T0293207

Invitation to Algebra

A Resource Compendium for Teachers, Advanced Undergraduate Students and Graduate Students in Mathematics

Invitation to Algebra

A Resource Compendium for Teachers, Advanced Undergraduate Students and Graduate Students in Mathematics

Vlastimil Dlab • Kenneth S Williams

Carleton University, Canada

NEW JERSEY • LONDON • SINGAPORE • BEIJING • SHANGHAI • HONG KONG • TAIPEI • CHENNAI • TOKYO

Published by

World Scientific Publishing Co. Pte. Ltd.
5 Toh Tuck Link, Singapore 596224
USA office: 27 Warren Street, Suite 401-402, Hackensack, NJ 07601
UK office: 57 Shelton Street, Covent Garden, London WC2H 9HE

Library of Congress Control Number: 2020937550

British Library Cataloguing-in-Publication Data
A catalogue record for this book is available from the British Library.

INVITATION TO ALGEBRA
A Resource Compendium for Graduate Students and Advanced Undergraduate Students in Mathematics

ISBN 978-981-121-997-9 (hardcover)
ISBN 978-981-121-998-6 (ebook for institutions)
ISBN 978-981-121-999-3 (ebook for individuals)

For any available supplementary material, please visit
https://www.worldscientific.com/worldscibooks/10.1142/11818#t=suppl

Desk Editor: Soh Jing Wen

To our wives Helena and Carole

Preface

Up until towards the end of the nineteenth century, algebra loosely comprised the study of a variety of concrete systems such as vectors, polynomials, quaternions, and matrices. However, in the decade or so before the end of that century, mathematicians recognized that by abstracting the common content of such systems these different realms of algebra could be brought efficiently together. The foremost abstraction was to treat them as sets of elements subject to an operation specified by certain abstract properties. Thus, for example, the set of equivalence classes of binary quadratic forms $ax^2 + bxy + cz^2$ treated by Gauss and the class of transformations

$$z \to \frac{az+b}{cz+d}$$

in the complex plane studied by Möbius were brought efficiently together under that part of abstract algebra known as the theory of groups.

Successful students know that learning an abstract subject like mathematics requires an appreciation of concrete examples which illustrate and motivate the underlying concepts. The learning process should imitate the way in which children learn new notions. A child does not learn what a table is through an abstract description of a certain construction but rather, after being shown many different tables, learns to single out tables from collections of furniture. In the same way an abstract notion should be built on well-chosen concrete examples. If a student is provided with a variety of concrete groups, then the definition of a group will arise naturally. The pedagogical principle which we shall follow is well expressed in a Chinese proverb attributed to **Confucius (551 - 479 BCE)**

I hear and I forget. I see and I remember. I do and I understand.

The usual desire of mathematicians to present their subject strictly logically from the beginning often obscures the learning process, kills the joy of discovery, and does not contribute to a lasting and deeper understanding of the substance of new knowledge.

In writing a book such as this it is vital to keep in mind who it is intended for. Our intention is that the book be a useful resource for graduate students and advanced undergraduates in mathematics. We hope too that it will be helpful to mathematics teachers at all levels as a source book and this is why so much emphasis has been placed on the development of the integers and rational numbers in the beginning chapters. Hopefully too a general reader of mathematics will find topics of interest. The material is presented in a challenging way so that the reader will be led to a deeper understanding of the subject. A familiarity with elementary linear algebra is assumed.

The origins of the book lie in a set of lecture notes prepared by the first author for students in an introductory algebra course given at Carleton University in Ottawa in the academic year 2005-2006. These notes were subsequently revised for courses given by the first author at Charles University in Prague during the years 2008 to 2012. More recently they have been revised again in conjunction with the second author taking into account the many valuable suggestions received from colleagues. They have also benefited from notes given in a variety of algebra courses by both authors. Tested in this way our book should provide useful guidance and reference to graduate students and advanced undergraduates in mathematics. No book lives in isolation from earlier publications and we have learned and benefited from textbooks by other authors. What we have learned from these books has helped improve our formulation of the theory and to hopefully avoid certain pitfalls in the exposition.

This book emphasizes the relationship between algebra and other parts of mathematics, especially geometry but also combinatorics and number theory, which are either missing from present algebra textbooks or are mentioned only marginally. It contains varied and diverse material that every mathematics graduate student and advanced undergraduate should be acquainted with. Its aim is to present beautiful, inspiring, and interesting topics, and to present them in such a way as to arouse interest in the reader to study and expand her/his horizons. This book roughly follows a prescribed scheme. Each chapter begins with some motivation and examples, followed by the central core material of the chapter with associated exercises, possibly followed in turn by more advanced topics and problems. The reader should understand that the exercises form an important part of the book. The reader should work through them to develop a proper understanding of the basic theoretical material. A summary of definitions of abstract concepts that have been introduced as examples earlier, is given in Chapter VI. There are also sections which deal with special topics to give a taste of contemporary algebra, some of which are left to Chapter X (Appendix). One of our objectives is to remove rigid boundaries between parts of mathematics and to stress the unity of mathematics. In particular, we have tried to show the very close relationship between algebra and geometry.

Finally, we make a few comments on the presentation. Exercises are integrated with the development of the theory in order to clarify the new concepts. Generally routine exercises are avoided. Solutions to some of the exercises appear at the end of the book. Sections marked by $\star$ serve primarily to extend our understanding of the relationship with other parts of mathematics. Material marked in this way will not be referred to later and can be skipped over at first reading. A number of notions are introduced informally in the first few chapters; the reader will find a brief more formal summary of these concepts in Chapter VI.

The presentation of the book received substantial assistance from Piroska Lakatos and Nadiya Gubareni. The book would not appear in the present form without their support and encouragement. The authors would like to thank them for many valuable remarks and corrections, and for their patience with frequent changes and amendments. We also wish to thank John Dixon for proofreading parts of the manuscript; his comments and advice lead to a number of improvements. We also thank Professor Masahisa Sato of Yamanashi University, Japan for his very careful reading of our manuscript which enabled us to correct many typos and errors. The first author is happy to acknowledge assistance of his son Daniel in the final form of presentation. Both authors thank the staff of World Scientific, particularly Rochelle Kronzek (USA) and Soh Jing Wen (Singapore), for their great care and cooperation in bringing our book into print.

Learning new concepts is not a simple process. A deeper understanding of new concepts is a process consisting in attaining a higher level of maturity. The course to follow is eloquently expressed by G. Chrystal in his classic textbook *Algebra. An Elementary Text-Book for the Higher Classes of Secondary schools and for Colleges* (Adam and Charles Black, Edinburgh, 1889):

> Every mathematical book that is worth reading must be read ''backwards and forwards'' if I may use the expression. I would modify Lagrange's advice a little and say, ''Go on, but often return to strengthen your faith.'' When you come on a hard and dreary passage, pass it over; and come back to it after you have seen its importance or found the need for it further on.

About the Book

An overwhelming majority of scientific writing, from popular books to specialized monographs, seems to follow current trends. And there is a good reason for that: a hot commodity is always in demand. But what we often forget is that the proliferation of fashionable topics is accompanied by less fashionable works slowly sinking into oblivion and, as a result of that, we often lose precious knowledge. (The "elimination of elimination theory" immediately comes to mind.) Historically, this has been countered by encyclopedic reference books or textbooks providing the research community with "snapshots" of the period's state of knowledge.

In the field of algebra, there has never been a shortage of well established classical textbooks. But in the last couple of decades, we are seeing a new trend – textbooks with significant deviation from what is normally described as the classical content. Some of these texts can be viewed as neoclassical, while others as outright postmodernist. With that in mind, I am happy to see the arrival of Invitation to Algebra by V. Dlab and K. S. Williams, with a distinctly interesting content. It would be difficult to pigeonhole this delightful book into any classification scheme. Its nine chapters (plus an appendix) have innocuous looking headings, like Natural Numbers, Integers, Divisibility, Complex numbers and Plane Geometry, Polynomials, Rings and Fields, etc., but the content of the chapters is full of interesting topics and surprising results. Have you ever thought of how many Venn diagrams for six subsets there are? Do you know what Petr's polygon is? Can you subdivide a square into an odd number of triangles of the same area? What did Napoleon actually prove? Do you know how to use continued fractions to realize a transmission by a gear train? Who is William Brouncker and what does he have to do with Pell's equation?

The primary audience for this book is graduate students and advanced undergraduates in mathematics, but it could also be useful for teachers of mathematics. Even an expert may surprise themselves by opening the book on any page and discovering something unexpected. An abundance of historical references and a number of rarely discussed topics serves an important purpose – to introduce the reader to results and ideas that are often missing from other books. There are also a large number of exercises, which are integrated into the exposition. From a more general perspective, a potential reader will notice the book's emphasis on the unity of mathematics, illustrated by the interplay between algebra, geometry, and combinatorics. For this reason, the book may also be called Invitation to Mathematics.

Alex Martsinkovsky, Associate Professor, Northeastern University, USA

Contents

CHAPTER I. INTRODUCTION

I.1 A few words on sets

Knowledge of the language of set theory is essential to understanding mathematics. In everyday life, we are often confronted by a concept which may be rather difficult to define and that can be described as a collection (family, aggregate, assembly, system, class). For instance, it is possible to talk about a collection of students present at a certain moment in a certain classroom; or, of a collection of books in a specific library at a specific time; or, of the collection of the letters "a" in a particular sentence, etc.

In each of these instances (situations, cases) the word "set" can be used instead of the word "collection".

In mathematics, we regularly deal with various sets. We deal with the set of all vertices of a given polygon, the set of all words of length 7 formed from the letters a, b, c, d and e, the set of all positive integers smaller than 5832, the set of all lines joining 5 fixed points in the plane, etc. The objects in a set are called its **elements**.

The sets in all these examples have one very important property in common: They all consist of a certain finite number of elements. In each case, we are either able to answer the question "how many" (number of students in a classroom, number of books in the library, number of letters "a" in the sentence, number of vertices, number of words, number of integers, number of lines) or point out that the natural number that is the answer to the question "how many" actually exists, although we may not at this moment and with our present knowledge, be able to determine it (for example, the number of books in the library is finite, however it depends on the number of books checked out at that particular moment in time).

Sets consisting of a finite number of elements are said to be **finite sets**. In the example of the letters "a" in the sentence, if no "a" occurs in the sentence then the set of such letters "a" has no elements and so the number of elements in it is zero and the set is called an **empty set**. As we will prove that there is only one empty set, we can denote it by $\emptyset$.

In mathematics, we regularly encounter other sets which do not have a finite number of elements. Such a set is called an **infinite set**. For instance, the set of all integers, the set of even integers, the set of prime numbers, and the set of all straight lines passing through a given point, are all examples of infinite sets.

We now introduce the following basic concepts and notation of set theory. We shall usually denote sets by capital letters $A, B, C, \ldots$, and their elements by lower case letters $a, b, c, \ldots$. Given a set A and an object x, the statement "x is an element of A" (or "x is in A" or "x is a member of A" or "x belongs to A") will be written $x \in A$. If x is not in A, we shall write $x \notin A$.

Any mathematical discourse occurs within a set containing all of the objects and sets which could possibly be needed in the discourse. Such a set is called a **universal** set and will be denoted by U (when it has not already been given a special symbol). In the example of the students in the classroom, we might take the set of all students as our universal set, while if we were doing arithmetic, we could take the integers as our universal set.

Suppose A and B are sets whose elements belong to some universal set U. If A and B contain exactly the same elements they are called **equal**, and we write $A = B$. If every element of A is also an element of B, then A is called a **subset** of B, and we write $A \subseteq B$. We see that $A \subseteq B$ if $x \in A$ logically implies $x \in B$ for all x. It follows that $A = B$ if and only if $A \subseteq B$ and $B \subseteq A$. This last characterization of set equality is useful because many problems of mathematics require us to show that two apparently different sets A and B are, in fact, equal. This can often be done by proving separately the two subset inclusions $A \subseteq B$ and $B \subseteq A$. We use this technique to show that the empty set is unique. Suppose that $\emptyset_1$ and $\emptyset_2$ are two empty sets. Logic tells us the empty set is a subset of every set A, that is, $\emptyset \subseteq A$. For, if not, then there is an element of the empty set which is not in A; but this is impossible since the empty set has no elements. Thus $\emptyset_1 \subseteq \emptyset_2$ and $\emptyset_2 \subseteq \emptyset_1$ proving $\emptyset_1 = \emptyset_2$. In what follows we always assume that any sets under consideration are subsets of some universal set U, but, for the sake of brevity, it will not always be mentioned explicitly.

Since every element in A is certainly in A, our definition gives $A \subseteq A$, and thus every set can be regarded as a subset of itself. (This small point will matter when we count the number of subsets of a finite set.) The set A and the empty set $\emptyset$ are called **improper** subsets of A; all of the other subsets (if any) of A are called **proper**. In the case where we want to emphasize that A is a subset of B with $A \neq B$ we will use the symbol $A \subset B$.

If a set A contains a small number of elements, say the numbers 1, 3 and 8, then it is convenient to exhibit the set by merely listing the elements, that is, $A = \{1, 3, 8\}$. However, often each element x of a set $A \subseteq U$ can be described by a certain property $\mathbf{P}(x)$ that it possesses. In this case the set A can be given as $A = \{x \in U \mid \mathbf{P}(x)\}$ (where the vertical bar is read "such that"). For example, if we take the set $\mathbb{Z}$ of integers as our universal set then the set S of all perfect squares can be described by

$$S = \{x \in \mathbb{Z} \mid x = y^2 \text{ for some integer } y\},$$

while if we take the set $\mathbb{N}$ of positive integers (also called natural numbers) as our universal set then the set E of all positive even numbers can be described by

$$E = \{x \in \mathbb{N} \mid x = 2y \quad \text{for some natural number } y\}.$$

This way of describing a set is sometimes called **set-builder notation** because the set is built-up from the condition $\mathbf{P}(x)$, which is satisfied if and only if x is in the set.

There are two important ways in which we can combine two subsets X and Y of some universal set U to yield a new subset of U. First, we define the **intersection** of X and Y, denoted $X \cap Y$, to be the subset of U consisting of all elements belonging both to X and to Y. Second, we define the **union** of X and Y, denoted $X \cup Y$, to be the subset of U of all elements belonging either to X or to Y (or to both). We say the sets X and Y are **disjoint** if their intersection is the empty set. The intersection and the union of any number (finite or infinite) of subsets are defined similarly. The reader is invited to define them.

The set consisting of all elements of X not belonging to Y is called the **difference** or **relative complement** of Y in X and is denoted by $X \setminus Y$; in particular, $X^* = U \setminus X$, for a subset X of the universe U is called the **(absolute) complement** of X.

The set of all subsets, proper or improper, of the universe U is called the **power set** of U and will be denoted by $\mathcal{P}(U)$. In the case where U is finite with n elements, it can be shown $\mathcal{P}(U)$ has 2^n elements, so the power set is often denoted by 2^U (including when U is infinite). This fact is proved in Section 2 of Chapter II: Fix an element $u_0 \in U$ and observe that the family of all subsets of U that contain u_0 and the family of all subsets of U that do not contain u_0 have the same number of elements. If this number is 2^{n-1}, then the number of elements of $\mathcal{P}(U)$ is $2 \cdot 2^{n-1} = 2^n$. Later in this chapter, we shall describe the structure of the power set in terms of the relation of **inclusion** of one subset in another.

Another important construction with two sets is the **Cartesian product** of X and Y, denoted $X \times Y$, of two subsets of possibly different universes. It is defined to be the set of all "ordered pairs" (x, y), where $x \in X$ and $y \in Y$. In set-builder notation, we have

$$X \times Y = \{(x, y) \,|\, x \in X, y \in Y\}.$$

More generally, the Cartesian product, $X_1 \times X_2 \times \cdots \times X_n$, of sets $X_1, X_2, \ldots, X_n$ is defined by

$$X_1 \times X_2 \times \cdots \times X_n = \{(x_1, x_2, \ldots, x_n) \,|\, x_t \in X_t, t = 1, 2, \ldots, n\}.$$

At times, we will use the short-hand $\prod_{t=1}^{n} X_t$ in place of $X_1 \times X_2 \times \cdots \times X_n$.

The mutual intersections and unions of a small number of sets can be easily illustrated in terms of Venn diagrams. **John Venn (1834 - 1923)** used a diagrammatic way of representing sets in his work on mathematical logic.

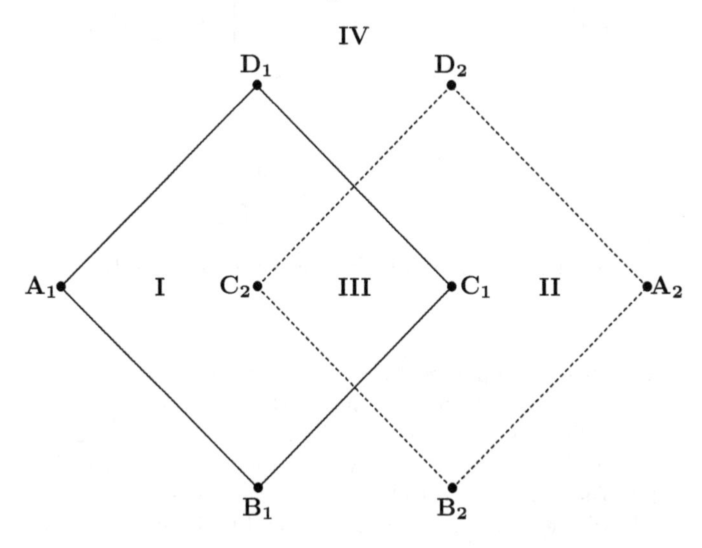

Figure I.1.1 Venn diagram for two sets

Already a picture of two different intersecting squares dividing the plane into four regions I, II, III, IV (see Figure I.1.1) can help solve some simple problems. Here is an example: Students are supposed to read two books this term: book A and book B. In a class of 25 students there are 3 students who did not read either book so far, 5 students who read only book A, and 7 students who read only book B. How many students read both book A and book B? Suppose that the square $A_1B_1C_1D_1$ of Figure I.1.1 represents the set of all students who read book A and the square $A_2B_2C_2D_2$ those who read book B. Then there are 3 students in region IV, 5 students in region I, and 7 students in region II. Thus there are 10 students in region III, and the required answer is 10.

The picture of three subsets (usually three circles) of a universe U (the plane) and all eight subsets formed by intersections and unions is well-known. In Figure I.1.2 we illustrate this situation by exhibiting three squares $A_tB_tC_tD_t$, $t = 1, 2, 3$ (in *"general position"*). To say that they are in general position means that there is nothing special (like an empty intersection of two of the subsets) about the placement of the squares, that is, that the universe is divided into 8 nonempty subsets, regions that we shall call the **components** of the Venn diagram. We note that a Venn diagram of n subsets A_t, $t = 1, \dots, n$, of U (in general position) will have 2^n components. Indeed, we may identify these components with the subsets of the set whose n elements are the subsets A_t.

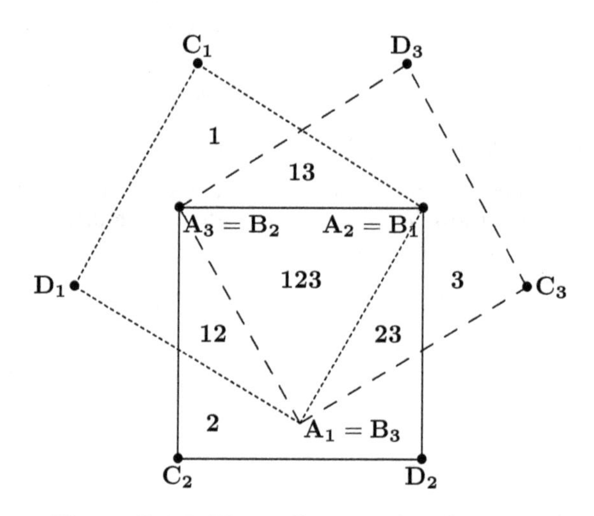

Figure I.1.2 Venn diagram for three sets

Exercise I.1.1. Students are supposed to read during the spring term 3 books: book A, book B and book C. At the end of the first month, there was still one student who did not read any of the books. There was no student who read only book A, 2 students read only book B, 3 students only book C. But there were 7 students who read books A and B, 6 students who read books A and C, 5 students who read books B and C, and there were 4 students who had already read all three books A, B and C. How many students were in the class?

Figure I.1.3 is a Venn diagram for four subsets represented by rectangles $\mathcal{Q}_t = A_tB_tC_tD_t$, $t = 1,2,3,4$. In the diagram, for instance, the area denoted by 124 describes the subset of all elements of the plane U that belong to $\mathcal{Q}_1, \mathcal{Q}_2$ and $\mathcal{Q}_4$ and do not belong to $\mathcal{Q}_3$, that is, the subset $(\mathcal{Q}_1 \cap \mathcal{Q}_2 \cap \mathcal{Q}_4) \setminus \mathcal{Q}_3$.

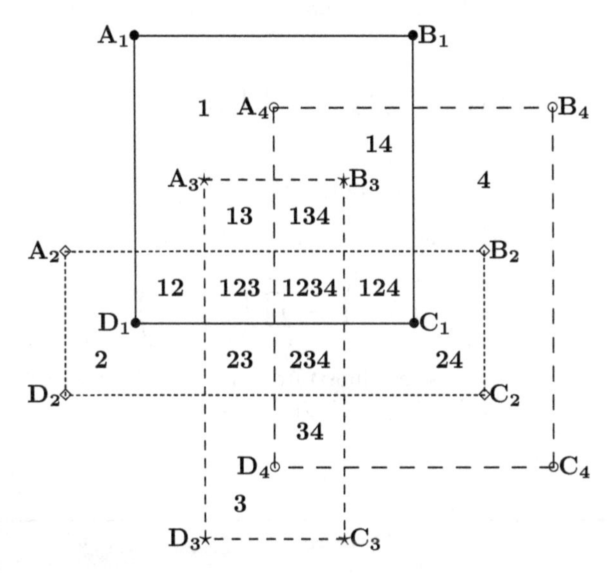

Figure I.1.3 Venn diagram for four sets

For a general definition of a Venn diagram, see Section 1 of Chapter X.

We use this opportunity to point out the enormity of the number of possible combinations of even a very few sets. A Venn diagram of **only 6 subsets of** U (in "general position") describes $\mathbf{2^{2^6} = 2^{64} \approx 1.8 \times 10^{19}}$ **subsets of** U. This is a number that dwarfs even the most generous estimate of the number of days since the Big Bang, that is, about 5×10^{12}, the number that is considered to be the age of our universe. In fact, the number of subsets defined by the Venn diagram of 8 subsets ($\approx 1.16 \times 10^{77}$) is comparable to the estimated number of atoms in the entire observable universe.

As already mentioned the definition of the intersection and of the union of subsets of U can be easily extended to an infinite family of subsets X_δ, indexed by some index set Δ ($\delta \in \Delta$). Thus, the intersection $\bigcap_\delta X_\delta$ consists of all elements that belong to X_δ for every $\delta \in \Delta$, while the union $\bigcup_\delta X_\delta$ consists of those elements that belong to X_{δ_0} for at least one index $\delta_0 \in \Delta$. We summarize the basic identities.

Theorem I.1.1. *Let $X, Y, Z, X_\delta, \delta \in \Delta$, be subsets of a set U. Then*

$$X \cup Y = Y \cup X \ \text{ and } \ X \cap Y = Y \cap X;$$

$$(X \cap Y) \cap Z = X \cap (Y \cap Z) \ \text{ and } \ (X \cup Y) \cup Z = X \cup (Y \cup Z);$$

$$(\bigcap_\delta X_\delta) \cup Y = \bigcap_\delta (X_\delta \cup Y) \;\; and \;\; (\bigcup_\delta X_\delta) \cap Y = \bigcup_\delta (X_\delta \cap Y).$$

Furthermore the following **de Morgan rules** *named after* **Augustus de Morgan (1806 - 1871)** *hold*

$$U \setminus \bigcap_\delta X_\delta = \bigcup_\delta (U \setminus X_\delta) \;\; and \;\; U \setminus \bigcup_\delta X_\delta = \bigcap_\delta (U \setminus X_\delta);$$

that is

$$(\bigcap_\delta X_\delta)^* = \bigcup_\delta X_\delta^* \;\; and \;\; (\bigcup_\delta X_\delta)^* = \bigcap_\delta X_\delta^*.$$

Exercise I.1.2. Denote the cartesian product $X_1 \times X_2 \times \cdots \times X_k$ by $\prod_{t=1}^{k} X_t$, that is the set $\{(x_1, x_2, \ldots, x_k) \mid x_t \in X_t\}$. For the sets $M_{rs}, r = 1, 2, \ldots, m, s = 1, 2, \ldots, n$, prove that

$$\bigcap_{s=1}^{n} \left(\prod_{r=1}^{m} M_{rs} \right) = \prod_{r=1}^{m} \left(\bigcap_{s=1}^{n} M_{rs} \right).$$

The situation for $m = n = 2$ is well illustrated in Figure I.1.4 with the sets represented by the line segments $M_{11} = A_1B_1, M_{12} = C_1D_1, M_{21} = A_2B_2$ and $M_{22} = C_2D_2$.

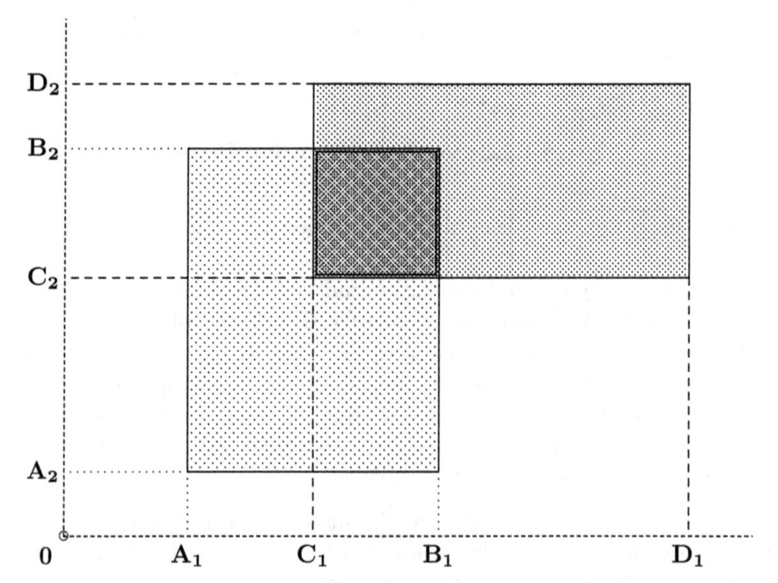

Figure I.1.4 $(\mathbf{M_{11}} \times \mathbf{M_{21}}) \cap (\mathbf{M_{12}} \times \mathbf{M_{22}}) = (\mathbf{M_{11}} \cap \mathbf{M_{12}}) \times (\mathbf{M_{21}} \cap \mathbf{M_{22}})$

In what follows, we shall denote the familiar important sets of numbers as follows:

$\mathbb{N} = \{1, 2, 3, \ldots, n, \ldots\}$ the *natural numbers* (positive integers),

$\mathbb{N}_0 = \{0, 1, 2, \ldots, n, \ldots\}$ the *non-negative integers*,

$\mathbb{Z} = \{0, \pm 1, \pm 2, \pm 3, \ldots, \pm n, \ldots\}$ the *integers*,

$\mathbb{Q} = \{\frac{a}{b} \mid a, b \in \mathbb{Z}, b \neq 0\}$ the *rational numbers* (rationals),

$\mathbb{R} = \{\text{all decimal expansions} \pm a_1a_2 \ldots a_k.b_1b_2b_3 \ldots\}$ the *real numbers* (reals),

$\mathbb{C} = \{a + bi \mid a, b \in \mathbb{R}, i^2 = -1\}$ the *complex numbers.*

These sets form a *tower* (or *chain*)

$$\mathbb{N} \subset \mathbb{N}_0 \subset \mathbb{Z} \subset \mathbb{Q} \subset \mathbb{R} \subset \mathbb{C}.$$

I.2 The important role of mappings

Given two sets A and B, we define a **mapping** (or **function**) f from A to B, denoted $f : A \longrightarrow B$, to be a rule which assigns to each element of A exactly one element of B. We shall write $f(a)$ for the unique element of B, assigned to a by f. The set A is called the **domain**, and the set B the **codomain** (or **range**) of f. The set $\{f(a) \mid a \in A\} \subseteq B$ is called the **image** of f and will be denoted by $\mathrm{Im} f$.

Two mappings f and g are **equal**, written $f = g$, if they have the same domain A, the same codomain, and for all $a \in A$, $f(a) = g(a)$.

The mapping f is said to be **one-to-one** (or **injective** or a **monomorphism**) if every element in the image comes from exactly one **pre-image** in the domain; that is, if $a_1 \neq a_2$ in A implies that $f(a_1) \neq f(a_2)$; or, equivalently, $f(a_1) = f(a_2)$ implies $a_1 = a_2$.

The mapping f is said to be **onto** B (or **surjective** or an **epimorphism**) if every element in B is the image of at least one element in A; that is, if for every $b \in B$, $b = f(a)$ for some $a \in A$, equivalently $\mathrm{Im} f = B$.

A one-to-one mapping of a set A onto a set B is called a **one-to-one correspondence** (or a **bijection**) of A onto B. We can see that, if there is a one-to-one correspondence f of A onto B, then there is a one-to-one correspondence g of B onto A. If $f(a) = b$ then $g(b) = a$. We shall record such a situation by the symbol $\simeq$, that is $A \simeq B$. When each of the sets A and B possesses an algebraic structure (such as some algebraic operations) and the one-to-one correspondence between A and B also preserves the structure (see Sections I.5 and I.7, and also Chapters VI, VII and VIII), the term **isomorphism** is used for the one-to-one correspondence.

The mapping from A to A that assigns to every $a \in A$, that same element $a \in A$ is called the **identity mapping** of the set A. It will be denoted by 1_A.

Exercise I.2.1. Let S be a finite set. Then $f : S \to S$ is surjective if and only if it is injective and thus if and only if it is bijective. Give an example where this is not true if S is an infinite set.

Exercise I.2.2. Define the mapping $f : \mathbb{Q} \to \mathbb{Q}$ by

$$f(x) = x^3 + 6x^2 + 12x + 18$$

and the mapping $g : \mathbb{R} \to \mathbb{R}$ by

$$g(x) = 1 + x \text{ for } x \in \mathbb{Q} \text{ and } g(x) = 1 - x \text{ for } x \in \mathbb{R} \setminus \mathbb{Q}.$$

Decide whether f is injective, surjective or bijective. Do the same for g.

The concept of a bijection is central to the notion of the **size** or **cardinality** of a set. Two sets are said to be **equivalent** (or to have the **same cardinality** or to have the same size) if there is a one-to-one correspondence between them. The reader can find more about the cardinality of sets in Section 2 of Chapter X.

In the case when two sets are equivalent and one of them is finite, then the other is also finite and both have the same number of elements, that is, they have the same size. The infinite sets equivalent to the set of all positive integers (natural numbers) are said to be **countable** or **denumerable**. The elements of a countable set can be listed as a (countable) sequence: $a_1, a_2, \ldots, a_n, \ldots$ with n running through the set of all natural numbers. We note that the set of all integers is countable (consider, for example, the listing $a_1 = 0, a_2 = 1, a_3 = -1, \ldots, a_{2k} = k, a_{2k+1} = -k, \ldots$) and the set of all even positive integers is also countable (consider the listing $a_1 = 2, a_2 = 4, a_3 = 6, \ldots$). It is less obvious that the set of all rational numbers (fractions of integers) is countable. In fact, the countable union (that is, the union of a countable family) of countable sets is countable. Thus all these sets have the same size! Also observe that **every subset of a countable set is either finite or countable**.

To see that the union of a countable family of countable sets is countable, we exhibit a scheme of a sequence (countable set) of sequences (countable sets)

$$\mathbf{a}_n = (a_{n1}, a_{n2}, \ldots, a_{nn}, \ldots),\ n = 1, 2, \ldots,$$

that allows a process which is useful in many other considerations, namely a process of forming a sequence comprising all entries "via diagonal paths" as indicated in Figure I.2.1. Here, the new sequence $\mathbf{b} = (b_1, b_2, \ldots b_k, \ldots)$ satisfies for $k = \frac{n(n-1)}{2} + 1 = \binom{n}{2} + 1,\ b_k = a_{1\,n}$ for n even and $b_k = a_{n1}$ for n odd. In Figure I.2.1 the index n is even. A particular version of this scheme sometimes appears in textbooks to show that the set of all rational numbers is countable.

Exercise I.2.3. Show that the preceding sequence $\mathbf{b}$ satisfies $b_1 = a_{11}$,
$b_{\binom{n}{2}+t} = a_{t\ n-t+1}$ for even $n \geq 2$ and every $t \in \{1, 2, \ldots, n\}$, and
$b_{\binom{n}{2}+t} = a_{n-t+1\ t}$ for odd $n \geq 3$ and every $t \in \{1, 2, \ldots, n\}$.

Exercise I.2.4. Use the preceding method to provide a detailed proof of the following statement: For an arbitrary natural number n, the set of all possible finite sequences $(a_0, a_1, \ldots, a_n)$ of integers (or rational numbers) is countable.

Example I.2.1. We show that not all sets are countable. For example, the set S_{01} of all sequences whose members are digits 0 or 1 is **not countable**. We apply the **diagonalization argument** of **Georg Cantor (1845 - 1918)** to give a proof of this statement by contradiction. We assume that the set S_{01} is countable. Then all its elements can be written as an enumeration

$$s_1, s_2, \ldots, s_n, \ldots,\quad \text{where } s_n = (a_{n1}, a_{n2}, \ldots, a_{nk}, \ldots) \text{ with } a_{nk} = 0 \text{ or } 1. \tag{I.2.1}$$

Now, the sequence

$$s = (a_1, a_2, \ldots, a_k, \ldots),\ \text{where } a_k = 0 \text{ if } a_{kk} = 1 \text{ and } a_k = 1 \text{ if } a_{kk} = 0,$$

does not belong to the enumeration (I.2.1). However, this contradicts s being an element of S_{01} and therefore belonging to the enumeration (I.2.1). This contradiction implies that the original assumption is false. Consequently, S_{01} is not countable.

Exercise I.2.5. Using the diagonalization argument, or otherwise, prove that the set $\mathbb{R}$ of the real numbers is not countable.

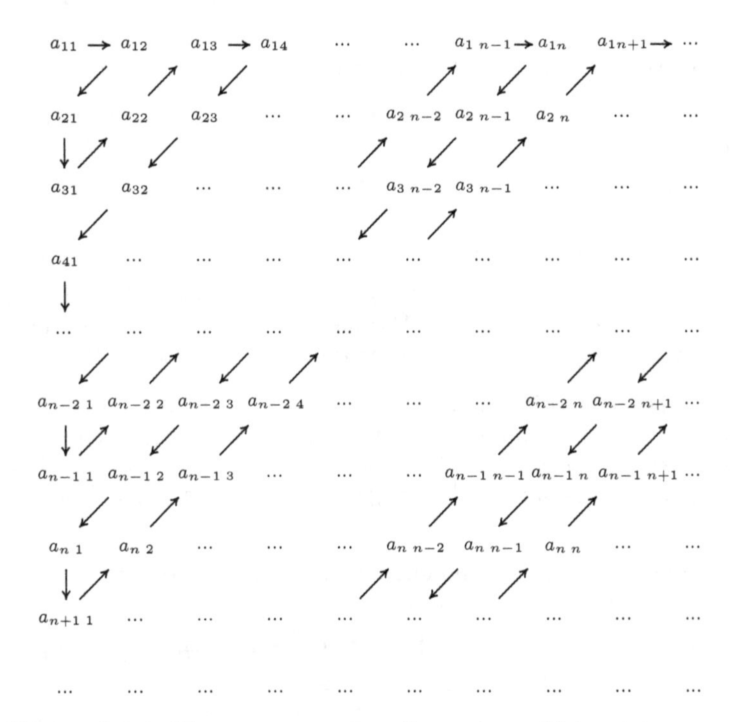

Figure I.2.1 The sequence $\mathbf{b} = (\mathbf{b}_{\binom{n}{2}+t}), \mathbf{n} \in \mathbb{N}, \mathbf{t} = 1, 2, \dots, \mathbf{n}$, where $\mathbf{b}_{\binom{n}{2}+t} = \mathbf{a}_{t\ n-t+1}$ (n even) or $\mathbf{b}_{\binom{n}{2}+t} = \mathbf{a}_{n-t+1\ t}$ (n odd)

A mapping $f : A \to B$ can be identified with its **graph**, that is, by the set of all (ordered) pairs (a, b), where $a \in A$, $b \in B$ and $f(a) = b$. A graph is thus a subset of the Cartesian product $A \times B$. Note that a subset $\Gamma \subseteq A \times B$ is a graph of a mapping of the set A into the set B if and only if

(1) there exists $(a, b) \in \Gamma$ for every $a \in A$ and

(2) if $(a, b_1) \in \Gamma$ and $(a, b_2) \in \Gamma$, then $b_1 = b_2$.

Exercise I.2.6. For real numbers x and y, consider the sets $S_1 = \{(x, y) \,|\, x + y = 7\}$, $S_2 = \{(x, y) \,|\, x^2 + y = 7\}$, $S_3 = \{(x, y) \,|\, x + y^2 = 7\}$, and $S_4 = \{(x, y) \,|\, x^2 + y^2 = 7\}$. Explain whether these sets are graphs of some functions or not.

The concept of a mapping plays a central role in contemporary mathematics. We shall witness its importance throughout this book.

The composition (or product) of two mappings $f : A \to B$ and $g : B \to C$ is the mapping $s = g \circ f : A \to C$ defined by

$$s(a) = (g \circ f)(a) = g(f(a)) \text{ for every } a \in A.$$

The composition of two injective, surjective, or bijective mappings is again injective, surjective, or bijective, respectively. Furthermore, composition has the properties stated in Theorem I.2.1.

Theorem I.2.1. *Let $f : A \to B$, $g : B \to C$ and $h : C \to D$ be given mappings. Then*

(1) $h \circ (g \circ f) = (h \circ g) \circ f$, *that is the composition of mappings is an* **associative (binary) operation**.

(2) *If $g \circ f$ is injective, then f is injective.*

(3) *If $g \circ f$ surjective, then g is surjective.*

(4) *A bijective mapping $f : A \to B$ defines a bijective mapping $f^{-1} : B \to A$ by setting $f^{-1}(b) = a$ if and only if $f(a) = b$. Moreover $f^{-1} \circ f = 1_A$ is the* **identity mapping** *of the set A and $f \circ f^{-1} = 1_B$ is the identity mapping of the set B.*

Remark I.2.1. The composition $(g \circ f) : A \to C$ of two mappings $f : A \to B$ and $g : B \to C$ may be bijective with neither f nor g being bijective. The preceding theorem only guarantees that f is injective and g is surjective. The following example illustrates this situation. Define $f : \mathbb{N}_0 \to \mathbb{Z}$ by $f(n) = n$ and $g : \mathbb{Z} \to \mathbb{N}_0$ by $g(z) = |z|$. Recall that $\mathbb{N}_0$ denotes the set of all non-negative numbers.

In fact, every mapping has an "opposite" decomposition: a surjective mapping followed by an injective one. Here is the formulation.

Theorem I.2.2. *Let $f : A \to B$ be a mapping. Then for a certain set C, $f = f_i \circ f_p$, where $f_p : A \to C$ is surjective and $f_i : C \to B$ is injective.*

Proof. Let $C = \text{Im} f = \{f(a) \mid a \in A\} \subseteq B$ and the assertions follow. □

Exercise I.2.7. The mapping $f : \mathbb{N} \to \mathbb{N}$ is defined by $f(x) = x + 1$. Determine all mappings $g : \mathbb{N} \to \mathbb{N}$ such that $g \circ f = 1_{\mathbb{N}}$. Is there a mapping satisfying $f \circ g = 1_{\mathbb{N}}$?

The set of all mappings from a set A to the set B will be denoted by $\text{Hom}(A, B)$. The notation Hom refers to the notion of homology. The word is of Greek origin (homos = same, logos = relation) and was introduced in the 17-th century to science to describe biological structures. It entered mathematics much later, probably through the work "Analysis Situs" of **Henri Poincaré (1854 - 1912)**. We note that for finite sets A and B of m and n elements, respectively, the set $\text{Hom}(A, B)$ has n^m elements. Indeed, each element of A can be mapped into n different elements of B, and thus the number of all the different mappings is $n \cdot n \cdots n = n^m$. We note that in the case $n = 2$, these mappings can be identified with the subsets of A.

We now consider the power set $\mathcal{P}(U)$ of U as a universal set together with what are called **morphism sets**, namely, $\text{Hom}(A, B)$ for all $A \subseteq U$ and $B \subseteq U$. We have a structure $\mathcal{C}$ containing **objects** (the elements of $\mathcal{P}(U)$) and **morphisms** (the elements of $\text{Hom}(A, B)$, where $A \subseteq U$ and $B \subseteq U$), which satisfy the following conditions:

(1) for any $f \in \text{Hom}(A, B), g \in \text{Hom}(B, C)$ and $h \in \text{Hom}(C, D)$,

$$h \circ (g \circ f) = (h \circ g) \circ f \ \in \text{Hom}(A, D), \text{ and}$$

(2) for any object $X \in \mathcal{C}$, there is the identity morphism $1_X \in \mathrm{Hom}(X, X)$, namely the identity mapping of the set X, satisfying the relations $1_B \circ f = f = f \circ 1_A$ for every morphism $f \in \mathrm{Hom}(A, B)$. $\mathcal{C}$ is an example of a **concrete category**; this concept is defined in Section 6 of Chapter X.

In subsequent sections of this book, we shall often encounter a type of mapping that maps the domain A into itself. Such a mapping $f : A \to A$ is called a **transformation** (of A). The set of all such transformations on the set A will be denoted by $T(A)$, that is $T(A) = \mathrm{Hom}(A, A)$. We note that the set $T(A)$ is endowed with an **associative binary operation**, denoted by the symbol $\circ$, of composition of transformations. In other words, the composition of two transformations of a set A is again a transformation of A. The identity mapping 1_A plays the role of the **neutral element** of $T(A)$ satisfying

$$1_A \circ f = f \circ 1_A = f \text{ for all } f \in T(A).$$

The set $T(A)$ endowed with the operation of composition is an example of an algebraic structure called a **monoid**, that is, a **semigroup with a neutral element** as defined more properly in Definition I.7.1.

A one-to-one mapping from a set to itself is called a **permutation**. The subset $S(A)$ of $T(A)$ consisting of all permutations of A is closed with respect to composition: If f and g belong to $S(A)$, then so does the composition $g \circ f$. Moreover, for any permutation $f \in S(A)$ the **inverse permutation** f^{-1}, satisfying $f \circ f^{-1} = f^{-1} \circ f = 1_A$, is also in $S(A)$. The set $S(A)$ together with the operation $\circ$ forms an algebraic structure called a **group**. The concept of a group is one of the fundamental concepts of algebra. We shall provide a formal definition in Chapter VI (Definition VI.1.3), and study this concept thoroughly in Chapter VIII (see also Theorem I.7.1).

If the set A is infinite, then both $T(A)$ and $S(A)$ are infinite, as well. If A is finite, then $T(A)$ and $S(A)$ are also finite. In fact, if A has n elements, then $T(A)$ has n^n elements (which is a special case of the number n^m of all mappings from a set of m elements to a set of n elements that we considered earlier) and $S(A)$ has $n!$ elements (the first element of A can be mapped into n elements, the second one into $n-1$ elements, and so on, and thus the number of all permutations of A is $n \cdot (n-1) \cdot (n-2) \cdot \cdots 2 \cdot 1 = n!$). It follows that even for small values of n the sets $T(A)$ have an enormous number of elements. Already, for $n = 10$, the monoid $T(A)$ has 10 billion elements. For $n = 2$ and $n = 3$, the size of $T(A)$ allows us an explicit listfing of its elements. Thus, for $A = \{1, 2, 3\}$, if we record the elements of $T(A)$ by triples $f(1)f(2)f(3)$, then

$$\begin{aligned} T(A) = \{&111, 112, 113, 121, 122, 123, 131, 132, 133,\\ &211, 212, 213, 221, 222, 223, 231, 232, 233,\\ &311, 312, 313, 321, 322, 323, 331, 332, 333\} \end{aligned}$$

and

$$S(A) = \{123, 132, 213, 231, 312, 321\}.$$

I.3 Three examples: Bijections in plane geometry

Example I.3.1 (Unrolling the circle). In the plane endowed with a Cartesian coordinate system, consider a circle $\mathbf{C}$ with center $(0, s)$ and radius $s(> 0)$. We associate with every point $(a,b) \neq (0,2s)$ of the circle the intersection point of the $x-$axis with the line connecting the points (a,b) and $(0,2s)$; that is, we "project" the point (a,b) onto the $x-$axis (see Figure I.3.1). The correspondence f defined in this way is a bijection between the points on the circle $\mathbf{C}$ except for the point $(0,2s)$, and the points on the $x-$axis.

As an exercise prove

$$f((a,b)) = \frac{2as}{2s-b} \text{ and } f^{-1}(x) = \left(\frac{4s^2x}{4s^2+x^2}, \frac{2sx^2}{4s^2+x^2}\right).$$

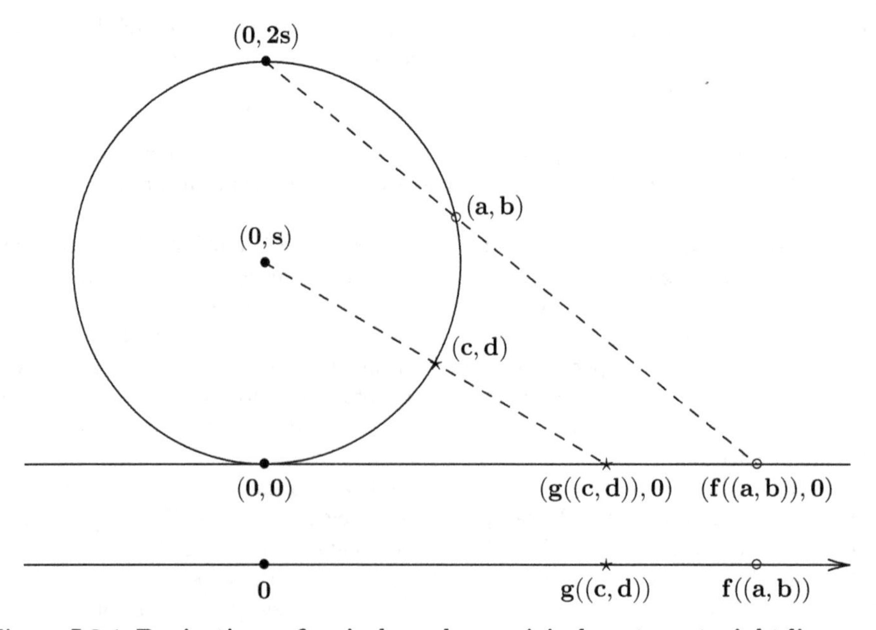

Figure I.3.1 Projections of a circle and a semicircle onto a straight line

Example I.3.2. In a similar manner, we can project the points of the lower semicircle (without the extreme points) of $\mathbf{C}$ from the center of the circle to the $x-$axis, see Figure I.3.1. In this way, we establish a bijection g between the semicircle and the $x-$axis. Similarly, as before, we can show that

$$g((c,d)) = \frac{cs}{s-d} \text{ and } g^{-1}(x) = \left(\frac{sx}{\sqrt{s^2+x^2}}, \frac{s(\sqrt{s^2+x^2}-s)}{\sqrt{s^2+x^2}}\right).$$

We note that the composition of f and g^{-1} defines a bijection between all the points different from $(0,2s)$ on the circle $\mathbf{C}$ and all the points on the lower semicircle of $\mathbf{C}$ without the extreme points:

$$(g^{-1} \circ f)((a,b)) = \left(2s\sqrt{\frac{b}{2s+3b}}, s - s\sqrt{\frac{2s-b}{2s+3b}}\right).$$

Example I.3.3. Barycentric coordinates. Let ABC be a triangle given by the cartesian coordinates $A = (a_1, a_2)$, $B = (b_1, b_2)$ and $C = (c_1, c_2)$. Let $A' = (a'_1, a'_2)$ be a point of the side BC. Then, treating the points A, B, C, A' as vectors, we obtain

$$A' = B + t_1(C - B), \quad \text{for some } 0 \le t_1 \le 1.$$

Similarly, a point X on the segment AA' satisfies

$$X = A + t_2(A' - A) = (1 - t_2)A + (t_2 - t_1t_2)B + t_2t_1C, \quad \text{where } 0 \le t_2 \le 1,$$

and thus

$$X = a_X\, A + b_X\, B + c_X\, C, \quad \text{where } a_X + b_X + c_X = 1,\ 0 \le a_X,\ 0 \le b_X,\ 0 \le c_X.$$

This establishes a bijection between all the points X inside or on the boundary of the triangle ABC and the triples (a_X, b_X, c_X) of real numbers satisfying

$$a_X + b_X + c_X = 1 \quad \text{and } 0 \le a_X,\ 0 \le b_X,\ 0 \le c_X.$$

The numbers a_X, b_X, c_X are called the **barycentric coordinates** of the point X and the point X is identified with the triple $(a_X,\ b_X,\ c_X)$. Thus $(1, 0, 0)$ are the barycentric coordinates of A, $(0, 1, 0)$ the barycentric coordinates of B, and $(0, 0, 1)$ the barycentric coordinates of C. The barycentric coordinates of the centroid T (the point of intersection of the medians of a triangle) of the triangle ABC are $(\frac{1}{3}, \frac{1}{3}, \frac{1}{3})$.

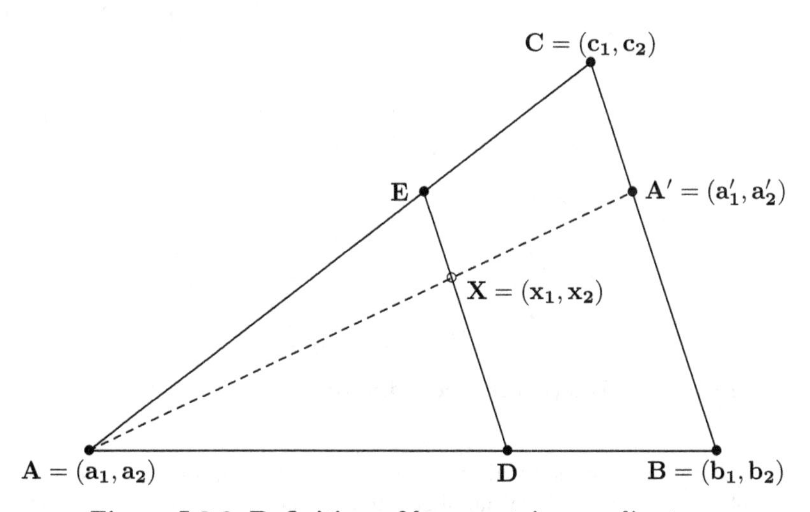

Figure I.3.2 Definition of barycentric coordinates

Exercise I.3.1. In Figure I.3.2 the line segment DE passing through the point X is parallel to BC. Prove that the set of points whose first barycentric coordinate is the same as that of the point X is equal to the set of the points of the segment DE. In particular, the points whose first barycentric coordinate is 0 are just the points of the side BC.

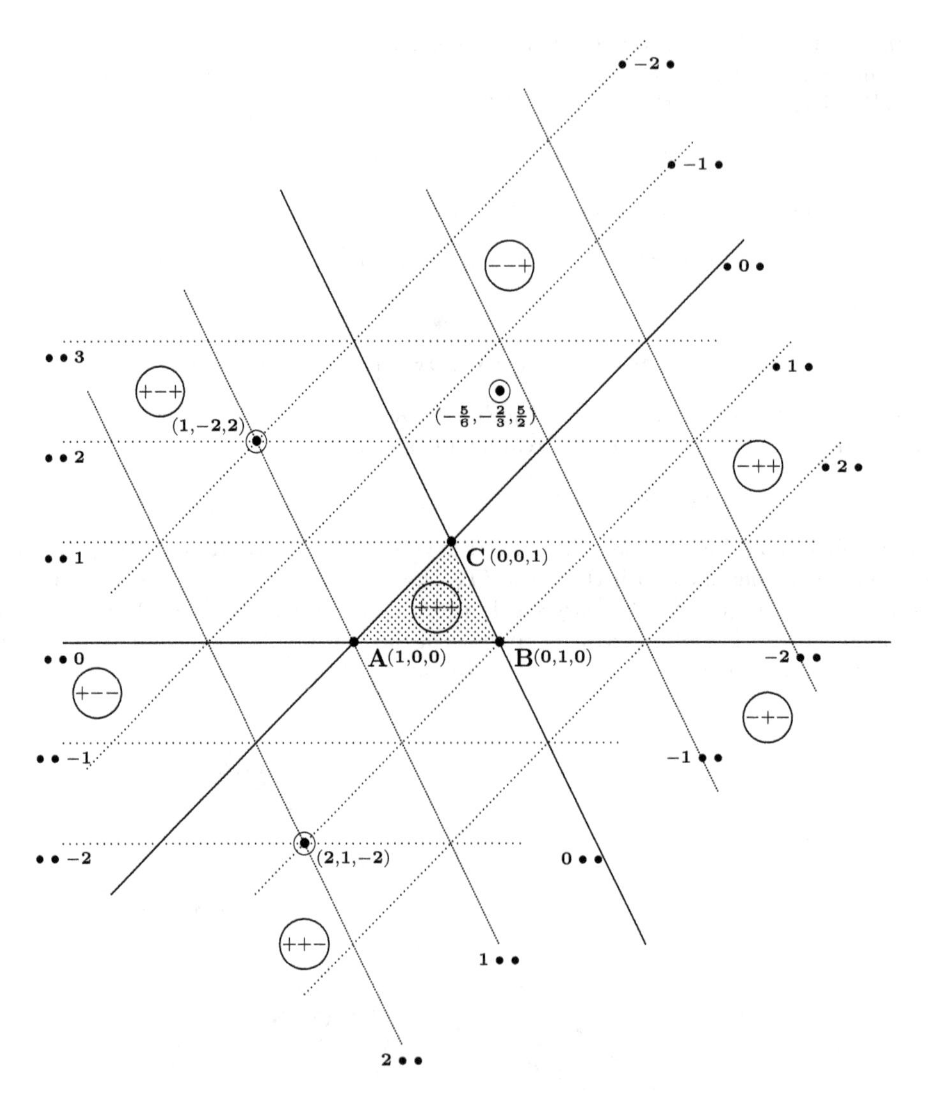

Figure I.3.3 Barycentric coordinates of points in the plane

Example I.3.3 can be modified and extended to describe (uniquely) every point X of the plane by a triple (a_X, b_X, c_X) of real numbers that are not necessarily non-negative: These are the barycentric coordinates of the point X:

$$X = a_X\,A + b_X\,B + c_X\,C, \text{ where } a_X + b_X + c_X = 1.$$

Figure I.3.3 illustrates the fact that some of the barycentric coordinates are zero or negative. The straight lines extending the sides of the triangle ABC divide the entire plane into seven regions. The points on these lines are just the points X whose barycentric coordinates (a_X, b_X, c_X) satisfy $a_X b_X c_X = 0$. The points A, B and C have two barycentric coordinates equal to zero, the other points on these lines have one zero coordinate. Each of the seven regions is characterized by the fact that all points in a given region have the same distribution of their positive and negative coordinates. This fact is indicated in Figure I.3.3 by a triple of symbols $+$ and $-$ in the little circles. Thus $+++$ describes the fact that the interior points of the triangle ABC are characterized by all three coordinates being positive. The region marked $--+$ contains all points X whose coordinates (a_X, b_X, c_X) satisfy $a_X < 0, b_X < 0$ and $c_X > 0$, etc.

As an illustration, we use barycentric coordinates to describe the points of a given line. We present this result as

Theorem I.3.1. *Let (a_R, b_R, c_R) and (a_S, b_S, c_S) be the barycentric coordinates of the points R and S. Let X be a point of the line defined by the points R and S. Then there is a unique number t such that*

$$(a_X, b_X, c_X) = ((1-t)a_R + ta_S, (1-t)b_R + tb_S, (1-t)c_R + tc_S)$$

are the barycentric coordinates of the point X.

Proof. Since

$$a_R\,\overrightarrow{RA} + b_R\,\overrightarrow{RB} + c_R\,\overrightarrow{RC} = \vec{0},\ a_S\,\overrightarrow{SA} + b_S\,\overrightarrow{SB} + c_S\,\overrightarrow{SC} = \vec{0} \text{ and } \overrightarrow{RX} = t\,\overrightarrow{RS},$$

the required equality $a_X\,\overrightarrow{XA} + b_X\,\overrightarrow{XB} + c_X\,\overrightarrow{XC} = \vec{0}$ can be obtained by making use of the simple relations

$$\overrightarrow{XA} = \overrightarrow{XR} + \overrightarrow{RA} = t\overrightarrow{SR} + \overrightarrow{RA},\quad \overrightarrow{XB} = t\overrightarrow{SR} + \overrightarrow{RB},\quad \overrightarrow{XC} = t\overrightarrow{SR} + \overrightarrow{RC} \text{ and}$$
$$\overrightarrow{RA} = \overrightarrow{RS} + \overrightarrow{SA},\quad \overrightarrow{RB} = \overrightarrow{RS} + \overrightarrow{SB},\quad \overrightarrow{RC} = \overrightarrow{RS} + \overrightarrow{SC}.$$

□

An immediate consequence of Theorem I.3.1 is the following assertion providing a geometric meaning of the positive barycentric coordinates of the points inside the triangle ABC which is very illustrative and leads to the fact that these coordinates are often called **planar coordinates**.

Theorem I.3.2. *Let X be an inner point of a triangle ABC. Denote by the symbol $|UVW|$ the area of the triangle UVW. Then the barycentric coordinates a_X, b_X, c_X of the point X satisfy*

$$a_X = \frac{|BCX|}{|ABC|},\ b_X = \frac{|CAX|}{|ABC|},\ c_X = \frac{|ABX|}{|ABC|}.$$

Proof. Let D be the point of intersection of the line through the points C and X and the side AB of the triangle ABC, see Figure I.3.4. By Theorem I.3.1, the barycentric coordinates of the points D and X are closely related:

$$a_D = \frac{a_X}{1-c_X},\ b_D = \frac{b_X}{1-c_X},\ c_D = 0.$$

Hence,

$$\frac{a_X}{b_X} = \frac{a_D}{b_D} = \frac{|DB|}{|AD|} = \frac{|DBC|}{|ADC|} = \frac{|DBX|}{|ADX|} = \frac{|DBC| - |DBX|}{|ADC| - |ADX|} = \frac{|BCX|}{|CAX|}.$$

In a similar manner, we obtain

$$\frac{b_X}{c_X} = \frac{|CAX|}{|ABX|} \text{ and } \frac{c_X}{a_X} = \frac{|ABX|}{|BCX|}.$$

Since $|BCX| + |CAX| + |ABX| = |ABC|$, Theorem I.3.2 follows. □

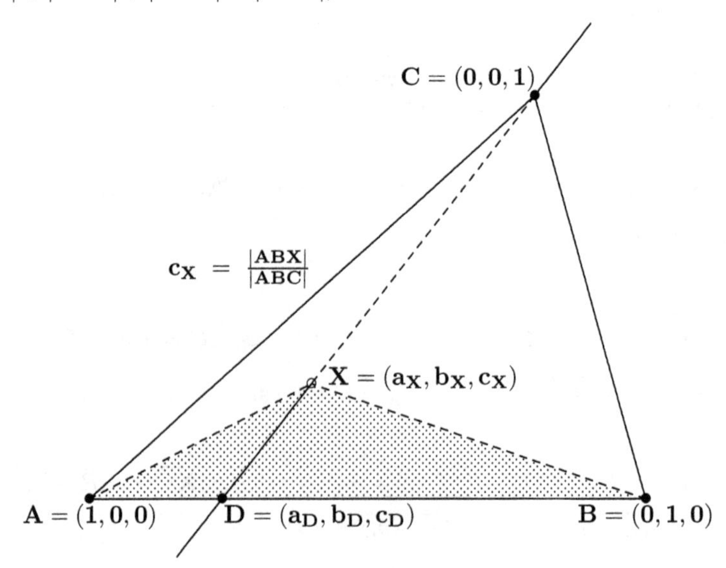

Figure I.3.4 Planar coordinates

Example I.3.4. We apply Theorem I.3.1 to prove the equalities displayed in Figure I.3.5. We consider a line passing through the centroid T of the triangle ABC. In Figure I.3.5 this line meets AB at F, BC at D, and CA extended at E. Other configurations can be treated similarly. The barycentric coordinates of the points A, B, C, T, D, E are $(1,0,0)$, $(0,1,0)$, $(0,0,1)$, $(\frac{1}{3}, \frac{1}{3}, \frac{1}{3})$, $(0, b_D, c_D = 1 - b_D)$, $(a_E, 0, c_E = 1 - a_E)$, respectively. By Theorem I.3.1, there is a number t such that

$$a_E = \frac{1}{3}t,\ 0 = (1-t)b_D + \frac{1}{3}t \text{ and } c_E = (1-t)c_D + \frac{1}{3}t.$$

Hence, $t = \frac{3b_D}{3b_D - 1}$, and therefore $a_E = \frac{b_D}{3b_D - 1}$ and $c_E = \frac{2b_D - 1}{3b_D - 1}$. From here,

$$\frac{|BD|}{|DC|} + \frac{|AE|}{|EC|} = \frac{1 - b_D}{b_D} + \frac{2b_D - 1}{b_D} = \frac{b_D}{b_D} = 1.$$

As an exercise, the reader should verify the remaining two equalities given in Figure I.3.5.

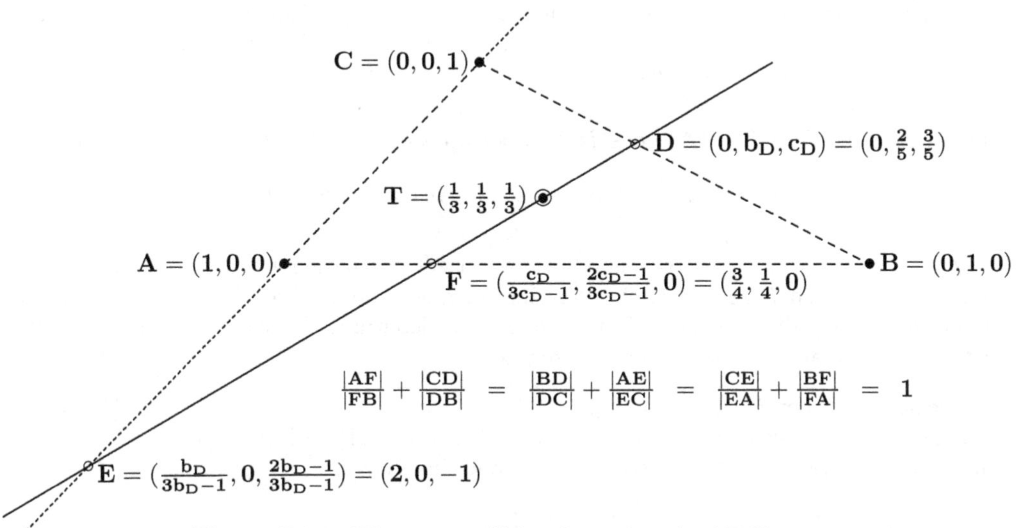

Figure I.3.5 Three equalities for triangle ABC

I.4 Partitions of sets, equivalence relations

We are now going to introduce the important concept of **a partition of a set**. This concept is intimately connected to the concept of a mapping in the following sense.

Let $f : A \to B$. For every element $b \in B$ denote by $f^{-1}(b)$ the subset of A consisting of those elements that are mapped by f to b, that is

$$f^{-1}(b) = \{a \in A \mid f(a) = b\}.$$

We observe that the collection of the nonempty subsets

$$\Sigma_f = \{f^{-1}(b) \mid b \in \mathrm{Im} f\}$$

of A possesses the following two properties:

(1) If $b_1 \neq b_2$, then $f^{-1}(b_1) \cap f^{-1}(b_2) = \emptyset$ and

(2) $\bigcup_{b \in B} f^{-1}(b) = A$.

Hence, Σ_f is a **partition** (or a **disjoint cover**) of the set A in accordance with the following definition.

Definition I.4.1. *A collection $\{A_\alpha \mid \alpha \in I\}$ of nonempty subsets of a set A is said to be a partition of A if*

(1) $A_{\alpha_1} \cap A_{\alpha_2} = \emptyset$ *for every pair of indices* $\alpha_1 \neq \alpha_2$ *and*

(2) $\bigcup_{\alpha \in I} A_\alpha = A$.

We have seen that every mapping from a set A defines a partition of A. In fact, the next theorem (the proof of which is left as an exercise) shows that this is the way in which **all partitions of a set** are obtained.

Theorem I.4.1. *Let $\mathcal{P} = \{A_\alpha \mid \alpha \in I\}$ be a partition of a set A. Then*

$$\mathcal{P} \ = \ \Sigma_f, \text{ where } f : A \to \mathcal{P} \text{ is defined by } f(a) = A_\alpha \text{ for all } a \in A_\alpha, \alpha \in I.$$

There is an obvious question related to the concept of a partition of a set. Although it is obvious that a map f from a set A determines uniquely the partition Σ_f, there are, on the other hand, many different maps from a set A that determine the same partition of A. To determine them all, solve the following exercise.

Exercise I.4.1. Let $f : A \to B$ and $g : A \to C$ be two maps from a set A. Then $\Sigma_f = \Sigma_g$ if and only if there exists a bijection $\varphi : \text{Im}f = \{f(a) \mid a \in A\} \to \text{Im}g = \{g(a) \mid a \in A\}$ such that $\varphi \circ f = g$. Equivalently, we say that the following diagram

$$\begin{array}{ccccc} A & \xrightarrow{f} & \text{Im}f & \subseteq & B \\ \| & & \downarrow \varphi & & \\ A & \xrightarrow{g} & \text{Im}g & \subseteq & C \end{array}$$

commutes, meaning the two paths from A to C in the diagram give the same result.

There is yet another very important connection originating from the concept of partitions of sets, or from the concept of a mapping for that matter. Considering a mapping $f : A \to B$, it is quite natural to regard $a_1 \in A$ and $a_2 \in A$ to be **f-equivalent** if they belong to the same element of the partition Σ_f, that is, if $f(a_1) = f(a_2)$. We denote this "new equality", this new binary relation (that is, a subset of the cartesian product $A \times A$) on the set A, by the symbol $a_1 \sim_f a_2$. It is quite obvious that $a_1 \sim_f a_1$, that $a_1 \sim_f a_2$ implies $a_2 \sim_f a_1$ and that $a_1 \sim_f a_2$ and $a_2 \sim_f a_3$ imply $a_1 \sim_f a_3$. In other words, the relation $\sim_f$ is an **equivalence** on the set A in the sense of Definition I.4.2. Recall that a (binary) relation $\sim$ on a set A is a subset of the cartesian product $A \times A$ and that it is customary to write $a_1 \sim a_2$ in place of $(a_1, a_2) \in \sim$.

Definition I.4.2. *A relation $\sim$ on a set A is said to be an equivalence if the following conditions are satisfied for all elements a_1, a_2, a_3 in A:*

(1) *$a_1 \sim a_1$ for all $a_1 \in A$.*

(2) *If $a_1 \sim a_2$, then $a_2 \sim a_1$.*

(3) *If $a_1 \sim a_2$ and $a_2 \sim a_3$, then $a_1 \sim a_3$.*

Properties (1), (2) *and* (3) *are called the* **reflexive, symmetric** *and* **transitive properties,** *respectively.*

We have seen that a partition $\mathcal{P} = \{A_\alpha \mid \alpha \in I\}$ of a set A defines the equivalence $\sim_\mathcal{P}$ on A:

$$a_1 \sim_\mathcal{P} a_2 \text{ if and only if } a_1 \text{ and } a_2 \text{ belong to the same } A_\alpha.$$

On the other hand, an equivalence $\sim$ on A defines the partition $\mathcal{P}_\sim = \{A_{\sim,\alpha} \mid \alpha \in I\}$:

Two elements a_1 and a_2 belong to the same $A_{\sim,\alpha}$ if and only if $a_1 \sim a_2$.

It is convenient to denote $A_{\sim,\alpha}$ simply by $[a]$, where $a \in A_{\sim,\alpha}$. Hence $[a_1] = [a_2]$ if and only if $a_1 \sim a_2$ and $[a]$ is the **equivalence class** (or, more precisely, $\sim$-equivalence class) of all elements $x \in A$ such that $x \sim a$. It is customary to denote by $A/\!\sim$ the set of all these $\sim$-equivalence classes. This set is called the **factor set** of A by the equivalence $\sim$. Given a mapping $f : A \to B$, we have $\Sigma_f = A/\!\sim_f$.

The close relationship between the partitions of a set A and the equivalences defined on a set A can be summarized in the following theorem.

Theorem I.4.2. *There is a bijection β between the set of all partitions $\mathcal{P}$ of a set A and the set of all equivalence relations $\sim$ on A. It is expressed by*

$$\beta : \mathcal{P} \to \sim_\mathcal{P} \quad \textit{and} \quad \beta^{-1} : \sim \to \mathcal{P}_\sim .$$

Moreover, there is a **natural (canonical) surjection** $\Phi : A \to A/\!\sim$ *defined by* $\Phi(a) = [a]$.

An important question concerning partitions is to determine the number $b(n)$ of all possible partitions of a set having n elements. In Section 3 of Chapter X we determine a recurrence relation for $b(n)$.

I.5 The concept of a lattice

We denote by $\mathcal{R}(A)$ the collection of all partitions of a given set A. There is more information about $\mathcal{R}(A)$ in Section 3 of Chapter X. A very important aspect of such a collection is the fact that there is a natural **partial order** of $\mathcal{R}(A)$.

Given two partitions $\mathcal{P} = \{A_\alpha \mid \alpha \in I\}$ and $\mathcal{Q} = \{B_\beta \mid \beta \in J\}$ of a set A, we shall say that $\mathcal{P}$ **is smaller than** $\mathcal{Q}$ and write $\mathcal{P} \preceq \mathcal{Q}$ if for every A_α, there is a B_β such that $A_\alpha \subseteq B_\beta$. We observe the following properties of the relation $\preceq$: For all $\mathcal{P}_1, \mathcal{P}_2$ and $\mathcal{P}_3$ in $\mathcal{R}(A)$ we have

(1) $\mathcal{P}_1 \preceq \mathcal{P}_1$.

(2) If $\mathcal{P}_1 \preceq \mathcal{P}_2$ and $\mathcal{P}_2 \preceq \mathcal{P}_1$, then $\mathcal{P}_1 = \mathcal{P}_2$.

(3) If $\mathcal{P}_1 \preceq \mathcal{P}_2$ and $\mathcal{P}_2 \preceq \mathcal{P}_3$, then $\mathcal{P}_1 \preceq \mathcal{P}_3$.

Thus the relation $\preceq$ is reflexive, antisymmetric and transitive, and such relations, in general, will be called partial orders.

We note that for a set A of more than two elements, there are partitions $\mathcal{P}$ and $\mathcal{Q}$ for which neither $\mathcal{P} \preceq \mathcal{Q}$ nor $\mathcal{Q} \preceq \mathcal{P}$ holds, that is, the order $\preceq$ of $\mathcal{R}(A)$ is not a **total (linear) order**. However, $\mathcal{R}(A)$ is a **lattice** in the sense of Definition I.5.3, as we shall see in the following section. We observe that the relation $\mathcal{P} \preceq \mathcal{Q}$ means that each $B_\beta \in \mathcal{Q}$ has a partition formed by certain subsets A_α belonging to the partition $\mathcal{P}$, that is, $\mathcal{P}$ is a **refinement** of the partition $\mathcal{Q}$.

In the previous remark, a structure given by "order" of elements was illustrated on a collection of all partitions of a given set. Similarly, an order was described and illustrated graphically on the collection of all subsets generated by a given family of sets in connection with Venn diagrams. Here, we shall note some properties of these relations.

We start with a definition.

Definition I.5.1. *A* **partially ordered set** $(P, \preceq)$ *is a set P together with a (binary) relation $\preceq$ which satisfies, for all elements x, y, z of P the following conditions:*

(1) $x \preceq x$ *(the relation $\preceq$ is* **reflexive***).*

(2) *If $x \preceq y$ and $y \preceq x$, then $x = y$ (the relation $\preceq$ is* **antisymmetric***).*

(3) *If $x \preceq y$ and $y \preceq z$, then $x \preceq z$ (the relation $\preceq$ is* **transitive***).*

If, in addition, for every x, y of P either $x \preceq y$ or $y \preceq x$, the order $\preceq$ is said to be **total** *(or* **linear***). Such a partially ordered set is often called a* **chain***, or a* **tower**.

The following two concepts play an important role in studying partially ordered sets.

Definition I.5.2. *Let $(P, \preceq)$ be a partially ordered set and $X \subseteq P$ a finite subset. Then an element $a \in P$ is said to be an* **upper bound** *of X if $x \preceq a$ for all $x \in X$. The* **least upper bound (supremum)** *of X is an upper bound u of X that satisfies $u \preceq a$ for any other upper bound a of X. Dually, an element $b \in P$ is said to be a* **lower bound** *of X if $b \preceq x$ for all $x \in X$. The* **greatest lower bound (infimum)** *of X is a lower bound l of X that satisfies $b \preceq l$ for any other lower bound b of X.*

Remark I.5.1. A subset X of a partially ordered set may or may not have a supremum or infimum. If a supremum exists, it is determined uniquely and often denoted by $\bigvee X$ or $\sup X$. Similarly, if an infimum exists, it is determined uniquely and often denoted by $\bigwedge X$ or $\inf X$. For a set of two elements, that is, if $X = \{x, y\}$, we shall write simply $x \vee y$ for $\bigvee X$ and $x \wedge y$ for $\bigwedge X$.

Definition I.5.3. *A partially ordered set $(P, \preceq)$ is said to be a* **lattice** *if every finite subset X of P has a least upper bound $\bigvee X$ (the* **join** *of X) and a greatest lower bound $\bigwedge X$ (the* **meet** *of X).*

Exercise I.5.1. (1) Let $X_1, X_2, \ldots, X_d$ be finite subsets of a lattice $(P, \preceq)$. Prove that

$$\bigvee \bigcup_{t=1}^{d} X_t = \bigvee \{\bigvee X_1, \bigvee X_2, \ldots, \bigvee X_d\}$$

and

$$\bigwedge \bigcup_{t=1}^{d} X_t = \bigwedge \{\bigwedge X_1, \bigwedge X_2, \ldots, \bigwedge X_d\}.$$

(2) A partially ordered set $(P, \preceq)$ is a lattice if and only if any two elements x, y of P have the least upper bound $x \vee y$ and the greatest lower bound $x \wedge y$.

(3) Show that $x \preceq y$ if and only if $x \vee y = y$ and that this equality holds if and only if $x \wedge y = x$.

Exercise I.5.2. Consider the collection $\mathcal{R}(A)$ of all partitions of a **finite** set A. Let

$$X = \{\mathcal{P}_1, \mathcal{P}_2, \dots, \mathcal{P}_n\} \subseteq \mathcal{R}(A), \text{ where } \mathcal{P}_t = \{A_{tk} \mid k = 1, \dots, n_t\} \text{ for } t \in \{1, 2, \dots, n\}.$$

Let $\mathcal{P}$ be the set of all nonempty intersections $A_{1k_1} \cap A_{2k_2} \cap \cdots \cap A_{nk_n}$, where $k_t \in \{1, 2, \dots, n_t\}$. Show that $\mathcal{P} = \bigwedge X$, that is, show that $\mathcal{P}$ is a partition of the set A which is the infimum of the set X. Hence every finite subset of $\mathcal{R}(A)$ has a greatest lower bound.

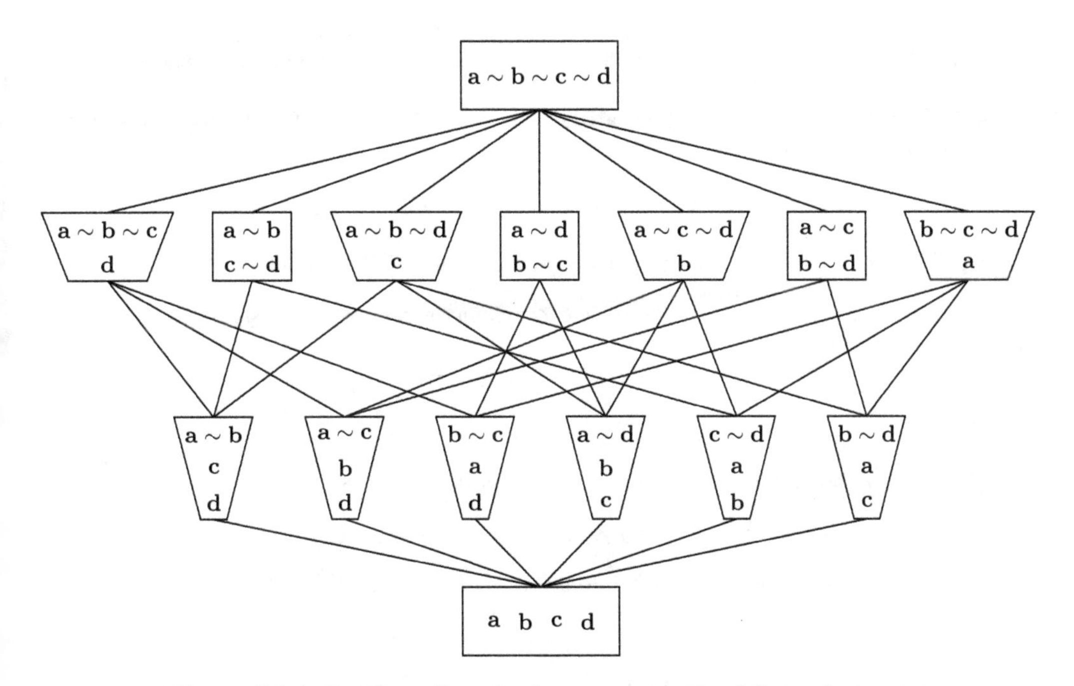

Figure I.5.1 Lattice of equivalences on a set of four elements

Exercise I.5.3. Given a subset Y of $\mathcal{R}(A)$, denote by X the subset of all partitions of a finite set A that are upper bounds of Y. The set X is nonempty as the one-element partition belongs to X and finite since $\mathcal{R}(A)$ is finite. Show that

$$\bigvee Y = \bigwedge X.$$

It follows that the partially ordered set $\mathcal{R}(A)$ of all equivalence relations on a given set A is a lattice. For sets A with a small number of elements, we can draw a diagram (graph) representing the lattice. Figure I.5.1 is the diagram of the lattice of all equivalences of a set of four elements $\{a, b, c, d\}$: $\mathcal{P}_1 \preceq \mathcal{P}_2$ if and only if there is a path in the graph from $\mathcal{P}_1$ to $\mathcal{P}_2$ of line segments directed upward.

In Section 4 of Chapter X we see that the lattice of equivalences is considerably larger for a set of five elements.

I.6 ⋆ Finite partially ordered sets, distributive lattices

In Chapter III, in dealing with divisibility of integers, we shall meet finite lattices L (of divisors of a given number) that will satisfy the condition

$$x \vee (y \wedge z) = (x \vee y) \wedge (x \vee z) \text{ for all } x, y, z \in L. \tag{I.6.1}$$

A lattice satisfying this condition will be called a **distributive lattice**. We shall see that the structure of distributive lattices reflects the divisibility structure of the integers.

Exercise I.6.1. Show that a lattice satisfies the condition (I.6.1) if and only if it satisfies the following "dual" condition

$$x \wedge (y \vee z) = (x \wedge y) \vee (x \wedge z) \text{ for all } x, y, z \in L.$$

The following statement is an important property of distributive lattices.

Theorem I.6.1. *Let L be a distributive lattice. Suppose $x, y, z \in L$ satisfy*

$$x \wedge z = y \wedge z \text{ and } x \vee z = y \vee z.$$

Then $x = y$.

Proof. To prove the statement, we apply a string of implications

$$\begin{aligned} x &= x \wedge (x \vee z) = x \wedge (y \vee z) = (x \wedge y) \vee (x \wedge z) \\ &= (y \wedge x) \vee (y \wedge z) = y \wedge (x \vee z) = y \wedge (y \vee z) = y, \end{aligned}$$

□

As a consequence of Theorem I.6.1, the following two lattices of five elements

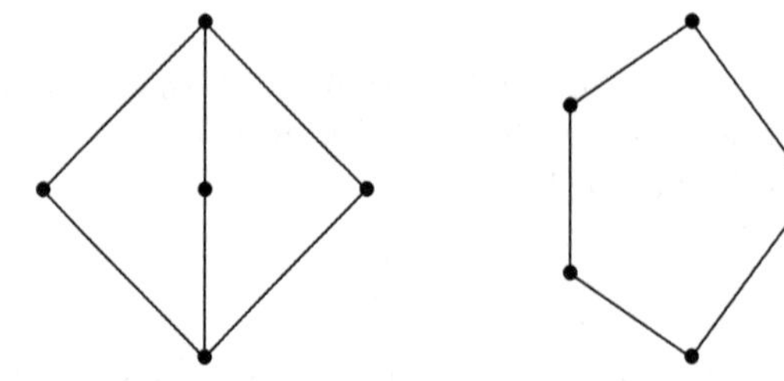

are not distributive.

Exercise I.6.2. Show that these two lattices are the only nondistributive lattices of five elements.

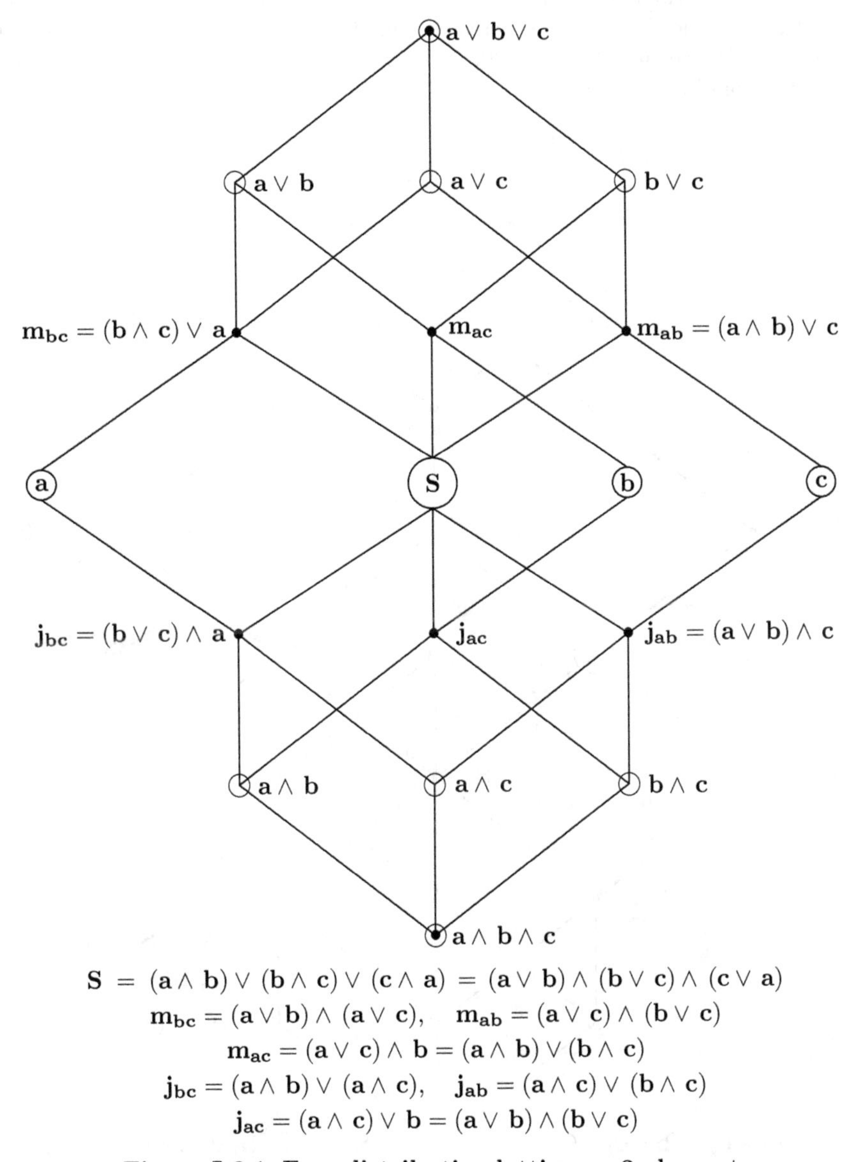

$$\mathbf{S} = (\mathbf{a} \wedge \mathbf{b}) \vee (\mathbf{b} \wedge \mathbf{c}) \vee (\mathbf{c} \wedge \mathbf{a}) = (\mathbf{a} \vee \mathbf{b}) \wedge (\mathbf{b} \vee \mathbf{c}) \wedge (\mathbf{c} \vee \mathbf{a})$$
$$\mathbf{m_{bc}} = (\mathbf{a} \vee \mathbf{b}) \wedge (\mathbf{a} \vee \mathbf{c}), \quad \mathbf{m_{ab}} = (\mathbf{a} \vee \mathbf{c}) \wedge (\mathbf{b} \vee \mathbf{c})$$
$$\mathbf{m_{ac}} = (\mathbf{a} \vee \mathbf{c}) \wedge \mathbf{b} = (\mathbf{a} \wedge \mathbf{b}) \vee (\mathbf{b} \wedge \mathbf{c})$$
$$\mathbf{j_{bc}} = (\mathbf{a} \wedge \mathbf{b}) \vee (\mathbf{a} \wedge \mathbf{c}), \quad \mathbf{j_{ab}} = (\mathbf{a} \wedge \mathbf{c}) \vee (\mathbf{b} \wedge \mathbf{c})$$
$$\mathbf{j_{ac}} = (\mathbf{a} \wedge \mathbf{c}) \vee \mathbf{b} = (\mathbf{a} \vee \mathbf{b}) \wedge (\mathbf{b} \vee \mathbf{c})$$

Figure I.6.1 Free distributive lattice on 3 elements

It can be shown that all distributive lattices that are generated by a finite number of elements are finite. This is true, in particular, about the (uniquely determined) **free distributive lattices**.

The free distributive lattice (without empty joins and meets) generated by two elements a and b has just 4 elements: $a, b, a \vee b, a \wedge b$. The free distributive lattice (without empty joins and meets) generated by three elements has already 18 elements. The size of these lattices grows dramatically with the increasing number of generators. It is known up to 8 generators: The lattice has then $56, 130, 437, 228, 687, 557, 907, 786$ elements. Figure I.6.1 shows the diagram of the free distributive lattice on 3 generators a, b, c.

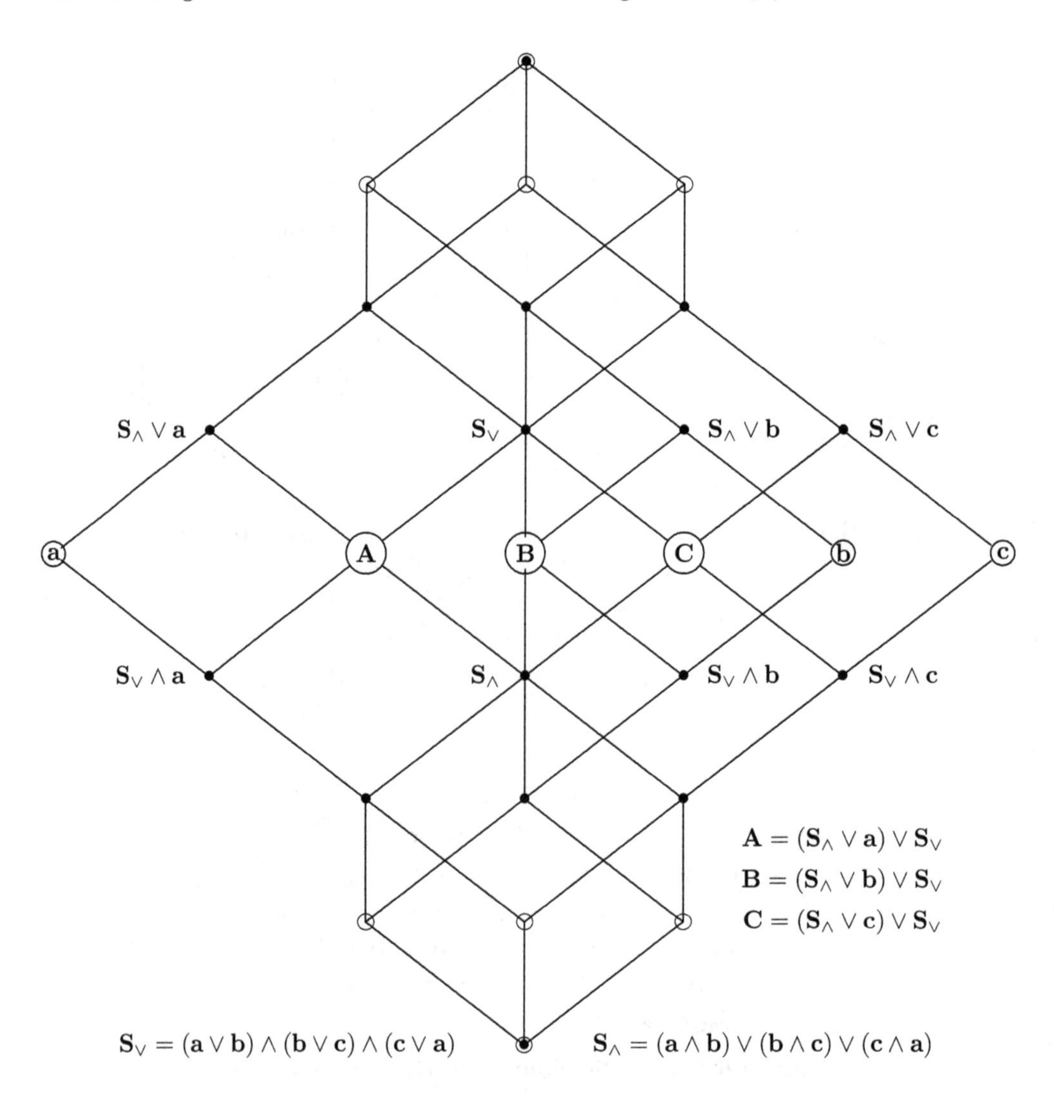

Figure I.6.2 Free modular lattice on 3 elements

Remark I.6.1. We mention that **Richard Dedekind (1831 - 1916)** showed that a **free modular lattice**, that is, a lattice satisfying the weaker condition

$$x \vee (y \wedge z) = (x \vee y) \wedge (x \vee z) \quad \text{for all } x \leq y \text{ or } x \leq z,$$

which is generated by three elements is finite and has 28 elements (see Figure I.6.2) while a free modular lattice on four or more generators is infinite.

Exercise I.6.3. Describe all elements of the lattice in Figure I.6.2 in terms of the generators a, b, c.

We are now going to exhibit a bijective relation between the family of finite distributive lattices and a family of particular finite partially ordered sets and show the way to represent this relation graphically. This will help us understand the bijective relation between the lattices that are disjoint unions of finite chains and certain subsets of natural numbers. For instance, we shall see that the lattice in Figure I.6.3 corresponds to the set of all natural numbers of the form $p_1 p_2^3 p_3^2$, where p_1, p_2, and p_3 are three different prime numbers.

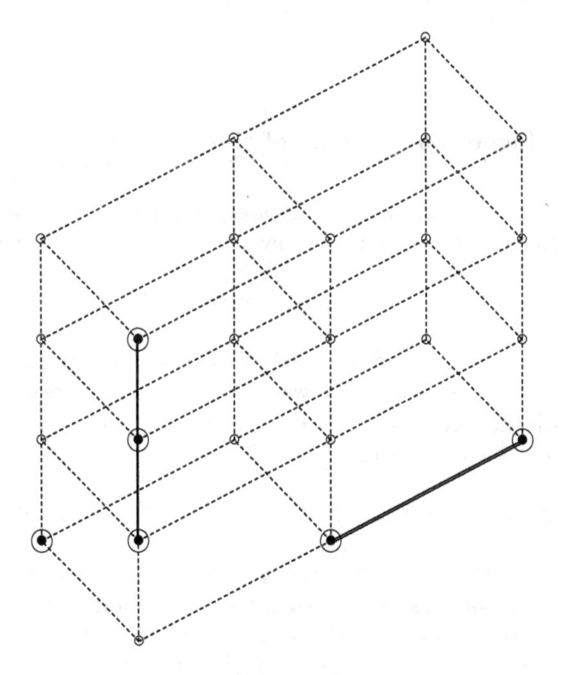

Figure I.6.3 Distributive lattice corresponding to the number $\mathbf{p_1 p_2^3 p_3^2}$

Exercise I.6.4. Describe the close relationship between Figures I.6.3 and I.6.4.

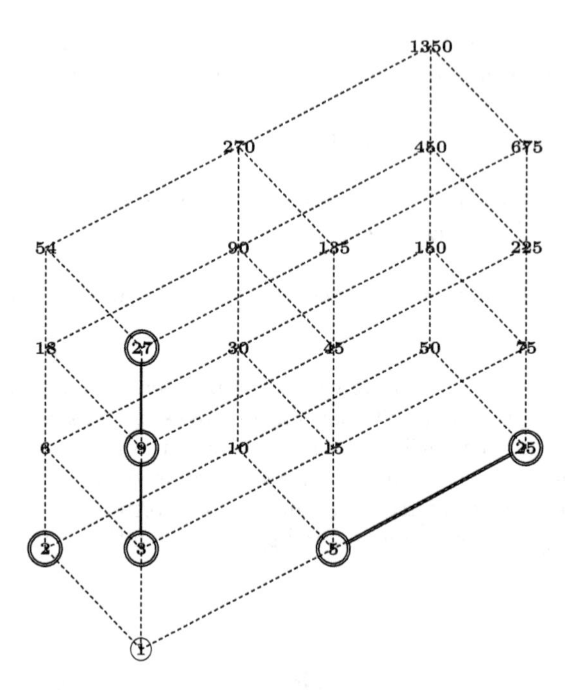

Figure I.6.4 Lattice of all divisors of 1350

We note that every finite lattice L has a "greatest" element $\mathbf{1} = \bigvee L$ and a "least" element $\mathbf{0} = \bigwedge L$. It will be convenient to define, for any $x \preceq y$ of a lattice L, the subset

$$L_{x,y} = \{z \in L \mid x \preceq z \preceq y\}.$$

We observe that $L_{x,y}$ is a lattice with the greatest element y and the least element x. When $L_{x,y}$ consists only of the elements x and y, we shall say that y **covers** x.

In accordance with the general use of the term, an **isomorphism** of two lattices L_1 and L_2 is a bijection $\varphi : L_1 \to L_2$ satisfying

$$\varphi(x \vee y) = \varphi(x) \vee \varphi(y) \quad \text{and} \quad \varphi(x \wedge y) = \varphi(x) \wedge \varphi(y) \quad \text{for all} \quad x, y \in L_1.$$

If such an isomorphism exists we say that the lattices L_1 and L_2 are **isomorphic.**

The terminology of two objects with the same structure **"being isomorphic"** will appear throughout the entire book. In each instance, be it for semigroups, groups or rings or any other objects, it will always be specified and properly defined.

We start with a simple statement

Theorem I.6.2. *Let $L = (L, \preceq)$ be a finite distributive lattice. Then, for any two elements $x, y \in L$, the lattices $L_1 = (\ L_{x,x\vee y}, \preceq\ )$ and $L_2 = (\ L_{x\wedge y,y}, \preceq\ )$ are isomorphic* (see Figure I.6.5).

Proof. The mappings $\varphi_y : L_1 \to L_2$ and $\varphi_x : L_2 \to L_1$ defined by

$$\varphi_y(z_1) = z_1 \wedge y \quad \text{and} \quad \varphi_x(z_2) = x \vee z_2 \quad \text{for} \quad z_1 \in L_1,\ z_2 \in L_2$$

are, with regard to the equalities

$$\varphi_x(\varphi_y(z_1)) = x \vee (z_1 \wedge y) = (x \vee z_1) \wedge (x \vee y) = z_1$$

and

$$\varphi_y(\varphi_x(z_2)) = (x \vee z_2) \wedge y = (x \wedge y) \vee (z_2 \wedge y) = z_2,$$

bijective. To derive the equality

$$\varphi_y(z_1 \vee z_1') = (z_1 \vee z_1') \wedge y = (z_1 \wedge y) \vee (z_1' \wedge y) = \varphi_y(z_1) \vee \varphi_y(z_1'),$$

as well as the other three equalities expressing the fact that φ_y and φ_x are isomorphisms, is left to the reader. □

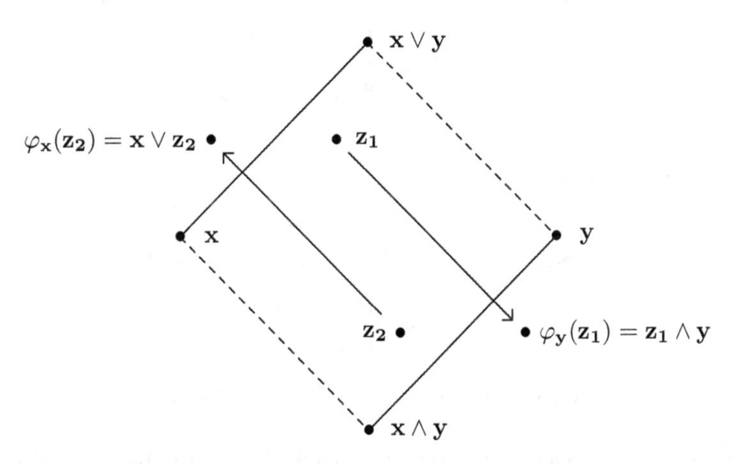

Figure I.6.5 Isomorphism of sectional lattices

In a finite lattice $L = (L, \preceq)$, there is, for any element $x \in L$ a chain

$$\mathbf{0} = x_0 \preceq x_1 \preceq x_2 \preceq \cdots \preceq x_r = x$$

such that x_{t+1} covers x_t for all $0 \leq t \leq r-1$. Such a chain (called the **Jordan-Hölder chain of** x) is not unique, and may have different lengths. This chain is named for **Ludwig Otto Hölder (1859 - 1937)** and **Marie Ennemond Camille Jordan (1838 - 1922)**. In a finite distributive lattice all these lengths are the same and we may speak about the **length** $d(x)$ **of the element** x.

The following two theorems are presented without detailed proofs. Both proofs make use of the principle of mathematical induction that will be discussed in Section 1 of Chapter II. These proofs can therefore be left as future exercises.

Theorem I.6.3. *Let L be a finite distributive lattice. Let*

$$\mathbf{0} = x_0 \preceq x_1 \preceq x_2 \preceq \cdots \preceq x_r = x$$

and

$$\mathbf{0} = y_0 \preceq y_1 \preceq y_2 \preceq \cdots \preceq y_s = x$$

be two Jordan-Hölder chains of the element x. Then $r = s$.

Figure I.6.6 indicates the idea of the proof.

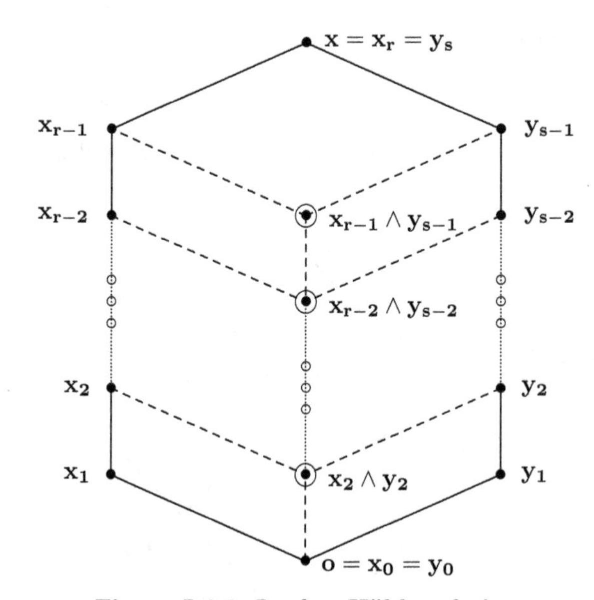

Figure I.6.6 Jordan-Hölder chains

The following concept plays a fundamental role for finite distributive lattices.

Definition I.6.1. *An element $g \neq \mathbf{0}$ is said to be* **irreducible** *(or, more precisely, $\vee$-irreducible) if, whenever $g = x \vee y$, necessarily either $g = x$ or $g = y$.*

Using distributivity, it is can be proved that whenever an irreducible element g satisfies $g \preceq \bigvee X$, then there is $x_0 \in X$ such that $g \preceq x_0$. Making use of this observation, we can state the following important assertion.

Theorem I.6.4. *Let $L = (L, \preceq)$ be a finite distributive lattice. Then, for every nonzero element $z \in L$, there is a uniquely defined subset $G_z \subseteq L$ of irreducible elements of L that satisfies*

$$z = \bigvee G_z \text{ and } z \neq \bigvee G \text{ for any proper subset } G \subset G_z.$$

Theorem I.6.4 has an immediate consequence that can be formulated as follows.

Corollary I.6.1. *There is a bijective mapping Φ from the family of finite distributive lattices onto the family of finite partially ordered sets, namely, $\Phi(L) = G_L$, where $G_L \subseteq L$ is the subset of all irreducible elements of L. The inverse mapping Φ^{-1} can be described as follows:*

If $(G, \preceq)$ is a finite partially ordered set, then $\Phi^{-1}[(G, \preceq)] = (L, \prec)$, where L is isomorphic to the set of all closed subsets I of G, that is, of subsets $I \subseteq G$ satisfying the following condition:

$$\textit{If } x \in I \textit{ and } y \preceq x, \textit{ then } y \in I.$$

Example I.6.1. In this example we illustrate the construction described in Corollary I.6.1. Let

$$(G, \preceq) = \{a, b, c, d, e \mid a \preceq d \preceq e,\ a \preceq c,\ b \preceq c,\ b \preceq e\}.$$

Then Figure I.6.7 describes the mapping Φ^{-1}:

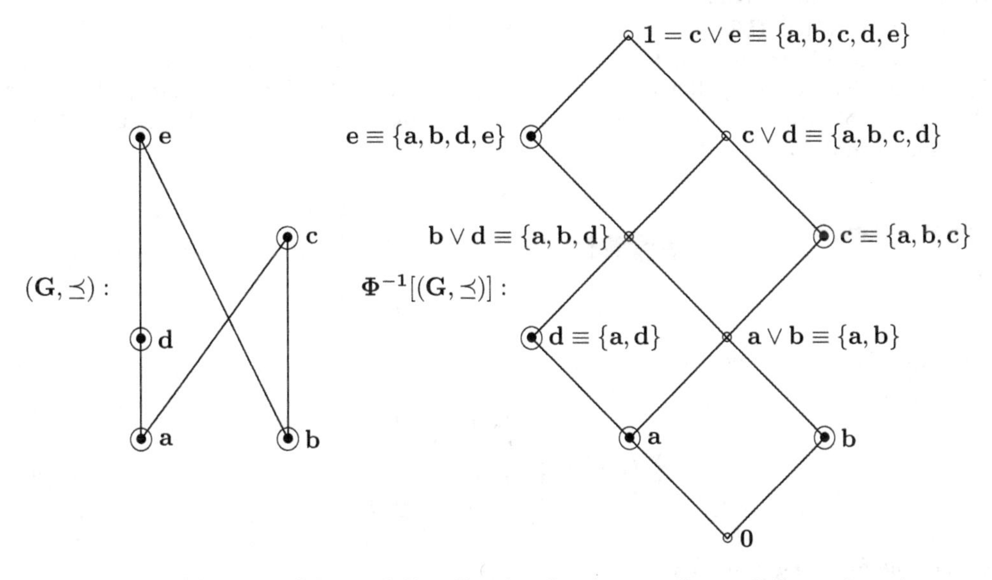

Figure I.6.7 Isomorphism of distributive lattices and partially ordered sets

Exercise I.6.5. Prove that a lattice is distributive if and only if, for any three elements a, b, c

$$a \wedge b = a \wedge c \text{ and } a \vee b = a \vee c \text{ implies } b = c.$$

I.7 Oriented graphs, path semigroups, monoids

The concepts to be introduced in this section serve as a background for further developments throughout the book. They unite various parts of mathematics and provide a solid tool for the introduction of polynomials in Chapter VII.

By a (finite) **oriented graph** (or **directed graph**) $\Gamma = (\mathbf{V}, \mathbf{H})$ we shall understand an arbitrary finite set $\mathbf{V} = \{V_1, V_2, \ldots, V_n\}$ of points called **vertices**, together with a finite set $\mathbf{H}$ of objects called **arrows**. It is customary to denote the vertices of a graph simply by $\{1, 2, \ldots, n\}$. Every arrow α has a **tail** (origin) $t(\alpha) = V_i$, and a **head** (end) $h(\alpha) = V_j$ for some $1 \leq i, j \leq n$. We shall record this fact by

$$\alpha : V(i) \to V(j) \quad \text{or by} \quad \alpha : t(\alpha) \to h(\alpha), \quad \text{or simply by} \quad i \xrightarrow{\alpha} j\,.$$

The **orientation** of the graph is given by two mappings $t : \mathbf{H} \to \mathbf{V}$ and $h : \mathbf{H} \to \mathbf{V}$, giving the direction of each arrow.

We observe that $\mathbf{H}$ is a disjoint union, $\mathbf{H} = \bigcup_{i,j} \mathrm{Hom}(V_i, V_j)$, where $\mathrm{Hom}(V_i, V_j)$ is the (finite) set of all arrows with $t(\alpha) = V_i$ and $h(\alpha) = V_j$; that is, all arrows originating at V_i and terminating at V_j. It may be the case that some of these sets are empty. Denoting the number of arrows in $\mathrm{Hom}(V_i, V_j)$ by t_{ij}, we can represent the graph Γ by its so-called **incidence matrix**

$$I(\Gamma) = \begin{pmatrix} t_{11} & t_{12} & \ldots & t_{1n} \\ t_{21} & t_{22} & \ldots & t_{2n} \\ \vdots & \vdots & \ddots & \vdots \\ t_{n1} & t_{n2} & \ldots & t_{nn} \end{pmatrix}.$$

Example I.7.1. The incidence matrix

$$I(\Gamma_0) = \begin{pmatrix} 0 & 1 & 1 & 0 \\ 1 & 1 & 2 & 0 \\ 0 & 0 & 0 & 1 \\ 0 & 0 & 3 & 2 \end{pmatrix}$$

defines the following graph Γ_0:

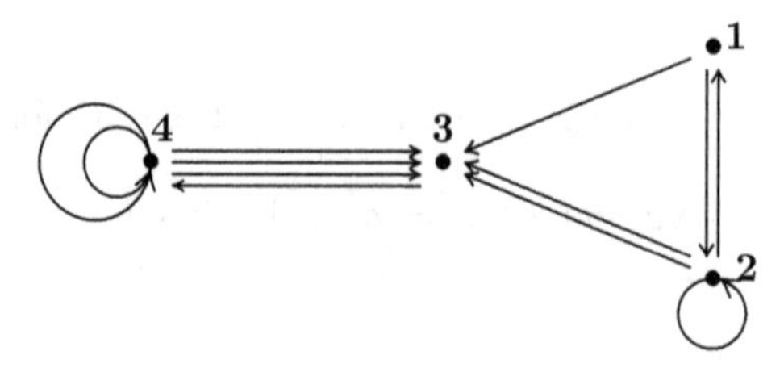

We leave it to the reader to describe the graphs whose incidence matrix is a symmetric matrix.

A **path** of an oriented graph is an ordered sequence of arrows $(\alpha_1\alpha_2\dots\alpha_n)$ such that the head of the arrow α_k coincides with the tail of the arrow α_{k+1}; that is, $h(\alpha_k)=t(\alpha_{k+1})$ for all $1\le k\le n-1$. The number k of the arrows is called the **length of the path.** The tail $t(\alpha_1)$ of the arrow α_1 is called the **tail of the path** and the head $h(\alpha_n)$ of the arrow α_n is called the **head of the path.** We shall say that such a path connects the vertex $i=t(\alpha_1)$ with the vertex $j=h(\alpha_n)$. A path for which the tail and the head coincide will be called an **oriented cycle**; it will be called a **loop** if its length equals 1. For every vertex i define formally a path ϵ_i of length 0; the vertex i is both its tail and head.

We denote the set of all paths of the graph Γ together with a symbol 0 (**zero**) by $\mathcal{S}=\mathcal{S}(\Gamma)$ and define on $\mathcal{S}$ the following (binary) operation $\circ$:

$$(\alpha_1\alpha_2\dots\alpha_n)\circ(\beta_1\beta_2\dots\beta_m)=(\alpha_1\alpha_2\dots\alpha_n\beta_1\beta_2\dots\beta_m),\text{ when } h(\alpha_n)=t(\beta_1),$$

$$(\alpha_1\alpha_2\dots\alpha_n)\circ(\beta_1\beta_2\dots\beta_m)=0 \text{ when } h(\alpha_n)\neq t(\beta_1) \text{ and}$$

$$0\circ\gamma=\gamma\circ 0=0\circ 0=0 \text{ for every path } \gamma \text{ of the graph } \Gamma.$$

In this way, we obtain what is called the **path semigroup** of Γ, denoted by $\mathcal{S}=(\mathcal{S},\circ)$. Note that the semigroup $\mathcal{S}$ is finite if and only if the graph Γ contains no oriented cycle.

Example I.7.2. The multiplication of the path algebra of the following graph

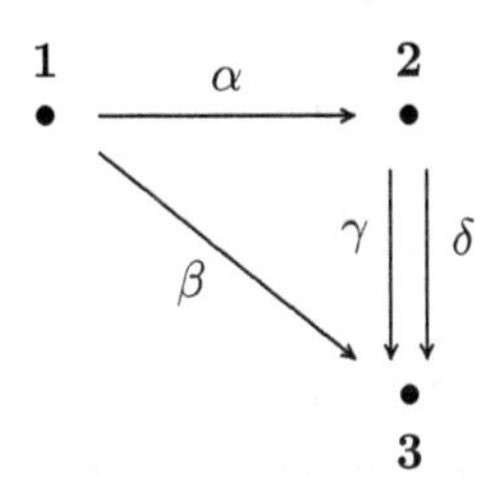

can be displayed in the **multiplication table**:

mult	0	ϵ_1	ϵ_2	ϵ_3	α	β	γ	δ	$\alpha\gamma$	$\alpha\delta$
0	0	0	0	0	0	0	0	0	0	0
ϵ_1	0	ϵ_1	0	0	α	β	0	0	$\alpha\gamma$	$\alpha\delta$
ϵ_2	0	0	ϵ_2	0	0	0	γ	δ	0	0
ϵ_3	0	0	0	ϵ_3	0	0	0	0	0	0
α	0	0	α	0	0	0	$\alpha\gamma$	$\alpha\delta$	0	0
β	0	0	0	β	0	0	0	0	0	0
γ	0	0	0	γ	0	0	0	0	0	0
δ	0	0	0	δ	0	0	0	0	0	0
$\alpha\gamma$	0	0	0	$\alpha\gamma$	0	0	0	0	0	0
$\alpha\delta$	0	0	0	$\alpha\delta$	0	0	0	0	0	0

Exercise I.7.1. Write down the multiplication table for the path semigroup of the graph

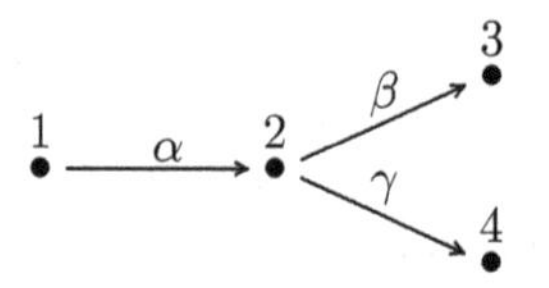

Compare this semigroup to the multiplicative semigroup consisting of ten 4×4 matrices

$$\{E_{11}, E_{22}, E_{33}, E_{44}, E_{12}, E_{23}, E_{24}, E_{13}, E_{14}, 0\};$$

Here, $E_{kl} = (e_{ij})$, where $e_{kl} = 1$ and $e_{ij} = 0$ otherwise. Therefore $E_{kl}E_{lm} = E_{km}$ and $E_{kl}E_{nm} = 0$ for $l \neq n$.

With this last example in mind, we define an **abstract semigroup** as follows.

Definition I.7.1. *A (nonempty) set S endowed with a binary associative operation $\circ$ is called a* **semigroup**, *and is denoted by $S = (S, \circ)$.*

A (nonempty) subset $T \subseteq S$ is said to be a **subsemigroup** *of $(S, \circ)$ if, for any two elements $s_1, s_2 \in T$, $s_1 \circ s_2$ also belongs to T.*

A semigroup $(M, \circ)$ containing a **neutral element** *$e \in M$, that is an element satisfying*

$$e \circ m = m \circ e = m \quad \text{for every element} \quad m \in M,$$

is called a **monoid**. *A* **submonoid** *of a monoid $(M, \circ)$ is a subsemigroup of $(M, \circ)$ containing the neutral element e.*

The neutral element e of a monoid $(M, \circ)$ can be proved to be unique, so we may denote it by 1. It is often called the **identity** element of $(M, \circ)$. We note that every semigroup $(S, \circ)$ can be "enlarged" to a monoid $(M, \circ)$ by setting $M = S \cup \{e\}$ and defining $e \circ e = e$ and $s \circ e = e \circ s = s$ for every element $s \in S$. We shall see that monoids appear in mathematics in a variety of situations.

An element $z \in S$ is called a **zero** of the semigroup S if

$$z \circ s = s \circ z = z \quad \text{for every element} \quad s \in S.$$

It can be proved that if a monoid $(M, \circ)$ possesses a zero element then there is only one such zero element in M, so we may speak of the zero of M, and denote it by 0. Again, as in the case of the path algebra, every semigroup can be extended to a semigroup with zero (by appending the zero element with appropriate operations $0 \circ 0 = 0 \circ s = s \circ 0 = 0$ for all $s \in S$).

In the familiar multiplicative monoid of all nonnegative integers $\mathbb{N}_0$, the identity and the zero are the numbers 1 and 0, respectively.

Another application of the concept of a monoid can be found in the theory of automata. A short description can be found in Section 5 of Chapter X.

Example I.7.3. We have seen that the path semigroups are semigroups with zero. In the case of the oriented graph

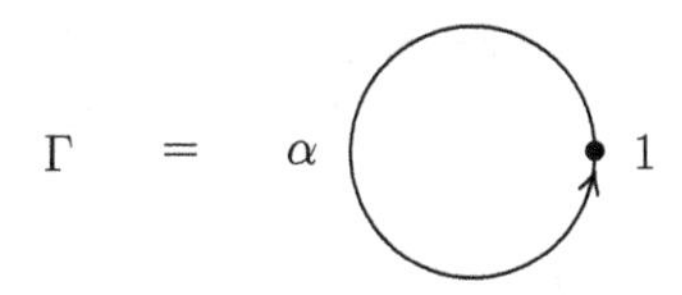

(whose incidence matrix is (1)), the head and tail of every path is the vertex 1. Consequently, the "composition" (or product) $\pi_1 \circ \pi_2$ is well-defined for any two paths π_1, π_2 of Γ. Thus, the set of all paths (without adjoining zero) is (with respect to the operation $\circ$) a semigroup S^1. This semigroup is **isomorphic** to the additive monoid of the non-negative numbers $\mathbb{N}_0$; that is, there is a bijection $f : S^1 \to \mathbb{N}_0$ such that

$$f(s_1 \circ s_2) = f(s_1) + f(s_2), f(\alpha) = 1, f(\epsilon_1) = 0.$$

We shall see that the graph Γ of Example I.7.3 will play a fundamental role in Chapter VII. Here are several further important examples of monoids and submonoids.

Example I.7.4. Define the monoid $F(A)$ of all "words" of finite length composed of letters (or symbols) of a finite set A (called the **alphabet**) along with the operation $\circ$ of **concatenation** (that is the stringing of words alongside each other). Thus

$$F(A) = \{a_{i_1} a_{i_2} \dots a_{i_k} \mid a_{i_t} \in A, 1 \leq t \leq k,\ k \in \mathbb{N}\} \cup \{e\},$$

where e is the empty word, and for $v = a_{i_1} a_{i_2} \dots a_{i_r}$ and $w = a_{j_1} a_{j_2} \dots a_{j_s}$ we have $v \circ w = a_{i_1} a_{i_2} \dots a_{i_r} a_{j_1} a_{j_2} \dots a_{j_s}$. We note that the letters in the strings $a_{i_1} a_{i_2} \dots a_{i_r}$ do not have to be distinct.

The composition of words is evidently associative. The **empty word** $e \in F(A)$, which has no letters, is the neutral element. Thus, $e \circ v = v \circ e = v$ for all words $v \in F(A)$. The monoid $F(A)$ is called the **free monoid over the basis** A.

Example I.7.5. (Definition of a **free cover of a monoid**). Let $(M, \circ)$ be an arbitrary monoid with the neutral element 1 and $B \subseteq M$ its **generating set**. This means that every element $m \in M$ can be expressed - not necessarily in a unique way - in the form

$$m = b_1 \circ b_2 \circ \dots \circ b_d \quad \text{where} \quad b_t \in B,\ t \in \{1, 2, \dots, d\}.$$

Let $F(B)$ be the free monoid over the basis B (defined in Example I.7.4). It is called a **free cover** of the monoid M if the following hold: The mapping $\varphi : F(B) \to M$ defined by

$$\varphi(v) = b_{i_1} \circ b_{i_2} \circ \dots \circ b_{i_r} \in M \quad \text{for every} \quad v = b_{i_1} b_{i_2} \dots b_{i_r} \in F(B) \quad \text{and} \quad \varphi(e) = 1$$

satisfies $\varphi(v \circ w) = \varphi(v) \circ \varphi(w)$ for all $v, w \in F(B)$ and is thus a homomorphism, a so-called **canonical** or **natural** homomorphism of $F(B)$ onto M. We note that its kernel $\mathrm{Ker}\varphi = \{a \in F(B) \mid \varphi(a) = 1\}$ is a submonoid of $F(B)$.

Example I.7.6. Consider the monoid $\mathcal{T}(X) = \mathrm{Hom}(X, X)$ of all mappings (transformations) $\alpha : X \to X$ of the set X with composition of mappings. We write $\alpha(x)$ for the image of $x \in X$, and note that $(\beta \circ \alpha)(x) = \beta(\alpha(x))$ for $\alpha, \beta \in \mathcal{T}(X)$. The **identity mapping** $1_X : X \to X$ of X is the neutral element of $\mathcal{T}(X)$.

The monoid $\mathcal{T}(X)$ contains the submonoid $\mathcal{S}(X)$ of all bijective transformations of X onto X. In this case, for every $\alpha \in \mathcal{S}(X)$, there is a (unique) **inverse element** $\beta \in \mathcal{S}(X)$ usually denoted by α^{-1} satisfying

$$\alpha \circ \beta = \beta \circ \alpha = 1_X.$$

The monoids with this property are called groups; they will be studied extensively in Chapter VIII. The set $\mathcal{S}(X)$ of all permutations of the set X with respect to the previously introduced algebraic structure is an important group called the **symmetric group** on the set X. **Arthur Cayley (1821 - 1895)** showed that any group is isomorphic to a subgroup of a symmetric group (see Theorem VIII.11.1).

The case when X is finite is important, particularly in the theory of counting (called **combinatorics** or **combinatorial theory**). If X has n elements, then $\mathcal{T}(X)$ has n^n elements and $\mathcal{S}(X)$ has $n!$ (n factorial) elements as already mentioned in Section I.2.

Next, we formulate two assertions concerning those semigroups with a certain desirable algebraic property called **cancellation.**

Definition I.7.2. *A semigroup S is called a* **semigroup with cancellation** *if, for every $s \in S, s \circ a = s \circ b$ implies that $a = b$ and similarly $a \circ s = b \circ s$ implies that $a = b$.*

A nontrivial semigroup with cancellation cannot contain the zero element since two distinct elements a and b in such a semigroup with 0 would satisfy $0 \circ a = 0 \circ b = 0$.

The first assertion shows that the cancellation property has an important consequence for finite monoids.

Theorem I.7.1. *In a finite monoid M with cancellation, for each element $x \in M$, there exists a (unique) element $y \in M$ such that $x \circ y = y \circ x = 1$. This unique element y will be denoted by x^{-1}. A monoid having this cancellation property is called a* **group**.

Proof. Let $\{x_1 = 1, x_2, \ldots, x_n\}$ be the set of all elements of the monoid M. Since M is a monoid with cancellation, the set $\{x \circ x_1, x \circ x_2, \ldots, x \circ x_n\}$ is again the set of all elements of the monoid M. Hence, there exists $i \in \{1, 2, \ldots, n\}$ such that $x \circ x_i = 1$. Similarly, there exists $j \in \{1, 2, \ldots, n\}$ such that $x_j \circ x = 1$. Now, since $x_j = x_j \circ (x \circ x_i) = (x_j \circ x) \circ x_i = x_i$, the assertions of the theorem follow. □

The second assertion involves construction of fractions. It concerns the commutative semigroups, that is, semigroups satisfying $s \circ t = t \circ s$ for all elements $s, t \in S$. The construction will be used a number of times in this book when constructing the integers, the rational numbers, the rational functions, and other systems. The important point of the next theorem is that S is not assumed to be a monoid, that is, S does not have to contain the identity element. Thus, it may be applied for instance to the multiplicative semigroup $S = k\mathbb{N}$ of all positive multiples of an (arbitrary) natural number k.

Theorem I.7.2. *Let $S = (S, \circ)$ be a commutative semigroup with cancellation. Then there exists a semigroup $G = (G, \circ)$ containing $(S, \circ)$ such that*

(1) *the semigroup G is commutative and contains the identity element 1;*

(2) *every* $g \in G$ *satisfies the equality* $s \circ g = r$ *for suitable elements* $s, r \in S$;

(3) *for every choice of elements* $s, r \in S$, *there is an element* $g \in G$ *satisfying* $s \circ g = r$.

As a consequence, for every element $g \in G$, *there exists its (uniquely defined) inverse* g^{-1} *satisfying* $g \circ g^{-1} = g^{-1} \circ g = 1$, *that is, the extension* $(G, \circ)$ *of the semigroup* $(S, \circ)$ *is a commutative group.*

Proof. We follow the standard construction: First, we construct the semigroup $(S \times S, \circ)$ of all ordered pairs of elements of S; here, $(s_1, s_2) \circ (r_1, r_2) = (s_1 \circ r_1, s_2 \circ r_2)$.

Next we define the following **equivalence relation** $\equiv$ on S, namely,

$$(s_1, s_2) \equiv (s_1', s_2') \text{ if and only if } s_1 \circ s_2' = s_2 \circ s_1'.$$

We denote the equivalence class containing a pair $(x_1, x_2) \in S \times S$ by $\overline{(x_1, x_2)}$ and the set of all $\equiv$–equivalence classes by G. For $(s_1, s_2) \equiv (s_1', s_2')$ and $(r_1, r_2) \equiv (r_1', r_2')$, we have

$$(s_1, s_2) \circ (r_1, r_2) \equiv (s_1', s_2') \circ (r_1', r_2'),$$

and we can define the binary operation $\circ$ on G by

$$\overline{(s_1, s_2)} \circ \overline{(r_1, r_2)} = \overline{(s_1 \circ r_1, s_2 \circ r_2)} = \overline{(s_1, s_2) \circ (r_1, r_2)}.$$

The operation $\circ$ on G is commutative and possesses the identity element $1 = \overline{(x, x)}$ (for any $x \in S$). Also, every element $s \in S$ can be identified with the element $\overline{(s \circ x, x)}$ of G. Thus $(S, \circ)$ is a subsemigroup of $(G, \circ)$.

Now, given $g = \overline{(s_1, s_2)} \in G$, we have

$$s \circ g = r \text{ for } s = \overline{(s_2 \circ x, x)} \text{ and } r = \overline{(s_1 \circ x, x)},$$

establishing the statement (2). Similarly, $g = \overline{(r, s)}$ satisfies $s \circ g = r$, verifying (3). Finally, given $g = \overline{(s_1, s_2)} \in G$, $g^{-1} = \overline{(s_2, s_1)}$, which completes the proof. □

Exercise I.7.2. Perform the construction described in Theorem I.7.2 for the multiplicative semigroup $S = k\mathbb{N}$ of all positive multiples of an (arbitrary) natural number k. Show that the resulting commutative group is the multiplicative group of all positive rational numbers.

Exercise I.7.3. Perform the construction described in Theorem I.7.2 for the multiplicative semigroup $S = \{p, p^2, p^3, \dots, p^k, \dots\}$ of all positive powers of a prime p. Is the assumption that p is a prime for the construction necessary?

Exercise I.7.4. Show that the group $(G, \circ)$ is "the smallest group extension" of the semigroup $(S, \circ)$ in the following sense: Denoting the injective mapping of $(S, \circ)$ into $(G, \circ)$ by α, then for any injective mapping β of $(S, \circ)$ into some group $(H, \circ)$, there is an injective mapping $\omega : G \to H$ such that $\omega\alpha = \beta$, that is, one can define ω so that the diagram

$$\begin{array}{ccc} S & \xrightarrow{\alpha} & G \\ \| & & \downarrow \omega \\ S & \xrightarrow{\beta} & H \end{array}$$

commutes.

Example I.7.7. $\star$ **(The monoid of arithmetic functions)** A mapping $f : \mathbb{N} \to \mathbb{C}$, from the natural numbers to the complex numbers is called an **arithmetic** function. Such functions arise naturally and frequently in number theory. In this example we establish an important result about arithmetic functions called the **Möbius inversion formula** in honor of **August Ferdinand Möbius (1790 - 1868)**.

We denote by A the set of all **arithmetic functions**. Let f and g belong to A. We define a **convolution** of f and g, $f \star g : \mathbb{N} \to \mathbb{C}$, by

$$(f \star g)\,(n) \;=\; \sum_{d|n} f(d)\, g(\frac{n}{d}),$$

where the sum runs through **all positive divisors** d **of the natural number** n. The function $f \star g$ is called the **Dirichlet product** of f and g in honor of **Lejeune Dirichlet (1805 - 1859)**. For example, we have

$$(f \star g)\,(12) \;=\; f(1)\,g(12) + f(2)\,g(6) + f(3)\,g(4) + f(4)\,g(3) + f(6)\,g(2) + f(12)\,g(1).$$

The concept of the convolution of two arithmetic functions arose in Dirichlet's work on infinite series of the type $\sum_{n=1}^{\infty} \frac{f(n)}{n^s}$, where f is a bounded arithmetic function and $s > 1$. Series of this type are called **Dirichlet series**. The arithmetic function f is called the Dirichlet coefficient of the Dirichlet series. If f and g are both bounded arithmetic functions then the product of their corresponding Dirichlet series is also a Dirichlet series with Dirichlet coefficient $f \star g$, that is

$$\sum_{n=1}^{\infty} \frac{f(n)}{n^s} \sum_{n=1}^{\infty} \frac{g(n)}{n^s} = \sum_{n=1}^{\infty} \frac{(f \star g)(n)}{n^s}.$$

Exercise I.7.5. Show that the binary operation $\star$ is commutative, that is, $f \star g = g \star f$ for all $f, g \in A$, and associative, that is, $(f \star g) \star\, h \;=\; f \star\, (g \star h)$ for all $f, g, h \in A$.

Show that the function $e : \mathbb{N} \to \mathbb{C}$ defined by $e(1) = 1$ and $e(n) = 0$ for $n > 1$ satisfies the relation

$$f \star e \;=\; e \star f \;=\; f \quad \text{for every } f \in A,$$

and conclude that $(A, \star)$ is a (commutative) monoid.

We consider two particular arithmetic functions and their $\star$-products.

We denote by $\varepsilon \in A$ the identity function (in contrast to the identity element e) defined by $\varepsilon(n) = 1$ for all $n \in \mathbb{N}$. Furthermore, we define the **Möbius function** μ by

$$\mu(n) = \begin{cases} 1 & \text{if } n = 1; \\ (-1)^k & \text{if } n = p_1\, p_2\, \ldots p_k, \text{where } p_t \text{ are distinct primes;} \\ 0 & \text{otherwise.} \end{cases}$$

We observe that the definition of the Möbius function requires knowing that a positive integer can be expressed uniquely as a product of primes, see Theorem III.10.2.

We denote by $\mathcal{S}_f$ the **sum function** of $f \in A$, which is defined by

$$\mathcal{S}_f(n) = \sum_{d|n} f(d).$$

We have that $\mathcal{S}_f = f \star \varepsilon = \varepsilon \star f$ for all $f \in A$. In Chapter III, see Remark III.10.3, we prove that

$$\mu \star \varepsilon \;=\; \varepsilon \star \mu \;=\; e,$$

that is, the $\star$-inverse of ε is μ. As an immediate consequence, we obtain the so-called **Möbius inversion formula**

$$f(n) \;=\; \sum_{d|n} \mu(d)\, \mathcal{S}_f(\frac{n}{d}), \quad \text{that is} \quad f = \mu \star \mathcal{S}_f.$$

Indeed, $\mu \star \mathcal{S}_f \;=\; \mu \star (\varepsilon \star f) \;=\; (\mu \star \varepsilon) \star f \;=\; e \star f \;=\; f$. We summarize this result:

$$g(n) \;=\; \sum_{d|n} f(d) \;\text{ if and only if }\; f(n) \;=\; \sum_{d|n} \mu(d)\, g(\frac{n}{d}).$$

Exercise I.7.6. Let $n \in \mathbb{N}$. Use the result

$$\sum_{d|n} \mu(d) = \begin{cases} 1 & \text{if } n = 1, \\ 0 & \text{if } n > 1, \end{cases}$$

to prove that if

$$g(n) = \sum_{d^2|n} f(n/d^2)$$

then

$$f(n) = \sum_{d^2|n} \mu(d) g(n/d^2).$$

Exercise I.7.7. Let $n \in \mathbb{N}$. We say that n is squarefree if $m^2 \mid n$ implies $m = 1$. Prove that

$$\sum_{d^2|n} \mu(d) = \begin{cases} 1 & \text{if } n \text{ is squarefree}, \\ 0 & \text{otherwise}. \end{cases}$$

Exercise I.7.8. Let $n \in \mathbb{N}$. Define $r(n) :=$ number of pairs (x, y) with $x \in \mathbb{Z}$ and $y \in \mathbb{Z}$ such that $x^2 + y^2 = n$ and $p(n) :=$ number of pairs (x, y) with $x \in \mathbb{Z}$ and $y \in \mathbb{Z}$ such that $x^2 + y^2 = n$ and $d(x, y) = 1$. Prove that

$$r(n) = \sum_{d^2|n} p(n/d^2).$$

Exercise I.7.9. Let $n \in \mathbb{N}$. It is known that

$$r(n) = 4 \sum_{\substack{d \mid n \\ d \equiv 1 \pmod 2}} (-1)^{(d-1)/2}.$$

Use Exercises I.7.6, I.7.7 and I.7.8 to show that

$$p(n) = 4 \sum_{\substack{d \mid n \\ d \equiv 1 \pmod 2 \\ n/d \text{ squarefree}}} (-1)^{(d-1)/2}.$$

What is the value of $p(n)$ if n is the product of k distinct primes $p \equiv 1 \pmod 4$?

CHAPTER II. NATURAL NUMBERS

II.1 Well-ordering and mathematical induction

The human mind's conception of the natural numbers as a never-ending list is in direct contrast with the apparent finiteness of human activities, experience, and life. It is one of the first moments when mathematics presents itself as something supernatural, perfect, and ideal, and as extending beyond individual or collective human experience.

The role of the natural numbers in mathematics is absolutely fundamental. This single set of objects possesses and allows notions of labeling, order, size, and especially arithmetic - the last of which is manifested in the familiar properties of addition and multiplication.

The basic structure of the system of natural numbers can be formally presented in the form of the **Peano axioms**, named in honor of **Giuseppe Peano (1858 - 1932)**.

The set $\mathbb{N}$ defined by the following properties is called the set of **natural numbers**.

(1) *There is a natural number* $1 \in \mathbb{N}$.

(2) *Every natural number* n *has a successor* $s(n) \in \mathbb{N}$.

(3) *There is no natural number whose successor is* 1.

(4) *If* $m \neq n$ *are (distinct) natural numbers, then* $s(n) \neq s(m)$.

(5) (**Principle of mathematical induction**) *Let* S *be any subset of the natural numbers satisfying the following two properties.*

   (i) $1 \in S$.

   (ii) *If* $n \in S$, *then* $s(n) \in S$.

   Then S *is the entire set* $\mathbb{N}$ *of all natural numbers.*

There are two binary algebraic operations, **addition** and **multiplication** of natural numbers denoted by $+$ and $\cdot$, respectively. It is worthwhile to point out that the Peano axioms involve only a **unary operation** $s : \mathbb{N} \to \mathbb{N}$ of assigning a successor. The binary operations of addition and multiplication (that is, in fact, just a "repeated" addition) are defined recursively in terms of s by

$$m + n = s^m(n) \text{ and } m \cdot n = s^{(m-1)\cdot n}(n).$$

We note that the inductive definition of multiplication requires the base step $1 \cdot n = n$ and can be expressed in the form

$$s(m) \cdot n = m \cdot n + n = s^{m \cdot n}(n).$$

The set $\mathbb{N}$ is endowed with a natural **order** denoted by $\leq$, determined by $n \leq s(n)$ and transitivity (for the definition of $<$ and further details see P. Halmos: Naive Set Theory, Springer-Verlag 1974). The principle of mathematical induction can be replaced by the following requirement placed on this order and often chosen as an axiom.

Well-ordering principle. *Every nonempty set of natural numbers contains a smallest number.*

There is an immediate consequence of this principle.

Theorem II.1.1 (The least offender (minimal counter-example)). *Let $\mathcal{S}(1), \mathcal{S}(2), \ldots, \mathcal{S}(n), \ldots$ be a sequence of statements, one for each natural number n. If some of these statements are false, then there is a first false statement, that is, there is $k \in \mathbb{N}$ such that $\mathcal{S}(k)$ is false while all $\mathcal{S}(i)$ for $1 \leq i < k$ are true.*

Proof. Let $S \subseteq \mathbb{N}$ be the set of all natural numbers n for which $\mathcal{S}(n)$ is false. By assumption, S is nonempty. Hence, by the well-ordering principle, there is a least integer $n_0 \in S$, and $\mathcal{S}(n_0)$ is the first false statement. □

Exercise II.1.1. Prove that for every $n \in \mathbb{N} \setminus \{1\}$ there is a unique $m \in \mathbb{N}$ (the predecessor of n) such that $n = s(m)$.

The principle of mathematical induction can be formulated in the following form suitable for applications.

Principle of mathematical induction. *Given statements $\mathcal{S}(n)$, one for each natural number n, suppose that*

(1) *$\mathcal{S}(1)$ is true* (**the base step**), *and*

(2) *if $\mathcal{S}(n)$ is true for a certain $n \geq 1$, then $\mathcal{S}(n+1)$ is true* (**the inductive step**).

Then $\mathcal{S}(n)$ is true for all natural numbers n.

Example II.1.1. *In this example we use the principle of mathematical induction to prove the inequality*

$$(1 + x)^n \geq 1 + nx \textit{ for all real numbers } x \geq 0 \textit{ and all natural numbers } n.$$

Let x be a fixed real number with $x \geq 0$. Let $\mathcal{S}(n)$ be the statement that $(1+x)^n \geq 1+nx$. As $(1+x)^1 = 1+1\cdot x$, the statement $\mathcal{S}(1)$ is true. Now suppose that $\mathcal{S}(n)$ is true for some natural number n. Then we have

$$(1+x)^{n+1} = (1+x)(1+x)^n \geq (1+x)(1+nx) = 1+(n+1)x+nx^2 \geq 1+(n+1)x$$

so that $\mathcal{S}(n+1)$ is true. Hence, by the principle of mathematical induction, $\mathcal{S}(n)$ holds for all natural numbers n.

Exercise II.1.2. Show that the preceding principle of mathematical induction can be formulated in an equivalent way as follows: *Given statements $\mathcal{S}(n)$, one for each natural number n, suppose that*

(i) *$\mathcal{S}(1)$ is true and*

(ii) *if $\mathcal{S}(k)$ is true for all $k \in \{1,2,\ldots,n\}$, then $\mathcal{S}(n+1)$ is true.*

Then $\mathcal{S}(n)$ is true for all natural numbers n.

The following theorem shows that the relation between the above formulated principles is very simple.

Theorem II.1.2. *The well-ordering principle and the principle of mathematical induction are equivalent.*

Proof. The proof involves the following two implications.

(a) The principle of mathematical induction follows from the principle of well-ordering. Let $S \subseteq \mathbb{N}$ be the set of all natural numbers for which $\mathcal{S}(n)$ is false. Let S be nonempty. Then, by Theorem II.1.1, there is a least integer n_0 in S such that $\mathcal{S}(n_0)$ is a false statement. By (1), $n_0 \neq 1$. Hence, there is a predecessor $n = n_0 - 1$, and $\mathcal{S}(n)$ is true. Now, by (2), $\mathcal{S}(n+1) = \mathcal{S}(n_0)$ is true, and this is a contradiction. We conclude that S is empty and thus $\mathcal{S}(n)$ holds for all natural numbers n.

(b) The principle of well-ordering follows from the principle of mathematical induction. Let S be a nonempty subset of $\mathbb{N}$ having no least number. We denote by C the set of all natural numbers that do not belong to S (that is, $C = \mathbb{N} \setminus S$). For every natural number n, we denote by $\mathcal{S}(n)$ the following statement

$$\{1,2,\ldots,n\} \subseteq C.$$

Clearly $\mathcal{S}(1)$ is true (otherwise 1 would be the least number in S). Also, if $\mathcal{S}(n)$ is true, so is $\mathcal{S}(n+1)$; for, otherwise $n+1$ would be the least number in S. Therefore, by the principle of mathematical induction, it follows that $C = \mathbb{N}$. Hence $S = \emptyset$, in contradiction to the assumption that S is nonempty. Consequently, S contains a smallest number.

□

Remark II.1.1. We clarify the notion of the principle of mathematical induction vis-a-vis the meaning of inductive reasoning. Consider the following example of a **Pell's equation** for natural numbers m, n

$$m^2 = 991n^2 + 1. \tag{II.1.1}$$

In Section 6 of Chapter IV, in dealing with applications of **continued fractions**, we shall see that this equation has an infinite number of solutions. The smallest solution of (II.1.1), that is the least value of n for which the number $991n^2 + 1$ is the square of a natural number, is

$$n = 12,055,735,790,331,359,447,442,538,767 \approx 1.2 \times 10^{28}.$$

For this n, $991n^2 + 1$ is the square of the number

$$m = 379,516,400,906,811,930,638,014,896,080.$$

Now, consider, for each $n \in \mathbb{N}$, the following statement

$$\mathcal{S}(n) : \ 991n^2 + 1 \ \text{ is not a square of a natural number.} \tag{II.1.2}$$

Suppose, starting from the very origin of our planet Earth (whose age is estimated to be less than 10^{10} years $< 10^{13}$ days), we started to verify the statement (II.1.2) for the value on the n-th day, then we would find that up to the present day the statement $\mathcal{S}(n)$ would be true every day. Inductive reasoning would then suggest that it should also be true tomorrow and thus forever, that is, the number $991n^2 + 1$ is never a square of a natural number. But that is NOT true. This surprising fact illustrates the profound difference between inductive reasoning and mathematical induction.

Of course, there are many less dramatic examples illustrating this feature. One can, for example, verify, step by step, that every odd number $1 < n < 5777$ can be written in the form $n = p + m^2$, where p is a prime number and $m \in \mathbb{N}$. Then, unexpectedly, this statement is not true for $n = 5777$.

Exercise II.1.3. Prove that for every natural number n,

(i) $11^{n+1} + 12^{2n-1}$ is a multiple of 133,

(ii) $1^2 - 2^2 + 3^2 - 4^2 + \cdots + (-1)^{n-1}n^2 \ = \ \frac{1}{2}\,(-1)^{n-1}n(n+1)$,

(iii) $1^3 + 3^3 + 5^3 + \cdots + (2n-1)^3 \ = \ n^2\,(2n^2 - 1)$,

(iv) $\frac{4^n}{n+1} \ < \ \frac{(2n)!}{(n!)^2}$.

II.2 Binomial coefficients and integer sequences

We start with two basic statements as exercises for the reader. Recall that, for a natural number n and an integer k satisfying $0 \le k \le n$, the **binomial coefficient** $\binom{n}{k}$ is defined by

$$\binom{n}{k} = \frac{n!}{(n-k)!\,k!} \tag{II.2.1}$$

with the understanding that $0! = 1$ (and thus $\binom{n}{0} = 1$ and $\binom{0}{0} = 1$). For $k > n$ we define $\binom{n}{k} = 0$. We note that, for $0 \le k \le n$ and $0 \le l \le n-k$ we have $\binom{n}{k} = \binom{n}{k+l}$ if and only if $n = 2k + l$.

Exercise II.2.1 (Binomial theorem). Prove, for **any numbers** a and b, and any natural number n,

$$(a+b)^n = \sum_{k=0}^{n} \binom{n}{k} a^{n-k} b^k.$$

Exercise II.2.2. Prove, for **any numbers** a and b, and any natural number n,

$$a^n - b^n = (a-b) \sum_{k=0}^{n-1} a^k\, b^{n-k-1}.$$

Both exercises can be proved using mathematical induction and, in the first statement, applying the formula of **al-Karaji (Abu Bakr al-Karaji (953 - about 1029))**

$$\binom{n}{k} + \binom{n}{k+1} = \binom{n+1}{k+1}. \tag{II.2.2}$$

The equality (II.2.2) follows immediately from the fact that $\binom{n}{k}$ denotes the number of choices of k elements from a set of n elements, the number of **combinations** of k elements. Selecting one particular element e from a set of $n+1$ elements, we can see right away that the number $\binom{n+1}{k+1}$ of combinations splits into $\binom{n}{k+1}$ combinations that **do not contain** the element e and $\binom{n}{k}$ combinations that **do contain** the element e. Of course, (II.2.2) can also be obtained by substituting the expressions for $\binom{n}{k}$, $\binom{n}{k+1}$ and $\binom{n+1}{k+1}$ from (II.2.1) into (II.2.2).

Taking the special case $a = b = 1$ in the result of Exercise II.2.1 we find

$$2^n = \sum_{k=0}^{n} \binom{n}{k}. \tag{II.2.3}$$

This describes the fact that a set with n elements has 2^n subsets. Similarly, for $a = 1$ and $b = -1$, we have

$$\sum_{k=0}^{n} (-1)^k \binom{n}{k} = 0.$$

This shows that the number of subsets with an odd number of elements equals the number of subsets with an even number of elements.

Exercise II.2.3. Prove the following general version of (II.2.2): For any triple $0 \le k \le r \le n$,

$$\sum_{s=r}^{n} \binom{s}{k} = \binom{n+1}{k+1} - \binom{r}{k+1}.$$

Example II.2.1. The general form of al-Karaji's formula (II.2.2) reads

$$\binom{m+n}{k} = \sum_{t=0}^{k} \binom{m}{k-t} \binom{n}{t}. \tag{II.2.4}$$

To see the equality (II.2.4), we express the number of combinations of k elements from a set S of $m+n$ elements in the following way: Consider the set S as the union of a subset M containing m elements and its complement $N = S \setminus M$ which has thus n elements. Now, the number of all subsets of S that have k elements with t of them belonging to the subset N is clearly $\binom{m}{k-t}\binom{n}{t}$. Adding all these numbers for $t = 0, 1, \ldots, k$, we obtain (II.2.4).

We note that for $m = n = k$ (II.2.4) gives the identity

$$\binom{2n}{n} = \sum_{t=0}^{n} \binom{n}{t}^2.$$

Exercise II.2.4. Prove (II.2.4) by considering the identity

$$(x+1)^m \, (x+1)^n \; = \; (x+1)^{m+n}.$$

Example II.2.2. Applying the definition of the binomial coefficient given in (II.2.1), we readily obtain

$$\binom{n}{s}\binom{n-s}{k-s} = \binom{n}{k}\binom{k}{s} \quad \text{for any natural numbers} \quad s \le k \le n. \tag{II.2.5}$$

In view of

$$2^k = \sum_{s=0}^{k} \binom{k}{s}$$

and (II.2.5), we obtain the equality

$$\binom{n}{k} 2^k = \sum_{s=0}^{k} \binom{n}{s}\binom{n-s}{k-s}. \tag{II.2.6}$$

Exercise II.2.5. Prove (II.2.6) directly by considering the set of all words composed of the letters X, Y and Z that have length n with exactly $n-k$ letters X and s letters Y.

Next, we exhibit some remarkable equalities which do not allow simple inductive explanations, but are rather easily justified in terms of representations as particular combinations.

Example II.2.3. For any integers k, l, n satisfying $0 \le k$, $0 \le l$ and $0 \le n \le k+l$, prove that

$$\sum_{r=0}^{k} \binom{n}{r}\binom{k+l-n}{k-r} 2^{n-r} = \sum_{s=0}^{l} \binom{n}{s}\binom{k+l-s}{k}. \tag{II.2.7}$$

Thus, choosing $k = l = n$, we deduce the identity

$$\sum_{t=0}^{n} \binom{n}{t}^2 2^t = \sum_{k=0}^{n} \binom{n}{k} \binom{2n-k}{n}.$$

In order to prove the equality (II.2.7) we observe that both sides of this formula represent the number of words composed from the letters X, Y and Z, and satisfying the following conditions: Every word has length $k+l$, contains exactly k letters X, and all letters Y must be among the first n letters from the left. We count these words in two distinct ways: First we count the $\binom{n}{r}$ ways of choosing r letters X among the first n letters with 2^{n-r} ways of choosing letters Y, and $\binom{k+l-n}{k-r}$ ways of choosing $k-r$ letters X among the last $k+l-n$ letters. Second we count the $\binom{n}{s}$ ways of choosing s letters Y among the first n letters, and $\binom{k+l-s}{k}$ ways of choosing k letters X among the remaining $k+l-s$ letters. These two counts must be equal, and the result follows.

Remark II.2.1. In order to get a better understanding of these identities, we can derive further equalities by choosing special values for k, l and n in (II.2.7).

Exercise II.2.6. Let m and n be natural numbers. Show that

$$\sum_{k=0}^{n} \binom{m+k}{m} = \binom{n+m+1}{m+1}.$$

Next, we introduce a concept that finds its motivation in the study of sequences of numbers. This concept will reveal the importance of binomial coefficients in the study of these sequences. All statements hold for sequences of real or complex numbers, however our interest will be confined to sequences of integers. The crucial idea of our approach is related to the concept of a "**differential**" of a given sequence. We illustrate it on some simple examples.

Given a sequence

$$\mathbf{a} = (a_1, a_2, \dots, a_n, \dots)$$

of numbers, we can create a new sequence, a "differential" of $\mathbf{a}$, by defining

$$d_1\mathbf{a} = (a_2 - a_1, a_3 - a_2, \dots, a_{n+1} - a_n, \dots).$$

We can keep repeating this process to obtain successively the sequences

$$d_2\mathbf{a}, d_3\mathbf{a}, \dots, d_m\mathbf{a}, \dots, \quad \text{where } d_m\mathbf{a} = d_1(d_{m-1}\mathbf{a}).$$

This process is closely related to a simple property of the sequence of squares, the sequence of cubes, in fact the sequence of the k-th powers of natural numbers for any $k = 1, 2, \dots$. One can see that for

$$\mathbf{a} \;=\; (\,1,\quad 4,\quad 9,\quad 16,\quad 25,\quad \dots\quad n^2,\quad \dots)$$

we have

$$\begin{array}{lcllllllll} d_1\mathbf{a} & = & (\,3, & 5, & 7, & 9, & 11, & \ldots, & 2n+1, & \ldots), \\ d_2\mathbf{a} & = & (\,2, & 2, & 2, & 2, & 2, & \ldots, & 2, & \ldots), \\ d_3\mathbf{a} & = & (\,0, & 0, & 0, & 0, & 0, & \ldots, & 0, & \ldots). \end{array}$$

Similarly, if

$$\mathbf{a} \;=\; (\,1, \;\; 8, \;\; 27, \;\; 64, \;\; 125, \;\; \ldots, \;\; n^3, \;\; \ldots),$$

we have

$$\begin{array}{lcllllllll} d_1\mathbf{a} & = & (\,7, & 19, & 37, & 61, & 91, & \ldots, & 3n^2+3n+1, & \ldots), \\ d_2\mathbf{a} & = & (\,12, & 16, & 24, & 30, & 36, & \ldots, & 6n+6, & \ldots), \\ d_3\mathbf{a} & = & (\,6, & 6, & 6, & 6, & 6, & \ldots, & 6, & \ldots), \\ d_4\mathbf{a} & = & (\,0, & 0, & 0, & 0, & 0, & \ldots, & 0, & \ldots). \end{array}$$

The fact that we end up with the zero sequence after a finite number of steps is an important property that allows us to express the n-th term of such sequences as an integral linear combination of binomial coefficients. For instance it applies to a sequence

$$\mathbf{a} = (P(1), P(2), \ldots P(n), \ldots)$$

consisting of the values of polynomials $P(x) = c_0 + c_1x + c_2x^2 + \cdots + c_dx^d$, where the coefficients c_i are integers and $c_d \neq 0$.

For example, for the polynomial

$$P(x) = -12 + x + 2x^2,$$

we obtain sequentially the following sequences

P(1)	P(2)	P(3)	P(4)	P(5)	P(6)	P(7)	P(8)	P(9)	P(10)	···	···
−9	−2	9	24	43	66	93	124	159	198	···	···
7	11	15	19	23	27	31	35	38	···	···	···
4	4	4	4	4	4	4	4	···	···	···	···
0	0	0	0	0	0	0	···	···	···	···	···

of which the general terms are

P(n−3) **P(n−2)** **P(n−1)** **P(n)**

P(n−2)−P(n−3) **P(n−1)−P(n−2)** **P(n)−P(n−1)**

P(n−1)−2P(n−2)+P(n−3) **P(n)−2P(n−1)+P(n−2)**

P(n)−3P(n−1)+3P(n−2)−P(n−3)=0

resulting after three steps in the zero sequence; that is, for all $n \geq 4$,

$$P(n) - 3P(n-1) + 3P(n-2) - P(n-3) = \sum_{k=0}^{3}(-1)^k\binom{3}{k}P(n-k) = 0.$$

Similarly, for the polynomial

$$P(x) = 5 - 8x - 3x^2 + 2x^3,$$

we obtain by differentiating the sequences

P(1)	P(2)	P(3)	P(4)	P(5)	P(6)	P(7)	P(8)	P(9)	P(10)	...	...
−4	−7	8	53	140	281	488	773	1148	1625	...	...
−3	15	45	87	141	207	285	375	477	...	...	...
18	30	42	54	66	78	90	102	...	...	...	...
12	12	12	12	12	12	12	...	...	...	...	...
0	0	0	0	0	0	...	...	...	...	...	...

and, for all $n \geq 5$, we have

$$P(n) - 4P(n-1) + 6P(n-2) - 4P(n-3) + P(n-4) = \sum_{k=0}^{4} (-1)^k \binom{4}{k} P(n-k) = 0.$$

In Section II.3, we show that the set of all sequences of numbers having the property that after a finite number of steps of differentiation they become the zero sequence are, in fact, in a bijection with the set of polynomials, whose values at integral points are integers. An example of such a polynomial whose coefficients are not all integers is the polynomial $\frac{1}{2}x + \frac{1}{2}x^2$.

We are now ready to formulate the general concept of a differential scheme, derive some of its basic properties and illustrate them on a number of well-known sequences.

Definition II.2.1. *By a* **differential scheme** $\mathbb{D}$ *of integers (possibly of real or complex numbers) we shall understand an infinite matrix* $(a_{m\,n})$ *for* $m \geq 1$ *and* $n \geq 1$ (see Figure II.2.1) *that satisfies, for all* m, n *the relations*

$$a_{m+1\,n} = a_{m\,n+1} - a_{m\,n}.$$

A differential scheme $\mathbb{D}$ is determined by its first row; that is, by the sequence $\mathbf{a} = \mathbf{a}_{\rightarrow} = (a_{11}, a_{12}, \ldots, a_{1n}, \ldots)$, as well as by its first column, that is, by the sequence $\mathbf{a} = \mathbf{a}_{\downarrow} = (a_{11}, a_{21}, \ldots, a_{m1}, \ldots)$. We note that, for every choice of $r, s, 1 \leq r, 1 \leq s$, the subscheme $(a_{r+k\,s+l})$, where $0 \leq k, 0 \leq l$ is again a differential scheme.

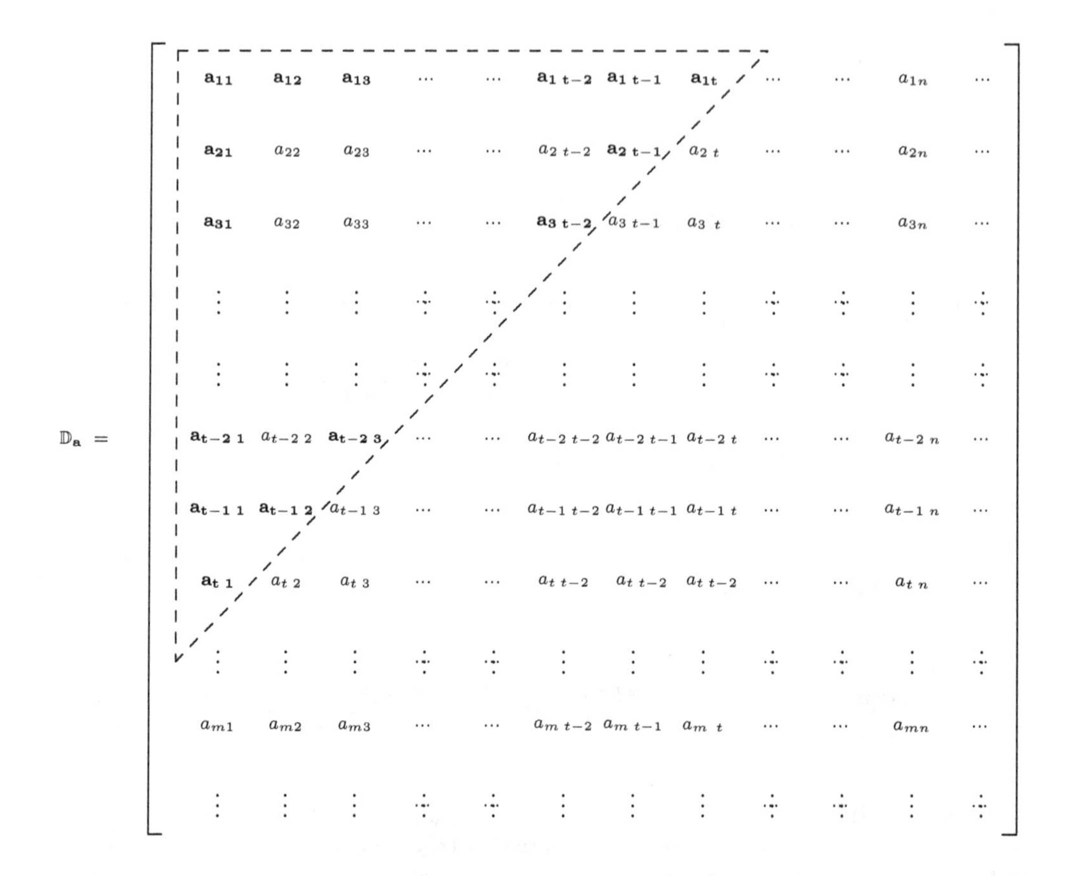

$$\mathbf{a_{m\,n+1}} = \mathbf{a_{m\,n}} + \mathbf{a_{m+1\,n}} \text{ for all } \mathbf{m}, \mathbf{n} \in \mathbb{N}$$

Figure II.2.1 Differential scheme with a marked triangle $\mathbb{T}_{11,t}$

In a given differential scheme, we shall consider triangles $\mathbb{T}_{11,\,t}$ as indicated in Figure II.2.1, as well as the "shifted" triangles $\mathbb{T}_{rs,\,t}$, and prove the relations

$$a_{1\,t} = \sum_{k=1}^{t} \binom{t-1}{k-1} a_{k1} \tag{II.2.8}$$

and

$$a_{t1} = \sum_{k=1}^{t} (-1)^{t-k} \binom{t-1}{k-1} a_{1k}. \tag{II.2.9}$$

We note that $\mathbb{T}_{rs,\,t}$ consists of all entries a_{ij} whose indices satisfy the inequalities $i \leq r, j \leq s$ and $i + j \leq r + s + t - 1$.

Our proofs of (II.2.8) and (II.2.9) are by mathematical induction on t. First, both equalities are valid for $t = 1$ and for $t = 2 : a_{12} \ = \ a_{11} + a_{21}$.

Next we assume that the first equality (II.2.8) holds in the triangle $\mathbb{T}_{11,t}$, and the equality

$$a_{2t} \ = \ \sum_{k=1}^{t} \binom{t-1}{k-1} a_{k+1\,1} \ = \ \sum_{k=2}^{t+1} \binom{t-1}{k-2} a_{k1}$$

holds in the triangle $\mathbb{T}_{21,t}$. Then $a_{1\,t+1} \ = \ a_{1t} + a_{2t}$ equals

$$\binom{t-1}{0} a_{11} + \sum_{k=2}^{t} \left[\binom{t-1}{k-1} + \binom{t-1}{k-2}\right] a_{k1} + \binom{t-1}{t-1} a_{t+1\,1}$$

$$= \binom{t}{0} a_{11} + \sum_{k=2}^{t} \binom{t}{k-1} a_{k1} + \binom{t}{t} a_{t+1\,1} \ = \ \sum_{k=1}^{t+1} \binom{t}{k-1} a_{k1},$$

which proves the required equality (II.2.8). We proceed similarly for the second equality (II.2.9). Here, we assume the validity of the equality (II.2.9) for the triangle $\mathbb{T}_{11,t}$ and the equality

$$a_{t2} \ = \ \sum_{k=1}^{t} (-1)^{t-k} \binom{t-1}{k-1} a_{1\,k+1} \ = \ \sum_{k=2}^{t+1} (-1)^{t-k+1} \binom{t-1}{k-2} a_{1k}.$$

for the triangle $\mathbb{T}_{12,t}$. Then the term $a_{t+1\,1} \ = \ a_{t2} - a_{t1}$ equals

$$\sum_{k=2}^{t} \left[\binom{t-1}{k-2} + \binom{t-1}{k-1}\right] a_{1k} + \binom{t-1}{t-1} a_{1\,t+1} + (-1)^{t-k+1} \binom{t-1}{0} a_{11},$$

which can be transformed, as in the previous case, to the required form

$$a_{t+1\,1} = \sum_{k=1}^{t+1} (-1)^{t-k+1} \binom{t}{k-1} a_{1k}.$$

Remark II.2.2. Applying the formula (II.2.8) to the m-th row of the differential scheme in Figure II.2.1, we obtain

$$a_{mn} \ = \ \sum_{k=1}^{n} \binom{n-1}{k-1} a_{k+m-1\,1}. \qquad \text{(II.2.10)}$$

We also point out that we can obtain, for every $s < r$, the equality

$$\sum_{k=s}^{r} (-1)^k \binom{r}{k} \binom{k}{s} \ = \ 0. \qquad \text{(II.2.11)}$$

We clarify the idea behind the proof of (II.2.11). The concept of a differential scheme yields a permutation of the set of all sequences $\mathcal{S}$; that is, the bijection $\Delta : \mathcal{S} \to \mathcal{S}$ defined by the relations (II.2.9): Every sequence $\mathbf{a}_{\rightarrow} = (a_{1t})$ is mapped to $\Delta\mathbf{a}_{\rightarrow} = \mathbf{a}_{\downarrow} = (a_{t1})$. The other way around, the inverse bijection ∇ assigns every sequence $\mathbf{a}_{\downarrow} = (a_{t1})$ the sequence $\nabla\mathbf{a}_{\downarrow} = \mathbf{a}_{\rightarrow} = (a_{1t})$ by means of (II.2.8). The bijection ∇ can be described by the infinite triangular matrix $M(\nabla) = (\nabla_{m\,n})$, where $\nabla_{m\,n} = \binom{m}{n}$ for $m \geq n$ and $\nabla_{m\,n} = 0$ for $m < n$. Similarly, the bijection Δ can be described by the triangular matrix $M(\Delta) = (\Delta_{m\,n})$, where $\Delta_{m\,n} = (-1)^{m-n}\binom{m}{n}$ for $m \geq n$ and $\Delta_{m\,n} = 0$ for $m < n$. The equality (II.2.11) is thus nothing more than an expression of the properties of the inverse matrices $M(\Delta)$ and $M(\nabla)$; that is, of the matrix relation

$$M(\Delta)M(\nabla) = \begin{bmatrix} 1 & 0 & 0 & 0 & 0 & \cdots \\ -1 & 1 & 0 & 0 & 0 & \cdots \\ 1 & -2 & 1 & 0 & 0 & \cdots \\ -1 & 3 & -3 & 1 & 0 & \cdots \\ 1 & -4 & 6 & -4 & 1 & \cdots \\ \cdots & \cdots & \cdots & \cdots & \cdots & \cdots \end{bmatrix} \begin{bmatrix} 1 & 0 & 0 & 0 & 0 & \cdots \\ 1 & 1 & 0 & 0 & 0 & \cdots \\ 1 & 2 & 1 & 0 & 0 & \cdots \\ 1 & 3 & 3 & 1 & 0 & \cdots \\ 1 & 4 & 6 & 4 & 1 & \cdots \\ . & \cdots & \cdots & \cdots & \cdots & \cdots \end{bmatrix} = I,$$

where I is the "identity matrix" $I = (\delta_{m\,n}) : \delta_{m\,n} = 1$ for $m = n$ and $\delta_{m\,n} = 0$ otherwise.

Example II.2.4. We illustrate several consequences of the important differential scheme of the **Hemachandra - Fibonacci sequence** (see Section II.5), which is named for **Acharya Hemachandra (1087 - 1172)** and **Leonardo Pisano Fibonacci (1170 - 1250)**.

To begin, set $\mathbf{a}_{\rightarrow} = (F_0, F_1, F_2, \ldots F_n, \ldots)$, where $F_0 = 0$, $F_1 = 1$ and $F_{t+1} = F_t + F_{t-1}$ for all $t \geq 1$ and $\mathbf{a}_{\downarrow} = \Delta\mathbf{a}_{\rightarrow} = (-F_0, F_1, -F_2, \ldots, (-1)^{n-1}F_n, \ldots)$.

There are two subschemes marked in Figure II.2.2, both of the same size $(n+1) \times (n+1)$. Using the equalities (II.2.8) and (II.2.9), we readily obtain

$$\sum_{k=1}^{n-1}(-1)^{k+1}\binom{n}{k}F_k = 0 \text{ for } n \text{ odd, and}$$

$$\sum_{k=1}^{n-1}(-1)^{k+1}\binom{n}{k}F_k = 2F_n \text{ for } n \text{ even.}$$

Furthermore,

$$F_{2n} = \sum_{k=0}^{n}\binom{n}{k}F_k,$$

and similarly

$$F_{2n+1} = \sum_{k=0}^{n}\binom{n}{k}F_{k+1}.$$

The reader can see that differential schemes of suitably selected sequences are a source of further discoveries and entertainment.

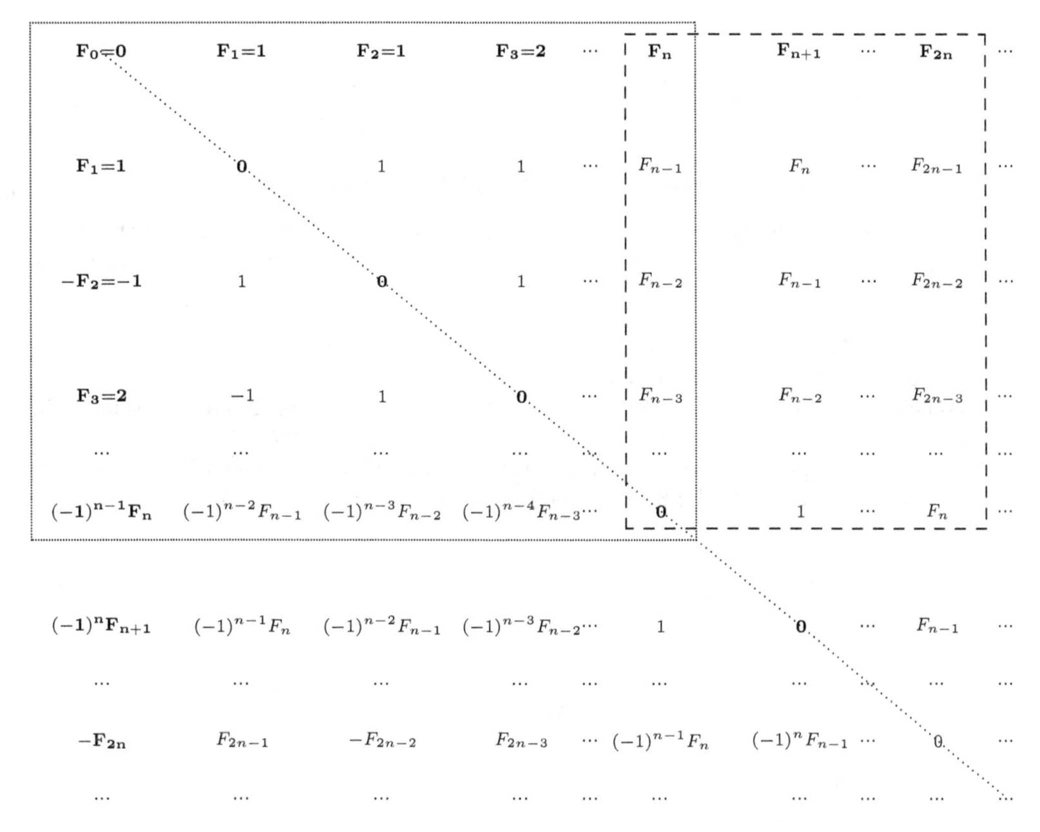

Figure II.2.2 Differential scheme of the Hemachandra - Fibonacci sequence

Theorem II.2.1 gives an important result concerning the sum s_{tn} of the first n terms of row t in a differential scheme.

Theorem II.2.1. *Let $t \geq 2$. The sum s_{tn} of the first n terms of row t in a differential scheme is equal to the difference of the $(n+1)$-th term and the first term of the row $t-1$, that is, (referring to* Figure II.2.1)

$$s_{tn} = \sum_{k=1}^{n} a_{tk} = a_{t-1\ n+1} - a_{t-1\ 1}. \tag{II.2.12}$$

Moreover, we have

$$s_{1n} = \sum_{v=1}^{n} a_{1v} = \sum_{v=1}^{n} \left(\sum_{k=1}^{v} \binom{v-1}{k-1} a_{k1} \right) = \sum_{k=1}^{n} \left(\sum_{v=k}^{n} \binom{v-1}{k-1} \right) a_{k1} = \sum_{k=1}^{n} \binom{n}{k} a_{k1} \tag{II.2.13}$$

and

$$s_{tn} = \sum_{k=1}^{n} a_{tk} = \sum_{k=1}^{n+1} \binom{n}{k-1} a_{k+t-2\;1} - a_{t-1\;1} = \sum_{k=1}^{n} \binom{n}{k} a_{k+t-1\;1}. \qquad \text{(II.2.14)}$$

Exercise II.2.7. Using mathematical induction, prove the identity (II.2.12).

Exercise II.2.8. Using (II.2.12) and (II.2.10), prove (II.2.14). Note that (II.2.13) derived earlier using al-Karaji's formula of Exercise II.2.2 is, in fact, the formula (II.2.14) for $t = 1$.

Example II.2.5. The Bell sequence $\mathbf{b} = (b(t))$ is introduced in Section 2 of Chapter X. The quantity $b(t)$ counts the number of equivalence relations on a given finite set of t elements. The equality (X.2.2) is just our equality (II.2.8) obtained from the differential scheme in which $b(t) = a_{1t}$ (see Figure II.2.3). In this scheme, $a_{t1} = b(t-1)$. Evidently, the differential scheme provides a very effective way of calculating the values of $b(t)$. Observe that, applying the equality (II.2.8), we obtain the recurrence relation for $b(n+1)$, namely,

$$b(n+1) = a_{1\;n+1} = \sum_{t=1}^{n-1} \binom{n}{t-1} a_{t\;1} = \sum_{t=0}^{n} \binom{n}{t} a_{t+1\;1} = \sum_{t=0}^{n} \binom{n}{t} b(t). \qquad \text{(II.2.15)}$$

b(0)=b(1) 1	**b(2)= 2**	**b(3)= 5**	**b(4)= 15**	**b(5)= 52**	**b(6)= 203**	**b(7)= 877**	**b(8)= 4140**	**b(9)= 21147**	⋯
b(1)=1	3	10	37	151	674	3263	17007	94828	⋯
b(2)=2	7	27	114	523	2589	13744	77821	467767	⋯
b(3)=5	20	87	409	2066	11155	64077	389946	2504665	⋯
b(4)=15	67	322	1657	9089	52922	325869	2114719	14418716	⋯
b(5)=52	255	1335	7432	43833	272947	1788850	12303997	88586135	⋯
b(6)=203	1080	6097	36401	229114	1515903	10515147	76282138	577461075	⋯
⋮	⋮	⋮	⋮	⋮	⋮	⋮	⋮	⋮	⋱

Figure II.2.3 Differential scheme of the Bell sequence $\mathbf{a}_{\rightarrow} = (\mathbf{b}(\mathbf{t}))$

Exercise II.2.9. Using (II.2.9), prove that

$$b(n+1) \;=\; \sum_{t=1}^{n}(-1)^{n-t}\binom{n}{t-1}b(t)+b(n).$$

Exercise II.2.10. We have seen that the Bell sequence

$$\mathbf{b} = (b(1), b(2), b(3), \dots, b(n), \dots)$$

satisfies

$$\Delta\mathbf{b} = (b(0), b(1), b(2), \dots, b(n-1), \dots).$$

Determine all sequences $\mathbf{a}_{\rightarrow} = (a_{11}, a_{12}, \dots, a_{1\,n}, \dots)$ having the property that $\mathbf{a}_{\downarrow} = \Delta\mathbf{a}_{\rightarrow} = (a_{11}, a_{21}, \dots, a_{n1}, \dots)$ satisfies $a_{1n} = a_{n+t\,1}$ for all $n \geq 1$ and a fixed $t \geq 1$. In particular, show that every nontrivial sequence $\mathbf{a}$ whose differential scheme satisfies $a_{1n} = a_{n+1\,1}$ for all $1 \leq n$ is a multiple of the Bell sequence.

We note that the rows and columns of this differential scheme form a number of sequences that have been studied in mathematics. For example, let $\{\mathcal{R}_k \mid 1 \leq k \leq b(t)\}$ be the set of all partitions of a set of t elements, and r_k the number of (non-empty) subsets that constitute the partition $\mathcal{R}_k$. Define the sequence

$$\mathbf{q} \;=\; (q(t)), \;\text{ where }\; q(t) \;=\; \sum_{k=1}^{b(t)} r_k.$$

Exercise II.2.11. Prove that the sequence $\mathbf{q}$ is the sequence of the second row of the differential scheme in Figure II.2.3.

Exercise II.2.12. Show that the t-th term of the second row, that is $q(t)$, equals the sum of all terms of the "diagonal" from the first term of the t-th row to the t-th term of the first row of the differential scheme in Figure II.2.3.

Exercise II.2.13. In a given differential scheme, consider the sequences of numbers in two adjacent columns. Denote the first sequence by $\mathbf{a} = (a_1, a_2, \dots, a_n, \dots)$ and the following one by $\mathbf{b} = (b_1, b_2, \dots, b_n, \dots)$.

(i) Show that $b_n = a_n + a_{n+1}$ for all $n \geq 1$.

(ii) Let the sequence $\mathbf{a}$ be defined recursively by the first two terms a_1, a_2 and by the relations

$$a_n \;=\; \alpha_n\, a_{n-1} + \beta_n\, a_{n-2} \;\text{ for all }\; n \geq 3,$$

where $(\alpha_n \mid n \geq 3)$ and $(\beta_n \mid n \geq 3)$ are arbitrary sequences of numbers such that

$$\delta_n \;=\; 1 + \alpha_n - \beta_n \neq 0 \;\text{ for all }\; n \geq 3.$$

Show that the sequence $\mathbf{b}$ is defined recursively in a similar manner, that is, it is given by the first two terms b_1, b_2 and the relations

$$b_n \;=\; \rho_n\, b_{n-1} + \sigma_n\, b_{n-2} \;\text{ for some sequences }\; (\rho_n \,|\, n \geq 3) \;\text{ and }\; (\sigma_n \,|\, n \geq 3).$$

(iii) Show that

$$\rho_n = 1 + \alpha_{n+1} - \frac{\delta_{n+1}}{\delta_n} \quad \text{and} \quad \sigma_n = \beta_n \frac{\delta_{n+1}}{\delta_n}.$$

(iv) Consider in detail the special case when the sequence $\{\delta_n \,|\, n \geq 3\}$ is the constant sequence $\{\delta, \delta, \ldots\}$.

As we have already seen, differential schemes allow a great many possibilities for the study of sequences and relations among them. We present some additional examples in Section 3 of Chapter II.

II.3 Arithmetic progressions of higher order

Here we deal with those sequences **a** having a differential scheme $\mathbb{D}_a$ which contains a zero row (so that all subsequent rows are zero rows). Surprisingly, such sequences are in a one-to-one correspondence with certain polynomials. This bijection is straightforward to write down explicitly using the mapping Δ, that is, using the equality (II.2.8). We describe this relationship in detail.

Definition II.3.1. *A sequence* $\mathbf{a} = \mathbf{a}_{\rightarrow} = (a_{11}, a_{12}, \ldots, a_{1\,n}, \ldots)$ *is called an* **arithmetic progression of higher order** *if there exists (in the differential scheme* $\mathbb{D}_a$ *as in* Figure II.2.1) *a number* r *such that*

$$a_{rt} = 0 \ \text{for all} \ 1 \leq t.$$

The least d *such that*

$$a_{d+2\,t} = 0 \ \text{for all} \ 1 \leq t.$$

is called the **order** *of the arithmetic progression* **a**.

Thus, in this terminology, the non-constant arithmetic progressions, as we know them from elementary algebra, are arithmetic progressions of order 1. The arithmetic progressions that satisfy $a_{1t} = c \neq 0$ for all $1 \leq t$ are the arithmetic progressions of order 0, while the zero progression is an arithmetic progression of order -1 (or we may say that it does not have an order).

Given an arithmetic progression $\mathbf{a} = (a_{1\,t})$ of order d, we obtain by (II.2.8) that

$$a_{1t} = \sum_{k=0}^{d} \binom{t-1}{k} a_{k+1\,1} = P(t), \tag{II.3.1}$$

where $P(t)$ is a polynomial of degree d. Indeed

$$P(t) = c_0 + c_1 t + c_2 t^2 + \cdots + c_d t^d, \ \text{where} \ c_d = \frac{a_{d+1\,1}}{d!} \neq 0.$$

On the other hand, given a polynomial $P(x) = c_0 + c_1 x + c_2 x^2 + \cdots + c_d x^d$ of degree d, $c_d \neq 0$, we may form the differential scheme with

$$a_{1\,t} = P(t), \ \text{and thus} \ a_{m1} = \sum_{k=0}^{m-1} (-1)^k \binom{m-1}{k} P(m-k) \ \text{for} \ 1 \leq k \leq d. \tag{II.3.2}$$

Exercise II.3.1. Prove the formula (II.3.2).

Exercise II.3.2. Prove that $a_{d1} = 0$ in the formula (II.3.1), that is, for a polynomial $P(x)$ of degree d,

$$\sum_{k=0}^{d-1} (-1)^k \binom{d-1}{k} P(d-k) = 0. \tag{II.3.3}$$

We may summarize the previous statements in the following theorem.

Theorem II.3.1. *There is a bijection between the set of all integral arithmetic progressions of order d and all integral polynomials of degree d, that is, the polynomials whose values at natural numbers are integers. This bijection is expressed by formulas* (II.3.1) *and* (II.3.2) *with* (II.3.3).

Let $d \in \mathbb{N}_0$. The sequence $\mathbf{a} = (1^d, 2^d, \ldots, n^d, \ldots)$ is an arithmetic progression of order d. It corresponds to the polynomial $P(x) = x^d$. Consequently, from the differential scheme for $\mathbf{a}$ and using the general summation formula (II.2.13), we can determine the sum

$$S_d(n) = \sum_{k=1}^{n} k^d.$$

For instance, for $d = 5$, we have the scheme

1	32	243	1024	3125	7776	16807	32768	59049	...
31	211	781	2101	4651	9031	15961	26281	...	...
180	570	1320	2550	4380	6930	10320	...	...	...
390	750	1230	1830	2550	3390	4350	...	...	...
360	480	600	720	840	960	1080	...	...	...
120	120	120	120	120	120	120	...	...	...
0	0	0	0	0	0	0	...	...	...

and thus

$$S_5(n) = \binom{n}{1} \times 1 + \binom{n}{2} \times 31 + \binom{n}{3} \times 180 + \binom{n}{4} \times 390 + \binom{n}{5} \times 360 + \binom{n}{6} \times 120$$

$$= \frac{n^2(n+1)^2}{12} (2n^2 + 2n - 1).$$

Exercise II.3.3. Prove $S_2(n) = \frac{1}{6}n(n+1)(2n+1)$, $S_3(n) = \frac{1}{4}n^2(n+1)^2$, $S_4(n) = \frac{1}{30}n(n+1)(6n^3 + 9n^2 + n - 1)$ and $S_6(n) = \frac{n(n+1)}{42}(6n^5 + 15n^4 + 6n^3 - 6n^2 - n + 1)$.

Exercise II.3.4. Prove that an arbitrary sequence of n numbers $\{a_1, a_2, \ldots, a_n\}$ can be (uniquely) extended to an arithmetic progression $\mathbf{a} = \{a_1, a_2, \ldots, a_n, a_{n+1}, \ldots\}$ of order $d \leq n-1$. Determine such a sequence of order ≤ 3 if $a_1 = 5, a_2 = -1, a_3 = 2$ and $a_4 = 0$.

Remark II.3.1. The sums $S_d(n) = \sum_{k=1}^n k^d$ can be found recursively in an elementary way. Assuming that we know already the sums $S_k(n)$ for all $1 \le k < d$ (which is a disadvantage of this approach), we have

$$S_d(n) = \frac{1}{d+1}\ \left((1+n)^{d+1} - 1 - \sum_{k=0}^{d-1} \binom{d+1}{k} S_k(n) \right). \tag{II.3.4}$$

This follows from the binomial theorem

$$(1+x)^{d+1} = 1 + \binom{d+1}{1}x + \cdots + \binom{d+1}{t}x^t + \cdots + \binom{d+1}{d+1}x^{d+1} = \sum_{t=0}^{d+1}\binom{d+1}{t}x^t.$$

If we add up the identities for $x = 0, 1, 2, \ldots, d+1$, we find

$$\begin{aligned}&S_{d+1}(n) - 1 + (1+n)^{d+1}\\ &= S_0(n) + \binom{d+1}{1}S_1(n) + \cdots + \binom{d+1}{t}S_t(n) + \cdots + \binom{d+1}{d}S_d(n) + \binom{d+1}{d+1}S_{d+1}(n),\end{aligned}$$

which is the equality (II.3.4).

Exercise II.3.5. Think over the following visual method of **Alhazen**, which is the latinized name of **Hasan Ibn al-Haytham (965 - 1040)**, for the summation of the series $\sum_{k=1}^n k^d$:

$$\sum_{k=1}^n k^{d+1} = (n+1)\sum_{k=1}^n k^d - \sum_{t=1}^n \sum_{k=1}^t k^d.$$

The following scheme has n rows with $n+1$ entries k^d for $k = 1, 2, \ldots, n$. For each k, the sum of the last k summands is written in the form $k \cdot k^d = k^{d+1}$. Thus the sum of all entries in this "rectangle" equals

$$(n+1)\sum_{k=1}^n k^d = \sum_{k=1}^n k^{d+1} + \sum_{t=1}^n \sum_{k=1}^t k^d.$$

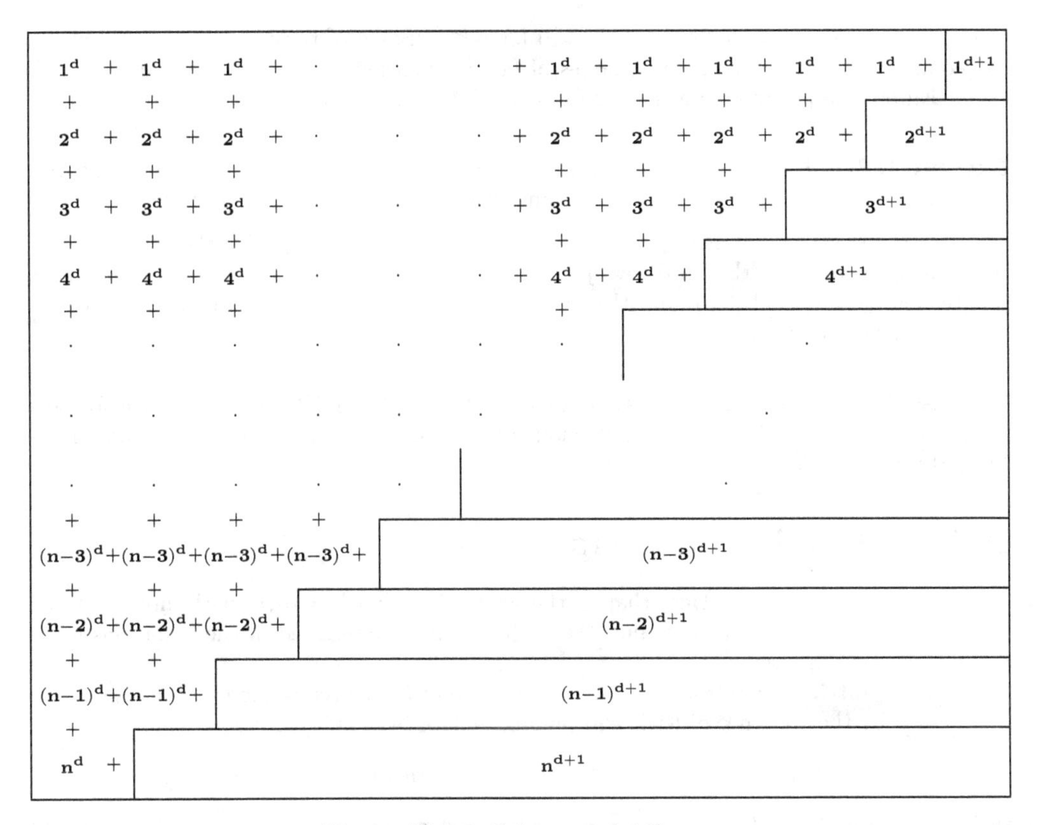

Figure II.3.1 Scheme of Alhazen

A natural question concerning special differential schemes arises in connection with geometric progressions. All non-trivial geometric progressions have, up to a non-zero multiple, a simple form, namely,

$$\mathbf{q} = (1, q, q^2, \ldots, q^t, \ldots) \text{ with } q \neq 0. \tag{II.3.5}$$

The following exercise shows that they can be easily identified.

Exercise II.3.6. Prove the following characterization of geometric progressions: Let $a \neq 0$ and $b \neq -a$ be two numbers. Then the first row of the differential scheme (of Figure II.2.1) such that $a_{11} = a$, $a_{21} = b$ and all rows are multiples of each other is the $a-$multiple of the progression $\mathbf{q}$ of (II.3.5) with $q = \frac{a+b}{a}$. Moreover, show that

$$\Delta\mathbf{q} = (1, q-1, (q-1)^2, \ldots, (q-1)^n, \ldots).$$

Now, write $\mathbf{q}_0 = \mathbf{q}$ of (II.3.5), and for every integer z, define the sequence $\mathbf{q}_z$ by

$$\mathbf{q}_z = \{(q_{zt}) \mid 1 \leq t\},\ q_{zt} = (q-1)^z \left(q^{t-1} - \sum_{k=1}^{-z} \binom{t-1}{k-1} (q-1)^{k-1} \right). \tag{II.3.6}$$

Here, as we have already agreed, $\binom{t-1}{k-1} = 0$ for $t < k$. Some of these sequences are related to geometrical problems and carry names of famous mathematicians. For instance, $\mathbf{q}_{-2}$ is called **Euler's sequence**, named after **Leonhard Euler (1707 - 1783)**.

Exercise II.3.7. Prove that the t-th term $q_{-2\,t}$ of Euler's sequence denotes the number of partitions of a set of $t-1$ elements containing just one subset having more than 1 element.

Now, for a natural number r, consider the (infinite) matrix $\mathbb{D}_{\mathbf{q}_{-r}}$ with the first row $\mathbf{q}_{-r}$, the second row $\mathbf{q}_{-(r-1)}$, the third row $\mathbf{q}_{-(r-2)}$, etc. Hence, the r-th row forms the geometric progression $\mathbf{q}_0 = \mathbf{q}$ of (II.3.5), and the $(r+3)$-th row forms the $(q-1)^s$-th multiple of the geometric progression $\mathbf{q}$.

Exercise II.3.8. Prove that $\mathbb{D}_{\mathbf{q}_{-r}}$ is a differential scheme. Using (II.2.8), recheck the formula for q_{zt} in (II.3.6) and describe the behaviour of the columns in relation to the rows of the Yang Hui-Pascal triangle.

II.4 ⋆ Catalan numbers

We consider a **plane lattice**, that is, the set of all points $P = (a, b)$ with integers a and b in a coordinate system of the plane. We shall be mainly interested in the situations when a and b are natural numbers.

A (finite) **lattice path** from a point (a, b) to a point (m, n) is a sequence of line segments determined by the sequence of their end points, that is, by a finite sequence of points

$$P_t = (m_t, n_t),\ (t = 0, 1, \ldots, d), P_0 = (m_0, n_0) = (a, b), P_d = (m_d, n_d) = (m, n),$$

such that, for each $t = 0, 1, \ldots, d-1$, either the first component or the second component of the neighboring points P_t and P_{t+1} are equal. In other words, the line segments P_tP_{t+1} are either "horizontal" or "vertical". There is an infinite number of paths between two given lattice points P_0 and P_d. We will however restrict our attention to the **increasing lattice paths** from P_0 to P_d. Such paths exist only if $a \le m$ and $b \le n$, and we shall assume that these inequalities hold from now on. Thus an increasing lattice path satisfies

$$a = m_0 \le m_1 \le \cdots \le m_d = m \ \text{ and } \ b = n_0 \le n_1 \le \cdots \le n_d = n.$$

Under these conditions, we see that there is only a finite number of such paths. In fact, the number $N((a, b), (m, n))$ of increasing lattice paths from P_0 to P_d is given by the formula

$$N((a,b),(m,n)) \;=\; \binom{m+n-a-b}{n-b}. \tag{II.4.1}$$

In particular, the number of increasing lattice paths from the origin to the point (n, n) is equal to $\binom{2n}{n}$.

We now prove (II.4.1). First of all, by means of a shift of coordinates, we may assume that $(a, b) = (0, 0)$. Then an increasing lattice path from $P_0 = (0, 0)$ to $P_d = (m, n)$ is fully determined by the points

$$P_1 = (x_1, 1), P_2 = (x_2, 2), \ldots, P_n = (x_n, n), \ \text{ where } 0 \le x_1 \le x_2 \le \cdots \le x_n \le m$$

in the following sense: The defining points of this path are

$$P_0 = (0,0),\ Q_1 = (x_1, 0),\ P_1 = (x_1, 1),\ Q_2 = (x_2, 1),\ P_2 = (x_2, 2),\ \dots$$

$$\dots,\ Q_n = (x_n, n-1),\ P_n = (x_n, n),\ P_d = (m, n).$$

Of course, P_0 and Q_1, as well as P_n and P_d may coincide (naturally, some of the P_i and Q_{i+1} may coincide as well, for $1 \le i \le n$). Consequently, there is a bijection between the set of increasing paths from $P_0 = (0,0)$ to $P_d = (m,n)$ and the set of combinations of n numbers from the set $\{1, 2, \dots, m+n\}$. Indeed, the above path is fully (and uniquely) described by the subset $\{x_1 + 1, x_2 + 2, \dots, x_n + n\}$. Therefore, the number of such paths is equal to $\binom{m+n}{n}$, and thus in the general case (shifting $(0,0)$ back to (a,b)), this number is equal to $\binom{m+n-a-b}{n-b}$.

We now define the sequence $(C(0), C(1), \dots, C(n), \dots)$ of **Catalan numbers**, named after **Eugène Charles Catalan (1814 - 1894)**. This is the sequence of numbers $1, 1, 2, 5, 14, 42, 132, 429, \dots$ that occurs in a variety of counting problems. Catalan defined $C(n)$ as the number of different ways a given convex $(n+2)-$gon can be divided into n triangles connecting its vertices with diagonals. Here, in (II.4.2), we describe $C(n)$ in two different ways: by a single formula as well as by a recurrent prescription.

Definition II.4.1. *Define $C(n)$ as the number of different increasing lattice paths from $(0,0)$ to (n,n) that do not pass above the diagonal $x = y$, that is, do not contain any point (r,s) with $r < s$. The Catalan sequence is the sequence of numbers*

$$(C(0), C(1), C(2), C(3), C(4), C(5), C(6), C(7) \dots) = (1, 1, 2, 5, 14, 42, 132, 429, \dots).$$

The Catalan sequence appears in a number of interesting problems, see for example R. P. Stanley, *Enumerative Combinatorics, Vol. 2*, Cambridge University Press, 1999.

Theorem II.4.1. *For every $n \in \mathbb{N}$, we have*

$$C(n) = \frac{1}{n+1}\binom{2n}{n} = \binom{2n}{n} - \binom{2n}{n-1} = \sum_{k=1}^{n} C(k-1)\, C(n-k). \tag{II.4.2}$$

Proof. Looking at Figure II.4.1 may help the reader in following the proof. Recall that we already know that the number of all increasing lattice paths from (a,b) to (n,n) is equal to

$$\binom{2n-a-b}{n-b}.$$

In particular, the number of all such paths from $(0,0)$ to (n,n) is $\binom{2n}{n}$.

Now, consider all paths from $(0,0)$ to (n,n) that cross the diagonal $x = y$. Of course, crossing the diagonal means getting to a point $(k, k+1)$ for some $k \in \{0, 1, \dots, n-1\}$. We denote by P_k $(k = 0, 1, \dots, n-1)$ the set of all increasing lattice paths from $(0,0)$ to (n,n) that pass through the point $(k, k+1)$, but do not pass through any point $(t, t+1)$ with $t < k$. These are the increasing lattice paths from $(0,0)$ to (k,k) that do not pass above the diagonal extended to $(k, k+1)$ and continued to (n,n) (see Figure II.4.1(a)). Therefore their total number is

$$C(k)\binom{2n-2k-1}{n-k-1}.$$

Thus, the total number of the increasing lattice paths from $(0,0)$ to (n,n) that cross the diagonal is

$$\sum_{k=0}^{n-1} C(k)\binom{2n-2k-1}{n-k-1}. \tag{II.4.3}$$

It is not difficult to determine this sum. Looking at Figure II.4.1(a) we can see that there is a bijection between the set P_k and the set Q_k of all increasing lattice paths from $(-1,1)$ to (n,n) that pass through the point $(k, k+1)$, but not through any point $(t, t+1)$ with $t < k$. The image q from Q_k of a path p from P_k in this one-to-one correspondence can be described as follows:

The initial segment from $(-1,1)$ to $(k, k+1)$ of q is a reflection of the initial segment from $(0,0)$ to $(k, k+1)$ of p in the line l while the remaining segments from $(k, k+1)$ to (n,n) coincide (see the shaded triangles and the shaded rectangle in Figure II.4.1(a).) Thus the total number of the increasing lattice paths from $(0,0)$ to (n,n) that cross the diagonal, that is, the sum (II.4.3) is equal to the number of all increasing lattice paths from $(-1,1)$ to (n,n):

$$\sum_{k=0}^{n-1} C(k)\binom{2n-2k-1}{n-k-1} = \binom{2n}{n-1}.$$

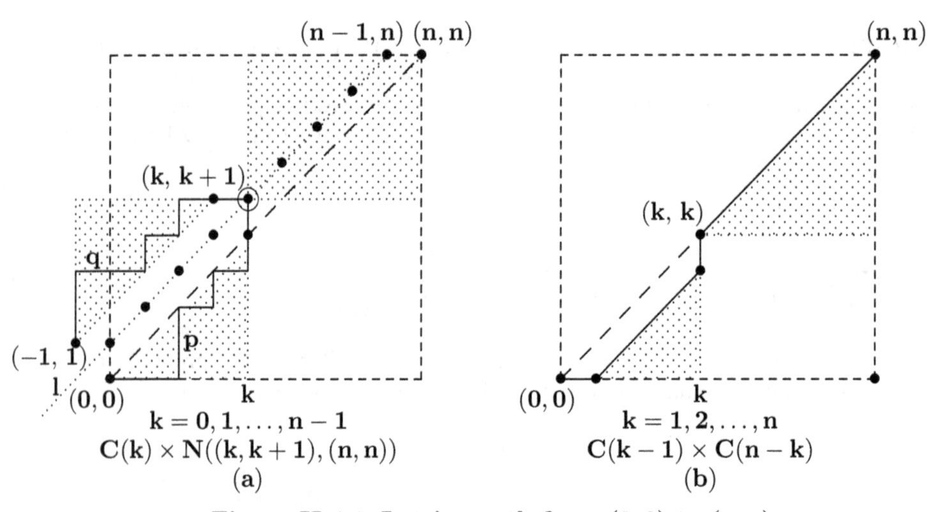

Figure II.4.1 Lattice path from (0, 0) to (n, n)

As a consequence, using the definition of $C(n)$, we have

$$C(n) = \binom{2n}{n} - \binom{2n}{n-1} = \frac{1}{n+1}\binom{2n}{n}.$$

Now we turn our attention to Figure II.4.1(b). Each of the increasing lattice paths from $(0,0)$ to (n,n) that does not pass above the diagonal consists of the following four segments: The path from $(0,0)$ to $(1,0)$, an increasing lattice path from $(1,0)$ to $(k,k-1)$, the pass from $(k,k-1)$ to (k,k), where $k>1$ is the least number so that (k,k) belongs to the path, and an arbitrary increasing lattice path from (k,k) to (n,n) that does not pass above the diagonal. For a fixed k, the number of such paths is $C(k-1)\,C(n-k)$ (see the shaded triangles in Figure II.4.1(b)). Since k runs through the set $\{1,2,\dots,n\}$, we readily obtain

$$C(n) \;=\; \sum_{k=1}^{n} C(k-1)\,C(n-k). \tag{II.4.4}$$

Theorem II.4.1 is now proved. □

Exercise II.4.1. Write down the differential scheme for the sequence of Catalan numbers. Denote the sequences defined by the first three columns by

$\mathbf{a} = (a_1, a_2, \dots, a_n, \dots)$ (**Motzkin sums**),

$\mathbf{b} = (b_1, b_2, \dots, b_n, \dots)$ (**Motzkin numbers**) and

$\mathbf{c} = (c_1, a_c, \dots, c_n, \dots)$ (**successive Motzkin sums**);

which are named after **Theodore Motzkin (1908 - 1970)**.

(i) Assuming that, for $n \geq 3$, $a_n = \frac{n-2}{n}\,(2a_{n-1} + 3a_{n-2})$, determine the functions $\varphi(n)$ and $\psi(n)$ such that

$$b_n \;=\; \varphi(n)\,b_{n-1} + \psi(n)\,b_{n-2} \;\text{ for }\; n \geq 3.$$

(ii) Prove that

$$c_n \;=\; \frac{1}{n+1}\,[2n\,c_{n-1} + 3(n-3)\,c_{n-2}] \;\text{ for }\; n \geq 3.$$

Exercise II.4.2. Interpret the sum (II.2.4) as the number of all different bracketings (with $n-2$ pairs of brackets). For example, there are $C(4)=5$ bracketings

$$\{((ab)c)d,\; (a(bc))d,\; (ab)(cd),\; a((bc)d),\; a(b(cd))\} \text{ of four symbols } a,b,c,d.$$

Exercise II.4.3. Let $G_{a,\,q} = \{x_1, \dots, x_t, \dots\}$ be the geometric progression defined recursively by $x_1 = a \neq 0$, $x_{t+1} = qx_t$ for $t = 1, 2, \dots$. Show that the m-th row of the differential scheme of the geometric progression $G_{a,\,q}$ forms the geometric progression $G_{a(q-1)^{m-1},\,q}$ and that the n-th column forms the geometric progression $G_{aq^{n-1},\,q-1}$, that is, the term a_{mn} of the scheme in Figure II.2.1 is

$$a(q-1)^{m-1}q^{n-1}.$$

II.5 Everybody has their own sequence

We consider **recursive sequences** of the type

$$\mathbf{X} = \{X_n \mid n = 1, 2, \ldots\}, \quad \text{where } X_{n+2} = AX_{n+1} + BX_n,$$

with given X_1, X_2, A and B. As before, we pay particular attention to sequences in which the chosen "initial conditions" are integers. We emphasize that the sequence $\mathbf{X}$ is uniquely determined by X_1, X_2, A and B. We have

$$X_3 = AX_2 + BX_1,\ X_4 = (A^2 + B)X_2 + ABX_1,\ X_5 = (A^3 + 2AB)X_2 + (A^2B + B^2)X_1, \ldots.$$

For centuries there has been considerable interest in the sequence with $A = B = X_1 = X_2 = 1$. This is the sequence $\mathbf{F} = (F_n)$ of numbers that **François Édouard Anatole Lucas (1842 - 1891)** proposed in May, 1876 to call **Fibonacci numbers**. This sequence was described by **Leonardo Pisano Fibonacci** in his book *Liber Abaci* in 1202 in a problem involving the growth of a hypothetical population of rabbits based on idealized assumptions. In fact, these numbers were already known long before in India in connection with Sanskrit prosody. The first documented occurrence of the series was in the work of an Indian writer by the name of **Pingala** in the 4-th century BCE. His prosaic study called *Chandas Shastra* presents the first known description of a binary numeral system in connection with the systematic enumeration of meters with fixed patterns of short and long syllables. His discussion of the combinatorics of meter contains the binomial theorem and the Fibonacci numbers, called **maatraameru**. They were later thoroughly described by **Gopala** and **Acharya Hemachandra** and are now known also as the **Gopala - Hemachandra numbers**. Their description is presented as the number of the rhythmical pattern and resembles the frequently used formulation of the numbers of ways of climbing a staircase consisting of n stairs by climbing either one or two stairs at a time. The recurrence rule is as follows: The number of ways to climb n stairs in this way is the sum of the number of ways to climb $n-1$ stairs (this is the case when we choose to start the climb by climbing a single stair) and the number of ways to climb $n-2$ stairs (which is the case when we choose to start the climb by climbing two stairs). The interest in the Fibonacci sequence is due to a considerable extent to its close relationship with the

$$\textbf{golden ratio } \varphi = \frac{1+\sqrt{5}}{2} \approx 1.61803398875$$

that appears in the formula of **Jacques Marie Binet (1786 - 1856)**

$$F_n = \frac{(1+\sqrt{5})^n - (1-\sqrt{5})^n}{2^n\sqrt{5}}, \tag{II.5.1}$$

This formula was probably already known to **Johannes Kepler (1571 - 1630)**. The proof of (II.5.1) was given in the year 1730 by **Abraham de Moivre (1667 - 1754)**, later also by **Daniel Bernoulli (1700 - 1782)** and in the year 1765 by **Leonhard Euler**.

We remark that some proofs of Binet's formula are quite involved. For instance, one can consider the ordinary generating function of the Fibonacci sequence $\{F_n \mid n = 1, 2, \ldots\}$:

namely $f(x) = \sum_{n=1}^{\infty} F_n x^{n-1} = 1 + x + 2x^2 + 3x^3 + 5x^4 + \cdots$ and show that, for $|x| < \frac{1}{2}$,

$$f(x) = \frac{-1}{x^2 + x - 1}.$$

Then, if we rewrite $f(x)$ as the sum of suitable partial fractions

$$\frac{a_1}{(x - b_1)} \quad \text{and} \quad \frac{a_2}{(x - b_2)},$$

then express these fractions in the form of infinite series, and finally compare these series with the original series, we obtain Binet's formula (II.5.1).

We wish to clarify the nature of such sequences and derive a general "Binet-like" formula for the n-th term using the method that has its origin in the work of **Abraham de Moivre**.

We choose numbers A, B with $B \neq 0$ and arbitrary X_1, X_2 to define the recursive sequence $\{X_n\}$ by

$$X_{n+2} = AX_{n+1} + BX_n \quad \text{for} \quad n = 1, 2, \ldots. \tag{II.5.2}$$

Our task is to find a formula for X_n. As the sequence is uniquely determined by the recurrence relation and the values of X_1 and X_2, it suffices to find a quantity $f_n(A, B, X_1, X_2)$ satisfying the recurrence relation and the initial conditions

$$f_1(A, B, X_1, X_2) = X_1 \quad \text{and} \quad f_2(A, B, X_1, X_2) = X_2,$$

as this establishes that $X_n = f_n(A, B, X_1, X_2)$. We use the chosen numbers A and B to form the quadratic equation

$$x^2 = Ax + B.$$

We denote the roots of this quadratic equation by α and β so that $A = \alpha + \beta$ and $B = -\alpha\beta$. As $B \neq 0$ we have $\alpha \neq 0$ and $\beta \neq 0$.

First suppose that $\alpha \neq \beta$. For any a and b the quantity $a\alpha^n + b\beta^n$ satisfies the recurrence relation (II.5.2) as

$$A(a\alpha^{n+1}+b\beta^{n+1})+B(a\alpha^n+b\beta^n) = (\alpha+\beta)(a\alpha^{n+1}+b\beta^{n+1})-\alpha\beta(a\alpha^n+b\beta^n) = a\alpha^{n+2}+b\beta^{n+2}.$$

We now determine the numbers a and b to satisfy the conditions

$$a\alpha + b\beta = X_1 \quad \text{and} \quad a\alpha^2 + b\beta^2 = X_2.$$

In this case, we obtain

$$a = \frac{1}{\alpha(\alpha - \beta)}(X_2 - \beta X_1); \quad b = \frac{1}{\beta(\beta - \alpha)}(X_2 - \alpha X_1),$$

and we deduce

$$X_n = \frac{1}{\alpha - \beta}[\alpha^{n-1}(X_2 - \beta X_1) - \beta^{n-1}(X_2 - \alpha X_1)]. \tag{II.5.3}$$

Next suppose that $\alpha = \beta$. In this case $A = 2\alpha$ and $B = -\alpha^2$ and for any a and b the quantity $(a + bn)\alpha^n$ satisfies the recurrence relation as

$$A(a+b(n+1))\alpha^{n+1}+B(a+bn)\alpha^n = 2\alpha(a+b(n+1))\alpha^{n+1}-\alpha^2(a+bn)\alpha^n = (a+b(n+2))\alpha^{n+2}.$$

We now determine the numbers a and b to satisfy the conditions

$$(a + b)\alpha = X_1 \text{ and } (a + 2b)\alpha^2 = X_2.$$

We obtain

$$a = (2\alpha X_1 - X_2)/\alpha^2 \text{ and } b = (X_2 - \alpha X_1)/\alpha^2.$$

Hence

$$X_n = \alpha^{n-2} [(n-1)X_2 - (n-2)\alpha X_1].$$

We summarize these results in the following theorem.

Theorem II.5.1. *Let* (II.5.2) *be a recursive sequence defined by* $A, B \neq 0, X_1$ *and* X_2. *Denote by* α *and* β *the roots of the quadratic equation* $x^2 = Ax + B$. *Then*

$$X_n = \frac{1}{\alpha - \beta} [\alpha^{n-1}(X_2 - \beta X_1) - \beta^{n-1}(X_2 - \alpha X_1)] \text{ if } \alpha \neq \beta$$

and

$$X_n = \alpha^{n-2} [(n-1)X_2 - (n-2)\alpha X_1] \text{ if } \alpha = \beta.$$

In Example II.5.1 we illustrate Theorem II.5.1 using some sequences that appear in the literature; in each case, the reader should check the formula. We have already mentioned Binet's formula for the Fibonacci sequence.

Example II.5.1. We give the n-th terms of four famous sequences.

Lucas sequence: $L_1 = 2$, $L_2 = 1$, $A = 1$, $B = 1$, and thus

$$\alpha = \frac{1+\sqrt{5}}{2}, \quad \beta = \frac{1-\sqrt{5}}{2}, \quad a = \frac{2}{1+\sqrt{5}}, \quad b = \frac{2}{1-\sqrt{5}};$$

$$L_n = \frac{(1+\sqrt{5})^{n-1} + (1-\sqrt{5})^{n-1}}{2^{n-1}}.$$

Pell sequence: $P_1 = 1$, $P_2 = 2$, $A = 2$, $B = 1$, and thus

$$\alpha = 1+\sqrt{2}, \quad \beta = 1-\sqrt{2}, \quad a = \frac{\sqrt{2}}{4}, \quad b = -\frac{\sqrt{2}}{4};$$

$$P_n = \sqrt{2}\, \frac{(1+\sqrt{2})^n - (1-\sqrt{2})^n}{4}.$$

Pell-Lucas sequence: $Q_1 = 1$, $Q_2 = 3$, $A = 2$, $B = 1$, and thus

$$\alpha = 1 + \sqrt{2},\ \beta = 1 - \sqrt{2},\ a = \frac{1}{2},\ b = \frac{1}{2};$$

$$Q_n = \frac{(1+\sqrt{2})^n - (1-\sqrt{2})^n}{2}.$$

Jacobsthal sequence: $J_1 = 1$, $J_2 = 1$, $A = 1$, $B = 2$, and thus

$$\alpha = 2,\ \beta = -1,\ a = \frac{1}{3},\ b = -\frac{1}{3};$$

$$J_n = \frac{2^n - (-1)^n}{3}.$$

Exercise II.5.1. Find a formula for the n-th member of each of the following sequences.

(i) Jacobsthal-Lucas: $K_1 = 2$, $K_2 = 1$, $A = 1$, $B = 2$;

(ii) Pell-Jacobstahl: $R_1 = 2$, $R_2 = 1$, $A = 1$, $B = 4$;

(iii) Chebyshev : $C_1 = 2$, $C_2 = 11$, $A = 11$, $B = -1$;

(iv) Pisot: $T_1 = 2$, $T_2 = 3$, $A = 3$, $B = -2$.

Exercise II.5.2. Determine X_n for the recursive sequence (II.5.2) defined by

(i) $X_1 = 2$, $X_2 = 22$, $A = 2$, $B = 9$.

(ii) $X_1 = A$, $X_2 = A^2 + 2B$, $B = -\frac{A^2}{4}$.

(iii) $X_1 = 1$, $X_2 = A$ and $X_{n+2} = (2A+1)X_{n+1} - A(A+1)X_n$.

Exercise II.5.3. Let c and d be two integers. Find X_1, X_2, A and B so that the recursive sequence (II.5.2) is the arithmetic progression $X_n = c + nd$.

Remark II.5.1. We note that even with integral choices of X_1, X_2, A and B the solution of the recurrence relation (II.5.2) may be given in terms of complex numbers. For example the choices $X_1 = 1$, $X_2 = 0$, $A = 2$, $B = -3$ give X_n in terms of complex numbers, namely,

$$X_n = \frac{-3i}{2\sqrt{2}} \left[(1 - i\sqrt{2})^{n-2} - (1 + i\sqrt{2})^{n-2}\right] \quad \text{for all } n \in \mathbb{N}.$$

II.6 An application: Mixing of two liquids

Two containers having the same volume C liters, contain solutions of a given percentage by volume. The first container contains A liters of p percent by volume of a certain liquid L and the other container B liters of q percent by volume of the same liquid L. We assume that $C < A+B < 2C$.

We describe the process of pouring the solutions from one container into the other.

1st step. Write $q = p_1$. Fill the first container by pouring solution from the second container. Thus, there will be C liters p_2 percent by volume of the liquid L, where

$$p_2 = \frac{Ap + (C-A)q}{C},$$

while in the second container remained $D = A+B-C$ liters of p_1 percent by volume of the liquid L.

2nd step. Fill the second container by pouring solution from the first container. We have thus poured in $E = 2C - (A+B)$ of the solution. The second container is now full of the p_3 percent by volume of the liquid L, where

$$p_3 = \frac{Dp_1 + Ep_2}{C}.$$

3rd step. We repeat the second step, this time by filling the first container by the solution from the second container. The first container contains now C liters of p_4 percent by volume of the liquid L, where

$$p_4 = \frac{Dp_2 + Ep_3}{C}.$$

Continuing in this process, we obtain a recursively defined sequence $\{p_n \mid n = 1, 2, \ldots\}$, where

$$p_1 = q,\ p_2 = \frac{Ap + (C-A)q}{C} \text{ and } p_{n+2} = \frac{D}{C}p_n + \frac{E}{C}p_{n+1} \text{ for } n = 1, 2, \ldots.$$

Exercise II.6.1. Show that, for every natural number n, we have

$$p_n = \frac{Ap + Bq}{A+B} + \frac{(-1)^n A(A+B-C)^{n-1}(p-q)}{(A+B)C^{n-1}}$$

and consequently, as expected,

$$\lim_{n\to\infty} p_n = \frac{Ap + Bq}{A+B}.$$

II.7 Characteristic functions

Let $\mathcal{P}(U)$ be the family of all subsets A of a universe U.

Definition II.7.1. *The* **characteristic** *function* $\mathbf{char}_A : U \to \{0,1\}$ *of the set* $A \subseteq U$ *is defined by*

$$\mathbf{char}_A(x) = 1 \text{ if } x \in A \text{ and } \mathbf{char}_A(x) = 0 \text{ if } x \notin A.$$

In general we can add and multiply real-valued or complex-valued functions defined on U. In the case of characteristic functions we have for $A \subseteq U$, $B \subseteq U$ and $x \in U$

$$(\mathbf{char}_A \pm \mathbf{char}_B)(x) = \mathbf{char}_A(x) \pm \mathbf{char}_B(x) \text{ and } (\mathbf{char}_A\, \mathbf{char}_B)(x) = \mathbf{char}_A(x)\, \mathbf{char}_B(x).$$

We now list some properties of characteristic functions, and leave it to the reader to prove them.

$\mathbf{char}_U = \mathbf{1}$ (that is, $\mathbf{char}_U(x) = 1$ for all $x \in U$), $\mathbf{char}_{A^*} = \mathbf{1} - \mathbf{char}_A$;

$A \subseteq B$ if and only if $\mathbf{char}_A \leq \mathbf{char}_B$ (that is, $\mathbf{char}_A(x) \leq \mathbf{char}_B(x)$ for all $x \in U$);

$A = B$ if and only if $\mathbf{char}_A = \mathbf{char}_B$;

$\mathbf{char}_{A\cap B} = \mathbf{char}_A\, \mathbf{char}_B$;

$\mathbf{char}_{A\cup B} = \mathbf{char}_A + \mathbf{char}_B - \mathbf{char}_{A\cap B} = \mathbf{char}_A + \mathbf{char}_B - \mathbf{char}_A\, \mathbf{char}_B$;

$\mathbf{char}_{A\setminus B} = \mathbf{char}_A\, (\mathbf{1} - \mathbf{char}_B)$;

$\mathbf{char}_{A\oplus B} = \mathbf{char}_A + \mathbf{char}_B - 2\, \mathbf{char}_A\, \mathbf{char}_B$.

Here, the set theoretical operation $\oplus$ is defined by $A \oplus B = (A \cup B) \setminus (A \cap B)$.

Exercise II.7.1. For A, B, C in $\mathcal{P}(U)$ prove that

$$\mathbf{char}_{A\cup B\cup C} = \mathbf{char}_A + \mathbf{char}_B + \mathbf{char}_C - \mathbf{char}_{A\cap B} - \mathbf{char}_{A\cap C} - \mathbf{char}_{B\cap C} + \mathbf{char}_{A\cap B\cap C}.$$

A generalization of the previous exercise can be formulated as the follows.

Theorem II.7.1. *Let* $A_1, A_2, \ldots, A_n$ *be subsets of* U. *Denote by* A *their union* $\bigcup_{k=1}^n A_k$. *Furthermore, denote by* $\mathcal{T}_k$ *the set of all possible* $k-$*tuples* $(i_1, i_2, \ldots, i_k)$ *satisfying* $i_1 < i_2 < \cdots < i_k$ *of the elements from* $\{1, 2, \ldots, n\}$. *Then*

$$\mathbf{char}_A = \sum_{k=1}^{n} \sum_{\mathcal{T}_k} (-1)^{k+1} \mathbf{char}_{A_{i_1}\cap A_{i_2}\cap\cdots\cap A_{i_k}}. \tag{II.7.1}$$

Exercise II.7.2. Write down the formula (II.7.1) explicitly for $n = 4$ to see the pattern. Then prove Theorem II.7.1 by induction using the above mentioned rules.

An elegant direct proof of (II.7.1) can be obtained from the following observation:

$$\mathbf{char}_{A_1\cup A_2\cup\cdots\cup A_n} = \mathbf{1} - \mathbf{char}_{(A_1\cup A_2\cup\cdots\cup A_n)^*} = \mathbf{1} - \mathbf{char}_{A_1^*\cap A_2^*\cap\cdots\cap A_n^*}$$

$$= \mathbf{1} - \mathbf{char}_{A_1^*}\,\mathbf{char}_{A_2^*}\,\ldots\,\mathbf{char}_{A_n^*} = \mathbf{1} - (\mathbf{1} - \mathbf{char}_{A_1})\,(\mathbf{1} - \mathbf{char}_{A_2})\,\ldots\,(\mathbf{1} - \mathbf{char}_{A_n})$$

$$= \sum_k \mathbf{char}_{A_k} - \sum_{k,l} \mathbf{char}_{A_k\cap A_l} + \sum_{k,l,m} \mathbf{char}_{A_k\cap A_l\cap A_m} - \cdots + (-1)^{n+1}\mathbf{char}_{A_1\cap A_2\cap\cdots\cap A_n}.$$

If U is a finite set, the **number of elements** $|A|$ of a subset $A \subseteq U$ is given by

$$|A| = \sum_{x\in U} \mathbf{char}_A(x).$$

Hence, we have the following important formula.

Theorem II.7.2. *Let $A_1, A_2, \ldots, A_n$ be subsets of a finite set U. Then*

$$| A_1 \cup A_2 \cup \cdots \cup A_n |= \sum_{k=1}^{n} \sum_{\mathcal{T}_k} (-1)^{k+1} | A_{i_1} \cap A_{i_2} \cap \cdots \cap A_{i_k} | .$$

Example II.7.1 (Inclusion-Exclusion Principle). Theorem II.7.2 can be reformulated in the form of the so-called Inclusion-Exclusion Principle:

Let U be a set containing u elements and let $P_1, P_2, \ldots, P_n$ be properties that the elements of U may or may not satisfy. Let $u_{i_1\, i_2 \ldots i_k}$ be the number of elements of U that have all the properties $P_{i_1}, P_{i_2}, \ldots P_{i_k}$. Then the number of elements of U that have none of the n properties is equal to

$$u - (u_1 + u_2 + \cdots + u_n) + (u_{12} + u_{13} + \cdots + u_{n-1\,n}) - (u_{123} + u_{124} + \cdots + u_{n-2\,n-1\,n})$$

$$+(-1)^k (u_{12\ldots k-1\,k} + u_{12\ldots k-1\,k+1} + \cdots + u_{n-(k-1)\,n-(k-2)\ldots n}) + \cdots + (-1)^n u_{12\ldots n}.$$

Exercise II.7.3. A survey of consumption of 3 different brands of coffee (**M**occa, **C**apuccino and **D**ecaffeinated) interviewed 1000 users and recorded the following findings: **M** was used by 790 people, **C** by 300 people; furthermore, 110 people used both **M** and **C**, 35 people both **M** and **D** and 15 people both **C** and **D**. Only 10 people have used all three brands of coffee. How many people used **D**?

II.8 A few final exercises

Exercise II.8.1. Prove, for all $m, n \in \mathbb{N}$, that

$$\sum_{k=m}^{n} \binom{n}{k}\binom{k}{m} = \binom{n}{m} 2^{n-m}.$$

Consequently, determine the sum

$$\sum_{k=1}^{n} k\binom{n}{k}.$$

Exercise II.8.2. Let $n \in \mathbb{N}$. Find an explicit formula for the sum

$$\frac{1}{1\cdot 4} + \frac{1}{4\cdot 7} + \frac{1}{7\cdot 10} + \cdots + \frac{1}{(3n-2)(3n+1)}$$

and for the product

$$\left(1 - \frac{1}{2^2}\right)\left(1 - \frac{1}{3^2}\right)\cdots\left(1 - \frac{1}{(n+1)^2}\right).$$

Exercise II.8.3. Find, for each $n \in \mathbb{N}$, the sum

$$\sum_{k=1}^{n} \frac{1}{k(k+2)}.$$

Exercise II.8.4. Show that for each $n \in \mathbb{N}$

$$\sum_{k=1}^{n} (-1)^k k^2 = (-1)^n \sum_{k=1}^{n} k \text{ and } \sum_{k=1}^{n} k \cdot k! = (n+1)! - 1.$$

Exercise II.8.5. Prove for each $n \in \mathbb{N}$ that

$$\sum_{k=0}^{n} \binom{2n}{2k} = \sum_{k=1}^{n} \binom{2n}{2k-1} = 2^{2n-1}.$$

Exercise II.8.6. Show that for each natural number n

$$\sum_{k=1}^{n} (2k-1)^2 = \binom{2n+1}{3} \text{ and } \sum_{k=1}^{n} (2k)^2 = \binom{2n+2}{3}.$$

Exercise II.8.7. Show that for each natural number n

$$\prod_{k=2}^{n} \left(1 - \frac{1}{1+2+\cdots+k}\right) = \frac{n+2}{3n}$$

and

$$\sum_{k=1}^{n} \frac{1}{\sqrt{k}} \leq 2\sqrt{n}.$$

Exercise II.8.8 (Lucas' Tower of Hanoi). Given three posts with n disks of decreasing diameters placed on one of the posts in such a way that the largest disk is at the bottom (thus making a stack of conical shape), determine the minimal number of steps required to move the entire stack to another post, obeying the following rules:

(i) Only one disk may be moved at a time.

(ii) Each move consists of taking the upper disk from one of the posts and sliding it onto another post, on top of the other disks that may already be present on that post.

(iii) No disk may be placed on top of a smaller disk.

Exercise II.8.9. Find the sum

$$S(n) = a_1 + a_2 + \cdots + a_n,$$

where $a_1 = 2$ and $a_k = 3a_{k-1} + 1$ for $k = 2, 3, \ldots$.

Exercise II.8.10. Define the sequence $\{u_n \mid n = 0, 1, 2, \ldots\}$ by

$$u_0 = c,\ u_n = au_{n-1} + b \ \text{ for } \ n \geq 1,\ a, b, c \in \mathbb{C}.$$

Find the explicit form of u_n. Does $\lim_{n\to\infty} u_n$ exist?

[You may like to give an application of this exercise to the evolution of some ecological groups.]

Exercise II.8.11. Writing the first $n+1$ rows of the Yang Hui - Pascal triangle in the form of the following triangular matrix

$$M_{n+1} = \begin{pmatrix} 1 & 0 & 0 & 0 & \cdots & 0 & 0 \\ 1 & 1 & 0 & 0 & \cdots & 0 & 0 \\ 1 & 2 & 1 & 0 & \cdots & 0 & 0 \\ 1 & 3 & 3 & 1 & \cdots & 0 & 0 \\ \cdots & \cdots & \cdots & \cdots & \cdots & \cdots & \cdots \\ \binom{n-1}{0} & \binom{n-1}{1} & \binom{n-1}{2} & \binom{n-1}{3} & \cdots & 1 & 0 \\ \binom{n}{0} & \binom{n}{1} & \binom{n}{2} & \binom{n}{3} & \cdots & \binom{n}{n-1} & 1 \end{pmatrix}$$

show that the triangular matrix $M_{n+1}^{-1} = (x_{rs})$ (that is, the matrix such that $M_{n+1}^{-1}\, M_{n+1}$ is the identity matrix) satisfies

$$x_{rs} \ = \ (-1)^{r-s}\binom{r-1}{s-1} \ \text{ for } r \geq s \text{ and } x_{rs} = 0 \text{ otherwise.}$$

Exercise II.8.12. Define **Stirling numbers** $\left[{n \atop k}\right]$ **of the first kind (James Stirling (1692 - 1770))** as the coefficients of the polynomial

$$x(x+1)(x+2)\cdots(x+n-1) = \begin{bmatrix} n \\ 1 \end{bmatrix} x + \begin{bmatrix} n \\ 2 \end{bmatrix} x^2 + \cdots + \begin{bmatrix} n \\ n-1 \end{bmatrix} x^{n-1} + \begin{bmatrix} n \\ n \end{bmatrix} x^n.$$

Prove that

$$\begin{bmatrix} n+1 \\ k \end{bmatrix} = \begin{bmatrix} n \\ k-1 \end{bmatrix} + n \begin{bmatrix} n \\ k \end{bmatrix} \quad \text{and}$$

$$\begin{bmatrix} n \\ 1 \end{bmatrix} = (n-1)!, \ \begin{bmatrix} n \\ n \end{bmatrix} = 1, \ \begin{bmatrix} n+1 \\ n \end{bmatrix} = \binom{n+1}{2}.$$

Exercise II.8.13. Show that the sequence $a_n = \left[{n+2 \atop n}\right]$ is an arithmetic progression of fourth order and use its differential scheme to show that

$$\begin{bmatrix} n+2 \\ n \end{bmatrix} = \binom{n+2}{3} \frac{3n+5}{4}.$$

Furthermore show that

$$\sum_{k=1}^{n} \begin{bmatrix} n \\ k \end{bmatrix} = \begin{bmatrix} n+1 \\ 1 \end{bmatrix} = n!$$

Exercise II.8.14. Write the first n rows of the "Stirling triangle" in the form of the following triangular matrix

$$N_n = \begin{pmatrix} 1 & 0 & 0 & 0 & \cdots & 0 & 0 \\ 1 & 1 & 0 & 0 & \cdots & 0 & 0 \\ 2 & 3 & 1 & 0 & \cdots & 0 & 0 \\ 6 & 11 & 6 & 1 & \cdots & 0 & 0 \\ \cdots & \cdots & \cdots & \cdots & \cdots & \cdots & \cdots \\ \left[{n-1 \atop 1}\right] & \left[{n-1 \atop 2}\right] & \left[{n-1 \atop 3}\right] & \left[{n-1 \atop 4}\right] & \cdots & 1 & 0 \\ \left[{n \atop 1}\right] & \left[{n \atop 2}\right] & \left[{n \atop 3}\right] & \left[{n \atop 4}\right] & \cdots & \left[{n \atop n-1}\right] & 1 \end{pmatrix}$$

and define (unsigned) **Stirling numbers** $\left\{{r \atop s}\right\}$ **of the second kind** using the entries of the inverse matrix $N_n^{-1} = (y_{rs})$ by

$$\begin{Bmatrix} r \\ s \end{Bmatrix} = (-1)^{r-s}\, y_{rs}.$$

Prove that

$$\begin{Bmatrix} n+1 \\ k \end{Bmatrix} = \begin{Bmatrix} n \\ k-1 \end{Bmatrix} + k \begin{Bmatrix} n \\ k \end{Bmatrix} \quad \text{and}$$

$$\begin{Bmatrix} n \\ 1 \end{Bmatrix} = 1, \ \begin{Bmatrix} n \\ n \end{Bmatrix} = 1, \ \begin{Bmatrix} n+1 \\ 2 \end{Bmatrix} = 2^n - 1, \ \begin{Bmatrix} n+1 \\ n \end{Bmatrix} = \binom{n+1}{2}.$$

Exercise II.8.15. Show that the sequence $a_n = \left\{ {n+2 \atop n} \right\}$ is an arithmetic progression of fourth order and use its differential scheme to show that

$$\left\{ {n+2 \atop n} \right\} = \binom{n+2}{3} \frac{3n+1}{4}$$

and thus

$$\left[{n+2 \atop n} \right] = \left\{ {n+2 \atop n} \right\} + \binom{n+2}{3}.$$

Furthermore, show that

$$\sum_{k=1}^{n} \left\{ {n \atop k} \right\} = b(n)$$

is the n-th Bell number (see Section 3 of Chapter X and Example II.2.5).

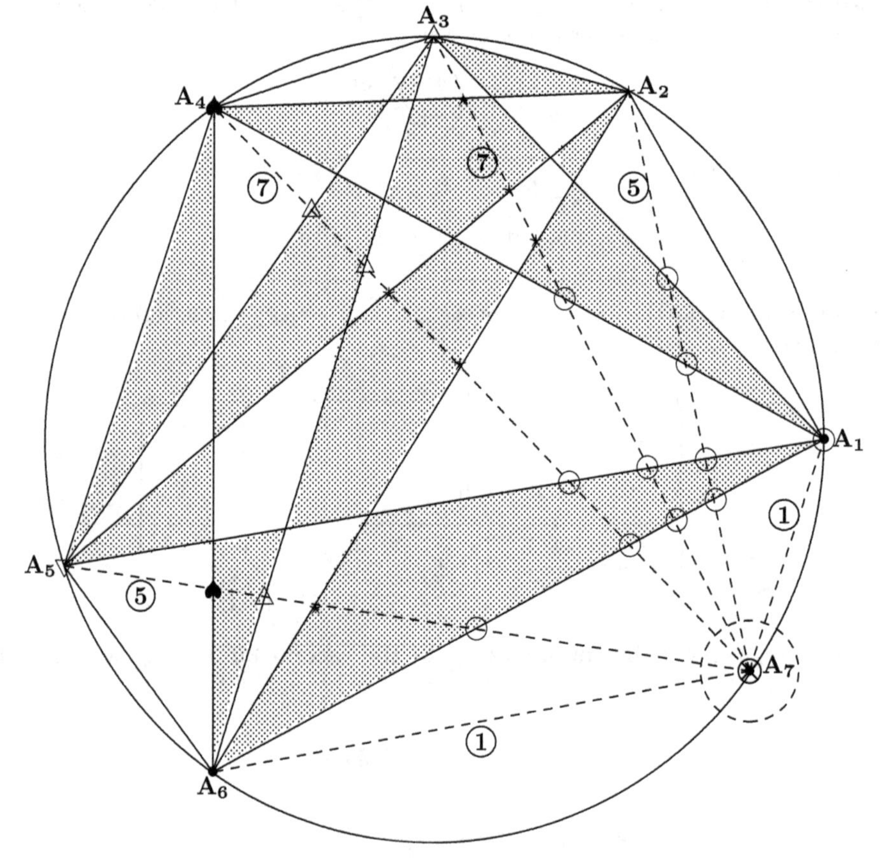

Figure II.8.1 $\mu(\mathbf{6}) = \mathbf{31}$, $\mu(\mathbf{7}) = \mu(\mathbf{6}) + \mathbf{1} + \mathbf{5} + \mathbf{7} + \mathbf{7} + \mathbf{5} + \mathbf{1} = \mathbf{57}$

Exercise II.8.16. Determine the maximal number of regions of a given circle that can be obtained by connecting n points of that circle by straight lines (secants).

n	μ(n)	Δn
1	**1**	
2	**2**	1
3	**4**	2
4	**8**	4
5	**16**	8
6	**31**	15
7	**57**	26
8	**99**	42
9	**163**	64
10	**256**	93
11	**386**	130
12	**562**	176
13	**794**	232
14	**1093**	299
15	**1471**	378
16	**1941**	470
17	**2517**	576
18	**3214**	697
19	**4048**	834
⋮	⋮	⋮

1	1	1	1	1	(1)	1	1	1	1	1	1	1	1	1	1	1	1	
1	2	3	4	(5)	6	7	8	9	10	11	12	13	14	15	16	17	……	
1	3	5	(7)	9	11	13	15	17	19	21	23	25	27	29	31	……		
1	4	(7)	10	13	16	19	22	25	28	31	34	37	40	43	……			
1	(5)	9	13	17	21	25	29	33	37	41	45	49	53	……				
(1)	6	11	16	21	26	31	36	41	46	51	56	61	……					
1	7	13	19	25	31	37	43	49	55	61	67	……						
1	8	15	22	29	36	43	50	57	64	71	78	……						
1	9	17	25	33	41	49	57	65	73	81	89	……						
1	10	19	28	37	46	55	64	73	82	91	100	109	……					
1	11	21	31	41	51	61	71	81	91	101	111	121	131	141	151	……		
1	12	23	34	45	56	67	78	89	100	111	122	133	144	155	……			
1	13	25	37	49	61	73	85	97	109	121	133	145	157	169	……			
1	14	27	40	53	66	79	92	105	118	131	144	157	170	183	……			
1	15	29	43	57	71	85	99	113	127	141	155	169	183	197	211	……		
1	16	31	46	61	76	91	106	121	136	151	166	181	196	211	226	……		
1	17	33	49	65	81	97	113	129	145	……								
1	⋮	⋮	⋮	⋮	⋮	⋮	⋮	⋮	⋮									

Figure II.8.2 Maximal number of regions $\mu(\mathbf{n})$.

Comments: Denote the maximal such number by $\mu(n)$. For $n = 1$ there is only one region, namely the entire interior of the circle, that is, $\mu(1) = 1$. Also we see that $\mu(2) = 2$, $\mu(3) = 4$, $\mu(4) = 8$, $\mu(5) = 16$. Figure II.8.1 shows that $\mu(6)$ **is not** the expected 32, but **31** and $\mu(7) = 57$. Figure II.8.2 lists in the column marked $\mu(n)$ the values for $1 \leq n \leq 19$ and indicates (by drawing the dotted diagonal lines) the way these values are obtained.

Exercise II.8.17. (i) Describe the way the values of $\Delta n = \mu(n) - \mu(n-1)$ in Figure II.8.2 (and thus of $\mu(n)$) are obtained and calculate $\mu(20)$.

(ii) Show that the sequence $(\mu(n))$ is an arithmetic progression of order 4 and using the corresponding differential scheme, find the expression for $\mu(n)$, that is, the polynomial of degree 4 whose value at n is $\mu(n)$.

(iii) Denote the points on the circle $A_1, A_2, \dots A_{n-1}, A_n$ (see Figure II.8.1 for $n = 7$). The connecting lines from the point A_n cut each existing region into two regions. For each individual line, the number of these new regions is determined by the number of the intersection points. Thus, the connecting line to A_1 creates one additional region, the connecting line to A_2 creates $n-1$ regions and the connecting line to A_3 creates $1+2(n-3)$ additional regions. In general, the line connecting A_n with A_k creates $1+(k-1)(n-k)$ regions. Adding the numbers of the new regions, show that $\Delta n = \frac{1}{6}(n-1)(n^2-5n+12)$.

(iv) Since Δn is a cubic polynomial in n, $\mu(n)$ is a polynomial of degree four in n. Use this fact and the result of (iii) to derive (in a different way) the formula for $\mu(n)$.

The last exercise of this section concerns Pythagorean triples. These triples are intimately related to the natural numbers, and, although named after **Pythagoras of Samos (569 - 475 BCE)**, have actually been known since ancient times. The oldest known record of Pythagorean triples appears on a Babylonian clay tablet Plimpton 322 from about 1800 BCE.

Exercise II.8.18 (Primitive Pythagorean Triples). A triple (x, y, z) is called a primitive Pythagorean triple, if x, y, z are natural numbers whose only positive common divisor is 1 and which satisfy the condition

$$x^2 + y^2 = z^2. \tag{II.8.1}$$

(a) Decide if the triple $(2004, 27853, 27925)$ is a primitive Pythagorean triple.

(b) Provide a geometrical representation of the Pythagorean triples as certain points on a unit circle.

(c) Show that if (x, y, z) is a primitive Pythagorean triple, then all pairs $(x, y), (x, z)$ and (y, z) are relatively prime, that is, these pairs of numbers have no positive common divisor but 1.

(d) Show that just one number of the primitive Pythagorean triple is even, and that it cannot be z. In what follows, we will assume that x is even.

(e) Prove that the equation (II.8.1) is equivalent to $\left(\frac{x}{2}\right)^2 = \left(\frac{z+y}{2}\right)\left(\frac{z-y}{2}\right)$, where $\frac{x}{2}$, $\frac{z+y}{2}$ and $\frac{z-y}{2}$ are integers and $\frac{z+y}{2}$ and $\frac{z-y}{2}$ are relatively prime.

(f) Deduce that there exist two integers u and v such that

$$u^2 = \frac{z+y}{2},\ v^2 = \frac{z-y}{2} \text{ and } 0 < v < u.$$

Express x, y and z as functions of u and v.

(g) Conversely, let u and v be two integers such that $0 < v < u$ and define

$$x = 2uv, y = u^2 - v^2 \text{ and } z = u^2 + v^2.$$

If the numbers u and v are relatively prime and of distinct parity, show that (x, y, z) is a primitive Pythagorean triple.

(h) Determine all primitive Pythagorean triples for $0 < v < u \leq 7$.

CHAPTER III. INTEGERS, DIVISIBILITY

III.1 The role of integers

The previous chapter has shown the significance of natural numbers. However, the full impact on the developments of society and science in particular lies with the integers. Their pivotal role in our lives can hardly be overstated. Already in our childhood, we assimilated quite naturally some of the properties of addition, multiplication and order which are generally taken for granted. In this context, one should mention the celebrated dictum of **Leopold Kronecker (1823 - 1891)** *"Die ganze Zahl schuf der liebe Gott; alles übrige ist Menschenwerk"* that can be freely translated as

"God created the integers; everything else is the work of man."

One should however recall that his contemporary, **Richard Dedekind** whose understanding of natural numbers was sketched in the previous chapter, expressed some reservations about this point of view.

The domain of the integers that inherently extends the domain of the natural numbers together with their addition, multiplication and order, is a basis for constructions of further number systems, of the entire algebra and modern analysis. This is a reason for proper understanding of its structure. And this is the goal of this chapter.

The notion of an integer has a lengthy history. Its developments extend to a relatively recent historical period when the concept of an integer included neither the number zero nor the negative numbers. Although one can already trace the negative numbers to China before our era, and although they are together with zero an indivisible part of Indian mathematics, the integers slowly entered European mathematics under the influence of Muslim science in the late middle ages. Even in the time of **Leonhard Euler** negative numbers do not seem to have been generally well understood. We recall the fact that as late as in the 18-th century some mathematicians held the view that the negative numbers are greater than infinity. The full understanding of the domain of integers includes both a proper understanding of the total ordering of this domain, as well as of the operation of subtraction. Many of these important facts will be expressed by saying that the integers form an **additive group**. The concept of a group will be one of the central concepts in our study. The ensuing presentation of the divisibility of the integers will be a valuable model for the study of abstract groups.

The notation for integer numerals also has a rich history. The origin of our present notation for integers by means of so-called **Arabic numerals** in the decimal system in Europe belongs to the late middle ages. This notation struggled for several centuries with the notation using **Roman numerals**. There was a contest between the **"algorists"** (proponents of the new system) and **"abacists"** (defenders of the older Roman numeral system). Both factions had staunch supporters and even as late as 1503 the well-known woodcut (see below) shows a competition between an abacist (a monk) and an algorist (a worldly scholar) seemingly favouring the algorist's progress. We will try to comprehend merits as well as shortcomings of these systems.

Equally, we will pay attention to the notation of the integers in different bases. Due to its importance in the contemporary era of computational technology we will pay particular attention to the binary numeral (that is, base-2 number) system. As a matter of interest, we present here the following table of the first 24 prime numbers recorded in different bases.

base 10	base 2	base 3	base 4	base 7	base 12
2	10	2	2	2	2
3	**11**	10	3	3	3
5	101	12	10	5	5
7	**111**	21	12	10	7
11	1011	102	21	14	**E**
13	1101	111	23	16	11
17	10001	122	32	23	15
19	10011	201	34	25	17
23	10111	212	43	32	**1E**
29	11101	1002	104	41	25
31	**11111**	1011	111	43	27
37	100101	1101	122	52	31
41	101001	1112	131	56	35
43	101011	1121	133	61	37
47	101111	1202	142	65	**3E**
53	110101	1222	203	104	45
59	111011	2012	214	113	**4E**
61	111101	2021	221	115	51
67	1000011	2111	232	124	57
71	1000111	2122	241	131	**5E**
73	1001001	2201	243	133	61
79	1001111	2221	10304	142	67

We point out that the notation in base-12 requires symbols for two new "digits" to express ten and eleven. Our table makes use of only one of them, namely the one that denotes eleven. We denote it by the letter E.

As we have already pointed out, understanding of the structure of integers consists above all in deeper comprehension of **divisibility** of integers. Here, we will stress the fundamental importance of the so-called **Euclidean division** and the closely related **Euclidean algorithm**, both named after **Euclid of Alexandria (about 325 - 265 BCE)**. Even a simple puzzle, such as the following one that **James Joseph Sylvester (1814 - 1897)** sent in 1884 to the *Educational Times* (and tells us of one of Sylvester's hobbies), can contribute to the initial understanding of the integers:

I have a large number of stamps to the value of 5d and 17d only. What is the largest denomination which I cannot make up with a combination of these two different values?

First, we point out that Sylvester reveals the fact that such a largest denomination exists, although this is not obvious from the start. However, it becomes clear and the puzzle is solved by making use of Figure III.1.1.

					0	1	2	3	4	5	6	7	8	9	10	11	12	13	14	...
0					**0**	17	34	51	**68**	**85**	102	119	136	**153**	**170**	187	204	221	**238**	...
1					**5**	22	39	56	**73**	**90**	107	124	141	**158**	.	.	.	.	.	...
2					**10**	27	44	61	**78**	**95**	112	129	146	**163**	.	.	.	.	.	...
3					**15**	32	49	**66**	**83**	100	117	134	**151**	**168**	.	.	.	.	.	...
4				**3**	20	37	54	**71**	88	105	122	139	**156**	173	.	.	.	.	.	...
5				**8**	25	42	59	**76**	93	110	127	144	**161**	178	.	.	.	.	.	...
6				**13**	30	47	**64**	**81**	98	115	132	**149**	**166**	183	.	.	.	.	.	...
7			**1**	18	35	52	**69**	86	103	120	137	**154**	171	188	.	.	.	.	.	...
8			**6**	23	40	57	**74**	91	108	125	142	**159**	176	193	.	.	244	.	.	...
9			**11**	28	45	62	**79**	96	113	130	147	**164**	181	198	.	.	.	.	.	...
10			**16**	33	50	**67**	**84**	101	118	135	**152**	**169**	186	203	.	.	.	.	.	...
11		**4**	21	38	55	**72**	89	106	123	140	**157**	174	191	208	.	.	.	.	.	...
12		**9**	26	43	60	**77**	94	111	128	145	**162**	179	196	213	.	.	.	.	.	...
13		**14**	31	48	**65**	**82**	99	116	133	**150**	**167**	184	201	218	**235**	.	.	.	303	...
14	**2**	19	36	53	**70**	87	104	121	138	**155**	172	189	206	223	.	.	.	.	.	...
15	**7**	24	41	58	**75**	92	109	126	143	**160**	177	194	211	228	.	.	.	.	.	...
16	**12**	29	46	**63**	**80**	97	114	131	158	**165**	182	199	216	233	**250**	.	.	.	318	...
17					**85**	102	119	136	**153**	**170**	187	204	221	**238**	**255**	272	289	306	**323**	...
18					**90**	107	124	141	**158**	.	.	.	.	.	.	277	.	.	.	...
19					**95**	112	129	146	**163**	.	.	.	.	.	.	.	.	.	.	...
20					100	117	134	**151**	**168**	.	.	.	.	.	.	.	.	.	.	...
21					105	122	139	**156**	173	.	.	.	.	.	.	.	.	.	.	...
22					110	127	144	**161**	178	.	.	.	.	.	.	.	.	.	.	...
23					115	132	**149**	**166**	183	.	.	.	.	.	.	.	.	336	.	...
24					120	137	**154**	171	188	.	.	.	.	.	.	.	.	.	.	...
25					125	142	**159**	176	193	.	.	244	.	.	.	.	.	.	.	...
26					130	147	**164**	181	198	.	.	.	.	.	.	.	.	.	.	...
27					135	**152**	**169**	186	203	.	.	.	.	.	.	.	.	.	.	...
28					140	**157**	174	191	208	.	.	.	.	.	.	.	.	.	.	...
29					145	**162**	179	196	213	.	.	.	.	.	.	.	.	.	.	...
30					**150**	**167**	184	201	218	**235**	.	.	.	303	.	.	.	.	.	...
31					**155**	172	189	206	223	.	.	.	.	.	.	.	.	.	.	...
32					**160**	177	194	211	228	.	.	.	.	.	.	.	.	.	.	...
33					**165**	182	199	216	233	**250**	.	.	.	318	.	.	.	.	.	...
34					**170**	187	204	221	**238**	**255**	272	289	306	**323**	...					...
35					.	.	.	.	.	.	277	.	.	.	...					...
36					.	.	.	.	.	.	.	.	.	.	...					
37					.	.	.	.	.	.	.	.	.	.	...					
38					.	.	.	.	.	.	.	.	.	.	...					
39					.	.	.	.	.	.	.	.	.	.	...					
40					.	.	.	.	.	.	.	.	336	.	...					
41					.	.	.	.	.	.	.	.	.	.	...					
42					.	.	244	.	.	.	.	.	.	.	...					
43					.	.	.	.	.	.	.	.	.	.	...					
44					.	.	.	.	.	.	.	.	.	.	...					
45					.	.	.	.	.	.	.	.	.	.	...					
46					.	.	.	.	.	.	.	.	.	.	...					
47					**235**	.	.	.	303	.	.	.	.	.	...					
48					.	.	.	.	.	.	.	.	.	.	...					
49					.	.	.	.	.	.	.	.	.	.	...					
50					**250**	.	.	.	318	.	.	.	.	.	...					
51					**255**	272	289	306	**323**	...					...					
52					.	277	.	.	.	...					...					

Figure III.1.1 Non-negative solutions of the linear diophantine equation $5x + 17y = c$

We return to this table in Section 6 of this chapter. Here we only remark that the table shows that the postage of

44d can be made up in a unique way $(2 \cdot 5d + 2 \cdot 17d)$,

144d can be made in two different ways $(5 \cdot 5d + 7 \cdot 17d = 22 \cdot 5d + 2 \cdot 17d)$,

244d can be made in three different ways $(8 \cdot 5d + 12 \cdot 17d = 25 \cdot 5d + 7 \cdot 17d = 42 \cdot 5d + 2 \cdot 17d)$,

277d can be made in four different ways $(1 \cdot 5d + 16 \cdot 17d = 18 \cdot 5d + 11 \cdot 17d = 35 \cdot 5d + 6 \cdot 17d = 52 \cdot 5d + 1 \cdot 17d)$ etc.

On the other hand, those 32 denominations listed outside the framing cannot be made with any combination of the available stamps and the largest such denomination is $63d$.

A similar slightly more involved scheme provides an explanation of the following story:

A company sells iPads in two versions: version "A" for \$ 299.– and a more advanced version "B" for \$ 472.–. One of the dealers reported a monthly sale of \$ 140,407.– with an apology that he had misplaced his records of how many iPads of each type he sold. For the company, knowing that for that amount there is only one possibility: sale of 277 iPads of type A and 122 iPads of type B, this was no problem. It is hard to believe how much we can read from the amount representing the monthly sale. We know, they informed the dealer, that you cannot gross \$ 100,000.–. On the other hand, we know that you can cash in any amount bigger than \$ 140,357.–. In case that you gross \$ 200,000.–, we know right away that you must have sold 440 iPads of type A and 145 iPads of type B. Unfortunately, such a definite answer we are unable to give in the case when you gross \$ 300,000.–. Are all these statements true? Shall we succeed in explaining and verifying the following table?

SALE	Number of iPads of type "A"	Number of iPads of type "B"
§ 140,400.–	408	39
\$ 140,401.–	187	179
\$ 140,402.–	438	20
\$ 140,403.–	217	160
\$ 140,404.–	468	1
\$ 140,405.–	247	141
\$ 140,406.–	26	281
\$ 140,407.–	277	122
\$ 140,408.–	56	262
\$ 140,409.–	307	103
\$ 140,410.–	86	243
\$ 141,000–	440	20
\$ 150,000.–	448	34
\$ 200,000.–	440	145
\$ 300,000.–	896 *or* 424	68 *or* 367
\$ 500,000.–	1336, *or* 864 *or* 392	213, *or* 512 *or* 811
\$ 1,000,000.–	3144, *or* 2672, *or* 2200, *or* 1728, *or* 1256, *or* 784 *or* 312	127, *or* 426, *or* 725, *or* 1024, *or* 1323, *or* 1622 *or* 1921

As we have already indicated, it is impossible to obtain exactly \$ 100,000.– or \$ 50,000.– in the "sale" column.

This problem will lead us to the theory of (integral) diophantine equations, to the theory of congruences, to Euclidean division, to the understanding of prime numbers, to classification of cyclic groups, to constructions of finite fields and a brief glance into cryptography.

III.2 Basic structure

With reference to Theorem I.7.2 we know that the (additive) monoid $(\mathbb{N}_0, +)$ of non-negative integers can be extended to the (additive) abelian (commutative) group $(\mathbb{Z}, +)$. We refresh this process that will lead to a formal construction of the integral domain $\mathbb{Z} = (\mathbb{Z}, +, ., \leq)$ of the integers.

Consider the cartesian product $\mathbf{W} = \mathbb{N}_0 \times \mathbb{N}_0$ of all pairs (a, b) with $a, b \in \mathbb{N}_0$. We can define addition on $\mathbf{W}$: $(a, b) + (c, d) = (a + c, b + d)$ and thus obtain an (additive) monoid. But this is not our goal. Our goal is to interpret the pair (a, b) as a **difference** between a and b, that is, for instance to interpret $(5, 2)$ as 3. Here are two problems. First, such an interpretation should be made unique: since $(7, 4), (103, 100)$, etc. should also be interpreted as 3, we need to make all such pairs "equal". Secondly, we will have to provide an interpretation for "new numbers" (a, b) with $a < b$. Both these obstacles can be overcome by the introduction of the equivalence

$$(a, b) \equiv (c, d) \text{ if and only if } \ a + d = b + c.$$

(Check that this **relation** is **reflexive**, that is, $(a, b) \equiv (a, b)$, **symmetric**, that is, $(a, b) \equiv (c, d)$ implies $(c, d) \equiv (a, b)$ and **transitive**, that is, $(a, b) \equiv (c, d)$ and $(c, d) \equiv (e, f)$ imply $(a, b) \equiv (e, f)$.) Thus $\mathbf{W}$ has been partitioned into an infinite set $\mathbb{Z}$ of $\equiv$-equivalence classes $\overline{(a, b)}$:

$$\overline{(a, b)} = \overline{(c, d)} \text{ if and only if } (a, b) \equiv (c, d).$$

All these classes are infinite. Observe that in each of these classes there is a **unique** pair of the form either $(a, 0)$ or $(0, d)$.

Now, define

$$\overline{(a, b)} + \overline{(c, d)} = \overline{(a + c, b + d)} \text{ and } \overline{(a, b)} \cdot \overline{(c, d)} = \overline{(ac + bd, ad + bc)}.$$

The addition is well-defined and the additive monoid $(\mathbb{Z}, +)$ is an additive group (see Theorem I.7.1). Check, in the same way, that the multiplication is also well-defined (that is, that the above definitions do not depend on the choice of the representatives of the $\equiv$-equivalent classes). Furthermore, we observe that the mapping

$$\varphi : \mathbb{N}_0 \longrightarrow \mathbb{Z} \text{ defined by } \varphi(a) = \overline{(a, 0)}$$

satisfies

$$\varphi(a + c) = \varphi(a) + \varphi(c) \text{ and } \varphi(a \cdot c) = \varphi(a) \cdot \varphi(c),$$

that is

$$\overline{(a + c, 0)} = \overline{(a, 0)} + \overline{(c, 0)} \text{ and } \overline{(ac, 0)} = \overline{(a, 0)} \cdot \overline{(c, 0)}.$$

Hence $(\mathbb{Z}, +, \cdot)$ is an **extension** of $(\mathbb{N}_0, +, \cdot)$. Moreover, we can define a linear (total) order on $\mathbb{Z}$ by

$$\overline{(a, b)} \leq \overline{(c, d)} \text{ if and only if } a + d \leq b + c.$$

(Check again that this order is well-defined!) We can readily see that

$$\begin{aligned} \overline{(0,b)} &\leq \overline{(c,0)} && \text{for all } b \text{ and } c, \\ \overline{(a,0)} &\leq \overline{(c,0)} && \text{if and only if } a \leq c \text{ and} \\ \overline{(0,b)} &\leq \overline{(0,d)} && \text{if and only if } b \geq d\,! \end{aligned}$$

Thus, simply writing $\overline{(a,0)} = a$ and $\overline{(0,b)} = -b$, we can summarize: There are two operations, **addition** and **multiplication** of the integers

$$\mathbb{Z} = \{0, \pm 1, \pm 2, \pm 3, \ldots, \pm n, \ldots\},$$

extending the respective operations of the natural numbers $\mathbb{N}$. Both are **commutative** and **associative**. Unlike $\mathbb{N}$, $\mathbb{Z} = (\mathbb{Z}, +)$ is, with respect to addition, an **abelian group**: There is a (**neutral element**) zero $0 \in \mathbb{Z}$ satisfying $0 + a = a$ for all $a \in \mathbb{Z}$ and, for any $a \in \mathbb{Z}$, there is an (**additive inverse**) opposite $-a \in \mathbb{Z}$ such that $a + (-a) = 0$. Hence, we may perform the operation of **subtraction** by

$$a - b = a + (-b). \tag{III.2.1}$$

Here, a warning may be in order: The symbol $-$ is used here in two different meanings: On one hand, to denote the opposite element $-b$ to the element b and on the other hand, to denote the binary operation on the set $\mathbb{Z}$. Thus, it is important to distinguish for example the distinct use in the widely used notation $5 - 3 = -3 + 5 = (-3) + 5$. To avoid this source of possible misunderstanding, **always** apply (III.2.1). Thus,

$$a - b - c = a + (-b) + (-c) \neq a - (b - c)\,!$$

We clarify this point in the following exercise.

Exercise III.2.1. Let $\ominus : \mathbb{Z} \times \mathbb{Z} \to \mathbb{Z}$ be a binary operation on the set $\mathbb{Z}$ of the integers satisfying the following properties:

(i) There is a neutral element $n \in \mathbb{Z}$ such that

$$a \ominus a = n \quad \text{and} \quad a \ominus n = a \quad \text{for all} \quad a \in \mathbb{Z}.$$

(ii) For every $c \in \mathbb{Z}$, there is an element $c^- \in \mathbb{Z}$ such that

$$(a \ominus b)^- = b \ominus a \quad \text{for all} \quad a, b \in \mathbb{Z}.$$

(iii) For all $a, b, c \in \mathbb{Z}$,

$$(a \ominus b) \ominus c = a \ominus (c \ominus b^-).$$

Prove that

(a) $n^- = n$ and for every $a \in \mathbb{Z}, n \ominus a = a^-$, $(a^-)^- = a$ and $n \ominus a^- = a$.
(b) The neutral element n is determined uniquely.
(c) For every number $c \in \mathbb{Z}$ the number c^- is determined uniquely.
(d) $(a \ominus b) \ominus b^- = b \ominus (a^- \ominus b^-) = a$.

(e) $a \ominus b = c \ominus b$ implies that $a = c$.
(f) $a \ominus b = n$ implies that $a = b$.
(g) The equation $a \ominus x = b$ has a unique solution $x = b^- \ominus a^-$.
(h) The equation $x \ominus a = b$ has a unique solution $x = b \ominus a^-$.
(i) The abelian group $(\mathbb{Z}, +)$ is isomorphic (in fact, identical) with the structure $(\mathbb{Z}, \star)$,

where

$$a \star b = a \ominus b^-.$$

If the operation $\ominus$ is the subtraction of integers, then the neutral element is $n = 0, b^- = -b$ and $a \star b = a + b$.

With respect to multiplication, integers ($\neq \pm 1$) have no (multiplicative) inverse, that is, $(\mathbb{Z}, \cdot)$ is **not** a (multiplicative) group. It is a **multiplicative monoid** (with a neutral element 1 satisfying $1 \cdot a = a$ for all $a \in \mathbb{Z}$).

The (**algebraic**) operations of addition and multiplication are (as in the case of $\mathbb{N}$) closely related by the **distributive property**

$$a(b + c) = ab + ac \text{ for all } a, b, c \in \mathbb{Z}.$$

We describe all these facts by saying that $\mathbb{Z}(+, \cdot)$ is a (**commutative**) **ring** (for more details about commutative rings see Chapter VII). In fact, the multiplication in $\mathbb{Z}(+, \cdot)$ has also the following **cancellation property**: If $a \neq 0$ and $ab = ac$, then $b = c$. Commutative rings with this property are called (**integral**) **domains**. Thus, $\mathbb{Z}(+, \cdot)$ is an integral domain.

There is also a natural extension of the (**linear**) **order** of $\mathbb{N}$ to $\mathbb{Z}$, namely,

$$\cdots < -4 < -3 < -2 < -1 < 0 < 1 < 2 < 3 < 4 < \ldots.$$

The integers a satisfying $0 < a$ are called **positive** (these are just all natural numbers) and those satisfying $a < 0$ are called **negative**. Recall that $a < b$ if and only if $0 < b - a$. It follows that for the integers a, b such that $a < b$,

$$a + c < b + c \text{ for all } c, \; ac < bc \text{ for all } c > 0 \text{ and } ac > bc \text{ for all } c < 0.$$

There is a geometrical representation of the integers on a **number line**

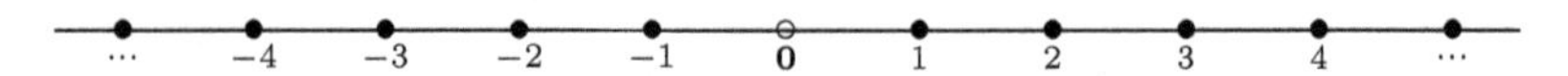

The distance of the number a from the origin is denoted by $|a|$ and is called the **absolute value** of a, that is

$$|a| = \begin{cases} a & \text{for } a \geq 0, \\ -a & \text{for } a \leq 0. \end{cases}$$

In this **linear order**, the set of all integers lost the property of the set of natural numbers of being well-ordered. Indeed, for example, the subset of all even numbers $2\mathbb{Z} = \{\ldots, -4, -2, 0, 2, 4, \ldots\}$ does not possess a least element.

III.3 Euclidean algorithm

We start with a rather "transparent" statement.

Theorem III.3.1 ("Envelopment" of a by multiples of b). *Let b be a natural number. Then, for any integer a, there is a unique integer q such that*

$$\mathbf{bq \leq a < b(q+1).}$$

Proof. Let $a > 0$ be a natural number. Then the statement is trivial if $a < b$ ($q = 0$) or $a = b$ ($q = 1$). Thus let $a > b$. Denote by A the subset of $\mathbb{N}$ defined as follows:

$$A = \{t \in \mathbb{N} \mid b(t+1) > a\}.$$

The set A is nonempty; for, $b(a+1) \geq ba+b \geq a+b > a$, and thus $a \in A$. By the well-ordering principle, there is a least $t = q$, and thus $bq \leq a < b(q+1)$; such q is unique.

If $a < 0$, then $-a \in \mathbb{N}$, and therefore there is q_1 such that $bq_1 \leq -a < b(q_1+1)$. Hence, $0 < b(q_1+1)+a$. Therefore, there is a unique integer q_2 such that $bq_2 \leq b(q_1+1)+a < b(q_2+1)$. Consequently, taking $q = q_2 - q_1 - 1$, $bq \leq a < b(q+1)$, as required. □

Remark III.3.1. The proof of Theorem III.3.1 can be reformulated as follows: The set $\mathbb{Z}$ of all integers is a disjoint union of the b subsets

$$S_t = \{qb + t \mid q \in \mathbb{Z}\}, \quad t = 0, 1, \ldots, b-1.$$

Then a has to belong to one of these sets. This property of the ordering of $\mathbb{Z}$ is often expressed by saying that the order is **Archimedean**. The word Archimedian is derived from **Archimedes of Syracuse (about 287 - 212 BCE)**.

Exercise III.3.1. Reformulate the proof in terms of the equivalence defined on $\mathbb{Z}$ whose equivalence classes are the subsets S_t.

Theorem III.3.1 provides a consequence that will become, once extended to an algorithm, a gateway to the study of divisibility of integers.

Theorem III.3.2 (Euclidean division). *Let a be an integer and b a natural number. Then there is a unique pair (q, r) of integers (the* **quotient** *q and the* **remainder** *r) such that*

$$a = bq + r \text{ and } 0 \leq r < b.$$

In some calculations, it is advantageous to use the following more expeditious form of "integral" division.

Exercise III.3.2. For any integer a and any natural number b there exist unique integers q and r such that

$$a = bq + r, \text{ where the absolute value } |r| \text{ satisfies } 2|r| < b \text{ or } 2r = b.$$

For instance, for $a = 19$ and $b = 5$, we have $19 = 5 \times 4 - 1$ while Euclidean division results in $19 = 5 \times 3 + 4$.

The case $r = 0$ in Theorem III.3.2 is important; we introduce the relevant definition.

Definition III.3.1. *Let a and b be two integers. We say that b* **is a divisor of** *a, or that a* **is a multiple of** *b, and write $b \mid a$ if there is an integer c such that $a = bc$. We write $b = \frac{a}{c}$ if $a = bc$.*

Remark III.3.2. If $a \neq 0$, then every divisor b of a satisfies the inequality

$$-|\mathbf{a}| \leq \mathbf{b} \leq |\mathbf{a}|.$$

Thus, **every non-zero integer has only a finite number of divisors.** Since $a \mid b$ and $b \mid a$ if and only if $a = \pm b$, we can restrict our formulations concerning divisibility to the positive integers (to the natural numbers) and in what follows we will, unless stated otherwise, do it. Observe that the relation "to be a divisor" is a **partial order** on $\mathbb{N}$; indeed, trivially, $a \mid a$, $a \mid b$ together with $b \mid a$ implies $a = b$, and $a \mid b$ together with $b \mid c$ result in $a \mid c$.

Here are some properties of the divisibility relation.

Theorem III.3.3. *Let a, b and c be integers. If c is a divisor of a and b, then c is also a divisor of* **the sum** $a+b$**, the difference** $a-b$*, and in general of* **any linear combination** *$ra + sb$, where r and s are integers.*

Exercise III.3.3. Prove all the statements of Theorem III.3.3, including the following more general formulation: If $c \mid a_t$ for $t \in \{1, 2, \ldots, n\}$, then $a \mid \sum_{t=1}^{n} r_t a_t$ for arbitrary integers r_t.

Now, for non-zero integers a_t, $t \in \{1, 2, \ldots, n\}$, we denote by $\mathcal{D}(a_1, a_2, \ldots, a_n)$ the set of **all their common positive divisors**: that is, $d \in \mathcal{D}(a_1, a_2, \ldots, a_n)$ if and only if $d > 0$ and $d \mid a_t$ for all $t \in \{1, 2, \ldots, n\}$. We already know that this set is finite. Theorem III.3.4 guarantees that this set has a **unique greatest element** $d = d(a_1, a_2, \ldots, a_n)$, which is called the **greatest common divisor** of the elements a_t, $t \in \{1, 2, \ldots, n\}$ and is characterized by the following two properties:

(1) $d \mid a_t$ for all $t \in \{1, 2, \ldots, n\}$;

(2) If some d' satisfies $d' \mid a_t$ for all $t \in \{1, 2, \ldots, n\}$, then $d' \mid d$.

Note that for $n = 1, \mathcal{D}(a)$ is just the set of all positive divisors of a and thus $d(a) = |a|$. The case $n = 2$ is particularly important. More general statements concerning the greatest common divisor for $n > 2$ can often be deduced from the case $n = 2$ using mathematical induction and the relation $d(a_1, a_2, \ldots, a_n) = d(a_1, d(a_2, \ldots, a_n))$. From now on we focus on the case $n = 2$.

Theorem III.3.4 (Euclidean algorithm). *Let a and b be two natural numbers. The following series of Euclidean divisions*

$$\begin{array}{rll} a = bq_0 + r_0, & 0 \leq r_0 < b, & \mathcal{D}(a,b) = \mathcal{D}(b,r_0); \\ b = r_0 q_1 + r_1, & 0 \leq r_1 < r_0 \ \ (\text{if } r_0 \neq 0), & \mathcal{D}(b,r_0) = \mathcal{D}(r_0,r_1); \\ r_0 = r_1 q_2 + r_2, & 0 \leq r_2 < r_1 \ \ (\text{if } r_1 \neq 0), & \mathcal{D}(r_0,r_1) = \mathcal{D}(r_1,r_2); \\ \cdots\cdots & \cdots\cdots & \cdots\cdots \\ r_{t-1} = r_t q_{t+1} + r_{t+1}, & 0 \leq r_{t+1} < r_t \ \ (\text{if } r_t \neq 0), & \mathcal{D}(r_{t-1},r_t) = \mathcal{D}(r_t,r_{t+1}); \\ \cdots\cdots & \cdots\cdots & \cdots\cdots \end{array}$$

is finite: it terminates when a remainder becomes zero. If this happens for $t = n$, that is,

$$\begin{array}{rll} r_{n-3} = r_{n-2}q_{n-1} + r_{n-1}, & 0 < r_{n-1} \leq r_{n-2}, & \mathcal{D}(r_{n-3},r_{n-2}) = \mathcal{D}(r_{n-2},r_{n-1}); \\ r_{n-2} = r_{n-1}q_n, & & \mathcal{D}(r_{n-2},r_{n-1}) = \mathcal{D}(r_{n-1},0); \end{array}$$

then $\mathcal{D}(a,b) = \mathcal{D}(r_{n-2},r_{n-1}) = \mathcal{D}(r_{n-1},r_n)$ and $d(a,b) = r_{n-1}$ (if $r_0 = 0$, then $d(a,b) = b$).

Proof. We have $\mathcal{D}(r_{n-1},r_n) = \mathcal{D}(r_{n-1},0) = \mathcal{D}(r_{n-1})$. Thus r_{n-1} is the greatest element in $\mathcal{D}(a,b)$ and is a divisor of both a and b. Moreover any divisor d' of a and b necessarily divides r_{n-1}. Hence $d(a,b) = r_{n-1}$. □

Exercise III.3.4. Prove that the set $\mathcal{M}(a,b)$ of all common positive multiples of the natural numbers a and b has a **unique smallest element.** This natural number is called the **least common multiple** of a and b and is denoted by $m(a,b)$. Prove that

$$m(a,b)\, d(a,b) = ab.$$

Exercise III.3.5. Show that $m = m(a,b)$ is characterized by the following properties:

(1) $a \mid m$ and $b \mid m$,

(2) if $a \mid m'$ and $b \mid m'$ for some m', then $m \mid m'$.

Example III.3.1. Using the Euclidean algorithm, show that the greatest common divisor of the numbers 31855 and 24099 is the number 277. Applying the algorithm we obtain

$$\begin{array}{ll} 31855 & = 24099 \times 1 + 7756, \\ 24099 & = 7756 \times 3 + 831, \\ 7756 & = 831 \times 9 + \underline{277}, \ \text{ and finally} \\ 831 & = 277 \times 3. \end{array}$$

Thus, $d(31855, 24099) = 277$ and the least common multiple of these two numbers is

$$m(31855, 24099) = \frac{31855 \times 24099}{277} = 2771385.$$

The Euclidean algorithm provides a method of finding the greatest common divisor and the least common multiple of given numbers. In addition, the greatest common divisor can be expressed as a **linear combination** of the numbers 31855 and 24099. Indeed,

$$\begin{aligned} 277 &= 7756 - 831 \times 9 \\ &= 7756 - (24099 - 7756 \times 3) \times 9 \\ &= 7756 \times 28 - 24099 \times 9 \\ &= (31855 - 24099) \times 28 - 24099 \times 9 \\ &= 31855 \times 28 - 24099 \times 37. \end{aligned}$$

We shall see that such an expression holds generally, see Bézout's theorem (Theorem III.5.1).

Example III.3.2. Here is another application of Euclidean division. Given two integers a and b such that $a^2 + b^2$ is a multiple of 3, we show that both a and b are multiples of 3.

Let $a = 3q_1 + r_1$ and $b = 3q_2 + r_2$, where r_1 and r_2 are the numbers 0, 1 or 2. Then

$$a^2 + b^2 = 3(3q_1^2 + 2q_1r_1 + 3q_2^2 + 2q_2r_2) + r_1^2 + r_2^2.$$

Hence, $r_1^2 + r_2^2$ is a multiple of 3. Since $r_1^2 + r_2^2 \in \{0, 1, 2, 4, 5, 8\}$, necessarily $r_1^2 + r_2^2 = 0$, and thus $r_1 = r_2 = 0$. Consequently, both a and b are multiples of 3.

Exercise III.3.6. Prove that if 3 is a divisor of $a^4 + b^4$, then 3 divides both a and b.

III.4 Lattice of divisors

We have already pointed out in Chapter I and again in the preceding section that the divisibility of integers endows the set of natural numbers $\mathbb{N}$ with a partial order $\preceq$, namely,

$$a \preceq b \text{ if and only if } a \mid b.$$

Now Theorem III.3.4 guarantees the existence of a least upper bound and a greatest lower bound for any finite set of elements of $\mathbb{N}$ and thus $(\mathbb{N}, \preceq)$ is a lattice (see Definition I.5.3).

Exercise III.4.1. Using Exercises I.5.2 and III.3.4, show that $(\mathbb{N}, \preceq)$ is a distributive lattice (see (I.6.1)).

Figure III.4.1 illustrates a part of the **Hasse diagram** (**Helmut Hasse (1898 - 1979)**), namely a part of the infinite lattice of all natural numbers (that does not possess a greatest element). A number x in a higher layer connected to a number y in a lower layer symbolizes that x is a multiple of y. Observe that in the first "layer" (that is, just above the smallest element 1) are the **prime numbers**, that is, those numbers p that have only two (trivial) divisors p and 1. This lattice contains the finite lattice $(\mathcal{D}(a), \preceq)$ for each element $a \in \mathbb{N}$. Figure III.4.2 depicts such a lattice, a parallelogram, for $a = 2160$. The numbers in parentheses in Figure III.4.2 refer to the next exercise.

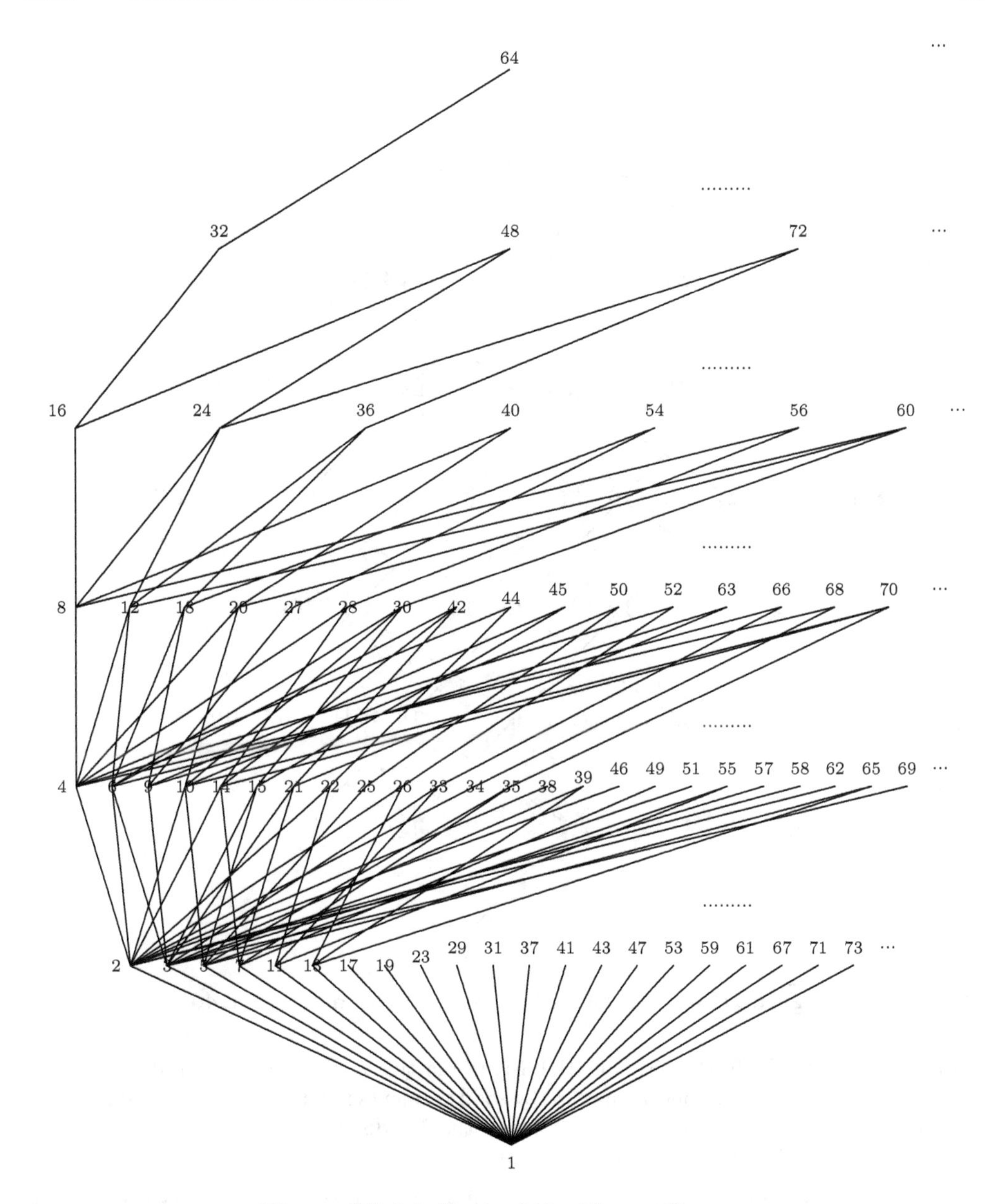

Figure III.4.1 Part of the Hasse diagram

Exercise III.4.2. Show that Figure III.4.2 also represents the lattice of all divisors of 3240. Fill in all the missing divisors and describe all numbers a such that Figure III.4.2 is the lattice of all divisors of a. Also, characterize the numbers a such that the set of all divisors of a is linearly ordered by $\preceq$.

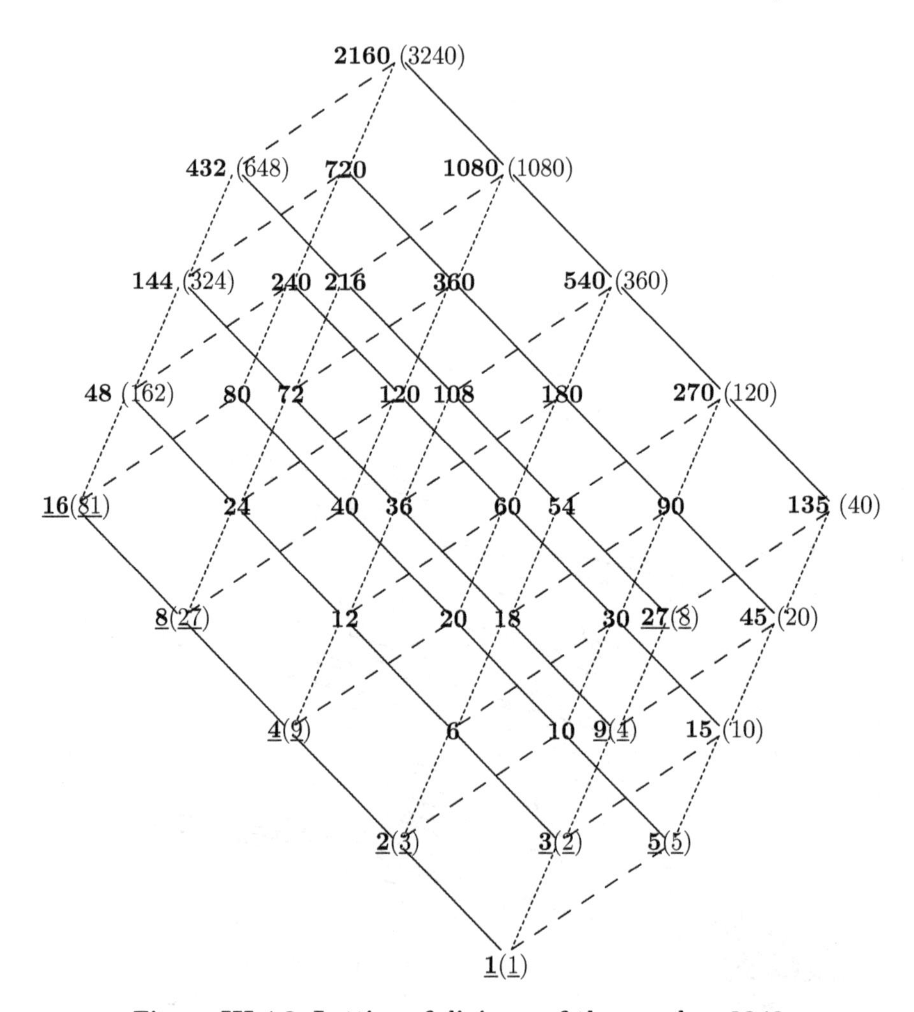

Figure III.4.2 Lattice of divisors of the number 3240

Exercise III.4.3. Verify that Figure III.4.3 is the lattice $\mathcal{L}$ of all divisors of 30030. Show that the sublattice of 18 numbers connected by solid segments is generated by the numbers 42, 110 and 195 and that this lattice is isomorphic to the free distributive lattice on 3 generators displayed in Figure I.6.1; define this isomorphism.

Recall that a **sublattice** of a lattice $\mathcal{L}$ is a subset $\mathcal{L}'$ of $\mathcal{L}$ that is a lattice such that the results of the operations $\vee$ and $\wedge$ in $\mathcal{L}'$ coincide with those in $\mathcal{L}$.

Exercise III.4.4. Show that the subsets

$$\mathcal{L}_1 = \{1, 2, 3, 5, 42, 110, 195, 2310, 2730, 4290, 30030\}$$

and

$$\mathcal{L}_2 = \{1, 2, 3, 5, 42, 110, 195, 30030\}$$

are distributive lattices and that neither $\mathcal{L}_1$ nor $\mathcal{L}_2$ is a sublattice of $\mathcal{L}$. Moreover, define an isomorphism of $\mathcal{L}_2$ and the lattice (**Boolean algebra**) (named after **George Boole (1815 - 1864)**) of all subsets of a 3−element set.

Figure III.4.3 is the lattice (6-dimensional cube) of all divisors of 30030.

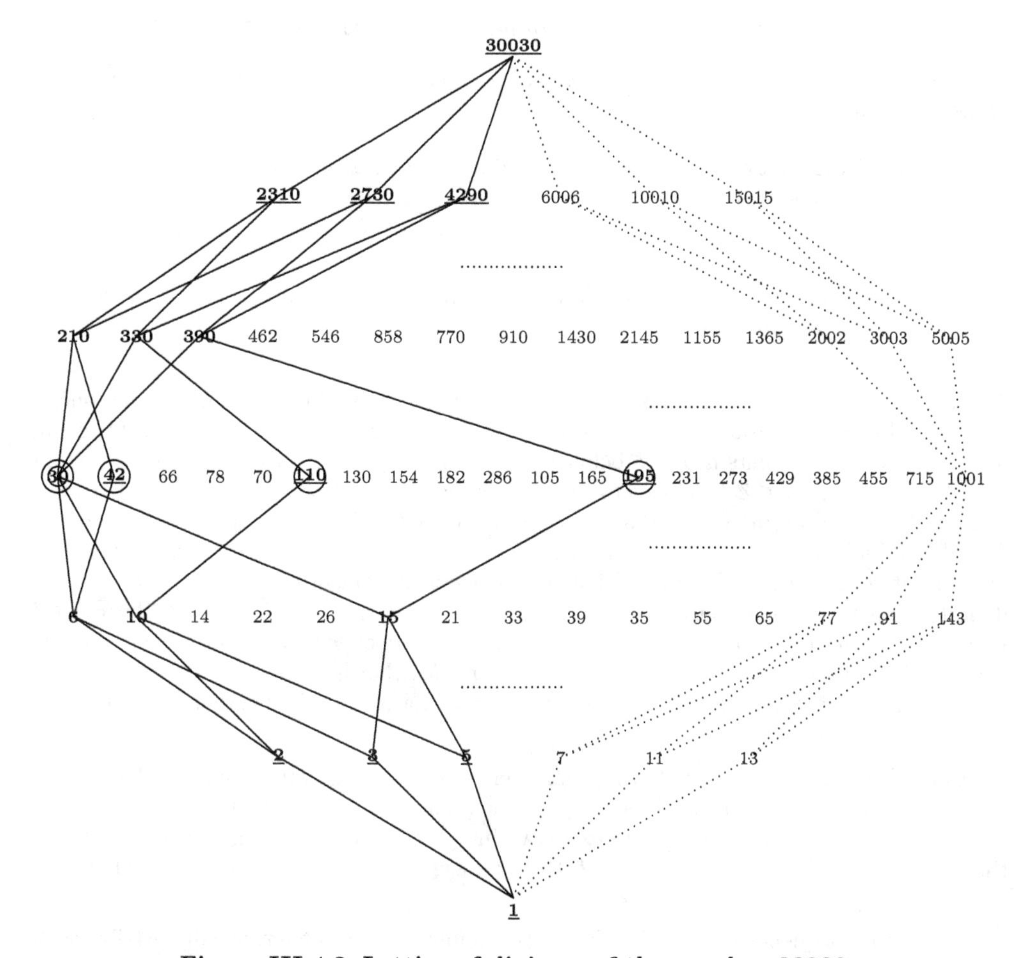

Figure III.4.3 Lattice of divisors of the number 30030

Exercise III.4.5. Prove that

$$d(m(a,b), m(a,c), m(b,c)) \;=\; m(d(a,b), d(a,c), d(b,c))$$

for any three natural numbers a, b and c.

Exercise III.4.6. For any natural numbers a, b, u and v, prove that

$$m(a,b) \times d(u,v) \;=\; d(m(a,b,u), m(a,b,v)) \times m(d(a,u,v), d(b,u,v)).$$

III.5 Diophantine equations

Theorem III.3.4 yields a number of consequences. Expressing (inductively)

$$\begin{aligned} d(a,b) &= r_{n-1} = r_{n-3} - r_{n-2}q_{n-1} = r_{n-3} - (r_{n-4} - r_{n-3}q_{n-2})q_{n-1} \\ &= r_{n-4}u_{n-3} + r_{n-4}v_{n-3} = \cdots = r_0u_1 + r_1v_1 = bu_0 + r_0v_0 = au + bv \end{aligned}$$

for suitable integers $u_{n-3}, v_{n-3}, \ldots, u_1, v_1, u_0, v_0, u, v$, we obtain a result due to **Étienne Bézout (1730 - 1783)**.

Theorem III.5.1 (Bézout's Theorem). *Let a and b be two non-zero integers and $d = d(a,b)$. Then there exist integers u and v such that*

$$au + bv = d.$$

Thus, in particular, if a and b are **relatively prime** *(that is if $d(a,b) = 1$), then there are integers u and v such that $au + bv = 1$.*

Remark III.5.1. An elegant proof of Bézout's theorem consists in making use of the concept of an **ideal** of the integral domain $\mathbb{Z}$. A subset $I \subseteq \mathbb{Z}$ is said to be an ideal of $\mathbb{Z}$ if the sum $a+b$ of any two elements $a, b \in I$ belongs to I and the product of an element $a \in I$ and an arbitrary integer $c \in \mathbb{Z}$ belongs to I. It is a consequence of Euclidean division that every ideal of $\mathbb{Z}$ is a **principal ideal**, that is, for every ideal I of $\mathbb{Z}$, there exists $d \in I$ such that $I = d\mathbb{Z} = \{dq \mid q \in \mathbb{Z}\}$. Indeed, using the well-ordering principle we can see that the set of all positive integers of the ideal I has a least element. Denoting this element by d and dividing an arbitrary element c of I by d, we obtain $c = dq + r$ with $0 \le r < d$. Since $r \in I$, necessarily we have $r = 0$, that is, $c = dq$. Now, for any two elements $a, b \in \mathbb{Z}$, consider the subset $J \subseteq \mathbb{Z}$ of all possible linear combinations $ax + by$, that is, $J = \{ax + by \mid x, y \in \mathbb{Z}\}$. It is easy to see that J is an ideal of $\mathbb{Z}$ and thus $J = d\mathbb{Z}$ for a suitable d and that $d = d(a,b)$.

Exercise III.5.1. Prove the following characterization of the greatest common divisor: Let $a, b \in \mathbb{Z}$ (not both 0). Then a positive integer c such that $c \mid a$ and $c \mid b$ is the greatest common divisor $c = d(a,b)$ if and only if there exist two integers u and v such that $au + bv = c$. Use this result to show that, for all integers t, the numbers $6t - 1$ and $1 - 4t$ are relatively prime.

Another consequence of Theorem III.5.1 is the following statement usually attributed to **Johann Carl Friedrich Gauss (1777 - 1855)**.

Theorem III.5.2 (Gauss' theorem). *Let a, b and c be three integers such that $a \mid bc$ and $d(a,b) = 1$. Then $a \mid c$.*

Proof. In view of our assumptions, there is $w \in \mathbb{Z}$ such that $bc = aw$. By Bézout's theorem, there are $u, v \in \mathbb{Z}$ such that $au + bv = 1$. Hence

$$c = auc + bvc = auc + avw = a(uc + vw),$$

that is, $a \mid c$, as required. □

Remark III.5.2. A particular case of Gauss' theorem is the following result, which is known as **Euclid's lemma**: *If a prime p divides a product ab, then p divides a or b (or both).*

Exercise III.5.2. Show that Gauss' theorem can be stated in the following (equivalent) form:

$$\text{If } d(a,b) = d(a,c) = 1, \text{ then } d(a,bc) = 1.$$

Both Bézout's and Gauss' theorems can be generalized to involve an arbitrary finite number of pairwise relatively prime (coprime) numbers. For use further on in this book, it is convenient to denote these numbers by $n_1, n_2, \ldots, n_r$ and their product by $n = n_1 n_2 \ldots n_r$. Without loss of generality (why?), we may assume that they are all positive.

Theorem III.5.3. *If $d(n_1, n_k) = 1$ for all $k \in \{2, 3, \ldots, r\}$, then $d(n_1, \frac{n}{n_1}) = 1$. Thus, the natural numbers $n_1, n_2, \ldots, n_r$ are relatively prime if and only if*

$$d(n_k, \frac{n}{n_k}) = 1 \ \ \text{for all} \ \ k \in \{1, 2, \ldots, r\}.$$

Proof. Assume that $d(n_1, n_k) = 1$ for all $k \in \{2, 3, \ldots, r\}$ and that $d(n_1, \frac{n}{n_1}) > 1$. Let $s+1 \geq 2$ be the smallest index for which $d(n_1, n_1 n_2 \ldots n_s n_{s+1}) > 1$. Since $d(n_1, n_1 n_2 \ldots n_s) = 1$ and $d(n_1, n_{s+1}) = 1$, we obtain a contradiction in view of the version of Gauss' theorem given in Exercise III.5.2. We leave the reverse assertion to the reader. □

Theorem III.5.4. *If the natural numbers $n_1, n_2, \ldots, n_r$ are pairwise relatively prime, then there exist integers w_k $(k = 1, 2, \ldots, r)$ such that*

$$\sum_{k=1}^{r} w_k \frac{n}{n_k} = 1. \tag{III.5.1}$$

Moreover, for each $k \in \{1, 2, \ldots, r\}$, there exists $z_k \in \mathbb{Z}$ such that

$$w_k \frac{n}{n_k} + z_k n_k = 1.$$

Proof. For $r = 2$, the statement is just Bézout's theorem (with $z_1 = w_2$ and $z_2 = w_1$). Proceeding by induction, assume that the statement is true for $r - 1$ integers. Then there exist integers v_k such that

$$\sum_{k=1}^{r-1} v_k \frac{n}{n_r n_k} = 1, \ \text{ and thus } \sum_{k=1}^{r-1} v_k \frac{n}{n_k} = n_r.$$

Since $d(n_r, \frac{n}{n_r}) = 1$, there are integers v_r and w_r such that

$$v_r n_r + w_r \frac{n}{n_r} = 1\,.$$

Hence, writing $w_k = v_r v_k$ for $k = 1, 2, \dots, r-1$, we obtain (III.5.1). Since, for every $k = 1, 2, \dots, r$ we have $n_k \mid \frac{n}{n_l}$ for all $l \neq k$, we also obtain

$$w_k \frac{n}{n_k} + z_k n_k = 1 \text{ with } z_k = \sum_{l \neq k} w_l \frac{n}{n_k n_l} \ .$$

This completes the proof. □

The previous results provide a solution of a **linear diophantine equation** in two unknowns.

Theorem III.5.5. *Let a, b and c be integers. The equation*

$$ax + by = c \tag{III.5.2}$$

has an integral solution if and only if the greatest common divisor $d = d(a,b)$ of the coefficients a and b divides c. If (x_0, y_0) is one of the solutions, that is, if $ax_0 + by_0 = c$, then

$$(x, y) = (x_0 + tb', y_0 - ta'), \quad \textit{where } a' = \frac{a}{d}, b' = \frac{b}{d} \quad \textit{and } t \in \mathbb{Z},$$

are all integral solutions of (III.5.2).

Proof. First, if (x_0, y_0) is a solution of (III.5.2), then $d \mid c$ by Theorem III.3.3. On the other hand, if $d \mid c$, that is, if $c = dc'$ with $c' \in \mathbb{Z}$, then, by Bézout's theorem, there are $u, v \in \mathbb{Z}$ such that $au + bv = d$ and thus $auc' + bvc' = c$. Consequently, $(x_0, y_0) = (uc', vc')$ is a solution of (III.5.2). For any other solution (x, y) of (III.5.2), we have $a(x_0 - x) + b(y_0 - y) = 0$, and thus

$$b'(y_0 - y) = a'(x - x_0), \text{ where } a = a'd, b = b'd \text{ with } d(a', b') = 1.$$

By Gauss' theorem, we have $b' \mid x - x_0$, that is,

$$x = x_0 + tb' \text{ for some } t \in \mathbb{Z}.$$

But then $b'(y_0 - y) = ta'b'$, and hence $y = y_0 - ta'$. For any $t \in \mathbb{Z}$

$$(x, y) = (x_0 + tb', y_0 - ta')$$

is a solution of (III.5.2). □

Remark III.5.3. We point our the geometrical meaning of Theorem III.5.5. In the plane with a given cartesian coordinate system, consider the lattice $\mathcal{L}$ of all points with integral coordinates; thus $\mathcal{L} \equiv \mathbb{Z} \times \mathbb{Z}$. Theorem III.5.5 asserts that any straight line of this plane either meets an infinite number of points of $\mathcal{L}$ (that form a sequence with a constant distance between any two neighboring points) or none of them.

Example III.5.1. We solve the diophantine equation

$$16377x + 231y = 48000.$$

First, we determine $d(16377, 231)$. Using the Euclidean algorithm, we obtain successively

$$\begin{aligned} 16377 &= 231 \times 70 + 207, \\ 231 &= 207 \times 1 + 24, \\ 207 &= 24 \times 8 + 15, \\ 24 &= 15 \times 1 + 9, \\ 15 &= 9 \times 1 + 6, \\ 9 &= 6 \times 1 + 3, \quad \text{and finally} \\ 6 &= 3 \times 2, \end{aligned}$$

so that $d(16377, 231) = 3$. Also

$$\begin{aligned} 3 &= 9 - 6 = 9 - (15 - 9) = 2 \times (24 - 15) - 15 = 2 \times 24 - 3 \times (207 - 8 \times 24) \\ &= (-29) \times (16377 - 70 \times 231) + 26 \times 231 = 16377 \times (-29) + 231 \times 2056. \end{aligned}$$

Consequently, we have

$$16377 \times (-464000 + 77t) + 231 \times (32896000 - 5459t) = 48000.$$

For $t = 6026$ we obtain the unique positive solution $(x, y) = (2, 66)$. This calculation can often be shortened by using the "accelerated" division version described in Exercise III.3.2:

$$\begin{aligned} 16377 &= 231 \times 71 - 24, \\ 231 &= 24 \times 10 - 9, \\ 24 &= 9 \times 3 - 3, \quad \text{and finally} \\ 9 &= 3 \times 3, \quad \text{that is,} \end{aligned}$$

$$3 = 9 \times 3 - 24 = \cdots = (231 \times 71 - 16377) \times 29 - 231 \times 3 = 16377 \times (-29) + 231 \times 2056.$$

We note that in Section 9 of this chapter we learn how to obtain the general solution $(x, y) = (2 + 77t, 66 - 5459t)$ more efficiently using congruences.

Exercise III.5.3. Find the solution of the diophantine equation

$$20221x + 5183y = 4453$$

for which the absolute value $\mid x + y \mid$ is minimal.

Exercise III.5.4. Two velodrome sprinters ride their bicycles at a constant speed. The first rider (A) covers one lap in a minutes, the second one (B) in b minutes. Assume that $a < b$. In how many minutes will the rider A overtake the rider B the first time?

[Hint: In the case that $d(a, b) = 1$, A will get ahead of B by one lap in $\frac{ab}{b-a}$ minutes.]

III.6 Non-negative solutions of diophantine equations

There are many problems, such as the one mentioned at the beginning of this chapter, that reduce to finding non-negative solutions of diophantine equations of the type

$$ax + by = c, \tag{III.6.1}$$

where a, b and c are positive integers. Since an integral solution of (III.6.1) exists if and only if the greatest common divisor of the numbers a and b is also a divisor of c, we only consider equations satisfying this condition. Hence, we can restrict our considerations to the case when the numbers a and b are relatively prime, that is, their greatest common divisor is 1.

Once the numbers a, b, c are given, it is not difficult to use the Euclidean algorithm (or, as we shall see in the next section, by using congruences) to solve the given equation. However, given a and b, to find the values c for which there is a **non-negative solution** of the equation (III.6.1) often requires rather lengthy computations. Here, by a non-negative solution we understand any solution (x_0, y_0) satisfying $x_0 \geq 0, y_0 \geq 0$. A **positive solution** satisfies $x_0 > 0$ and $y_0 > 0$.

We are going to prove the following, perhaps surprising, result.

Theorem III.6.1. *For given relatively prime positive integers a and b, there is only a finite number $n_{a,b}$ of positive integers c such that there is no non-negative solution of the equation* (III.6.1). *In fact,*

$$n_{a,b} = \frac{1}{2}\left(ab - (a+b) + 1\right) = \frac{1}{2}(a-1)(b-1)$$

and the largest c, for which there is no non-negative solution of (III.6.1) *is $ab-(a+b)$. Thus, for every $c > ab-(a+b)$ there exist non-negative solutions. Moreover, one of them (x_0, y_0) satisfies the condition $0 \leq x_0 < b$. Also we have $(t-1)a \leq y_0 < ta$ for a certain integer t and there are precisely t different non-negative solutions of* (III.6.1).

Proof. The proof follows in a straightforward manner from a proof of the statements for a particular example, namely for the puzzle of Joseph Sylvester mentioned at the beginning of this chapter. It leads to the diophantine equation

$$5x + 17y = c. \tag{III.6.2}$$

We write down all the possible (non-negative linear) combinations $a_{i,j} = 5i + 17j, 0 \leq i, 0 \leq j$, of the numbers 5 and 17 in the form of an infinite matrix $\mathbf{M}(5,17) = (a_{i,j})$ (see Figure III.1.1).

All values of c for which the equation (III.6.2) possesses a non-negative solution, appear (precisely once) in the first 17 rows of the matrix $\mathbf{M}(5,17)$. Figure III.1.1 indicates the division of the matrix $\mathbf{M}(5,17)$ into an infinite number of submatrices $\mathbf{M}_{k,l}, 0 \leq k, 0 \leq l$ of the type 17×5. Hereby

$$\begin{gathered}\mathbf{M}_{0,1} = \mathbf{M}_{1,0},\\ \mathbf{M}_{0,2} = \mathbf{M}_{1,1} = \mathbf{M}_{2,0},\\ \mathbf{M}_{0,3} = \mathbf{M}_{1,2} = \mathbf{M}_{2,1} = \mathbf{M}_{3,0},\\ \mathbf{M}_{0,4} = \mathbf{M}_{1,3} = \mathbf{M}_{2,2} = \mathbf{M}_{3,1} = \mathbf{M}_{4,0}, \text{ etc.}\end{gathered}$$

We can see that there are values of c that do not appear in this matrix. These are the values that appear in Figure III.1.1 left outside the matrix $\mathbf{M}_{0,0}$. To be more precise, they appear outside the matrix left of the fifth till seventeenth rows. Here the largest value is $63 = 5 \times 17 - 5 - 17$. This means that for **all(!)** values $c \geq 64$, the equation (III.6.2) has non-negative solutions. Translating this process to the general case of the equation (III.6.1), the largest value for which there is no non-negative solution is $a \times b - a - b = a \times b - (a+b)$.

In every submatrix $\mathbf{M}_{k,l}$, there is a certain duality. We describe it in detail in the matrix $\mathbf{M}_{0,0}$. We see that $a_{i,j} + a_{16-i,4-j} = 148$ $[= 2 \times 5 \times 17 - (5+17)]$. All (sporadic) values smaller than 63 are in a bijective correspondence with all (sporadic) elements greater than $85 = 148 - 63$. Their number $n_{5,17}$ is the same. However, the elements $a_{i,j}$ of the submatrix $\mathbf{M}_{0,0}$ that satisfy $85 \leq a_{i,j}$ are also in a bijective correspondence with the values of c that are not elements of the matrix $\mathbf{M}(5,17)$ (that is, with those that are outside the matrix $\mathbf{M}(5,17)$). Their values are $a_{i,j} - 85$. Since all 21 values $64 \leq c \leq 84$ belong to $\mathbf{M}_{0,0}$, we see that the number of those values of c for which there is no non-negative solution of (III.6.1) equals $n_{5,17} = \frac{1}{2}((5-1) \times (17-1)) = 32$ (as we could already see from Figure III.1.1). Again, translating this process to the general case of the equation (III.6.1), we have $n_{a,b} = \frac{1}{2}\left((a-1) \times (b-1)\right) = \frac{1}{2}\left(ab - (a+b) + 1\right)$.

Now, given c, by making use of the matrix $\mathbf{M}_{5,17}$ we can not only find whether (III.6.2) has a non-negative solution, but also **how many** such solutions exist. Their number is always finite and it is determined by the position of the number c in the submatrix

$$\mathbf{M}_0 = \begin{pmatrix} \mathbf{M}_{0,0} & \mathbf{M}_{0,1} & \mathbf{M}_{0,2} & \mathbf{M}_{0,3} & \cdots \end{pmatrix}.$$

If $c \in \mathbf{M}_{0,t} = \mathbf{M}_{1,t-1} = \cdots = \mathbf{M}_{t,0}$, then there are $t+1$ such solutions. For example the equation

$$5x + 17y = 133$$

has just one non-negative solution $(x,y) = (13,4)$, because $c = 133 \in \mathbf{M}_{0,0}$. The equation

$$5x + 17y = 122$$

has just two such solutions $(x,y) = (4,6)$ and $(21,1)$, because $c = 122 \in \mathbf{M}_{1,0}$. The equation

$$5x + 17y = 244$$

has just three such solutions $(x,y) = (8,12), (25,7)$ and $(42,2)$, because $c \in \mathbf{M}_{2,0}$. The equation

$$5x + 17y = 277$$

has just four such solutions $(x,y) = (1,16), (18,11), (35,6)$ and $(52,1)$, because $c \in \mathbf{M}_{3,0}$ etc.

We recapitulate and reformulate the preceding considerations for $a = 5$ and $b = 17$ to the general case. The infinite matrix $\mathbf{M}(a,b) = (a_{i,j})$, where $a_{i,j} = ai + bj$ for $i = 0,1,2,\ldots$ and $j = 0,1,2,\ldots$, is divided into the submatrices $\mathbf{M}_{k,l} = (a_{i,j}), i \in \{kb, kb+1, \ldots, (k+1)b - 1\}, j \in \{ka, ka+1, \ldots, (k+1)a - 1\}$, of size $b \times a$ and thus has a form of an infinite matrix whose elements are the matrices $\mathbf{M}_{k,l}$:

$$\mathbf{M}(a,b) = \begin{pmatrix} \mathbf{M}_{0,0} & \mathbf{M}_{0,1} & \mathbf{M}_{0,2} & \cdots & . & \cdots \\ \mathbf{M}_{1,0} & \mathbf{M}_{1,1} & \mathbf{M}_{1,2} & \cdots & . & \cdots \\ \mathbf{M}_{2,0} & \mathbf{M}_{2,1} & \mathbf{M}_{2,2} & \cdots & . & \cdots \\ \vdots & \vdots & \vdots & \vdots\vdots\vdots & \vdots & \vdots\vdots\vdots \\ . & . & . & \cdots & \mathbf{M}_{k,l} & \cdots \\ \vdots & \vdots & \vdots & \vdots\vdots\vdots & \vdots & \vdots\vdots\vdots \end{pmatrix}.$$

Every matrix $\mathbf{M}_{k,l}$ of type $b \times a$ has the form

$$\begin{pmatrix} m_{k,l} & m_{k,l}+b & m_{k,l}+2b & \cdots & m_{k,l}+(a-1)b \\ m_{k,l}+a & m_{k,l}+a+b & m_{k,l}+a+2b & \cdots & m_{k,l}+a+(a-1)b \\ m_{k,l}+2a & m_{k,l}+2a+b & m_{k,l}+2a+2b & \cdots & m_{k,l}+2a+(a-1)b \\ \vdots & \vdots & \vdots & \cdots & \vdots \\ m_{k,l}+(b-1)a & m_{k,l}+(b-1)a+b & m_{k,l}+(b-1)a+2b & \cdots & m_{k,l}+2ab-(a+b) \end{pmatrix},$$

where $m_{k,l} = (k+l)ab$. Moreover, define for every $0 \leq k$ the submatrix

$$\mathbf{N}_k = \begin{pmatrix} \mathbf{M}_{k,0} & \mathbf{M}_{k,1} & \mathbf{M}_{k,2} & \mathbf{M}_{k,3} & \cdots \end{pmatrix}$$

and observe that **all** values of c for which there is a non-negative solution of (III.6.2) are just the elements of the submatrix $\mathbf{M}_o$. Moreover, we point out that $\mathbf{M}_{k_1,l_1} = \mathbf{M}_{k_2,l_2}$ if and only if $k_1 + l_1 = k_2 + l_2$. The proof of the general statement follows. □

Now, there is a related question asking to determine the conditions for the existence of the **positive** integral solutions of (III.6.1). Here is the relevant statement.

Theorem III.6.2. *For given relatively prime positive integers a and b, there is only a finite number $n^+_{a,b}$ of positive integers c such that there is no positive solution of the equation* (III.6.1). *In fact,*

$$n^+_{a,b} = \frac{1}{2}\,(ab + (a+b) - 1) = \frac{(a+1)(b+1)}{2} - 1$$

and the largest c, for which there is no positive solution of (III.6.1) *equals ab. Thus, for every $c > ab$ there exist positive solutions. Moreover, one of them (x_0, y_0) satisfies the condition $0 < x_0 \leq b$. At the same time, $(t-1)a < y_0 \leq ta$ for a certain integer t and this very t determines that there are precisely t different positive solutions of* (III.6.1).

To prove Theorem III.6.2, construct, in analogy to $\mathbf{M}_{a,b}$ in Figure III.1.1, the matrix $\mathbf{M}^+_{a,b}$.

Exercise III.6.1. Using the matrix $\mathbf{M}^+_{5,17}$ in Figure III.6.1, prove Theorem III.6.1.

					0	1	2	3	4	5	6	7	8	9	10	11	12	13	14	15	16	...
0					0	17	34	51	68	85	102	119	136	153	170	187	204	221	238	255	272	...
1					**5**	22	39	56	73	**90**	107	124	141	148	**175**	192	209	226	243	**260**	277	...
2					**10**	27	44	61	78	**95**	112	129	146	163	**180**	.	.	.	.	.	.	...
3					**15**	32	49	66	83	**100**	117	134	151	168	**185**	.	.	.	.	.	.	...
4				**3**	**20**	37	54	71	**88**	**105**	122	139	156	**173**	**190**	.	.	.	.	.	.	...
5				**8**	25	42	59	76	**93**	110	127	144	161	**178**	195	.	.	.	.	.	.	...
6				**13**	30	47	64	81	**98**	115	132	149	166	**183**	200	.	.	.	.	.	.	...
7			**1**	**18**	35	52	69	**86**	**103**	120	137	154	**171**	**188**	205	.	.	.	.	.	.	...
8			**6**	23	40	57	74	**91**	108	125	142	159	**176**	193	210	.	244	.	.	.	.	...
9			**11**	28	45	62	79	**96**	113	130	147	164	**181**	198	215	.	.	.	.	.	.	...
10			**16**	33	50	67	84	**101**	118	135	152	169	**186**	203	220	.	.	.	.	.	.	...
11		**4**	**21**	38	55	72	**89**	**106**	123	140	157	**174**	**191**	208	225	.	.	.	.	.	.	...
12		**9**	26	43	60	77	**94**	111	128	145	162	**179**	196	213	230	.	.	.	.	.	.	...
13		**14**	31	48	65	82	**99**	116	133	150	167	**184**	201	218	235	.	.	.	303	.	.	...
14	**2**	**19**	36	53	70	**87**	**104**	121	138	155	**172**	**189**	206	223	240	.	.	.	.	.	.	...
15	**7**	24	41	58	75	**92**	109	126	143	160	**177**	194	211	228	245	.	.	.	.	.	.	...
16	**12**	29	46	63	80	**97**	114	131	148	165	**182**	199	216	233	250	.	.	.	318	.	.	...
17	**17**	34	51	68	**85**	**102**	119	136	**153**	170	**187**	204	221	238	255	**272**	289	306	323	340	.	...
18						107	124	141	148	**175**	192	209	216	243	260	277	.	.	.	.	.	...
19						112	129	146	163	**180**	.	.	.	.	.	.	.	.	.	...		
20						117	134	151	168	**185**	.	.	.	.	.	.	.	.	.	...		
21						122	139	156	**173**	**190**	.	.	.	.	.	.	.	.	.	...		
22						127	144	161	**178**	195	.	.	.	.	.	.	.	.	.	...		
23						132	149	166	**183**	200	.	.	.	.	.	.	.	336	.	...		
24						137	154	**171**	**188**	205	.	.	.	.	.	.	.	.	.	...		
25						142	159	**176**	193	210	.	244	.	.	.	.	.	.	.	...		
26						147	164	**181**	198	215	.	.	.	.	.	.	.	.	.	...		
27						152	169	**186**	203	220	.	.	.	.	.	.	.	.	.	...		
28						157	**174**	**191**	208	225	.	.	.	.	.	.	.	.	.	...		
29						162	**179**	196	213	230	.	.	.	.	.	.	.	.	.	...		
30						167	**184**	201	218	235	.	.	.	303	.	.	.	.	.	...		
31						**172**	**189**	206	223	240	.	.	.	.	.	.	.	.	.	...		
32						**177**	194	211	228	245	.	.	.	.	.	.	.	.	.	...		
33						**182**	199	216	233	250	.	.	.	318	.	.	.	.	.	...		
34						**187**	204	221	238	255	**272**	289	306	323	340	.	.	.	.	...		
35						.	.	.	.	.	277	.	.	.	.	.	.	.	.	...		
36						.	.	.	.	.	.	.	.	.	...							
37						.	.	.	.	.	.	.	.	.	...							
38						.	.	.	.	.	.	.	.	.	...							
39						.	.	.	.	.	.	.	.	.	...							
40						.	.	.	.	.	.	.	336	.	...							
41						.	.	.	.	.	.	.	.	.	...							
42						.	244	.	.	.	.	.	.	.	...							
43						.	.	.	.	.	.	.	.	.	...							
44						.	.	.	.	.	.	.	.	.	...							
45						.	.	.	.	.	.	.	.	.	...							
46						.	.	.	.	.	.	.	.	.	...							
47						.	.	.	303	.	.	.	.	.	...							
48						.	.	.	.	.	.	.	.	.	...							
49						.	.	.	.	.	.	.	.	.	...							
50						.	.	.	318	.	.	.	.	.	...							
51						**272**	289	306	323	340	.	.	.	.	...							
52						277	.	.	.	.	.	.	.	.	...							

Figure III.6.1 Positive solutions of the diophantine equation $5x + 17y = c$. $M^{+}_{5,17}$

Remark III.6.1. We can now apply Theorem III.6.1 and Theorem III.6.2 to the story about selling the iPads mentioned at the beginning of this chapter. In this case, we deal with the equation

$$299x + 472y = c,$$

where the numbers 299 and 472 are relatively prime. Here, the number of values of c for which there is no non-negative solution is $\frac{1}{2} \times 298 \times 471 = 70179$ and the largest such value is $298 \times 471 - 1 = 140357$, while the number of values for which there is no positive solution is $\frac{1}{2} \times 300 \times 473 - 1 = 70949$ and the largest such value is $299 \times 472 = 141128$.

Exercise III.6.2. A company produces cables in two fixed lengths: 7 meters and 47 meters. These cables cannot be altered. A row of houses is to be connected by continuous length of cable made by joining up cables of these two lengths. Each house is at a distance of more than 300 meters from its nearest neighbor. Under what conditions this can be done?

(This exercise requires finding all natural numbers c for which there exist non-negative solutions of the diophantine equation $7x + 47y = c$.)

Exercise III.6.3. Let m be a given natural number. Determine the largest n such that for all $b \in \{1, 2, \ldots, n\}$ the diophantine equation $49x + by = c$ has a non-negative solution for all $m \leq c$. Determine n^+ such that for all $b \in \{1, 2, \ldots, n^+\}$ the diophantine equation $49x + by = c$ has a positive solution for all $m \leq c$.

III.7 ⋆ Systems of linear diophantine equations

In this short section, we describe an algorithm for solving systems of linear diophantine equations. It concerns the following problem: *Given an integral* $m \times n$ *matrix* $\mathbf{A} = (a_{ij}), a_{ij} \in \mathbb{Z}$ *and an integral* $m-$*vector* $\mathbf{b} = (b_i) \in \mathbb{Z}^m$, *find all integral solutions* $\mathbf{x} = (x_j) \in \mathbb{Z}^n$ *such that*

$$\mathbf{Ax} = \mathbf{b}.$$

Example III.7.1. Find all integral solutions of the linear system

$$\begin{aligned} 4x_1 + x_2 + x_3 + 4x_4 &= 11 \\ 3x_1 - 3x_2 + 2x_3 + 3x_4 &= 7 \\ 7x_1 + 10x_2 - 5x_3 + 9x_4 &= 16 \end{aligned} .$$

The above mentioned algorithm for solving such a problem is based on the following theorem.

Theorem III.7.1. *Given two integers* a *and* b *with the greatest common divisor* $d = d(a, b)$, *there exists an integral* 2×2 *matrix* $\mathbf{A}$ *such that* $\det \mathbf{A} = 1$ *and*

$$\mathbf{A}\begin{pmatrix} a \\ b \end{pmatrix} = \begin{pmatrix} d \\ 0 \end{pmatrix}, \quad \text{that is,} \quad (a \;\; b)\mathbf{A}^T = (d \;\; 0),$$

where $\mathbf{A}^T$ *denotes the transpose of the matrix* $\mathbf{A}$.

Proof. By Bézout's theorem, there are integers u, v such that

$$ua + vb = d;$$

if $a = a'd$ and $b = b'd$, we get $ua' + vb' = 1$. Thus, the matrix

$$\mathbf{A} = \begin{pmatrix} u & v \\ -b' & a' \end{pmatrix}$$

satisfies the required conditions. We observe that $\mathbf{A}^T = \left(\begin{smallmatrix} u & -b' \\ v & a' \end{smallmatrix}\right)$. □

Next, by repeated use of Theorem III.7.1, we arrive at the following theorem.

Theorem III.7.2. *Let $\mathbf{A} = (a_{ij})$ be an $m \times n$ integral matrix. Let d be the greatest common divisor of all numbers $\{a_{ij} \mid i = 1$ or $j = 1\}$, that is, of all members of the first row and the first column. Then there exist matrices $\mathbf{L} = \mathbf{L}_{m \times m}$ and $\mathbf{R} = \mathbf{R}_{n \times n}$ such that*

$$\mathbf{B} = \mathbf{L}\,\mathbf{A}\,\mathbf{R} = (b_{ij}),$$

where $b_{11} = d, b_{1j} = 0$ for $j \in \{2, 3, \ldots, n\}$, $b_{i1} = 0$ for $i \in \{2, 3, \ldots, m\}$. Moreover, the matrices $\mathbf{L}$ and $\mathbf{R}$ can be chosen to satisfy $\det \mathbf{L} = \det \mathbf{R} = 1$, in which case they are called unimodular matrices.

An effective solution of a given system does not require applying the full statement of the theorem. It is sufficient to use the matrix $\mathbf{L}$ only, as indicated in the following solution to Example III.7.1.

In Example III.7.1 we successively multiply the augmented matrix of the system on the left by the matrices

$$\begin{pmatrix} 1 & -1 & 0 \\ -3 & 4 & 0 \\ 0 & 0 & 1 \end{pmatrix}, \begin{pmatrix} 1 & 0 & 0 \\ 0 & 1 & 0 \\ -7 & 0 & 1 \end{pmatrix}, \begin{pmatrix} 1 & 0 & 0 \\ 0 & 1 & -1 \\ 0 & 6 & -5 \end{pmatrix} \text{ that is, by } \begin{pmatrix} 1 & -1 & 0 \\ 4 & -3 & -1 \\ 17 & -11 & -5 \end{pmatrix},$$

(which is the product of these three matrices) so as to obtain the equivalent system

$$\begin{array}{rcrcrcrcr} x_1 & + & 4x_2 & - & x_3 & + & x_4 & = & 4 \\ & & 3x_2 & + & 3x_3 & - & 2x_4 & = & 7 \\ & & & & 20x_3 & - & 10x_4 & = & 30 \end{array}.$$

First we solve the equation $20x_3 - 10x_4 = 30$, that is, $2x_3 - x_4 = 3$. Then, we substitute the general solution $x_3 = k \in \mathbb{Z}$, $x_4 = -3 + 2k$ into the second equation to obtain $3x_2 - k = 1$. Setting $x_2 = t \in \mathbb{Z}$ we deduce $k = 3t - 1$ so that

$$x_2 = t,\ x_3 = -1 + 3t,\ x_4 = -5 + 6t.$$

Substituting these values into the first equation, we obtain $x_1 + 7t = 8$, and thus the general solution of the system is

$$x_1 = 8 - 7t,\ x_2 = t,\ x_3 = -1 + 3t,\ x_4 = -5 + 6t.$$

For example, for $t = 1$, we obtain the following solution: $x_1 = x_2 = x_4 = 1, x_3 = 2$.

Exercise III.7.1. Solve the following diophantine system

$$\begin{array}{rcrcrcr} 6x_1 & + & 6x_2 & + & 25x_3 & = & -1 \\ 14x_1 & + & 15x_2 & + & 30x_3 & = & 27 \end{array} .$$

Remark III.7.1. In the preceding cases, $n = m + 1$. Such systems occur very often. In the general case, we will always require $n > m$. Indeed the systems with $n \leq m$ have integral solutions only in very exceptional cases. If $n > m$, then the general solution depends on $n - m$ integral parameters. The following example illustrates the way to proceed.

Example III.7.2. In order to find all integral solutions of the equation

$$105x_1 + 70x_2 + 42x_3 + 30x_4 = 100.$$

we proceed as follows:

(i) Solve the equation $42x_3 + 30x_4 = 6u$, that is, $7x_3 + 5x_4 = u$ (note that $d(42, 30) = 6$). The general solution is $x_3 = \ - 2u + 5a, x_4 = 3u - 7a$.

(ii) Substitute these values into the original equation: $105x_1 + 70x_2 - 84u + 210a + 90u - 210a = 100$, that is,

$$105x_1 + 70x_2 + 6u = 100.$$

Hence, we solve $70x_2 + 6u = 2v$, that is, $35x_2 + 3u = v$ (note that $d(70, 6) = 2$). The general solution is $x_2 = -v + 3b, u = 12v - 35b$ and by additional substitution, we get

$$x_2 = -v + 3b,\ x_3 = -24v + 70b + 5a,\ x_4 = 36v - 105b - 7a.$$

(iii) Substitute again: $105x_1 - 70v + 210b - 42 \times 24v + 42 \times 70b + 210a + 30 \times 36v - 30 \times 105b - 210a = 100$, that is,

$$105x_1 + 2v = 100.$$

The general solution $x_1 = 0 + 2p, v = 50 - 105p$ finally yields

$$x_1 = 2p, x_2 = -50 + 105p + 3b, x_3 = -1200 + 105 \times 24p + 70b + 5a,$$
$$x_4 = 1800 - 105 \times 36p - 105b - 7a.$$

(iv) Putting $b = c + 17$, we get

$$x_1 = 2p, x_2 = 1 + 105p + 3c, x_3 = -10 + 105 \times 24p + 70c + 5a, x_4 = 15 - 105 \times 36p - 105c - 7a.$$

(v) Finally, let $c = q - 36p$, and then $a = r - 15q$ in order to get the general solution in a simple form

$$x_1 = 0 + 2p,\ x_2 = 1 - 3p + 3q,\ x_3 = 0 - 5q + 5r,\ x_4 = 1 - 7r.$$

Exercise III.7.2. Solve the following diophantine system of two linear equations

$$\begin{array}{rcrcrcrcr} x_1 & + & 2x_2 & - & 6x_3 & + & 4x_4 & = & 30 \\ 5x_1 & + & 10x_2 & + & 4x_3 & - & 6x_4 & = & 90 \end{array} ,$$

and show that there is a solution with $x_1 = a$ for an arbitrary even integer a and a solution with $x_2 = b$ for an arbitrary integer b.

An important application of the described method of solution of systems of linear diophantine equations is the **Chinese remainder theorem**. It is usually formulated in terms of congruences as we shall see in the next section. It concerns the following system of r equations in $r+1$ unknowns:

$$\begin{array}{lllllll} x & -n_1x_1 & & & & & = a_1 \\ x & & -n_2x_2 & & & & = a_2 \\ x & & & -n_3x_3 & & & = a_3 \\ \dots & \dots\dots & \dots\dots & \dots\dots & \dots\dots & \dots\dots & \dots \\ x & & & & -n_kx_k & & = a_k \\ \dots & \dots\dots & \dots\dots & \dots\dots & \dots\dots & \dots\dots & \dots \\ x & & & & & -n_rx_r & = a_r \end{array} \qquad \text{(III.7.1)}$$

Theorem III.7.3. *If the natural numbers $n_1, n_2, \dots, n_r$ are pairwise relatively prime, then the system* (III.7.1) *of r linear diophantine equations in $r+1$ unknowns $x, x_1, x_2, \dots, x_k, \dots, x_r$ has an infinite number of solutions. If $(x_0, x_{10}, x_{20}, \dots, x_{k0}, \dots, x_{r0})$ is one of the solutions, then*

$$\{(x_0 + nt,\ x_{10} + \frac{n}{n_1}\,t,\ x_{20} + \frac{n}{n_2}\,t,\ \dots,\ x_{k0} + \frac{n}{n_k}\,t,\ \dots, x_{r0} + \frac{n}{n_r}\,t) \mid t \in \mathbb{Z}\},$$

where $n = n_1n_2\cdots n_r$, is the set of all solutions of the system (III.7.1).

Proof. The system of equation (III.7.1) is equivalent to the system

$$\begin{array}{llllllll} x & -n_1x_1 & & & & & & = a_1 = b_1 \\ & n_1x_1 & -n_2x_2 & & & & & = a_2 - a_1 = b_2 \\ & & n_2x_2 & -n_3x_3 & & & & = a_3 - a_2 = b_3 \\ \dots\dots & \dots\dots & \dots\dots & \dots\dots & \dots\dots & \dots\dots & \dots\dots & \\ & & & n_{k-1}x_{k-1} & -n_kx_k & & & = a_k - a_{k-1} = b_k \\ \dots\dots & \dots\dots & \dots\dots & \dots\dots & \dots\dots & \dots\dots & \dots\dots & \\ & & & & n_{r-1}x_{r-1} & -n_rx_r & & = a_r - a_{r-1} = b_r \end{array}.$$

Due to the assumption $d(n_{r-1}, n_r) = 1$, the last equation has the general solution

$$x_{r-1} = x^*_{r-1} + n_rt_{r-1},\ x_r = x^*_r + n_{r-1}t_{r-1} \ \text{ with arbitrary integer } \ t_{r-1}.$$

Substituting this value of x_{r-1} into the previous equation, we obtain the equation

$$n_{r-2}x_{r-2} - n_{r-1}n_rt_{r-1} = b_{r-1} + n_{r-1}x^*_{r-1} \ \text{ for } \ x_{r-2} \ \text{ and } \ t_{r-1}.$$

Again, due to $d(n_{r-2}, n_{r-1}n_r) = 1$, this equation has the general solution

$$x_{r-2} = x^*_{r-2} + n_{r-1}n_rt_{r-2},\ t_{r-1} = t^*_{r-1} + n_{r-2}t_{r-2} \ \text{ with arbitrary integer } \ t_{r-2}.$$

Continuing this process, we derive for every $k \in \{1, 2, \dots, r\}$ the general solution

$$x_k = x^*_k + n_{k+1}n_{k+2}\dots n_rt_r \ \text{ and } \ t_{k+1} = t^*_{k+1} + n_kt_k$$

of the equation

$$n_k x_k - n_{k+1} n_{k+2} \dots n_r \, t_{k+1} = b_{k+1} + n_{k+1} x_{k+1}^*.$$

The first equation of the system (III.7.1), after substituting for x_1 the solution

$$x_1 = x_1^* + n_2 n_3 \dots n_r \, t_1, \text{ where } t_1 \in \mathbb{Z},$$

of the equation

$$n_1 x_1 - n_2 n_3 \dots n_r \, t_2 = b_2 + n_2 x_2^*,$$

becomes

$$x - n_1 n_2 n_3 \dots n_r t_1 = b_1 + n_1 x_2^* \text{ for } x \text{ and } t_1.$$

Thus, the general solution is

$$x = x_0 + n\, t, \; t_1 = t_1^* + t \text{ with arbitrary integer } t.$$

Hence

$$x_1 = x_1^* + n_2 n_3 \dots n_r \, (t_1^* + t) = (x_1^* + n_2 n_3 \dots n_r \, t_1^*) + n_2 n_3 \dots n_r \, t = x_{10} + \frac{n}{n_1} \, t.$$

Continuing in substituting back, we obtain, for every $k \in \{1, 2, \dots, r\}$,

$$\begin{aligned} x_k = {} & (x_k^* + n_{k+1} n_{k+2} \dots n_r \, t_k^* + n_{k-1} n_{k+1} n_{k+2} \dots n_r \, t_{k-1}^* + \cdots \\ & + n_1 n_2 \dots n_{k-1} n_{k+1} n_{k+2} \dots n_r \, t_1^*) + \frac{n}{n_k} \, t, \end{aligned}$$

that is, $x_k = x_{k_0} + \frac{n}{n_k} \, t$ for an arbitrary integer t. □

Remark III.7.2. The previous proof is rather tedious. However, it provides a method of solving systems of the form (III.7.1). We note that the customary formulation of the Chinese remainder theorem (see Theorem III.9.12) requires only the determination of the value of the unknown x. To that end, we may apply Theorem III.5.4: For each $k \in \{1, 2, \dots, r\}$ find w_k satisfying the diophantine equation $w_k \frac{n}{n_k} + z_k n_k = 1$ and get the solution in the form

$$x = \sum_{k=1}^{r} w_k a_k \frac{n}{n_k}.$$

Indeed, we can readily check that

$$\sum_{k=1}^{r} w_k a_k \frac{n}{n_k} - n_k \sum_{l \neq k} w_l (a_l - a_k) \frac{n}{n_k n_l} = a_k.$$

Exercise III.7.3. Solve the diophantine system

$$\begin{array}{lllll} x & - 10x_1 & & & = -1 \\ x & & - 3x_2 & & = \;\; 0 \\ x & & & - 7x_3 & = \;\; 1 \end{array}.$$

We conclude this section with a few exercises that emphasize the importance of Bézout's and Gauss' theorems.

Exercise III.7.4. Prove the following statements (for integers a, b, c, n):

(i) Any odd number $2n + 1$ is relatively prime to n. In general, any number $ab + 1$ is relatively prime to a.

(ii) For arbitrary integers a, b, c, the numbers $ab+1$ and $(ab+1)c+a$ are relatively prime.

(iii) The numbers a and b are relatively prime if and only if ab and $a+b$ are relatively prime.

(iv) If $a_1b_2 - a_2b_1 = \pm 1$, then the numbers $a_1 + a_2$ and $b_1 + b_2$ are relatively prime.

Exercise III.7.5. Prove the following statements :

(i) If $d(a, b) = 1, a \mid c$ and $b \mid c$, then also $ab \mid c$.

(ii) If $d(a, b_t) = 1$ for $t = 1, 2, \ldots, n$, then $d(a, b_1b_2 \ldots b_n) = 1$.

(iii) If $d(a, b) = 1$, then also $d(a^m, b^n) = 1$ for all $m, n \in \mathbb{N}$.

(iv) If $d(a, b) = 1$ and $u \mid a, v \mid b$, then also $d(u, v) = 1$.

III.8 Notation - representation of integers

The subject of this short section has a rich history. The familiar decimal numeral system, that is, base ten numeral system, the most widely used system of recording numbers by modern civilizations today, has been brought to Europe through Hindu-Arabic civilization and fully accepted in Europe only by the late Middle Ages. It was the ancient Sumerians sexagesimal, that is, base sixty numeral system, developed by Babylonian culture that was used in Europe until then, and is still used in a modified form for measuring time and angles. We remark that the Babylonians recorded an approximation of the irrational number $\sqrt{2}$ by

$$(1; 24, 51, 10) = 1 + \frac{24}{60} + \frac{51}{60^2} + \frac{10}{60^3} = \frac{30547}{216000} = 1.414212963 \approx 1.414213562\ldots$$

nearly four thousand years ago! The binary numeral system, that is, base-2 numeral system, acquired during the past several decades a fundamental importance due to the fact that it is presently used by almost all modern computers and computer-based devices. The basic operations in such a system are almost trivial, however any use in our daily arithmetic is handicapped by lengthy presentation. To make this point, compare the Babylonian approximation of $\sqrt{2}$ with a comparable expression in the binary system:

$$0.011\,010\,100\,000\,100\,111\,011\,100\,011 = \frac{1}{2^2} + \frac{1}{2^3} + \frac{1}{2^5} + \cdots + \frac{1}{2^{28}}.$$

The underlying reason for the existence of a numerical representation of any natural number in a base B, where B is a natural number greater than 1, is the Euclidean division. Here is the statement.

Theorem III.8.1 (Base B numeral system). *Let $B \geq 2$ be a natural number. Then, for any natural number a, there are integers a_i with $0 \leq a_i < B$ such that*

$$a = a_n B^n + a_{n-1} B^{n-1} + \cdots + a_2 B^2 + a_1 B + a_0. \tag{III.8.1}$$

Moreover, this expression of $a = (a_n\, a_{n-1}\, \ldots a_2\, a_1\, a_0)_B$ **in base** *B with $a_n \neq 0$ is unique.*

Proof. We obtain the above B-adic expansion of a by means of the following repeated applications of Euclidean division:

$$\begin{aligned} a &= Bq_1 + a_0,\ 0 \leq a_0 < B, \\ q_1 &= Bq_2 + a_1, 0 \leq a_1 < B, \\ q_2 &= Bq_3 + a_2, 0 \leq a_2 < B, \\ &\ldots\ldots\ldots\ldots\ldots\ldots\ldots\ldots \\ q_{n-1} &= Bq_n + a_{n-1}, 0 \leq a_{n-1} < B, \end{aligned}$$

where $a > q_1 > q_2 > \cdots > q_{n-1} > b > q_n = a_n$; (if $a < B$, then $q_1 = 0$ and $a = a_0$). We leave the proof of uniqueness to the reader. □

The expression of the number a in (III.8.1) will be denoted as

$$a = (a)_B = (a_n a_{n-1} \ldots a_2 a_1 a_0)_B.$$

Example III.8.1. The number $2282 = (2282)_{10}$ (in our decimal system) has the following forms:
$(100011101010)_2$ in the binary (2-adic) system,
$(10010112)_3$ in the tertiary (3-adic) system,
$(4352)_8$ in the 8-adic system and
$(13T2)_{12}$ in the 12-adic system (here, we denote the digits by $1, 2, \ldots, 8, 9, T = 10, E = 11$):

$$2 \times 10^3 + 2 \times 10^2 + 8 \times 10 + 2 \ = \ 1 \times 12^3 + 3 \times 12^2 + T \times 12 + 2.$$

Exercise III.8.1. Verify the expressions for the prime numbers in various bases as they are presented in Section 1 of Chapter III. Using a factorization of the expression $B^4 + B^2 + 1$, prove that $(10101)_B$ cannot be a prime for any choice of the base B.

The representation of numbers in the binary system provides an explanation of the following **"ancient (egyptian) multiplication"**.

Example III.8.2 (Ancient Multiplication Scheme). Let a and b be natural numbers. In the binary system, let

$$a = a_n \times 2^n + a_{n-1} \times 2^{n-1} + \cdots + a_2 \times 2^2 + a_1 \times 2 + a_0,$$

so that $a_t = 0$ or 1 for each $t \in \{0, 1, \ldots, n\}$. We now show how $\mathbf{a} \times \mathbf{b}$ can be calculated by **"doubling"** and **"halving"**. We write

b	$a = 2q_1 + a_0$	$a_0 = 1$	if and only if	$a = q_0$ is odd
$2b$	$q_1 = 2q_2 + a_1$	$a_1 = 1$	if and only if	q_1 is odd
2^2b	$q_2 = 2q_3 + a_2$	$a_2 = 1$	if and only if	q_2 is odd
2^3b	$q_3 = 2q_4 + a_3$	$a_3 = 1$	if and only if	q_3 is odd
....	....	....	....	....
2^tb	$q_t = 2q_{t+1} + a_t$	$a_t = 1$	if and only if	q_t is odd
....	....	....	....	....
$2^{n-1}b$	$q_{n-1} = 2q_n + a_{n-1}$	$a_{n-1} = 1$	if and only if	q_{n-1} is odd
2^nb		$a_n = q_n = 1.$		

Thus,

$$ab = a_n(2^nb) + a_{n-1}(2^{n-1}b) + \cdots + a_1(2^2b) + a_1(2b) + a_0 = \sum_{\substack{t=0 \\ q_t \text{ odd}}}^{n} 2^tb.$$

This method of multiplication had some advantage with Roman numerals because it did not require knowledge of any multiplication tables. This may be one of the reasons that the contest between abacists and algorists mentioned in Section 1 of this chapter took such a long time. We point out that the curriculum of the universities in the middle ages contained, in addition to addition and subtraction, just this very **doubling** and **halving method**. We now provide an illustration of using this multiplication scheme. In addition we apply it to Roman numerals. Take $\mathbf{a} = \mathbf{163}$ and $\mathbf{b} = \mathbf{289}$. Then we have

289	**163**	$a_0 = 1$	•
578$(= 2 \times 289)$	**81**$(= \frac{1}{2} \times 162)$	$a_1 = 1$	•
1156$(= 2 \times 578)$	**40**$(= \frac{1}{2} \times 80)$		
2312$(= 2 \times 1156)$	**20**$(= \frac{1}{2} \times 40)$		
4624$(= 2 \times 2312)$	**10**$(= \frac{1}{2} \times 20)$		
9248$(= 2 \times 4624)$	**5**$(= \frac{1}{2} \times 10)$	$a_5 = 1$	•
18496$(= 2 \times 9248)$	**2**$(= \frac{1}{2} \times 4)$		
36992$(= 2 \times 18496)$	**1**$(= \frac{1}{2} \times 2)$	$a_7 = 1$	•

Consequently, $\mathbf{a} \times \mathbf{b}$ equals $\mathbf{289} + \mathbf{578} + \mathbf{9248} + \mathbf{36992} = \mathbf{47107}$.

Now, consider a = XXVI = 26 and b = LVII = 57. Then

LVII	*XXVI*	
CXIIII	*XIII*	•
CCXXVIII	*VI*	
CCCCLVI	*III*	•
DCCCCXII	*I*	•

and thus

$$\begin{aligned}26 \times 57 &= XXVI \times LVII \\ &= CXIIII + CCCCLVI + DCCCCXII = MCCCCLXXXII = 1482.\end{aligned}$$

Exercise III.8.2. Devise a multiplication scheme based on the idea that numbers are expressed in base-3 numeral system. In this case, you have to take into account that the remainder in the respective Euclidean division will be $0, 1$ or 2.

The expression of integers in the decimal numeral system enables us to recognize easily some divisors of a given number. For example numbers ending in 0 are multiples of 10 and numbers ending in 0 or 5 are multiples of 5. We now give some other divisibility criteria for numbers expressed in the decimal system.

Remark III.8.1. Let $a = (a_n\, a_{n-1}\, \dots\, a_1\, a_0)_{10}$. Then

(1) The number a is divisible by **2**, resp. **5**, if and only if the last digit a_0 is divisible by 2, resp. 5.

(2) The number a is divisible by **4**, resp. **10**, resp. **25**, if and only if the number $a_1 10 + a_0$ is divisible by 4, resp. 10, resp. 25.

(3) The number a is divisible by **8**, resp. **125**, if and only if the number $a_2 10^2 + a_1 10 + a_0$ is divisible by 8, resp. 125.

(4) The number a is divisible by **3**, resp. **9**, if and only if the sum of the digits $\sum_{k=0}^{n} a_k$ is divisible by 3, resp. 9. This follows from the fact that $10^k - 1$ is divisible by 9 and thus $a - \sum_{k=0}^{n} a_k$ is divisible by 9.

(5) The number a is divisible by **11** if and only if

$$\sum_{t=0}^{[\frac{n}{2}]} a_{2t} - \sum_{t=0}^{[\frac{n-1}{2}]} a_{2t+1} = (a_0 + a_2 + a_4 + \cdots) - (a_1 + a_3 + a_5 + \cdots) = 11c$$

for some integer c. Here we use the fact that $10^k - (-1)^k$ is divisible by 11.

We provide further criteria in the next section dealing with congruences.

III.9 Congruences in $\mathbb{Z}$

The Euclidean division allows yet another reformulation of the previous results in terms of the fundamental concept of a **congruence relation**. In brief, we will consider a and r in the division $a = bq + r$ with $0 \le r < b$ for a natural number $b > 1$ "equal" and rewrite the relation $a = bq + r$ as $a \equiv r \pmod b$: **a is congruent to r modulo b.**

Here is the basic definition.

Definition III.9.1. *Let $n \geq 2$ be a (fixed) natural number, and a and b arbitrary integers. We say that a is* **congruent to** *b* **modulo** *n and write*

$$a \equiv b \pmod{n}$$

if $a - b$ is a multiple of n.

Exercise III.9.1. Let $n \geq 2$ be a natural number, and a and b arbitrary integers. If q_1, r_1, q_2 and r_2 are integers such that

$$a = nq_1 + r_1, \;\; 0 \leq r_1 < n \;\text{ and }\; b = nq_2 + r_2, \;\; 0 \leq r_2 < n,$$

prove that

$$a \equiv b \pmod{n} \text{ if and only if } r_1 = r_2.$$

Have a look at Figure III.1.1 and see that all numbers in the same row are congruent to each other modulo 17 and all numbers in the same column are congruent to each other modulo 5. In fact, the figure provides a simple description of these **congruence classes.**

Exercise III.9.2. Show that $[a]_n = \{a + tn \mid t \in \mathbb{Z}\}$ is the set of all integers that are congruent to a modulo n. Thus, there are n such subsets of $\mathbb{Z}$. For example, for $n = 5$ these subsets are

$$\begin{aligned}
[0]_5 &= \{\ldots, -15, -10, -5, \mathbf{0}, 5, 10, 15, \ldots\},\\
[1]_5 &= \{\ldots, -14, \;\; -9, \;\; -4, \mathbf{1}, 6, 11, 16, \ldots\},\\
[2]_5 &= \{\ldots, -13, \;\; -8, \;\; -3, \mathbf{2}, 7, 12, 17, \ldots\},\\
[3]_5 &= \{\ldots, -12, \;\; -7, \;\; -2, \mathbf{3}, 8, 13, 18, \ldots\},\\
[4]_5 &= \{\ldots, -11, \;\; -6, \;\; -1, \mathbf{4}, 9, 14, 19, \ldots\}.
\end{aligned}$$

Since the subsets $[a]_n \subset \mathbb{Z}$ form a (disjoint) covering of $\mathbb{Z}$, the congruence $\equiv$ modulo n is an equivalence and the subsets $[a]_n$ are $\equiv$ - equivalent classes. Of course this fact, that is, that the (binary) relation $\equiv$ modulo n is **reflexive, symmetric and transitive**, can be easily proven directly. Now, to simplify the notation, in the case that there will be no danger of misunderstanding, we will simply denote the $\equiv$ - equivalent class $[a]_n$ of the number a by $[a]$. Hence $[a] = [b]$ if and only if $a \equiv b \pmod{n}$; the numbers a and b are **representatives** of the class $[a] = [b]$. The term "congruence" reflects the following very fundamental property of this equivalence relation.

Theorem III.9.1 (Compatibility theorem). *Let $n \geq 2$ be a natural number. Let a, b, c and d be integers such that $a \equiv b \pmod{n}$ and $c \equiv d \pmod{n}$. Then*

$$a+c \equiv b+d \pmod{n}, \;\; ac \equiv bd \pmod{n} \;\text{ and }\; a^k \equiv b^k \pmod{n} \text{ for any natural number } k.$$

Proof. The first congruence follows from the identity $a + c = (a - b) + (c - d) + b + d$, the second from the identity $ac - bd = (a - b)c + b(c - d)$, and the last one from $a^k - b^k = (a^{k-1} + a^{k-2}b + \cdots + b^{k-1})(a - b)$. □

Remark III.9.1. Be aware that $ac \equiv bc \pmod{n}$ **does not** imply that $a \equiv b \pmod{n}$ **!** For example we see that $9 \equiv 15 \pmod{6}$, but $3 \not\equiv 5 \pmod{6}$.

In fact, we can formulate the following relevant result that follows from Gauss' theorem (Theorem III.5.2).

Theorem III.9.2. *If $ac \equiv bc \pmod{n}$ and $d = d(c,n)$ is the greatest common divisor of c and n, then $a \equiv b \pmod{n'}$, where $n = dn'$. Thus, if c and n are relatively prime, then $a \equiv b \pmod{n}$.*

Theorem III.9.1 allows us to define addition and multiplication on the set

$$\mathbb{Z}_n = \{[0], [1], [2], \dots, [n-1]\}$$

of all $\equiv$ - equivalence classes modulo n. Indeed the following addition $\oplus$ and multiplication $\odot$ is well-defined, that is, does not depend on the choice of the representatives of individual congruence classes:

$$[a] \oplus [b] = [a+b] \text{ and } [a] \odot [b] = [a \cdot b]. \tag{III.9.1}$$

Exercise III.9.3. Prove that the definitions of $\oplus$ and $\odot$ **do not depend** on the choice of the representatives of the congruence classes.

The behavior of multiplication in $\mathbb{Z}_n$ depends in an essential way on the number n. If n is **composite**, that is, if

$$n = a \cdot b \text{ with } a > 1 \text{ and } b > 1,$$

then $[a] \odot [b] = [0]$ for non-zero elements $[a]$ and $[b]$, that is, $\mathbb{Z}_n$ contains non-zero elements whose product is zero, the so-called (nontrivial) **divisors of zero.** On the other hand, if n is not composite, that is, if n is a **prime number**, then for any $[a] \neq [0]$, there is an element $[b]$ such that $[a] \odot [b] = [1]$ and thus there are no (nontrivial) divisors of zero. Indeed, for such $[a] \neq [0]$ necessarily $d(a,n) = 1$ and thus, in view of Bézout's theorem,

$$ab + nc = 1 \text{ for suitable integers } b \text{ and } c.$$

Hence, $[a] \odot [b] = [1]$.

Now, in view of (III.9.1), the (canonical) mapping

$$f : \mathbb{Z} \longrightarrow \mathbb{Z}_n \text{ defined by } f(a) = [a] \tag{III.9.2}$$

satisfies

$$f(a+b) = f(a) \oplus f(b) \text{ and } f(a \cdot b) = f(a) \odot f(b).$$

A map with these two properties will be called a **homomorphism** and we can formulate the following theorem.

Theorem III.9.3. *The mapping $f : \mathbb{Z} \longrightarrow \mathbb{Z}_n$ defined in* (III.9.2) *is a (canonical) homomorphism of the integral domain $(\mathbb{Z}, +, .)$ onto the finite commutative ring $(\mathbb{Z}_n = \mathbb{Z}_n, \oplus, \odot)$. The* **kernel** *$Ker\ f = \{a \mid f(a) = [0]\}$ of this homomorphism is the subset $n\mathbb{Z} = \{n \cdot x \mid x \in \mathbb{Z}\}$ of $\mathbb{Z}$. If n is composite, the ring $(\mathbb{Z}_n, \oplus, \odot)$ contains (non-trivial) divisors of zero; if n is a prime, then $(\mathbb{Z}_n, \oplus, \odot)$ is a field, that is, a commutative ring in which for every non-zero element $[a]$ there exists $[b]$ satisfying $[a] \cdot [b] = [1]$.*

Proof. In order to prove that $(\mathbb{Z}_n, \oplus, \odot)$ is a commutative ring, we must show that for any elements $[a], [b], [c] \in \mathbb{Z}_n$ the following equalities hold:

$$[a] \oplus [b] = [b] \oplus [a], \quad ([a] \oplus [b]) \oplus [c] = [b] \oplus ([a] \oplus [c]), \quad [a] \oplus [0] = [a] \quad \text{and} \quad [a] \oplus [-a] = [0],$$

$$[a] \odot [b] = [b] \odot [a], \quad ([a] \odot [b]) \odot [c] = [b] \odot ([a] \odot [c]), \quad [a] \odot [1] = [a] \quad \text{and}$$

$$[a] \odot ([b] \oplus [c]) = ([a] \odot [b]) \oplus ([a] \odot [c]).$$

The first line says that $(\mathbb{Z}_n, \oplus)$ is an additive group, the second line that $(\mathbb{Z}_n, \odot)$ is a commutative monoid, and the third line describes the interaction between addition and multiplication, namely, that multiplication distributes over addition. This interaction is called the **distributive law.** All three lines follow from (III.9.2).

Furthermore, we have already seen that in the case that n is a prime, then for any $[a] \neq [0]$, there is $[b] \in \mathbb{Z}_n$ (which will be denoted by $[a]^{-1}$ in what follows) such that $[a] \odot [b] = [a] \odot [a]^{-1} = [1]$. Thus the set of all non-zero elements forms a multiplicative group. A commutative ring with this property is called a **field**. □

Remark III.9.2. If $(R, +, \cdot)$ is a commutative ring, a non-empty subset I of R is called an ideal if

$$i, j \in I \text{ imply } i + j \in I$$

and

$$r \in R, i \in I \text{ imply } r \cdot i \in I.$$

In this sense the subring $n\mathbb{Z}$ of the ring $\mathbb{Z}$ is an **ideal** and $\mathbb{Z}_n$ (often denoted $\mathbb{Z}/\ n\mathbb{Z}$) is a quotient ring of $\mathbb{Z}$ (see Chapters VII and IX).

Exercise III.9.4. Write down the addition and multiplication tables for

(i) the ring $\mathbb{Z}_6 = (\mathbb{Z}_6, \oplus, \odot)$;

(ii) the field $\mathbb{Z}_5 = (\mathbb{Z}_5, \oplus, \odot)$.

Since the canonical mapping f defined by (III.9.2) is a **ring homomorphism** (that is, preserves the algebraic operations of addition and multiplication), f carries identities valid in the integral domain $(\mathbb{Z}, +, .)$ to the ring $(\mathbb{Z}_n, \oplus, \odot)$. Thus, for example the binomial theorem (of Exercise II.2.1) results for $n = 10$ in the identity

$$(a + b)^{10} \equiv a^{10} + 5a^8b^2 + 2a^5b^5 + 5a^2b^8 + b^{10} \pmod{10}.$$

There is no obvious pattern for a "binomial theorem" in $\mathbb{Z}_n$ for a general n. However, due to the relation

$$k\binom{p}{k} = p\binom{p-1}{k-1} \quad \text{for a prime } p \text{ and } k \in \{1, 2, \ldots, p-1\},$$

which means that, by Gauss' theorem, p is a divisor of $\binom{p}{k}$, we obtain the following simple form of the binomial theorem in $\mathbb{Z}_p$.

Theorem III.9.4. *For any two integers a and b and a prime p,*

$$(a+b)^p \equiv a^p + b^p \pmod{p}, \quad \textit{that is,} \quad ([a]+[b])^p = [a]^p + [b]^p \ \textit{for} \ [a],[b] \in \mathbb{Z}_p.$$

Theorem III.9.4 yields for any natural number a the relation

$$(a+1)^p - (a+1) \equiv a^p - a \pmod{p}.$$

Thus, since $1^p - 1 \equiv 0 \pmod{p}$, we obtain by induction that $a^p - a \equiv 0 \pmod{p}$ for any integer a and we can formulate the following theorem attributed to **Pierre de Fermat (1601 - 1665)**.

Theorem III.9.5 (Fermat's little theorem). *Let p be a prime and a an integer such that $d(a,p)=1$. Then*

$$a^{p-1} \equiv 1 \pmod{p}, \quad \textit{that is,} \quad [a]^{p-1} = [1] \ \textit{for every} \ [0] \neq [a] \in \mathbb{Z}_p.$$

Remark III.9.3. The adjective "little" stands in the previous theorem in contrast to **Fermat's last theorem** that states that the equation

$$x^n + y^n = z^n$$

has no non-zero integer solutions for x, y and z when $n > 2$. Fermat wrote, in the margin of Diophantus's Arithmetica *"I have discovered a truly remarkable proof which this margin is too small to contain"*, a claim that is believed to be unjustified. After countless attempts, the theorem was finally proved by **Andrew John Wiles (1953 -)** in November 1994. However we remark that the many unsuccessful attempts to prove the theorem over a 300 year period led to a wealth of mathematical discoveries. The proof of the theorem has many important consequences for mathematics.

Fermat's little theorem is, in fact, a special case of a more general assertion that we are going to present. We have already seen that, for a composite n, there are non-trivial divisors of zero in $\mathbb{Z}_n = (\mathbb{Z}_n, \oplus, \odot)$. A natural question asks what are the remaining elements like. We shall derive a simple answer, using the concept of an **invertible element** (or a **unit**) of $\mathbb{Z}_n$, that is, an element $[a] \in \mathbb{Z}_n$ such that there is an element $[b] \in \mathbb{Z}_n$ with the property $[a] \odot [b] = [1]$. Such an element $[b]$ with this property is uniquely determined; for, if $[a] \odot [c] = [c] \odot [a] = [1]$, then $[c] = [c] \odot [a] \odot [b] = [b]$. Thus, in $\mathbb{Z}_n$, $[a]$ is invertible if and only if $[a]^{-1}$ exists. Now, since the product of two invertible elements is an invertible element and $[1]$ is an invertible element, we conclude that the subset

$$\mathbf{U}_n = \mathbf{U}(\mathbb{Z}_n) = \{[a] \in \mathbb{Z}_n \mid [a] \text{ is an invertible element of } \mathbb{Z}_n\}$$

is a finite multiplicative commutative group. We are now ready to formulate the answer to our earlier question, which asked for a characterization of the integers a such that $[a]_n$ is not a divisor of zero in $\mathbb{Z}_n$.

Theorem III.9.6. *An element $[a]$ belongs to the group $\mathbf{U}_n$ if and only if the greatest common divisor $d(a,n) = 1$. The number of elements of the group $\mathbf{U}_n$, that is, the* **order** *of the group $\mathbf{U}_n$, is thus the number of $a \in \{1, 2, \ldots, n\}$ such that $d(a,n) = 1$ and is denoted by $\varphi(n)$ and is called the* **Euler function**. *If $d = d(a,n) > 1$, then $[a]$ is a zero divisor.*

Proof. We first point out a rather simple fact that $a \equiv b \pmod{n}$ implies that $d(a,n) = d(b,n)$ which is implicitly used in the formulation of the theorem. Only the last statement needs justification. Let $[a] \neq [0], a = db$ and $n = dc$ with $1 < d < n$. Then $[a] \odot [c] = [b] \odot [n] = [0]$ with $[c] \neq [0]$. □

Example III.9.1. Here is the multiplication table of $\mathbf{U}_{12}$:

$\cdot$	$[1]$	$[5]$	$[7]$	$[11]$
$[1]$	$[1]$	$[5]$	$[7]$	$[11]$
$[5]$	$[5]$	$[1]$	$[11]$	$[7]$
$[7]$	$[7]$	$[11]$	$[1]$	$[5]$
$[11]$	$[11]$	$[7]$	$[5]$	$[1]$

Thus, defining the mapping $f : \mathbf{U}_{12} \to V$, where $V = \{(x,y) \mid x, y \in \mathbb{Z}_2\}$ is the additive group of the 2-dimensional vector space over the field $\mathbb{Z}_2 = \{[0]_2, [1]_2\}$ of two elements by

$$f([1]_{12}) = ([0]_2, [0]_2),\ f([5]_{12}) = ([1]_2, [0]_2),\ f([7]_{12}) = ([0]_2, [1]_2) \text{ and } f([11]_{12}) = ([1]_2, [1]_2),$$

we can see that

$$f([a] \odot [b]) = f([a]) + f([b]) \text{ for all } [a], [b] \in \mathbf{U}_{12}.$$

Hence, f is an isomorphism of the group $\mathbf{U}_{12}$ and the **Klein four-group** (Vierergruppe) V_4 **Felix Klein (1849 - 1925)**. We learn more about this group in Chapter VIII.

Exercise III.9.5. Making use of the multiplication tables of the groups $\mathbf{U}_7$ and $\mathbf{U}_9$, or otherwise, show that $\mathbf{U}_7 \simeq \mathbf{U}_9$ is a group generated by a single element of order 6. Denote by $\mathbb{Z}_6 = (\mathbb{Z}_6, +)$ the additive group of the ring $\mathbb{Z}_6$ and verify that the mappings $f_1 : \mathbf{U}_7 \to \mathbb{Z}_6$ and $f_2 : \mathbf{U}_9 \to \mathbb{Z}_6$ defined by $f_1([5]_7^k) = f_2([5]_9^k) = [k]_6, 1 \leq k \leq 6$, are isomorphisms.

We now present a result that will be given in Chapter VIII as **Lagrange's theorem**. It concerns finite groups. A finite group is a finite monoid G with the property that every element $g \in G$ possesses the inverse $g^{-1} : gg^{-1} = g^{-1}g = 1$. Observe that, due to finiteness, in the sequence of the "powers" of an element $g : g, g^2, g^3, \ldots,$ there must be two exponents $r > s$ such that $g^r = g^s$. But then $g^{r-s} = g^r(g^{-1})^s = g^s(g^{-1})^s = 1$. Let $n > 0$ be the **least** exponent such that $g^n = 1$; call it the **order** of the element g and of the **cyclic group** consisting of the elements $\{g, g^2, g^3, \ldots, g^n = 1\}$ **generated by the element** g.

Theorem III.9.7. *Let G be a finite group of r elements. Then the order of every element $g \in G$ is a divisor of r and thus $g^r = 1$.*

Proof. Denote by H the cyclic subgroup of order s of G generated by the element g, that is $H = \{g, g^2, \ldots, g^s = 1\}$. If $s = r$, we are done. Otherwise, there is an element $h_1 \in G \setminus H$. Now, the set $h_1 H = \{h_1 g, h_1 g^2, \ldots, h_1 g^s = h_1\}$ also has s elements. This follows from the fact that $h_1 g^i = h_1 g^j$ if and only if $g^i = g^j$. Moreover, $H \cap h_1 H = \emptyset$; for, $g^i = h_1 g^j$ implies $h_1 = g^{i-j} \in H$. If $H \cup h_1 H = G$, then $r = 2s$ and thus $g^r = 1$. Otherwise, there is $h_2 \in G \setminus (H \cup h_1 H)$. Again, the subset $h_2 H = \{h_2 g, h_2 g^2, \ldots, h_2 g^s = h_2\}$ has s elements and is disjoint with $H \cup h_1 H$. If $H \cup h_1 H \cup h_2 H = G$, $r = 3s$ and $g^r = 1$. Otherwise, we can continue in this finite process until we reach a disjoint decomposition of G into k subsets $H, h_1 H, h_2 H, \ldots, h_k H$ of s elements each. Thus, $r = ks$, the order s of the element g is a divisor of r and $g^r = 1$. □

We are now ready to formulate the following generalization of Fermat's little theorem proved in 1736 by **Leonhard Euler**. The value of **Euler's totient function** $\varphi : \mathbb{N} \to \mathbb{N}$ at n is the order of the group $\mathbf{U}_n$, that is, the number of positive integers k less than or equal to n which are relatively prime with n. In particular, $\varphi(1) = 1$ and $\varphi(p) = p - 1$ for every prime p.

Theorem III.9.8 (Euler's totient theorem). *If n and a are relatively prime positive integers, then*

$$a^{\varphi(n)} \equiv 1 \pmod{n}.$$

Proof. Euler's totient theorem is a consequence of Theorem III.9.7. □

Exercise III.9.6. Formulate and prove a converse of Euler's totient theorem.

In the next section, we describe a simple way of calculating the values of the function φ.

We have already seen that, for a prime p, $\mathbf{U}_p = \mathbb{Z}_p \setminus \{[0]\}$. We have $[1]^2 = [p-1]^2 = [1]$ in $\mathbf{U}_p$.

Exercise III.9.7. Let p be a prime. Show that, for $a \in \mathbf{U}_p, a \neq [1], a \neq [p-1]$ always $a^2 \neq [1]$. Thus, in the group $\mathbf{U}_p$ there are $\frac{1}{2}(p-3)$ pairs (a, b) of elements $a, b \in \mathbf{U}_p$ such that $a \neq b$ and $a \odot b = [1]$.

Using the statement of Exercise III.9.7, we obtain a result attributed to **John Wilson (1741 - 1793)**, although it was already stated in the 7-th century by **Bhaskara I** and later by **Ibn Al-Haytham (about 1000 AD)**, and proved by **Joseph-Louis Lagrange (1736 - 1813)**.

Theorem III.9.9 (Wilson's theorem). *If p is a prime, then*

$$(p-1)! \equiv -1 \pmod{p}.$$

Remark III.9.4. We present some more divisibility criteria for numbers written in decimal notation $a = (a_n\, a_{n-1}\, \dots\, a_1\, a_0)_{10}$. We have already seen in Remark III.8.1 that the number a is divisible by 9, if and only if the sum of the digits $\sum_{k=0}^{n} a_k$ is divisible by 9. This criterion is just a reflection of the fact that $10 \equiv 1 \pmod 9$ and thus $10^k \equiv 1 \pmod 9$ for every natural number k. Justify, in a similar manner the divisibility of a by 11 (making use of $10^k \equiv (-1)^k \pmod{11}$) as mentioned in Remark III.8.1. We describe some further criteria.

(1) Since $10^2 \equiv 2 \pmod 7$,

$$a \equiv a_1 10 + a_0 + 2\,(a_n 10^{n-2} + a_{n-1} 10^{n-3} + \cdots + a_3 10 + a_2).$$

Repeated use of this congruence yields a criterion for the divisibility of an integer by the number 7. For example,

$$37790347 \equiv 755806 + 47 \equiv 15116 + 53 \equiv 302 + 69 \equiv 6 + 71 \equiv 0 \pmod 7,$$

and thus 37790347 is divisible by 7. We can also make use of the congruence $10^3 \equiv -1 \pmod 7$; then

$$a \equiv a_2 10^2 + a_1 10 + a_0 - \;(a_n 10^{n-3} + a_{n-1} 10^{n-4} + \cdots + a_3) \pmod 7.$$

Applying this method to the previous example, we get now $37790347 \equiv 347 - 37790 \equiv 443 - 37 \equiv 0 \pmod 7$.

Since $10 \equiv 3 \pmod 7$, $10^2 \equiv 2 \pmod 7$, $10^3 \equiv -1 \pmod 7$, $10^4 \equiv -3 \pmod 7$, $10^5 \equiv -2 \pmod 7$, $10^6 \equiv 1 \pmod 7$, etc. we obtain the general result

$$a \equiv (a_0 + 3a_1 + 2a_2) - (a_3 + 3a_4 + 2a_5) + (a_6 + 3a_7 + 2a_8) - \cdots \pmod 7,$$

(2) Since $10^3 \equiv -1 \pmod{13}$,

$$a \equiv a_2 10^2 + a_1 10 + a_0 - \;(a_n 10^{n-3} + a_{n-1} 10^{n-4} + \cdots + a_3) \pmod{13}.$$

(3) Since $10^2 \equiv -2 \pmod{17}$,

$$a \equiv a_1 10 + a_0 - 2\,(a_n 10^{n-2} + a_{n-1} 10^{n-3} + \cdots + a_3 10 + a_2) \pmod{17}.$$

(4) Similarly, for numbers $z = 27$, or $z = 37$, or $z = 111$ we can use the congruences $10^3 \equiv 1 \pmod z$. Hence

$$a \equiv a_2 10^2 + a_1 10 + a_0 + \;(a_n 10^{n-3} + a_{n-1} 10^{n-4} + \cdots + a_3) \pmod z,$$

where $z = 27$, or $z = 37$, or $z = 111$.

Exercise III.9.8. Show that

$$a \equiv a_1 10 + a_0 + 5\,(a_n 10^{n-2} + a_{n-1} 10^{n-3} + \cdots + a_3 10 + a_2) \pmod{19},$$

and

$$a \equiv a_1 10 + a_0 + 3\,(a_n 10^{n-2} + a_{n-1} 10^{n-3} + \cdots + a_3 10 + a_2) \pmod{97}$$

Exercise III.9.9. (i) Making use of the congruence $50 \equiv 1 \pmod 7$, determine the remainders of Euclidean division by 7 of the following numbers 50^{100}, 100, 100^3 and $50^{100}+100^{100}$.

(ii) Show that $671^{800} - 1$ is divisible by 6.

(iii) Show that the number $10a + b$, where a and b are non-negative integers, is divisible by 13 if and only if $a + 4b$ is divisible by 13. Hence, decide which of the numbers 1234567; 569556; 987654; 8888; 6666; 666666; 50801932; 6567 are divisible by 13.

Now dear reader relax and play a **simple game** (a version of the old Chinese game Jianshizi) with us the authors. Before us we have a heap of $n \geq 5$ match sticks. You begin by taking 1, 2 or 3 match sticks from the pile. Then we take 1, 2 or 3 match sticks from the remaining pile. Then you take 1, 2 or 3 match sticks from those left, and so on. If you take the last match stick you lose, whereas if we take the last match stick we lose. Can you devise a stategy for you to win?

We now reformulate Theorem III.5.5 concerning the solution of a linear diophantine equation in terms of congruences.

Theorem III.9.10 (Solution of $ax \equiv b \pmod n$). *The congruence*

$$ax \equiv b \pmod n \quad \text{with } a, n \in \mathbb{N} \text{ and } b \in \mathbb{Z} \tag{III.9.3}$$

has a solution if and only if b is a multiple of the greatest common divisor $d = d(a, n)$ of a and n. Thus, in particular, (III.9.3) *always has a solution if a and n are relatively prime.*

If x_0 is one of the solutions of (III.9.3), *then all the other solutions have the form*

$$x = x_0 + tn', \quad \text{where } n = dn' \text{ and } t \in \mathbb{Z}.$$

Proof. The equation (III.9.3) is equivalent to the diophantine equation

$$ax - ny = b$$

and thus Theorem III.9.10 follows from Theorem III.5.5. It is nevertheless useful to underline the importance of the Euclidean algorithm and, in this instance, of Bézout's theorem that lies in the core of the proof. □

Example III.9.2. We solve the equation

$$129x \equiv 213 \pmod 6.$$

Here $(129, 6) = 3$ and $3|213$. Hence a solution exists. The equation $129x \equiv 213 \pmod 6$ is equivalent to

$$3x \equiv 3 \pmod 6$$

and therefore to $x \equiv 1 \pmod 2$. Thus all odd numbers x satisfy the equation $129x \equiv 213 \pmod 6$. The previous example solves the diophantine equation

$$129x + 6y = 213.$$

We get $x = 1+2t$ and $y = 14-43t$ with $t \in \mathbb{Z}$. In fact, this translation of linear diophantine equations to congruence equations, and thus to simple equations in $\mathbb{Z}_n$, is a very effective way to solve them.

Yet another reformulation of Theorem III.5.5, this time as a simple equation in the ring $\mathbb{Z}_n$, is as follows.

Theorem III.9.11. *The equation*

$$[a]x = [b], \quad \textit{where } [a] \neq [0], \tag{III.9.4}$$

has a solution in a ring $\mathbb{Z}_n$ if and only if b is a multiple of the greatest common divisor $d = d(a,n)$. If $d \mid b$, then there are d different solutions of the equation (III.9.4) *in $\mathbb{Z}_n$; they differ by multiples of $[n']$, where $n' = \frac{n}{d}$.*

If, in particular, $d = 1$, then the equation has a unique solution. In this case, the inverse $[a]^{-1}$ of $[a]$ exists and $x = [a]^{-1} \odot [b]$.

Proof. Theorem III.9.11 is just a translation of Theorem III.5.5. We go through the process of translation once more. The equation (III.9.4) is equivalent to the diophantine equation

$$ax - ny = b.$$

If $d \mid b$ this equation is equivalent to

$$a'x - n'y = b', \quad \text{where } a = da', \ n = dn' \text{ and } b = db'.$$

If (x_0, y_0) is one of the solutions, then all the solutions satisfy $x = x_0 + n't$ with $t \in \mathbb{Z}$. Thus there are precisely d different congruence classes x represented by numbers

$$x_0, \ x_0 + n', \ x_0 + 2n', \ \ldots, \ x_0 + (d-1)n'$$

that solve the equation (III.9.4).

The inverse $[a]^{-1}$ of $[a]$ is, of course, the congruence class $[x_0]$ represented by a solution $x = x_0$ of the diophantine equation $ax - ny = 1$. □

Example III.9.3. Solve the equation $[6]x = [21]$ in $\mathbb{Z}_{33}$. Here, $d(6,33) = 3 \mid 21$, so we will have 3 different solutions. Indeed, the corresponding diophantine equation $6x - 21 = 33y$ has the solution $x = 9 + 11t$, $t \in \mathbb{Z}$ and thus $[9], [20]$ and $[31]$ are the three solutions of the given equation.

The equation $[6]x = [21]$ in $\mathbb{Z}_{31}$ has a unique solution because $d(6,31) = 1$. This time, we deal with the diophantine equation $6x - 21 = 31y$ and thus $3 \equiv y \pmod 6$. The solutions are $y = 3 + 6t$ and $x = 19 + 31t$. Therefore $x = [19]$ is the (unique) solution of the given equation. We also see that the solution of $6x \equiv 1 \pmod{31}$ yields the inverse $[6]^{-1} = [26]$.

Remark III.9.5. For a prime n, every $[a] \neq [0]$ in $\mathbb{Z}_n$ has an inverse element and every equation $[a]x = [b]$ has a unique solution $x = [a]^{-1} \odot [b]$. We observe that the non-zero elements of $\mathbb{Z}_n$ form a multiplicative group.

We conclude this section with the following promised formulation of the Chinese remainder theorem in terms of congruences.

Theorem III.9.12. (Chinese remainder theorem) *Let $m_1, \ldots, m_k$ be pairwise relatively prime positive integers. Let $a_1, \ldots, a_k \in \mathbb{Z}$. Let*

$$m = m_1 m_2 \cdots m_k = m_1 M_1 = m_2 M_2 = \cdots = m_k M_k.$$

As $d(m_i, M_i) = 1$, there is a unique integer n_i modulo m_i such that

$$n_i M_i \equiv 1 \pmod{m_i}.$$

Then the system of congruences

$$x \equiv a_i \pmod{m_i} \quad (i = 1, \ldots, k)$$

has a unique solution x modulo m given by

$$x = n_1 M_1 a_1 + \cdots + n_k M_k a_k.$$

Proof. This theorem follows from Theorem III.7.3 or alternatively by directly determining $n_1 M_1 a_1 + \cdots + n_k M_k a_k$ modulo m using $n_i M_i \equiv 1 \pmod{m_i}$ and $n_j M_j \equiv 0 \pmod{m_i}$ for $j \neq i$. □

Exercise III.9.10. Solve the following simultaneous congruences:
(i) $3x \equiv 1 \pmod 5$ and $2x \equiv 1 \pmod 3$;
(ii) $5x \equiv 14 \pmod{17}$ and $4x \equiv 7 \pmod 5$.

Exercise III.9.11. Let $m_1, \ldots, m_k \in \mathbb{N}$ (not necessarily coprime in pairs). Let $a_1, \ldots, a_k \in \mathbb{Z}$.
(i) Prove that the system of congruences

$$x \equiv a_i \pmod{m_i}, \quad i = 1, 2, \ldots, k,$$

has a solution if and only if

$$d(m_i, m_j) \text{ divides } a_i - a_j \text{ for all } i, j \in \{1, \ldots, k\}.$$

(ii) Prove that there is a sequence $n_1, \ldots, n_k$ with the following properties

$$n_i (\in \mathbb{N}) \text{ divides } m_i, \quad d(n_i, n_j) = 1 \text{ if } i \neq j, \quad N = l(m_1, \ldots, m_k) = n_1 \cdots n_k.$$

(iii) Define $N_i = N/n_i$. Suppose that the system of congruences

$$x \equiv a_i \pmod{m_i}, \quad i = 1, 2, \ldots, k,$$

is solvable. Prove that the integer x given by

$$(N_1 + N_2 + \cdots + N_k)x \equiv a_1N_1 + a_2N_2 + \cdots + a_kN_k \pmod{N}$$

is a solution of the system.

This exercise is based on a generalization of the Chinese remainder theorem due to **Fredric T. Howard**, see The College Mathematics Journal 33 (2002), pp. 279–282.

Example III.9.4. Solve the following simultaneous system of congruences

$$x \equiv 4 \pmod{9}, \; x \equiv 2 \pmod{10}, \;\text{ and }\; x \equiv 7 \pmod{11}.$$

We write the general solution of the first congruence in the form $x = 4 + 9k, k \in \mathbb{Z}$. Since it has to satisfy the second congruence, $9k \equiv 8 \pmod{10}$. Thus $k = 2+10l$ which in turn means that $x = 22 + 90l, l \in \mathbb{Z}$. In order that the third congruence is satisfied, $90l \equiv 7 \pmod{11}$ and thus $2l \equiv 18 \pmod{11}$ and therefore $l = 9 + 11t, t \in \mathbb{Z}$. Hence $x = 22 + 810 + 990t = 832 + 990t, t \in \mathbb{Z}$ is the general solution of the system.

Exercise III.9.12. Solve the following simultaneous congruences:

$$x \equiv 5 \pmod{13}, \; x \equiv 10 \pmod{14} \;\text{ and }\; x \equiv 15 \pmod{107}.$$

Exercise III.9.13. Find the smallest natural number which leaves remainders

(i) 1, 3 and 5 when divided by 9, 11 and 13, respectively.

(ii) r, $r + a$ and $r + b$ when divided by pairwise relatively prime natural numbers q, $q + a$ and $q + b$, respectively.

[Hint: Look for a specific solution to the respective system of congruences.]

Exercise III.9.14. Prove

$$\{3l + 1 \mid l \in \mathbb{N}_0\} = \{2^{2k}(6l + 1) \mid k, l \in \mathbb{N}_0\} \cup \{2^{2k+1}(6l + 5) \mid k, l \in \mathbb{N}_0\}.$$

What is the analogous result for $\{3l + 2 \mid l \in \mathbb{N}_0\}$?

III.10 Prime numbers

We begin by recalling the definitions of a prime number and a composite number.

Definition III.10.1. *A natural number $p > 1$ is called a* **prime number**, *or a* **prime**, *if the only positive divisors of p are 1 and p. The natural numbers (> 1) that are not prime are called* **composite** *numbers.*

Thus, **2** is the least (and the only even) prime, followed (in increasing order) by

$$\mathbf{3, 5, 7, 11, 13, 17, 19, 23, 29, 31, 37, 41, 43, 47, 53, 59, 61, 67, 71, 73, 79, 83, 89, 97,}$$

$$\mathbf{101, 103, 107, 109, 113, 127, 131, 137, 139, 149, 151, 157, 163, 167, 173, 179}, \ldots.$$

If the natural number $n > 1$ is composite it must have divisors other than 1 and n since otherwise it would be a prime. Hence $n = rs$ for some natural numbers r and s satisfying $1 < r < n$ and $1 < s < n$.

If n is a natural number greater than 1 it must be divisible by some prime. Suppose not. Then, by the well-ordering principle, there is a least positive integer $m > 1$ which is not divisible by a prime. Clearly m cannot be a prime and so it must be composite. Hence $m = rs$, where r and s are natural numbers satisfying $1 < r < m$ and $1 < s < m$. Then, by the minimality of m, r must be divisible by some prime p. Hence m must be divisible by p. This is a contradiction so every natural number $n > 1$ must have a prime divisor.

The lattice $\mathcal{D}(p)$ of the positive divisors of a prime p has only two elements. In the lattice $\mathcal{D}(n)$ of all divisors of a natural number $n > 1$, the **atoms** of $\mathcal{D}(n)$ are the elements $x \neq 1$ of the lattice such that, for any $y \in \mathcal{D}(n)$, $1 < y \leq x$ implies $y = x$. These are precisely all the **prime divisors** of n. If a natural number $n > 1$ is not a prime and p is the least prime divisor of n, then $n = pq$ with $p \leq q$ for some positive integer q. Therefore there is always a prime divisor p of n satisfying the inequality $p^2 \leq n$ and thus

$$2 \leq p \leq \sqrt{n}.$$

This inequality allowed **Eratosthenes of Cyrene (276 - 194 BCE)** to give a simple algorithm for finding all prime numbers up to any given limit, the so-called **sieve of Eratosthenes**. The algorithm consists in iteratively marking all composite numbers, starting with multiples of 2, then multiples of 3, multiples of 5 etc. Thus, to decide if a given number n is a prime or a composite number, we write down all numbers $1, 2, 3, \ldots, n$ and sequentially mark 1, all multiples of 2, all multiples of $3, \ldots,$ up to all multiples of the largest prime p that satisfies $p^2 \leq n$. The numbers which are left unmarked in this process are exactly all primes less than or equal to n. Hence, to determine whether n is prime, it is sufficient to check if any of the primes p satisfying $p^2 \leq n$ is a divisor of n.

Example III.10.1. Using the sieve of Eratosthenes we find all the primes less than or equal to 133, and in particular, whether 133 is a prime or not. We observe that 11 is the largest prime satisfying $11^2 \leq 133$. We mark by underlining in the sequence $1, 2, 3, 4, 5, 6, 7, \ldots, 131, 132, 133$ the integer 1 and all (non-trivial) multiples of $2, 3, 5, 7$ and 11, see Figure III.10.1. Since $\underline{133}$ is underlined, 133 is a composite number. The set

$$\{2, 3, 5, 7, 11, 13, 17, 19, 23, 29, 31, 37, 41, 43, 47, 53, 59, 61, 67, 71, 73,$$

$$79, 83, 89, 97, 101, 103, 107, 109, 113, 127, 131\}$$

of numbers that are not underlined, is the set of all primes smaller than 133. We note that marking multiples of $2, 3, 5, 7$ and 11 we could have determined all primes smaller than 144.

$\underline{1}$	**2**	**3**	$\underline{4}$	**5**	$\underline{6}$	**7**	$\underline{8}$	$\underline{9}$	$\underline{10}$	**11**
$\underline{12}$	13	$\underline{14}$	$\underline{15}$	$\underline{16}$	17	$\underline{18}$	19	$\underline{20}$	$\underline{21}$	$\underline{22}$
23	$\underline{24}$	$\underline{25}$	$\underline{26}$	$\underline{27}$	$\underline{28}$	29	$\underline{30}$	31	$\underline{32}$	$\underline{33}$
$\underline{34}$	$\underline{35}$	$\underline{36}$	37	$\underline{38}$	$\underline{39}$	$\underline{40}$	41	$\underline{42}$	43	$\underline{44}$
$\underline{45}$	$\underline{46}$	47	$\underline{48}$	$\underline{49}$	$\underline{50}$	$\underline{51}$	$\underline{52}$	53	$\underline{54}$	$\underline{55}$
$\underline{56}$	$\underline{57}$	$\underline{58}$	59	$\underline{60}$	61	$\underline{62}$	$\underline{63}$	$\underline{64}$	$\underline{65}$	$\underline{66}$
67	$\underline{68}$	$\underline{69}$	$\underline{70}$	71	$\underline{72}$	73	$\underline{74}$	$\underline{75}$	$\underline{76}$	$\underline{77}$
$\underline{78}$	79	$\underline{80}$	$\underline{81}$	$\underline{82}$	83	$\underline{84}$	$\underline{85}$	$\underline{86}$	$\underline{87}$	$\underline{88}$
89	$\underline{90}$	$\underline{91}$	$\underline{92}$	$\underline{93}$	$\underline{94}$	$\underline{95}$	$\underline{96}$	97	$\underline{98}$	$\underline{99}$
$\underline{100}$	101	$\underline{102}$	103	$\underline{104}$	$\underline{105}$	$\underline{106}$	107	$\underline{108}$	109	$\underline{110}$
$\underline{111}$	$\underline{112}$	113	$\underline{114}$	$\underline{115}$	$\underline{116}$	$\underline{117}$	$\underline{118}$	$\underline{119}$	$\underline{120}$	$\underline{121}$
$\underline{122}$	$\underline{123}$	$\underline{124}$	$\underline{125}$	$\underline{126}$	127	$\underline{128}$	$\underline{129}$	$\underline{130}$	131	$\underline{132}$
$\underline{133}$										

Figure III.10.1 Sieve of Eratosthenes

Now, for a given natural number $n \geq 2$, we define

$$n^* = n! + 1.$$

If a prime p divides n^*, then $p > n$. Hence we obtain the following well-known theorem assigned to **Euclid of Alexandria**.

Theorem III.10.1 (Euclid's theorem). *The set of all prime numbers is infinite.*

Exercise III.10.1. For a given prime p_0, denote by P_0 the set of all primes less than or equal to p_0. Show that every prime that divides the product

$$\left(\prod_{p \in P_0} p\right) + 1$$

is greater than p_0. Derive Euclid's theorem.

Exercise III.10.2. Show that for any natural number $n > 1$ the set of $n - 1$ consecutive numbers $N^* = \{n^* + t \mid 1 \leq t \leq n - 1\}$ contains no prime numbers.

We now show that every natural number (> 1) is a product of primes in essentially only one way. This is the essence of the so-called **Fundamental theorem of arithmetic**. A more general theorem about integral domains in which every non-zero element is a unique product of prime elements (such domains are called unique factorization domains) is proved in Chapter IX.

Theorem III.10.2 (Fundamental theorem of arithmetic). *Every natural number $n > 1$ is a product of primes, and this product is unique except for the order of the factors.*

Proof. Suppose that there exists a natural number $n(> 1)$, which is not a product of primes. Let S be the non-empty set of all such natural numbers. By the well-ordering principle S contains a least element m. As $m \in S$, m is not a prime and $m > 1$. Thus m is composite, so $m = rs$, where r and s are natural numbers satisfying $1 < r < m$ and $1 < s < m$. By the minimality of m in S, we have $r, s \notin S$ so each of r and s is a product of primes. Thus $m = rs$ is a product of primes. This is a contradiction. Hence every natural number $n > 1$ must be a product of primes.

We now prove the uniqueness. Suppose that there exists an integer $n > 1$ which has two different representations as a product of primes. Let T be the non-empty set of all such natural numbers. By the well-ordering principle T has a least element m. Then $m > 1$ and

$$m = p_1 \cdots p_t = q_1 \cdots q_u,$$

where $p_1, \ldots, p_t, q_1, \ldots, q_u$ are primes such $\{q_1, \ldots, q_u\}$ is not a reordering of $\{p_1, \ldots, p_t\}$. Now $p_1 \mid m$ so $p_1 \mid q_1 \cdots q_u$. Hence, by Gauss' theorem, we deduce $p_1 \mid q_h$ for some $h \in \{1, \ldots, u\}$. By relabeling, if necessary, we may suppose that $h = 1$. Thes $p_1 \mid q_1$. But p_1 and q_1 are both primes so we must have $p_1 = q_1$. Then

$$\frac{m}{p_1} = p_2 \cdots p_t = q_2 \cdots q_u.$$

Moreover $\{q_2, \ldots, q_u\}$ is not a reordering of $\{p_2, \ldots, p_t\}$. Thus m/p_1 has two different representations as a product of primes. Since $m/p_1 < m$ this contradicts the minimality of m in T. Hence every $n > 1$ must have a unique (up to order) representation as a product of primes. □

We usually write this **prime decomposition** of a natural number n in the form

$$n = p_1 p_2 \ldots p_r \text{ with } p_1 \leq p_2 \leq \cdots \leq p_r, \text{ or}$$

$$n = \prod_{t=1}^{s} p_t^{k_t} = p_1^{k_1} p_2^{k_2} \ldots p_s^{k_s}, \text{ where } p_1 < p_2 < \cdots < p_s \text{ (are all distinct).}$$

The exponents are called the **multiplicities** of the respective primes. Note that m is a divisor of n if and only if

$$m = p_1^{l_1} p_2^{l_2} \ldots p_s^{l_s}, \text{ where } l_t \in \{0, \ldots, k_t\} \text{ for all } t \in \{1, \ldots, s\}.$$

This results in $1 + k_t$ possible choices of l_t for each prime p_t and so we may formulate the following result.

Theorem III.10.3. *The number of (positive) divisors of the natural number*

$$n = p_1^{k_1} p_2^{k_2} \ldots p_s^{k_s}, \textit{ where the primes } p_t \ (t = 1, \ldots, s) \textit{ are distinct}$$

is equal to

$$(1 + k_1)(1 + k_2) \cdots (1 + k_s).$$

Although enumerating the products of a given number of primes is effortless, the "opposite operation" of factorization of large composite numbers into their prime divisors is very hard. No efficient algorithm to perform this operation is known. In fact, in Section 12 of this chapter, we shall see that the problem of factoring a number, which is the product of two large primes, is at the very heart of a coding procedure.

Once we know all the prime factors of two given natural numbers a and b, we can express their factorizations in terms of a common set $\{p_1, p_2, \ldots, p_t\}$ of distinct primes:

$$a = p_1^{k_1} p_2^{k_2} \ldots p_t^{k_t}, \quad b = p_1^{l_1} p_2^{l_2} \ldots p_t^{l_t}, \quad \text{with all } k_i \geq 0, \ l_i \geq 0. \tag{III.10.1}$$

Exercise III.10.3. Using the factorizations of a and b in the form (III.10.1), prove that the greatest common divisor

$$d(a,b) = p_1^{u_1} p_2^{u_2} \ldots p_t^{u_t}, \text{ where } u_i = \min(k_i, l_i) \ (i = 1, \ldots, t)$$

and the least common multiple

$$m(a,b) = p_1^{v_1} p_2^{v_2} \ldots p_t^{v_t}, \text{ where } v_i = \max(k_i, l_i) \ (i = 1, \ldots, t).$$

Remark III.10.1. In this remark we return to the Möbius function defined in Section 7 of Chapter 1. The reader should review the definitions of μ, e and φ given there. If $p_1, \ldots, p_t$ are t distinct primes, and $v_1, \ldots, v_t$ are non-negative integers, we deduce from the definition of μ that

$$\mu(p_1^{v_1} \cdots p_t^{v_t}) = \begin{cases} 0 & \text{if at least one } v_i \geq 2, \\ (-1)^{v_1 + \cdots + v_t} & \text{otherwise.} \end{cases}$$

Let $n (\in \mathbb{N}) > 1$ have the prime factorization $n = p_1^{k_1} \cdots p_t^{k_t}$, where $t \in \mathbb{N}$ and each $k_i \in \mathbb{N}$. Then

$$\sum_{d|n} \mu(d) = \sum_{v_1=0}^{k_1} \cdots \sum_{v_t=0}^{k_t} \mu(p_1^{v_1} \cdots p_t^{v_t}) = \sum_{v_1=0}^{1} \cdots \sum_{v_t=0}^{1} (-1)^{v_1 + \cdots + v_t} = 0.$$

Thus for all $n \in \mathbb{N}$ we have

$$\sum_{d|n} \mu(d) = \begin{cases} 1 & \text{if } n = 1, \\ 0 & \text{if } n > 1. \end{cases}$$

Hence, for all $n \in \mathbb{N}$, we have

$$(\mu \star \varepsilon)(n) = \sum_{d|n} \mu(d) \varepsilon(n/d) = \sum_{d|n} \mu(d) = e(n)$$

and thus

$$\mu \star \varepsilon = e,$$

which was stated but not proved in Section 7 of Chapter 1.

We are now going to determine an arithmetic formula for Euler's totient function φ, which was introduced in the last section. We recall that for each natural number n, $\varphi(n)$ counts the number of integers $k \in \{1, \ldots, n\}$ that satisfy $d(k,n) = 1$. We have already noted that

$\varphi(1) = 1$ and that for every prime p, $\varphi(p) = p-1$. As a preliminary step we determine $\varphi(p^k)$ for every prime p and every natural number k. We have

$$\begin{aligned}
\sum_{r=0}^{k} \varphi(p^r) &= \sum_{r=0}^{k} \text{number of } n \in \{1, 2, \ldots, p^r\} \text{ with } d(n, p^r) = 1 \\
&= \sum_{r=0}^{k} \text{number of } n \in \{1, 2, \ldots, p^{k-r}\} \text{ with } d(n, p^{k-r}) = 1 \\
&= \sum_{r=0}^{k} \text{number of } n \in \{1, 2, \ldots, p^k\} \text{ with } d(n, p^k) = p^r \\
&= \text{number of } n \in \{1, 2, \ldots, p^k\} \\
&= p^k.
\end{aligned}$$

Thus

$$\varphi(p^k) = \sum_{r=0}^{k} \varphi(p^r) - \sum_{r=0}^{k-1} \varphi(p^r) = p^k - p^{k-1} = p^{k-1}(p-1) = p^k \left(1 - \frac{1}{p}\right). \qquad \text{(III.10.2)}$$

We observe that $\varphi(p^k)$ is always an even number when p is an odd prime. The same is true when $p = 2$ provided $k \geq 2$. We now come to the main step in determining a formula for $\varphi(n)$ for all $n \in \mathbb{N}$.

Theorem III.10.4. *For any pair of relatively prime natural numbers a and b, we have*

$$\varphi(ab) = \varphi(a)\varphi(b). \qquad \text{(III.10.3)}$$

Proof. We display all the integers $n \in \{0, 1, \ldots, ab-1\}$ in a rectangular array of size $b \times a$ in which the (k, l)-entry $(k = 0, 1, \ldots, b-1;\ l = 0, 1, \ldots, a-1)$ is

$$c(k, l) \equiv ka + lb \pmod{ab}, \quad 0 \leq c(k, l) \leq ab - 1.$$

As $d(a, b) = 1$, Gauss' theorem (Theorem III.5.2) ensures that all the entries are distinct. The entries in the first column are the b integers $0a, 1a, \ldots, (b-1)a$. For $k = 0, 1, \ldots, b-1$, if $d(k, b) = 1$ the entries in the k-th row are all coprime with b whereas if $d(k, b) > 1$ all the entries of the k-th row have a common factor (> 1) with b. Similarly the entries in the first row are the a integers $0b, 1b, \ldots, (a-1)b$. For $l = 0, 1, \ldots, a-1$, if $d(l, a) = 1$ the entries in the l-th column are all coprime with a whereas if $d(l, a) > 1$ all the entries of the l-th column have a common factor (> 1) with a. The number of n in the array satisfying $d(n, ab) > 1$ is $ab - \varphi(ab)$. These n are exactly those such that $d(n, a) > 1$ or $d(n, b) > 1$ or both. The numbers n of the array such that $d(n, ab) > 1$ are precisely the numbers in the rows whose first entry $c_{k,1}$ satisfies $d(c_{k,1}, b) > 1$ and in the columns whose first entry satisfies $d(c_{1,j}, a) > 1$. The number of the former type is $(b - \varphi(b))a$ and the number of the latter is $(a - \varphi(a))b$. Those numbers that are at the intersection of these rows and columns are counted twice. Hence

$$ab - \varphi(ab) = (b - \varphi(b))a + (a - \varphi(a))b - (b - \varphi(b))(a - \varphi(a))$$

from which (III.10.3) follows. □

The formula (III.10.3) is illustrated in Figure III.10.2 for the values $a = 4,\ b = 15$.

Exercise III.10.4. Reproduce the previous proof for $n = 60$ illustrating each step in the Figure III.10.1.

From (III.10.2) and (III.10.3) we obtain the determination of $\varphi(n)$ for an arbitrary natural number n.

$\mathbb{Z}_{15}$:	$\mathbb{Z}_4$:	[0] ⇓	[3]	[2] ⇓	[1]
[0]	⇒	0	15	30	45
[4]		4	(19)	34	(49)
[8]		8	(23)	38	(53)
[12]	⇒	12	27	42	57
[1]		16	(31)	46	(1)
[5]	⇒	20	35	50	5
[9]	⇒	24	39	54	9
[13]		28	(43)	58	(13)
[2]		32	(47)	2	(17)
[6]	⇒	36	51	6	21
[10]	⇒	40	55	10	25
[14]		44	(59)	14	(29)
[3]	⇒	48	3	18	33
[7]		52	(7)	22	(37)
[11]		56	(11)	26	(41)

Figure III.10.2 The numbers $0 \le k \le 60 = 4 \times 15$ coprime to 60 are indicated by ◯

Theorem III.10.5. *If $n = p_1^{k_1} p_2^{k_2} \dots p_s^{k_s}$, where $p_1, p_2, \dots, p_s$ are distinct primes, then*

$$\varphi(n) = p_1^{k_1-1} p_2^{k_2-1} \dots p_s^{k_s-1}(p_1-1)(p_2-1)\dots(p_s-1) = n\left(1-\frac{1}{p_1}\right)\left(1-\frac{1}{p_2}\right)\cdots\left(1-\frac{1}{p_s}\right).$$

Example III.10.2. We can easily calculate

$$\varphi(504) = \varphi(2^3 \cdot 3^2 \cdot 7) = 504 \cdot \frac{1}{2} \cdot \frac{2}{3} \cdot \frac{6}{7} = 144,$$

and

$$\varphi(720) = \varphi(2^4 \cdot 3^2 \cdot 5) = 720 \cdot \frac{1}{2} \cdot \frac{2}{3} \cdot \frac{4}{5} = 192.$$

Thus $\mathbf{U}_{504}$ has 144 elements, while $\mathbf{U}_{720}$ has 192 elements. For small values of n it is rather easy to describe the structure of $\mathbf{U}_n$. The next section will provide some tools for such a description.

Exercise III.10.5. Determine the order of the groups $\mathbf{U}_{60}, \mathbf{U}_{72}, \mathbf{U}_{96}, \mathbf{U}_{100}$ and $\mathbf{U}_{120}$.

Exercise III.10.6 (Matiyasevich's parabola: This parabola is named after Yuri Vladimirovich Matiyasevich (1947 -)). Prove that the set of prime numbers can be defined as the integral "holes" (see Figure III.10.3) on the y axis of planar cartesian coordinates obtained in the following way: The integral points $(a, b), a \geq 2, b \geq 2$ of the parabola $y = x^2$ are connected by lines. Their points of intersection have integral coordinates and lie on the y axis. Mark them. The unmarked points $(0, p), p \in \mathbb{Z}$ are all the prime numbers.

Note that in order to determine all primes $p \leq n$, it is sufficient to consider the straight lines defined by the points $(\pm a, a^2)$ for $a \leq \sqrt{n}$. This reflects a relationship to the sieve of Eratosthenes (see Example III.10.1).

III.11 Direct products of cyclic groups

We start with a definition of a concept that we have already met informally.

Definition III.11.1. *A monoid $(G, \cdot)$ (with the identity element e) is said to be a* **cyclic group** *if there is an element (a* **generator***) $g \in G$ such that every element $x \in G$ is a power $x = g^r$ for a certain integer r and moreover, $x^{-1} \in G$ exists such that $x \cdot x^{-1} = e$.*

In this case, we shall simply write $G = \langle g \rangle$. Of course, $g^0 = e$ and $g^r \cdot g^s = a^{r+s}$ and $(g^r)^{-1} = g^{-r}$. Consequently, G is a commutative (abelian) group : $x \cdot y = y \cdot x$. Formulate these rules in terms of a theorem.

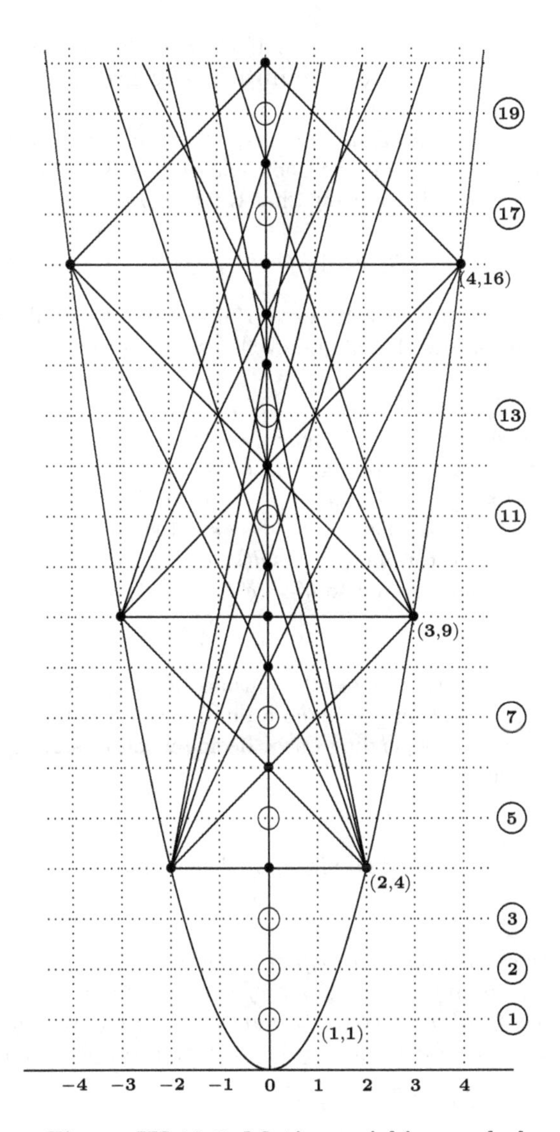

Figure III.10.3 Matiyasevich's parabola

Theorem III.11.1. *Let $G = \langle g \rangle$ be a cyclic group. Then there exists a surjective homomorphism (an epimorphism) $f : \mathbb{Z} \to G$ of the additive group of all integers onto G.*

If $\mathrm{Ker} f = \{0\}$, *that is, if f is an isomorphism, then $(G, \cdot) \simeq (\mathbb{Z}, +)$.*

If $\mathrm{Ker} f \neq \{0\}$, *then* $\mathrm{Ker} f = n\mathbb{Z}$ *for certain natural number $n \geq 2$ and $(G, \cdot) \simeq (\mathbb{Z}_n, \oplus)$.*

Proof. Simply define $f(t) = g^t$ for all $t \in \mathbb{Z}$. If $g^r \neq g^s$ for all integers $r \neq s$, f is an isomorphism. In this case, we speak of a **infinite cyclic group** that will be denoted by $C(\infty)$.

Otherwise, let $g^r = g^s$ for $r > s$. Thus $k = r - s > 0$ and $g^k = e$. Let $n > 0$ be the smallest natural number such that $g^n = e$. Using the Euclidean division $k = nq + r$ with $0 \leq r < n$, we infer that $g^r = e$ and thus $r = 0$. It turns out that $\mathrm{Ker} f = \{k \mid g^k = e\} = \{nq \mid q \in \mathbb{Z}\} = n\mathbb{Z}$. Hence

$$(G, \cdot) = \mathrm{Im} f \simeq (\mathbb{Z}_n, \oplus).$$

In this case, we speak of a **finite cyclic group of order** n that will be denoted by $C(n)$. Note that, up to an isomorphism, the infinite cyclic group, as well as the finite cyclic group of any order n, is unique. □

Suppose that the cyclic group $C(n)$ is generated by g. Thus $g^n = e$ and if $g^l = e$ then n divides l. We show that if h is another generator of $C(n)$ then $h = g^s$ for some integer s with $d(s, n) = 1$. As $C(n) = \langle g \rangle = \langle h \rangle$ we have $g = h^r$ and $h = g^s$ for some integers r and s. Hence $g = g^{rs}$ and so $g^{rs-1} = e$. Thus n divides $rs - 1$ so $d(s, n) = 1$. Now we show that g^s, where $d(s, n) = 1$, is a generator of $C(n)$. As $d(s, n) = 1$, for any integer a there exists an integer b such that $a \equiv bs \pmod{n}$ and so $g^a = g^{bs} = (g^s)^b$, proving that g^s is a generator of $C(n)$.

Example III.11.1. Let G be the set of all rotations of the plane around the origin that are obtained by repeatedly rotating the plane around the origin by a fixed angle α (both clockwise and anticlockwise). Depending on α, decide when this group is infinite. Can it be a cyclic group of any order?

The structure of a cyclic group is determined by the arithmetic of integers in the following sense.

Theorem III.11.2. *Any non-trivial subgroup of the infinite cyclic group is the infinite cyclic group. The subgroups of the finite cyclic group of order n are all cyclic and form a lattice isomorphic to the lattice of all divisors of the number n: If $C(n) = \langle g \rangle$ then the subgroup corresponding to the divisor a of n is the subgroup generated by g^b, where $n = ab$.*

Proof. In view of the isomorphisms described in Theorem III.11.1, we will simply describe the subgroups of the additive groups $\mathbb{Z}$ and $\mathbb{Z}_n$. We point out that the powers of elements are, in the additive notation, replaced by integral multiples of elements:

$$n \times a = a + a + \cdots + a \text{ (n times)}, \ (-n) \times a = n \times (-a) \text{ etc.}$$

Let $H \neq \{0\}$ be a subgroup of the additive group of $\mathbb{Z}$ and let a be the least positive number in the group H (remember the well-ordering). For any other $b \in H$, write $b = qa + r$, where $0 \leq r < a$. Since $r = b - q \times a \in H$, we must have $r = 0$. Consequently, we deduce $H = a\mathbb{Z} = \langle a \rangle = \{q \times a \mid q \in \mathbb{Z}\}$. □

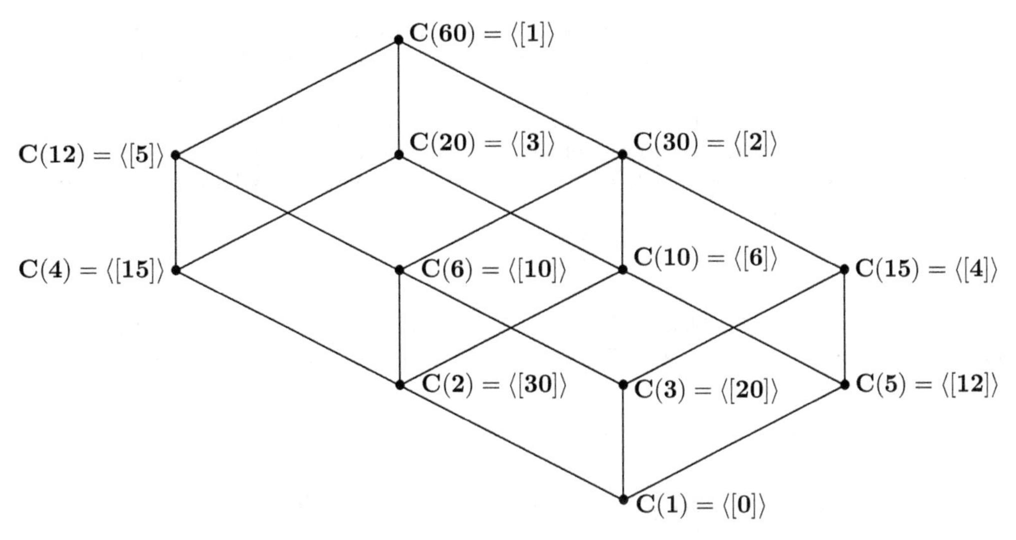

Figure III.11.1 Lattice of subgroups of $\mathbf{C(60)} \simeq (\mathbb{Z}_\mathbf{n}, \oplus)$.

Hence, all non-trivial subgroups of the additive group of integers (that is, of the infinite cyclic group) are infinite cyclic groups and a part of their lattice is illustrated in Figure III.4.1.

Now, the description of all subgroups of the additive group $(\mathbb{Z}_n, \oplus)$ follows from the description of all subgroups of $(\mathbb{Z}, +)$ via the canonical homomorphism f introduced in Theorem III.11.1. Have a look at the following commutative diagram (where $\hookrightarrow$ denotes embeddings).

$$\begin{array}{ccccccc}
\langle n\rangle & \hookrightarrow & \langle b = aq\rangle & \hookrightarrow & f^{-1}(H) = \langle a\rangle & \hookrightarrow & \mathbb{Z} \\
\downarrow f & & \downarrow f & & \downarrow f & & \downarrow f \\
\{[0]\} & \hookrightarrow & \langle [b] = [a][q]\rangle & \hookrightarrow & H = \langle f(a) = [a]\rangle & \hookrightarrow & \mathbb{Z}_n
\end{array}$$

If H is a subgroup of $\mathbb{Z}_n$, then the preimage ("inverse image")

$$f^{-1}(H) = \{z \in \mathbb{Z} \mid f(z) \in H\}$$

is a subgroup of $\mathbb{Z}$ and thus $f^{-1}(H) = \langle a\rangle \supseteq \langle n\rangle = \text{Ker} f$ for some natural number a. This means that $n = ar$ and $H = \langle f(a) = [a]\rangle \simeq C(r)$. On the other hand, for any divisor r of n, there is a (unique) cyclic subgroup generated by $[a] \in \mathbb{Z}_n$, where $n = ra$. In fact, $\langle [a]\rangle = \{[0], [a], 2 \times [a], \ldots, (r-1) \times [a]\}$. Since $\langle [b]\rangle \subseteq \langle [a]\rangle$ in $\mathbb{Z}_n$ if and only if $a \mid b$, we obtain the required isomorphism of the lattice of all subgroups of the cyclic group $C(n)$ and the lattice $\mathcal{D}(n)$ of all divisors of n (see the illustration in Figure III.11.1).

Exercise III.11.1. Determine the lattice of all subgroups of the cyclic groups $C(90)$ and $C(150)$ and compare it with the lattice in Figure III.11.1.

We turn our attention to Euler's totient function φ. Consider again all subgroups of the additive group $(\mathbb{Z}_n, \oplus)$; each of them is described by a divisor of n. Consider the mapping $h : \mathbb{Z}_n \to \mathcal{D}(n)$ (the lattice of all divisors of n) defined by $h([a]) = d$, where d is the order of the subgroup of $\mathbb{Z}_n$ generated by $[a]$. Thus the number of elements in the preimage $h^{-1}(d)$ equals to the number of the generators of the (unique) subgroup of order d, that is, to $\varphi(d)$. Now, the preimages of the elements of $\mathcal{D}(n)$ define, as we know from Chapter I, a (disjoint) partition of $\mathbb{Z}_n$. Therefore

$$\sum_{d|n} \varphi(d) = n.$$

The special case of this identity when n is a power of a prime was proved arithmetically in Section 10 of this chapter. Applying the Möbius inversion formula to this equation (see Example I.7.7), we obtain

$$\varphi(n) = \sum_{d|n} \mu(d) \frac{n}{d}.$$

We leave it to the reader to deduce that Theorem III.10.5 from this result.

Exercise III.11.2. For any two natural numbers a and b, prove

$$\varphi(ab) = \frac{\varphi(a)\varphi(b)}{\varphi[d(a,b)]}\, d(ab).$$

In Example III.9.1, we have described the group $\mathbf{U}_{12}$ as a "cartesian product" of two cyclic groups $\langle[5]\rangle \simeq C(2)$ and $\langle[7]\rangle \simeq C(2)$. We now formalize this construction.

Definition III.11.2. *Let $C(n_1), C(n_2), \ldots, C(n_k)$ be a set of k cyclic groups. The group G of all $k-$tuples $(g_1, g_2, \ldots, g_k)$ with coordinate-wise multiplication will be denoted by*

$$G = C(n_1) \times C(n_2) \times \cdots \times C(n_k)$$

and called the **direct product** *of the groups $C(n_1), C(n_2), \ldots, C(n_k)$.*

Thus, in particular, $(e, e, \ldots, e)$ is the identity element and $(g_1^{-1}, g_2^{-1}, \ldots, g_k^{-1})$ is the inverse element of the element $(g_1, g_2, \ldots, g_k)$. Of course, once we will use the additive notation (as it is often the case for commutative groups), we will speak about the **direct sum** of groups. The term non-trivial direct product or sum will be reserved for the situation when at least two components (factors or summands) are non-trivial groups.

Exercise III.11.3. Let the cyclic subgroups $C(n_1), C(n_2), \ldots, C(n_k)$ of a **commutative** group G generate the group G (that is, every element of the group G is a product of some elements from the subgroups $C(n_1), C(n_2), \ldots, C(n_k)$). Prove that the following three statements are equivalent:

(1) Every element $g \in G$ possesses a **unique** expression as a product $g = g_1 g_2 \ldots g_k$ with $g_t \in C(n_t)$ for $1 \leq t \leq k$.

(2) For every $1 \leq t \leq k$, the intersection of the subgroup G_t generated by $C(n_1), C(n_2), \ldots,$ $C(n_{t-1}), C(n_{t+1}), \ldots, C(n_k)$ and the subgroup $C(n_t)$ is trivial:

$$G_t \cap C(n_t) = \{e\}.$$

(3) The group G is isomorphic to the direct product of the cyclic groups

$$C(n_1), C(n_2), \ldots, C(n_k).$$

Cyclic groups can often be decomposed into a direct product of smaller components (subgroups). Such a situation is described in the following theorem.

Theorem III.11.3. *Let*

$$n = p_1^{k_1} p_2^{k_2} \ldots p_s^{k_s}, \quad \textit{where all primes } p_t, 1 \leq t \leq s, \quad \textit{are distinct.}$$

Then

$$C(n) \simeq C(p_1^{k_1}) \times C(p_2^{k_2}) \times \cdots \times C(p_s^{k_s}).$$

Proof. Let $C(n) = \langle g \rangle$. For each $1 \leq t \leq k$ consider the subgroup G_t of $C(n)$ generated by the element g^{n_t}, where $p_t^{k_t} n_t = n$. Thus

$$G_t = \{g^{n_t}, g^{2n_t}, \ldots, g^{p_t^{k_t} n_t} = g^n = e\}$$

is a cyclic subgroup of order $p_t^{k_t}$. Now, a simple argument, or a reference to Theorem III.5.4 guarantees that every element of g is a **unique** product of elements from $G_t, 1 \leq t \leq k$. $\square$

Exercise III.11.4. Show that the cyclic group $C(p^n)$, where p is a prime, cannot be decomposed into a non-trivial direct product. What is the lattice of all subgroups of $C(p^n)$?

Example III.11.2. The group $C(6) = \langle g \rangle$ is a direct product of the subgroup $C(3) = \langle g^2 \rangle$ and the subgroup $C(2) = \langle g^3 \rangle$. The group $\mathbf{U}_{30}$ has $\varphi(30) = 8$ elements and contains two cyclic groups of order 4 and three groups of order 2. It is easy to check that $\langle [7] \rangle \simeq C(4)$, $\langle [11] \rangle \simeq C(2)$ and $\mathbf{U}_{30} = \langle [7] \rangle \times \langle [11] \rangle \simeq C(4) \times C(2)$. For large n such a decomposition of $\mathbf{U}_n$ is demanding.

Exercise III.11.5. Verify:

(i) $\mathbf{U}_{504} = \langle [5] \rangle \times \langle [13] \rangle \times \langle [55] \rangle \times \langle [71] \rangle = \langle\ [101]\ \rangle \times \langle\ [349]\ \rangle \times \langle\ [433]\ \rangle \times \langle\ [503]\ \rangle \simeq$ $C(6) \times C(6) \times C(2) \times C(2)$.

(ii) $\mathbf{U}_{720} = \langle\ [241]\ \rangle \times \langle\ [19]\ \rangle \times \langle\ [37]\ \rangle \times \langle\ [71]\ \rangle \times \langle\ [89]\ \rangle \simeq C(3) \times C(4) \times C(4) \times C(2) \times C(2)$.

Exercise III.11.6. Verify that $\mathbf{U}_{100} = \langle [21] \rangle \times \langle [7] \rangle \times \langle [51] \rangle \simeq C(5) \times C(4) \times C(2)$ and determine the structure of $\mathbf{U}_7, \mathbf{U}_9, \mathbf{U}_{72}$ and $\mathbf{U}_{120}$.

In Chapter VIII we shall reformulate Theorem III.11.3 for any finite commutative group G showing that such a group has a unique direct decomposition into its primary components (maximal subgroups of prime power order). There we shall prove that any finite commutative group is a direct product of (finite) cyclic groups (and that such a decomposition is unique up to an isomorphism). Furthermore, we shall extend this result in two directions: On the one hand, to the commutative group that are generated by a finite number of elements, and on the other hand to commutative groups G of bounded order, that is, to groups (not necessarily finite) for which there exists a number n such that $g^n = e$ for all $g \in G$.

III.12 Codes and cryptography

In this section, we are going to present a few applications of modular arithmetic (congruences).

1. Check digits

These are simple applications, some of which we meet almost daily. They can be called collectively check-digit procedures. The aim is to detect possible errors in a digit sequence carrying some information (codes) or to check validity of certain data encoded in a sequence of letters and digits.

UPC (Universal Product Code) is an error detecting code that is broadly used to track trade items in stores. It consists of 11–tuple digit code $a_1a_2\ldots a_{11}$ and the check digit a_{12} (designed to detect an error in the 11–digit code) such that

$$3 \times \sum_{t=0}^{5} a_{2t+1} + \sum_{t=1}^{6} a_{2t} \equiv 0 \pmod{10}.$$

In Europe the **UPC** symbols $a_1a_2\ldots a_{12}a_{13}$ (with the check digit a_{13}) are such that

$$\sum_{t=0}^{6} a_{2t+1} + 3 \times \sum_{t=1}^{6} a_{2t} \equiv 0 \pmod{10}.$$

Exercise III.12.1. Verify the correctness of the **UPC** code of a product near you. Determine the check digit a_{12} for the code 88517002781.

ISBN (International Standard Book Number) is a sequence of 10 digits $b_1b_2\ldots b_{10}$ identifying in the first 9 digits the country, the publisher and registration of a book, with the last digit b_{10} as the check key. Here, a valid **ISBN** satisfies

$$\sum_{t=1}^{10} t \times b_{11-t} \equiv 0 \pmod{11}.$$

Exercise III.12.2. Check validity of the **ISBN** of some books around you. Find all possible values of b_8 and b_9 such that $2842250b_8b_91$ is an **ISBN** code.

IBAN (International Bank Account Number) is used in some countries to identify bank accounts across national borders. It consists of a string of up to 34 letters and digits such that the first two digits identify the country and a sequence of digits includes the bank account number, branch identifier and further relevant information. To validate **IBAN**, one checks first the total **IBAN** length (which differs from country to country), moves the original 4 first characters to the end of the string and replace each letter of the string by a pair of digits: $A \to 10, B \to 11, \cdots, Z \to 35$. Then the string of digits is interpreted as a decimal integer and checked whether this number is congruent to 1 modulo 97.

Here is an **IBAN** formatting example for Switzerland, where the length of the **IBAN** codes is 21:

$$CH9300762108720862657$$

This code translates into the decimal integer $n = 762108720862657121793$. As $762108 \equiv 76 \pmod{97}$, $76720862 \equiv 70 \pmod{97}$, and $70657121 \equiv 90 \pmod{97}$, we have

$$\begin{aligned}
n &= 762108 \times 10^{15} + 720862 \times 10^9 + 657121 \times 10^3 + 793 \\
&\equiv 76 \times 10^{15} + 720862 \times 10^9 + 657121 \times 10^3 + 793 \pmod{97} \\
&= 76720862 \times 10^9 + 657121 \times 10^3 + 793 \\
&\equiv 70 \times 10^9 + 657121 \times 10^3 + 793 \pmod{97} \\
&= 70657121 \times 10^3 + 793 \\
&\equiv 90 \times 10^3 + 793 \pmod{97} \\
&= 90793 \\
&\equiv \mathbf{1} \pmod{97},
\end{aligned}$$

and thus the code is valid.

Exercise III.12.3. Check if the following **IBAN** identifies a fictitious United Kingdom bank account (the length is correct):

$$GB23BBSC60971331926819.$$

2. Affine cryptography

A very old simple-minded cipher, based on modular arithmetics and allegedly used by **Julius Caesar (100 - 44 BCE)**, is the **affine cipher** using the function

$$f : \mathbb{Z}_n \to \mathbb{Z}_n \text{ defined by } f[x] = [a] \odot [x] \oplus [b] \text{ with } d(a, n) = 1.$$

Equivalently, f is an affine bijection of the set S_n of n non-negative numbers smaller than n onto itself given by the formula

$$f(x) = y \equiv ax + b \pmod{n}$$

and thus depending on n, a, b. A message is first translated, letter by letter, into a string of numbers from S_n which are then mapped by the function f to obtain an encrypted message. To read this message, we just use, in the same fashion, the inverse function

$$f^{-1}[y] = [a]^{-1} \odot [y] \ominus [a]^{-1} \odot [b].$$

So the "key" for this affine cipher consists of three natural numbers n and $0 \leq a, b \leq n-1$, $d(a,n) = 1$. Since the English alphabet has 26 letters, we need $n \geq 26$. Hence $n = 26$ is usually chosen. Care should be taken in this case, because $\mathbb{Z}_{26}$ is not a field and so not all numbers could be "inverted". It is therefore advisable to add another three numbers to represent, say the symbols {. , !}, and calculate in the field $\mathbb{Z}_{29}$. In any case, we will systematically use the translation of the letters $\{A, B, \dots, Z\}$ of the English alphabet according to this table

A	B	C	D	E	F	G	H	I	J	K	L	M	...	V	W	X	Y	Z
0	1	2	3	4	5	6	7	8	9	10	11	12	...	21	22	23	24	25

Example III.12.1. Consider the affine cipher given by $n = 26, a = 3$ and $b = 10$. Encipher the message " Peace". First PEACE is translated to 15 4 0 2 4, and thus mapped by the affine mapping to 3 22 10 16 22, which reads "DWKQW". If you get a reply "ZAJWVWJ", use the inverse cipher $f^{-1}[y] = [9] \odot [y] \oplus [14]$ to map 25 0 9 22 21 22 9 to 5 14 17 4 21 4 17 and read the message "FOREVER".

Exercise III.12.4. Associate the letters of the alphabet $\{A, B, C, \dots, Z\}$ with the (ordered) set of elements $\{[0], [1], [2], \dots, [25]\}$ of $\mathbb{Z}_{26}$.
(i) An (affine) code $f : \mathbb{Z}_{26} \to \mathbb{Z}_{26}$ is defined by $f(x) = [11] \odot x \ominus [18]$. Encipher the word MATHEMATICS.
(ii) Find the inverse mapping to f and decipher the word JAEHVGZGWM.

Exercise III.12.5. Associate the letters of the alphabet $\{A, B, C, \dots, Z\}$ with the (ordered) set of elements $\{[0], [1], [2], \dots, [25]\}$ of $\mathbb{Z}_{26}$.
(i) Find the affine code $f : \mathbb{Z}_{26} \to \mathbb{Z}_{26}$ such that O is encoded as A and U is encoded as Y.
(ii) Show that, in this code, there are two letters that coincide with their codes. Determine them.
(iii) Under which conditions is there an affine cipher that encodes a pair of chosen letters into two prescribed letters. Under what condition is such a cipher unique, if it exists.

Now, you can see a fundamental drawback in using this cipher. It really means just a permutation of the letters; in Example III.16 it is according to the table:

A	B	C	D	E	F	G	H	I	J	K	L	...	V	W	X	Y	Z
K	N	Q	T	W	Z	C	F	I	L	O	R	...	V	Y	B	E	H

Therefore, if we already know the language that is used for communication, it is easy to decipher this code. For, each letter appears in a given language with a definite frequency and therefore this statistical consideration is sufficient to make this cipher practically useless. We shall resolve this weakness by introducing genuine "more-dimensional" affine ciphers in which the encoding of a given letter will be changing depending on the position in the message.

We make a short interlude into a subject close to topics of Linear Algebra, viz. to the subject of **modules over the rings $\mathbb{Z}_n$.**

In what follows, we will simplify the notation: The elements $[a]$ of $\mathbb{Z}_n$ will simply be denoted by a, $[a] \oplus [b]$ by $a+b$ and $[a] \odot [b]$ by ab. Furthermore, $(\mathbb{Z}_n)^h$ will denote the module of all h-tuples $\mathbf{v}$ of elements of $\mathbb{Z}_n$:

$$\mathbf{v} = (v_1, v_2, \dots, v_h)^T, \quad \text{where } v_1, v_2, \dots, v_h \in \mathbb{Z}_n.$$

The module $(\mathbb{Z}_n)^h$ is, with respect to coordinate-wise addition, an abelian group and the elements $\mathbf{v}$ of $(\mathbb{Z}_n)^h$ can be "multiplied" by elements of $\mathbb{Z}_n$:

$$a\mathbf{v} = a(v_1, v_2, \ldots, v_h) \;=\; (av_1, av_2, \ldots, av_h),$$

whereby all "scalar multiplication" rules hold: For $\mathbf{v}, \mathbf{w} \in (\mathbb{Z}_n)^h$ and $a, b \in \mathbb{Z}_n$,

$$a(\mathbf{v}+\mathbf{w}) = a\mathbf{v} + a\mathbf{w},\ (a+b)\mathbf{v} = a\mathbf{v} + b\mathbf{v},\ a(b\mathbf{v}) = (ab)\mathbf{v} \text{ and } 1\mathbf{v} = \mathbf{v}.$$

Thus, if $n = p$ is a prime number, $(\mathbb{Z}_p)^h$ is just a $h-$dimensional vector space of all h-tuples over $\mathbb{Z}_p$.

Definition III.12.1. *Given an $h \times h$ matrix $\mathbf{R} = (r_{i,j})$ over the ring $\mathbb{Z}_n$ and an h-tuple $\mathbf{v}$ of $(\mathbb{Z}_n)^h$, the mapping $f : (\mathbb{Z}_n)^h \to (\mathbb{Z}_n)^h$ defined by $f(\mathbf{x}) = \mathbf{R}\mathbf{x} + \mathbf{v}$ will be called an* **affine mapping**.

Remark III.12.1. If $\mathbf{R} = \mathbf{0}_{h\times h}$, f is a **constant mapping**. If $\mathbf{v}$ is the zero h-tuple $\mathbf{0}$, then f is a **linear transformation** of $(\mathbb{Z}_n)^h$. In particular, a linear transformation f is an **identity mapping** if $\mathbf{R} = \mathbf{I}_{n\times n}$. It is easy to see that, if $g : (\mathbb{Z}_n)^h \to (\mathbb{Z}_n)^h$ is another affine mapping, defined by $g(\mathbf{x}) = \mathbf{S}\mathbf{x} + \mathbf{w}$, then the **composition mapping** $f(g(\mathbf{x})) = \mathbf{RS}\mathbf{x} + \mathbf{R}\mathbf{w} + \mathbf{v}$ is also an affine mapping.

Definition III.12.2. *An affine mapping f of $(\mathbb{Z}_n)^h$ will be called* **regular** *if there exists a mapping f^{-1}, an* **inverse mapping** *such that $f(f^{-1})$ is the identity mapping of $(\mathbb{Z}_n)^h$.*

Remark III.12.2. We have that f is regular if and only if the matrix $\mathbf{R}$ has an inverse. In that case, every h-tuple of $(\mathbb{Z}_n)^h$ is an image $f(\mathbf{x})$ of some $\mathbf{x}$. For $h = 1$, this means, as we have seen, that n and r in the mapping $f(x) = rx + s$ of $\mathbb{Z}_n$ are relatively prime. If, in particular, $n = p$ is a prime, then $\text{Im}f = \mathbb{Z}_p$ for every affine mapping that is not constant.

Remark III.12.3. Before we return to the affine ciphers, refresh some topics from linear algebra, such as solution of a system of linear equations, determinants, inverse matrices etc. with the following exercises over the field $\mathbb{Z}_p$:

(1) Solve the following system of linear equations over $\mathbb{Z}_7$:

$$\begin{array}{rcrcrcl} 5\,x_1 & + & 2\,x_2 & + & 2\,x_3 & = & 1, \\ 2\,x_1 & + & 1\,x_2 & + & 1\,x_3 & = & 0, \\ 1\,x_1 & + & 1\,x_2 & + & 6\,x_3 & = & 0. \end{array}$$

(2) Define linear transformations

$$f : {\mathbb{Z}_5}^3 \;\to\; {\mathbb{Z}_5}^3$$

by

(i) $f(x_1, x_2, x_3) = (x_1 + x_2, x_2 + x_3, x_3)$;

(ii) $f(x_1, x_2, x_3) = (x_1 + x_2, x_2 + x_3, x_3 + x_1)$.

Find the inverse transformations.

(3) Find the kernel and the range of the linear transformation

$$F : {\mathbb{Z}_2}^5 \rightarrow {\mathbb{Z}_2}^3$$

defined by $F(\mathbf{x}) = \mathbf{A}\mathbf{x}$, where

$$\mathbf{A} = \begin{pmatrix} 1 & 0 & 0 & 1 & 0 \\ 1 & 1 & 0 & 0 & 1 \\ 0 & 0 & 1 & 1 & 0 \end{pmatrix}.$$

To perform an h–dimensional affine encipher, we divide the entire message into block of h letters. After transforming the letters into the elements of $\mathbb{Z}_n$, we map the individual h–block, that is, the elements of $(\mathbb{Z}_n)^h$ by means of an agreed affine transformation $f : (\mathbb{Z}_n)^h \rightarrow (\mathbb{Z}_n)^h$ into encrypted h–block that we then translate back into letters, that is, the message that will be sent out. We illustrate this process with the following example, using the prime number $n = 37$ this time.

Example III.12.2. Associate the letters of the alphabet $\{A, B, C, \ldots, Z\}$, the set of the digits $\{0, 1, 2, \ldots, 9\}$ and the "blank" symbol $\oslash$ with the (ordered) set of elements

$$\{0, 1, 2, \ldots, 25, 26, 27, \ldots, 35, 36\} \text{ of } \mathbb{Z}_{37}.$$

Define the affine mapping $f : \mathbb{Z}_{37} \rightarrow \mathbb{Z}_{37}$ by $f(\mathbf{x}) = \mathbf{R}\mathbf{x} + \mathbf{v}$, where

$$\mathbf{R} = \begin{pmatrix} 1 & 0 & 3 \\ 0 & 1 & 2 \\ 4 & 3 & 0 \end{pmatrix} \quad \text{and} \quad \mathbf{v} = \begin{pmatrix} 1 \\ 3 \\ 0 \end{pmatrix}.$$

To encipher ABSTRACT ALGEBRA, consider the 3-blocks

$$\oslash AB \;\; STR \;\; ACT \;\; \oslash AL \;\; GEB \;\; RA\oslash \;,$$

that is, the triples

$$36\;0\;1 \quad 18\;19\;17 \quad 0\;2\;19 \quad 36\;0\;11 \quad 6\;4\;1 \quad 17\;0\;36.$$

These are mapped (using $f(\mathbf{x}) = \mathbf{R}\mathbf{x} + \mathbf{v}$) to

$$3\;5\;33 \quad 33\;19\;18 \quad 21\;6\;6 \quad 33\;25\;33 \quad 10\;9\;36 \quad 15\;1\;31,$$

that is, to the following ciphertext:

$$DF77TSVGG7Z7KJ \;\; PB5.$$

To decipher the message, find

$$\mathbf{S} = \mathbf{R}^{-1} = \begin{pmatrix} 25 & 18 & 31 \\ 16 & 13 & 33 \\ 29 & 31 & 2 \end{pmatrix} \quad \text{and} \quad \mathbf{w} = -\mathbf{S}\mathbf{v} = \begin{pmatrix} 32 \\ 19 \\ 26 \end{pmatrix};$$

and then use $f^{-1}(\mathbf{x}) = \mathbf{S}\mathbf{x} + \mathbf{w}$.

Exercise III.12.6. As in the previous example, associate the letters of the alphabet $\{A, B, C, \ldots, Z\}$, the set of the digits $\{0, 1, 2, \ldots, 9\}$ and the "blank" symbol $\oslash$ with the (ordered) set of elements $\{0, 1, 2, \ldots, 25, 26, 27, \ldots, 35, 36\}$ of $\mathbb{Z}_{37}$.
(i) Decipher the message

$$FSIHZE3WOZDLVJ7OC0BFZOTLWM8D3AR8X \oslash WF$$

enciphered using an affine code $f : (\mathbb{Z}_{37})^2 \to (\mathbb{Z}_{37})^2$, knowing that the first two letters of the message are TH and that the message ends with letters MAIL.
(ii) Answer the message, using the same code, by THANK YOU.

Exercise III.12.7. Associate the letters of the alphabet $\{A, B, C, \ldots, Z\}$, the set of the digits $\{0, 1, 2, \ldots, 9\}$ and the "blank" symbol $\oslash$ with the (ordered) set of elements

$$\{0, 1, 2, \ldots, 25, 26, 27, \ldots, 35, 36\} \text{ of } \mathbb{Z}_{37}$$

as in the previous exercise.
(i) Determine the linear code $f : (\mathbb{Z}_{37})^3 \to (\mathbb{Z}_{37})^3$ given by

$$f\left(\begin{pmatrix} x \\ y \\ z \end{pmatrix}\right) = \begin{pmatrix} 1 & 0 & a \\ 0 & 0 & b \\ 3 & 1 & c \end{pmatrix} \times \begin{pmatrix} x \\ y \\ z \end{pmatrix},$$

where $a, b, c \in \mathbb{Z}_{37}$ are to be determined knowing that ABC is enciphered as $C9B$.

(ii) Write down the form of the inverse mapping $f^{-1}\begin{pmatrix} x \\ y \\ z \end{pmatrix}$ and decipher the message UX6O0I41ZZLZ.

3. Cryptographic system RSA

The basic idea of public-key cryptography is that the **key**, that is, the number scheme, the encoding algorithm, that one needs to encode messages are **publicly known.** Thus, anyone can encrypt and send a message. However, only the recipient, the one who has installed the key, knows how to decode the received message.

One of the widely used public-key systems is the so-called **RSA** cryptosystem. This system is named after **Ronald L. Rivest (1947 -)**, **Adi Shamir (1952 -)** and **Leonard Max Adleman (1945 -)**, who on the basis of the 1976 work of **Bailey Whitfield Diffie (1944 -)**, **Martin Edward Hellman (1945 -)** and **Ralph C. Merkle (1952 -)** made the cryptographic system public in 1977. In fact an equivalent system was developed in 1973 by **Clifford Cocks (1950 -)**. However his work was not declassified until 1997. The RSA system is based on elementary number theory. Its security is supported by the fact that there is known no algorithm for factorization of large composite numbers and presently the factorization of large numbers even for the most advanced computers is infeasible.

The number theoretical basis of the **RSA** system is Fermat's little theorem. We formulate the entire scheme as the following theorem.

Theorem III.12.1. *Let p, q be distinct prime numbers, $N = pq$ and E a natural number satisfying*

$$1 < E < (p-1)(q-1) \text{ and } d(E,(p-1)(q-1)) = 1.$$

Then, there is a unique $D, 1 < D < (p-1)(q-1)$ such that

$$ED \equiv 1 \pmod{(p-1)(q-1)} \text{ and for any natural } a,\ a^{ED} \equiv a \pmod{N}.$$

Thus

$$a^E \equiv b \pmod{N} \text{ if and only if } b^D \equiv a \pmod{N}.$$

Proof. The existence of a unique D follows from Bézout's theorem. After all, D is just the inverse of the element E in the group of units $\mathbf{U}_{(p-1)(q-1)}$. Since $ED = 1 + t(p-1)(q-1)$ for some integer t, we get, using Fermat's Little Theorem, that

$$a^{ED} = a\,(a^{(p-1)})^{t(q-1)} \equiv a \pmod{p}$$

and similarly $a^{ED} \equiv a \pmod{q}$. Hence, $a^{ED} \equiv a \pmod{pq}$. The theorem follows. □

Example III.12.3 (naive illustration). Let $p = 3, q = 17, e = 11$. Thus $N = 51, (p-1)(q-1) = 32$ and $d(11,32) = 1$. For $a = 2$, $a^{11} = 2048 \equiv 8 \pmod{51}$ and $8^3 = 512 \equiv 2 \pmod{51}$. For $a = 3$, $3^{11} = 177147 \equiv 24 \pmod{51}$ and $24^3 = 13824 \equiv 3 \pmod{51}$, etc.

We describe the entire process of **transmission using RSA:**

A(lice) wants to send a string of digits (message) to B(ob).

- ⋆**1** **B** chooses p, q and E satisfying the prescribed conditions, $N = pq$;
- ⋆**2** **B** calculates $D:\ 1 \le D < (p-1)(q-1)$ and $ED \equiv 1 \pmod{(p-1)(q-1)}$;
- ⋆**3** **B** announces the key (N, E) publicly;
- ⋆**4** **A** converts the string of digits into a series of numbers a smaller than N;
- ⋆**5** For each a, **A** calculates

$$b \equiv a^E \pmod{N},\ 1 \le b \le N \text{ and sends } b \text{ to } \mathbf{B};$$

- ⋆**6** **B** deciphers the message by calculating

$$b^D = a^{ED} \equiv a \pmod{N}.$$

Here is an illustration of this procedure for a naive choice of Bob: $p = 7,\ q = 11,\ E = 13$; thus $N = 77$.

Bob calculates $D:\ 13D \equiv 1 \pmod{60}$; thus, $D = 37$ (for, $13 \times 37 - 1 = 480$)

Bob announces the public key $(77, 13)$.

Alice encodes the word ALGEBRA into the following string: 00 11 06 04 01 17 00 of pair of digits, using the translation $A = 00, B = 01, C = 03, \ldots, Z = 25$, and calculates, for each number a in the 2-blocks, $a^E \equiv b \pmod{77}$:

$$00 \mapsto 00, 11 \mapsto 11, 06 \mapsto 62, 04 \mapsto 53, 01 \mapsto 01, 17 \mapsto 73, 00 \mapsto 00;$$

Alice sends to Bob the following message 00116253017300.

Here, e.g. $6^{13} \equiv 62 \pmod{77}$ and back $62^{37} \equiv 6 \pmod{77}$.

Remark III.12.4. Our choices were naive indeed. In real life situations the primes selected would be much larger, each prime would have several hundred digits in the decimal system.

Exercise III.12.8. Formulate and prove Theorem III.12.1 considering the least common multiple m of $p-1$ and $q-1$ instead of their product.

Exercise III.12.9. Suppose that in an RSA Public Key Cryptosystem, $p = 37, q = 73$ and $E = 5$. Thus $N = 2701$. Using Exercise III.12.5, verify that $m = 72$ and $D = 29$. Show that the letter $V = 21$ is encoded to 189 and that $189^{29} \equiv 21 \pmod{2701}$.

Exercise III.12.10. Suppose that in an RSA Public Key Cryptosystem, $p = 11, q = 17$ and $E = 107$. Encrypt the word *LOVE* using two-digits blocks and the usual translation of the 26-letter alphabet into numbers. What is the "secret" key D?

CHAPTER IV. RATIONAL AND REAL NUMBERS

We are going to treat both rational and real numbers in the same chapter. This is quite unusual. The reason is rather simple; although both concepts are widely diverse, there are arithmetical reasons to treat them together at the level of introductory algebra. As much as the Euclidean algorithm forms a basis of integral arithmetic, when expressed in terms of continued fractions, it brings the real numbers into three disjoint subsets: the set of rational numbers, the set of irrational roots of quadratic equations and the set of remaining irrational numbers (that will include all transcendental numbers, that is, real numbers that are not roots of any algebraic equation).

At this moment, we should hasten to point out the fundamental, immense difference between rational numbers (fractions) and irrational numbers. While the former are given by "God" as **Leopold Kronecker** put it, the latter are a purely human abstract construction. Indeed, the rational numbers are derived quite naturally from the natural numbers, or integers; this does not mean that the history of this development was straightforward. The acceptance of zero and negative numbers was not without obstacles. However, today, as we have seen, the passage from natural numbers to fractions does not bring any conceptual difficulties and has a clear physical interpretation.

The matter with the real numbers (and as we shall see, with the complex numbers) is quite different. The difficulties lie right in the concept: There is no physical representation, the real irrational number is an abstract human construction, be it Dedekind's cuts or Cauchy-Weierstrass fundamental sequences. A common formulation that the real numbers correspond to the points of a line is rather artificial: It is true that they can be represented by the points of a line, but that line is an "ideal" line, equally an abstract construction as the real numbers themselves. So there is nothing "real" about these numbers. Once we accept this fact, we readily realize that the concept of a complex number (including those "imaginary") is of the same level of abstraction. They just correspond to the points of the (abstractly constructed "ideal") plane.

In fact, the abstract construction of the real numbers is an ideal opportunity to underline the fact that mathematics is indeed a human endeavor that, in spite of its abstractness serves well in dealing with concrete problems. The more abstract it is, strangely enough, the better it serves. It is rather unfortunate that this point is not already underlined enough during secondary school instruction. It would certainly make the transition from secondary school to university easier.

IV.1 Rational numbers

The extension of the integral domain of integers to the field of rational numbers follows a very simple algebraic path, one which is now known to us. We have already introduced it formally in the Introduction (Theorem I.7.2) and used it by extending the domain of natural numbers to integers. In the same way, as pairs of natural numbers (a, b) represented under a certain equivalence the "new" integral numbers, we may introduce the rational numbers as pairs of integers (a, b) with $b \neq 0$.

Exercise IV.1.1. In the set Q of all formal pairs $a/b, a \in \mathbb{Z}, b \in \mathbb{Z}, b \neq 0$, define the relation $\sim$ by $a/b \sim c/d$ if and only if $ad = bc$; it is an equivalence and denote the equivalence class of a/b by $[a/b]$. Furthermore, we denote the set of all equivalence classes by $\mathbb{Q}$. Also show that the definition of addition $[a/b] + [c/d] = [(ad + bc)/bd]$ does not depend on the choice of the representatives of the respective classes and conclude that $(\mathbb{Q}, +)$ is a commutative group with zero $[0/b]$. Show that also the definition of multiplication $[a/b] \cdot [c/d] = [ac/bd]$ does not depend on the choice of the representatives of the classes and that $(\mathbb{Q} \setminus \{[0/b]\}, \cdot)$ is a multiplicative commutative group with the identity element $[a/a]$. Verify the validity of the distributive law. Hence $(\mathbb{Q}, +, \cdot)$ is a field.

The field $\mathbb{Q}$ of Exercise IV.1.1 is the field of rational numbers (of fractions of integers). The elements $u = [a/b]$ are freely written as $\frac{a}{b}$; thus

$$\frac{a}{b} + \frac{c}{d} = \frac{ad + bc}{bd} \quad \text{and} \quad \frac{a}{b} \cdot \frac{c}{d} = \frac{ac}{bd}.$$

Now, one can achieve a unique notation (a unique representative) for the number $[a/b]$ by requiring additional conditions on a and b. This is generally done by asking that $b > 0$ and that a and b be relatively prime. We underline the fact that for $u = [a/b] \neq 0 = [0/b]$ and $v = [c/d]$ the equation

$$ux = v$$

has a **unique solution** $x = [bc/ad]$. Of course, by abuse of notation the equation can be written as $\frac{a}{b}x = \frac{c}{d}$ and its unique solution as $\frac{bc}{ad}$.

We also point out that there is an **injection** $f : \mathbb{Z} \to \mathbb{Q}$ defined by $f(a) = [a/1]$ that induces an isomorphism of the integral domain $\mathbb{Z}$ of integers and the image of f, that is, the integral domain of all rational numbers $[a/1], a \in \mathbb{Z}$. We will identify these two sets and say that the integral domain $\mathbb{Z}$ is a subring of the field $\mathbb{Q}$.

The order in $\mathbb{Z}$ can easily be extended to $\mathbb{Q}$. Indeed, define $u = [a/b] > v = [c/d]$ (or simply $\frac{a}{b} > \frac{c}{d}$) if and only if $abd^2 > b^2cd$.

Exercise IV.1.2. Verify that the relation $>$ does not depend on the choice of representatives of u and v and show that $u = [a/b]$ is positive (that is, $u > 0 = [0/d]$) if and only if the integer $ab > 0$. Show that always either $u > v$, or $v > u$ or $u = v$, that is, the ordering of $\mathbb{Q}$ is linear (total). Moreover, for the rational numbers $u, v, u < v$ and w prove that

$$u + w < v + w\,,\ uw < vw \text{ if } w > 0 \text{ and } vw < uw \text{ if } w < 0.$$

There are non-empty subsets of $\mathbb{Q}$ which have no least element: For instance, the set of all positive rational numbers does not have a least element. For, with every $u > 0$ also $u > \frac{u}{2} > 0$. In other words, $\mathbb{Q}$ is not well-ordered.

Since every non-zero number has a multiplicative inverse in $\mathbb{Q}$, divisibility of integers becomes a triviality in $\mathbb{Q}$: every non-zero integer (and, indeed, every non-zero rational number) is a divisor of any other rational number. However, the Euclidean division of integers, and the corresponding Euclidean algorithm, formulated in terms of rational numbers, brings a new important concept of a **continued fraction**.

Example IV.1.1. The fact that to every pair of integers a and $b > 0$ integers q and $0 \leq r < b$ satisfying the equality $a = bq + r$ are assigned in a unique way in $\mathbb{Z}$, means that an arbitrary rational number $u = \frac{a}{b}$ defines uniquely an integer q (its "integral part" denoted usually by $q = \lfloor u \rfloor$) and a remainder $0 \leq \frac{r}{b} = u - \lfloor u \rfloor < 1$. This interpretation reflects the basic character of the Euclidean division in a very simple (and natural) way. Moreover it indicates the process of expressing every rational number in the form of a **simple continued fraction**. For example, using the Euclidean algorithm with $a = 832$ and $b = 357$, one obtains successively

$$\begin{aligned} 832 &= 357 \cdot \mathbf{2} + 118, \\ 357 &= 118 \cdot \mathbf{3} + 3, \\ 118 &= 3 \cdot \mathbf{39} + 1, \\ 3 &= 1 \cdot \mathbf{3} + 0, \end{aligned}$$

and thus

$$\frac{832}{357} = \mathbf{2} + \frac{118}{357} = \mathbf{2} + \frac{1}{\frac{357}{118}} = \mathbf{2} + \frac{1}{\mathbf{3} + \frac{3}{118}} = \mathbf{2} + \frac{1}{\mathbf{3} + \frac{1}{\frac{118}{3}}} = \mathbf{2} + \frac{1}{\mathbf{3} + \frac{1}{\mathbf{39} + \frac{1}{\mathbf{3}}}}.$$

In this way, we have expressed the rational number $\frac{832}{357}$ in the form of a continued fraction. In general, we can formulate this procedure as follows.

Theorem IV.1.1. *Let $u = [a/b] \in \mathbb{Q}$. Then there is a unique integer $q = \lfloor u \rfloor$ satisfying $u \leq q < u + 1$. Thus, for $b > 0$,*

$$\frac{a}{b} = q + \frac{r}{b} = q + \frac{1}{\frac{b}{r}} \quad \textit{with } 0 \leq r < b.$$

If

$$\begin{aligned} &a = bq_0 + r_0, && 0 < r_0 < b, \\ &b = r_0 q_1 + r_1, && 0 < r_1 < r_0, \\ &r_0 = r_1 q_2 + r_2, && 0 < r_2 < r_1, \\ &\ldots\ldots\ldots\ldots && \ldots\ldots\ldots\ldots \\ &r_{t-1} = r_t q_{t+1} + r_{t+1}, && 0 < r_{t+1} < r_t, \\ &\ldots\ldots\ldots\ldots && \ldots\ldots\ldots\ldots \\ &r_{N-3} = r_{N-2} q_{N-1} + r_{N-1}, && 0 < r_{N-1} < r_{N-2}, \\ &r_{N-2} = r_{N-1} q_N, && \end{aligned}$$

we can write u in the form of a simple continued fraction

$$u = [q_0; q_1, q_2, \ldots, q_{N-1}, q_N] = q_0 + \cfrac{1}{q_1 + \cfrac{1}{q_2 + \cfrac{1}{\cdots + \cfrac{\cdots}{\cdots + \cfrac{\cdots}{\cdots + \cfrac{1}{q_{N-1} + \cfrac{1}{q_N}}}}}}}, \quad \text{(IV.1.1)}$$

where $q_t > 0$ for $t = 1, 2, \ldots, N$ and $q_N \neq 1$.

Proof. This is just a translation of the Euclidean algorithm. Indeed, the Euclidean algorithm (IV.1.1) can simply be rewritten line by line as

$$\frac{a}{b} = q_0 + \frac{1}{\frac{b}{r_0}}, \quad \frac{b}{r_0} = q_1 + \frac{1}{\frac{r_0}{r_1}}, \quad \frac{r_1}{r_1} = q_2 + \frac{1}{\frac{r_1}{r_2}}, \quad \text{etc.}$$

Note that **all** $q_t > 0$ for $t = 1, 2, \ldots, N$ and that $q_N \neq 1$. Observe that the condition $q_N \neq 1$ guarantees the uniqueness of the expression. For, otherwise, $[q_0; q_1, q_2, \ldots, q_{N-1}, 1] = [q_0; q_1, q_2, \ldots, q_{N-1} + 1]$. □

Definition IV.1.1. *The expression* (IV.1.1)*, where $q_t \in \mathbb{Z}$ for $t = 0, 1, 2, \ldots, N$, $q_t > 0$ for $t = 1, 2, \ldots, N$ and $q_N \neq 1$, is called a* **finite simple continued fraction** *and is denoted by $[q_0; q_1, q_2, \ldots, q_{N-1}, q_N]$. For every $k \in \{0, 1, \ldots, N\}$, we call the partial continued fraction $[q_0; q_1, q_2, \ldots, q_{k-1}, q_k]$ the* **k-th convergent** *of the continued fraction* (IV.1.1).

Note that $[q_0; q_1, q_2, \ldots, q_{N-1}, q_N] = q_0 + \frac{1}{[q_1; q_2, \ldots, q_{N-1}, q_N]}$. This will provide a convenient step in some inductive proofs. The convergents represent "the best approximations" of the value of the continued fractions and will be one of the important tools in approximating real numbers in the next section.

An application of Theorem IV.1.1 yields a basic result concerning finite continued fractions.

Theorem IV.1.2. *Any finite simple continued fraction represents a rational number. Conversely any rational number can be expressed in* **a unique way** *as a finite simple continued fraction.*

From a theoretical point of view this is a very important fact. Every rational number is uniquely expressed in a finite way. This is certainly not the situation when we express rational numbers in the decimal system (or a system with another base). First, there is no uniqueness of expressions in the decimal system: 1 and 0.99999... represent the same number and there are numbers like $\frac{1}{3}$ that do not have a finite representation in the decimal system: $\frac{1}{3} = 0.3333...$

As we have already mentioned, given a continued fraction $[q_0; q_1, q_2, \dots, q_{N-1}, q_N]$, then for every $k \in \{0, 1, \dots, N\}$, the "partial" continued fraction

$$[q_0; q_1, q_2, \dots, q_{k-1}, q_k] = q_0 + \cfrac{1}{q_1 + \cfrac{1}{q_2 + \cfrac{1}{\cdots + \cfrac{\cdots}{\cdots + \cfrac{1}{q_{k-1} + \cfrac{1}{q_k}}}}}}$$

is said to be the k-th **convergent** of the continued fraction (IV.1.1). The first four convergents are

$$[q_0;] = \frac{q_0}{1}, \quad [q_0; q_1] = \frac{q_0q_1 + 1}{q_1}, \quad [q_0; q_1, q_2] = \frac{q_0q_1q_2 + q_0 + q_2}{q_1q_2 + 1},$$

and

$$[q_0; q_1, q_2, q_3] = \frac{q_0q_1q_2q_3 + q_0q_1 + q_0q_3 + q_2q_3 + 1}{q_1q_2q_3 + q_1 + q_3}.$$

The next theorem shows how the convergents of the simple continued fraction $[q_0; q_1, \dots, q_N]$ can be determined.

Theorem IV.1.3. *Given a simple continued fraction $[q_0; q_1, \dots, q_N]$, we define the two sequences of integers $\{a_k \mid k = 0, 1, \dots, N\}$ and $\{b_k \mid k = 0, 1, \dots, N\}$ recursively by*

$$a_0 = q_0, \quad b_0 = 1,$$

$$a_1 = q_0q_1 + 1, \; b_1 = q_1,$$

and

$$a_k = a_{k-1}q_k + a_{k-2}, \; b_k = b_{k-1}q_k + b_{k-2}, \; k = 2, 3, \dots, N.$$

Then, for $k = 0, 1, \dots, N$, we have

$$b_k \in \mathbb{N}, \quad d(a_k, b_k) = 1, \quad [q_0; q_1, \dots, q_k] = \frac{a_k}{b_k}.$$

Proof. By Theorem IV.1.1 we have $q_k \in \mathbb{N}$ for $k \in \{1, 2, \dots, N\}$ so recursively from $b_0 = 1$, $b_1 = q_1$, $b_k = b_{k-1}q_k + b_{k-2}$ $(k = 2, 2, \dots, N)$, we see that $b_k \in \mathbb{N}$ for $k = 0, 1, \dots, N$.

We have $d(a_0, b_0) = d(q_0, 1) = 1$ and $d(a_1, b_1) = d(q_0q_1 + 1, q_1) = 1$. For $k = 2, 3, \dots, N$, we have

$$\begin{aligned}
a_kb_{k-1} - a_{k-1}b_k &= (a_{k-1}q_k + a_{k-2})b_{k-1} - a_{k-1}(b_{k-1}q_k + b_{k-2}) \\
&= -(a_{k-1}b_{k-2} - a_{k-2}b_{k-1}) \\
&= (-1)^2(a_{k-2}b_{k-3} - a_{k-3}b_{k-2}) \\
&= \dots \\
&= (-1)^{k-1}(a_1b_0 - a_0b_1) \\
&= (-1)^{k-1}((q_0q_1 + 1)1 - q_0q_1) \\
&= (-1)^{k-1},
\end{aligned}$$

so that

$$d(a_k, b_k) = 1, \quad k = 0, 1, 2, \ldots, N.$$

We have therefore established that the fraction $\frac{a_k}{b_k}$ is in lowest terms for $k = 0, 1, 2, \ldots, N$. Now we define for $k \in \{1, 2, \ldots, N\}$ and a positive real number x

$$\langle q_0; q_1, q_2, \ldots, q_{k-1}, x\rangle = q_0 + \cfrac{1}{q_1 + \cfrac{1}{q_2 + \cfrac{1}{\cdots + \cfrac{\cdots}{\cdots + \cfrac{1}{q_{k-1} + \cfrac{1}{x}}}}}}.$$

We make the inductive hypothesis that

$$\langle q_0; q_1, q_2, \ldots, q_{k-1}, x\rangle = \frac{xa_{k-1} + a_{k-2}}{xb_{k-1} + b_{k-2}}$$

holds for $k = 2, 3, \ldots, m$ for some $m < N$. This hypothesis holds for $k = 2$ as

$$\langle q_0; q_1, x\rangle = q_0 + \frac{1}{q_1 + \frac{1}{x}} = \frac{x(q_0 q_1 + 1) + q_0}{xq_1 + 1} = \frac{xa_1 + a_0}{xb_1 + b_0}.$$

Now, by the inductive hypothesis for $k = m$ and the recursion relations for a_m and b_m, we obtain

$$\begin{aligned}\langle q_0; q_1, \ldots, q_{m-1}, q_m, x\rangle &= \langle q_0; q_1, \ldots, q_{m-1}, q_m + \tfrac{1}{x}\rangle \\ &= \frac{(q_m + \frac{1}{x})a_{m-1} + a_{m-2}}{(q_m + \frac{1}{x})b_{m-1} + b_{m-2}} \\ &= \frac{x(q_m a_{m-1} + a_{m-2}) + a_{m-1}}{x(q_m b_{m-1} + b_{m-2}) + b_{m-1}} \\ &= \frac{xa_m + a_{m-1}}{xb_m + b_{m-1}}.\end{aligned}$$

This establishes that the inductive hypothesis holds for $k = m + 1$ and thus by the principle of mathematical induction for all $k \in \{2, 3, \ldots, N\}$.

Finally, with $m = k - 1$ and $x = q_k$, we have

$$[q_0; q_1, \ldots, q_{k-1}, q_k] = \langle q_0; q_1, \ldots, q_{k-1}, q_k\rangle = \frac{q_k a_{k-1} + a_{k-2}}{q_k b_{k-1} + b_{k-2}} = \frac{a_k}{b_k},$$

as asserted. □

Example IV.1.2. The recursions in Theorem IV.1.3 make it simple to compute the convergents of a given continued fraction. For instance, if $u = [1; 2, 3, 4, 5]$, then

q_k	a_k	b_k
$q_0 = 1$	1	1
$q_1 = 2$	$2 \cdot 1 + 1 = 3$	2
$q_2 = 3$	$3 \cdot 3 + 1 = 10$	$3 \cdot 2 + 1 = 7$
$q_3 = 4$	$4 \cdot 10 + 3 = 43$	$4 \cdot 7 + 2 = 30$
$q_4 = 5$	$5 \cdot 43 + 10 = 225$	$5 \cdot 30 + 7 = 157$

and the convergents are successively

$$u_0 = 1, \quad u_1 = \frac{3}{2}, \quad u_2 = \frac{10}{7}, \quad u_3 = \frac{43}{30} \quad \text{and} \quad u_4 = u = \frac{225}{157}.$$

The assertions of Theorem IV.1.4 are simple consequences of Theorem IV.1.3 and their proofs are left as exercises for the reader. (The reader will recognize the first equality from the proof of Theorem IV.1.3.) However, they are very important. In a later section, they will enable us to extend the theory of continued fractions to "infinite" continued fractions.

Theorem IV.1.4. *Let $[q_0; q_1, q_2, \ldots, q_{N-1}, q_N]$ be a finite simple continued fraction and $\frac{a_k}{b_k}$, $k = 0, 1, \ldots, N$, its convergents. Then, for every $k \in \{1, 2, \ldots, N\}$, we have*

$$\frac{a_k}{b_k} - \frac{a_{k-1}}{b_{k-1}} = \frac{(-1)^{k-1}}{b_k b_{k-1}}, \quad \textit{equivalently,} \quad a_k b_{k-1} - a_{k-1} b_k = (-1)^{k-1};$$

for $k \in \{2, \ldots, N\}$, we have

$$\frac{a_k}{b_k} - \frac{a_{k-2}}{b_{k-2}} = \frac{(-1)^k q_k}{b_k b_{k-2}}, \quad \textit{equivalently,} \quad a_k b_{k-2} - a_{k-2} b_k = (-1)^k q_k;$$

and thus

$$\frac{a_0}{b_0} < \frac{a_2}{b_2} < \cdots < u = \frac{a_N}{b_N} < \cdots < \frac{a_3}{b_3} < \frac{a_1}{b_1}.$$

Exercise IV.1.3. Expand $\frac{277}{47}$ and $-\frac{57}{32}$ into simple continued fractions.

Exercise IV.1.4. Formulate an expression for $a_k b_{k-3} - a_{k-3} b_k$.

Here is an application of continued fractions to the solution of linear diophantine equations.

Corollary IV.1.1. *If $u = [q_0; q_1, q_2, \ldots, q_{N-1}, q_N] = \frac{a}{b}$, where $b \in \mathbb{N}$ and $d(a, b) = 1$, then $x = b_{N-1}, y = -a_{N-1}$ is a solution of the diophantine equation*

$$ax + by = (-1)^{N-1}.$$

Proof. Observing that $u = \frac{a_N}{b_N}$, the statement follows from the first equality in Theorem IV.1.4. □

Example IV.1.3. To illustrate the merits of Corollary IV.1.1, refer back to Example IV.1.2: We can see that

$$x = 30,\ y = -43 \text{ is a solution of the equation } 225x + 157y = -1,$$

$$x = 7,\ y = -10 \text{ is a solution of the equation } 43x + 30y = 1, \text{ etc.}$$

We shall return to the concept of a continued fraction in the section on real numbers. Extending the concept to "infinite" continued fractions, we shall concentrate on the **approximation** of real numbers by means of rational numbers.

Remark IV.1.1. In Chapter I, we defined Cantor's concept of **cardinality** of a set. In the case of a finite set, this concept formalized the concept of "the number of elements". We also pointed out the fact that a proper subset can have the same cardinality as the entire set and provided illustrations of this phenomena. Here again, we have an opportunity to demonstrate this fact by showing that the cardinality of the set of all positive rational numbers equals the cardinality of the set of natural numbers, that is, that *the set of all rational numbers is countable.* The proof is simple (comp. Figure I.2.1 in Section I.2): Order the rational numbers $\frac{a}{b}$, where a and b are relatively prime positive integers into the sequence satisfying

$$\frac{a}{b} \prec \frac{c}{d} \text{ if and only if } a+b < c+d \text{ or if } a+b = c+d \text{ and } a < c.$$

Such a relation produces a well-ordering.

Exercises IV.1.5 and IV.1.6, as well as Theorem IV.1.5, again show the close relationship between rational numbers and integers. Let a and b be integers with $b > 0$. We recall that Euclidean division asserts that there are integers q and r satisfying

$$a = bq + r, \quad \text{where } r \in \{0, 1, \ldots, b-1\}.$$

We mention that this can be expressed in the following equivalent form

$$a = bq' - r', \quad \text{where } q' \in \mathbb{Z} \text{ and } r' \in \{0, 1, \ldots, b-1\}$$

by taking $q' = q + 1$ and $r' = b - r$ if $r > 0$ and $q' = q$ and $r' = 0$ if $r = 0$.

Exercise IV.1.5. Prove that every rational number $u = \frac{a}{b}$ can be expressed **uniquely** in the form

$$u = u_0 + \frac{a_1}{b_1} + \frac{a_2}{b_2} + \cdots + \frac{a_n}{b_n},$$

where u_0 is an integer, $a_1, a_2, \ldots, a_n$ are positive integers, $b_1, b_2, \ldots, b_n$ are powers of mutually distinct primes, $a_t < b_t$ for $t = 1, 2, \ldots, n$, and b_t is a divisor of b for $t = 1, 2, \ldots, n$.

Exercise IV.1.6. Prove that every rational number u satisfying the inequality $0 < u < 1$ can be written as a sum of unitary fractions, that is, in the form

$$u = \frac{1}{n_1} + \frac{1}{n_2} + \cdots + \frac{1}{n_k}, \tag{IV.1.2}$$

where $n_1, n_2, \ldots, n_k$ are positive integers satisfying $n_1 < n_2 < \cdots < n_k$.

Remark IV.1.2. The statement of the last exercise is sometimes attributed to **James Joseph Sylvester**, although expressing fractions as sums of unitary fractions can be traced back to ancient Egyptian times. The expression (IV.1.2) is not unique. For example, we have

$$\frac{191}{505} = \frac{1}{7} + \frac{1}{8} + \frac{1}{9} = \frac{1}{3} + \frac{1}{22} + \frac{1}{5544},$$

and, for any integer $k \geq 1$,

$$\frac{4}{4k+3} = \frac{1}{k+1} + \frac{1}{(k+1)(4k+3)} = \frac{1}{k+2} + \frac{1}{(k+1)(k+2)} + \frac{1}{(k+1)(4k+3)}.$$

An expression of the form (IV.1.2) can be found using an algorithm that starts as follows. If $u = \frac{a}{b}$, where a and b are integers with $0 < a < b$, write $b = an_1 - r_1$, where n_1 and r_1 are integers satisfying $a > r_1 \geq 0$, and then $bn_1 = r_1 n_2 - r_2$, where n_2 and r_2 are integers satisfying $r_1 > r_2 \geq 0$, etc. Formulate this algorithm in detail and then derive the expressions

$$\frac{16}{17} = \frac{1}{2} + \frac{1}{3} + \frac{1}{10} + \frac{1}{128} + \frac{1}{32640}$$

and

$$\frac{357}{832} = \frac{1}{3} + \frac{1}{11} + \frac{1}{207} + \frac{1}{75779} + \frac{1}{13050962496}.$$

Theorem IV.1.5. *Let the rational numbers u_1 and u_2 satisfy the equation*

$$u_1^2 + u_2^2 = n,$$

where n is a natural number. Then there exists integers a and b such that $a^2 + b^2 = n$.

Proof. The statement follows by induction from the following auxiliary result:
Let $r_1, s_1 \in \mathbb{Z}$ and $d_1 \in \mathbb{N}, d_1 > 1$, satisfy the equation

$$\left(\frac{r_1}{d_1}\right)^2 + \left(\frac{s_1}{d_1}\right)^2 = n.$$

Then there exists $r_2, s_2 \in \mathbb{Z}$ and $d_2 \in \mathbb{N}, 0 < d_2 < d_1$ such that

$$\left(\frac{r_2}{d_2}\right)^2 + \left(\frac{s_2}{d_2}\right)^2 = n.$$

To prove this auxiliary statement, denote by r and s the integers such that

$$\frac{r_1}{d_1} = r + \frac{r_1'}{d_1}, \quad \frac{s_1}{d_1} = s + \frac{s_1'}{d_1},$$

where

$$\left|\frac{r_1'}{d_1}\right| \leq \frac{1}{2} \text{ and } \left|\frac{s_1'}{d_1}\right| \leq \frac{1}{2},$$

and thus

$$\left(\frac{r_1'}{d_1}\right)^2 + \left(\frac{s_1'}{d_1}\right)^2 \leq \frac{1}{2}.$$

Now, the quadratic equation

$$\left(r + x\left(\frac{r_1}{d_1} - r\right)\right)^2 + \left(s + x\left(\frac{s_1}{d_1} - s\right)\right)^2 = n$$

has, in view of our assumption, a root $x_1 = 1$. A little elementary algebra shows that

$$\left(r + x\left(\frac{r_1}{d_1} - r\right)\right)^2 + \left(s + x\left(\frac{s_1}{d_1} - s\right)\right)^2 - n = (x-1)(v\,x + n - r^2 - s^2),$$

where

$$v = \left(\frac{r_1}{d_1} - r\right)^2 + \left(\frac{s_1}{d_1} - s\right)^2 = r^2 + s^2 + n - \frac{2r_1 r + 2s_1 s}{d_1}$$

is a rational number $\frac{d_2}{d_1}$ satisfying the inequality $\frac{d_2}{d_1} \leq \frac{1}{2}$ and therefore the second root of the equation x_2 can be expressed in the form

$$x_2 = \frac{(r^2 + s^2 - n)d_1}{d_2} = \frac{t\,d_1}{d_2}, \quad \text{where } t \in \mathbb{Z},\ d_2 \in \mathbb{N} \text{ and } 0 < d_2 \leq \frac{1}{2}d_1.$$

Hereby

$$r + \frac{td_1}{d_2}\left(\frac{r_1}{d_1} - r\right) = \frac{r_2}{d_2}, r_2 \in \mathbb{Z}$$

and

$$s + \frac{td_1}{d_2}\left(\frac{s_1}{d_1} - s\right) = \frac{s_2}{d_2}, s_2 \in \mathbb{Z}$$

satisfy the equality

$$\left(\frac{r_2}{d_2}\right)^2 + \left(\frac{s_2}{d_2}\right)^2 = n,$$

as required.

Theorem IV.1.5 can now be obtained by a finite application of the auxiliary statement. Indeed, after a finite number of applications, the denominator of the fractions $\frac{r_t}{d_t}$ and $\frac{s_t}{d_t}$ will reach the value $d_t = 1$. □

Remark IV.1.3. Here is a geometrical interpretation of Theorem IV.1.5. If no lattice point (that is, a point whose coordinates are integers) lies on such a circle (that is, such that the square of its radius is an integer) with center at the origin and radius whose square is an integer, then that circle passes through no point whose coordinates are rational numbers. This has a close connection with the Pythagorean triples of Exercise II.8.17(b).

IV.2 Rational numbers in a B-base numeral system

Let B be a natural number greater than 1. As we have seen earlier, it is quite simple to express any natural (or integral) number k in the **base** B, that is, in the form

$$k = a_h B^h + a_{h-1} B^{h-1} + \cdots + a_1 B^1 + a_0 B^0. \tag{IV.2.1}$$

This is achieved by using Euclidean division:

$k = B\, q_0 + a_0,$ where $0 \le a_0 < B$ and $q_0 < k$,

$q_0 = B\, q_1 + a_1,$ where $0 \le a_1 < B$ and $q_1 < q_0 < k$,

......

$q_{h-2} = B\, q_{h-1} + a_{h-1}$, where $0 \le a_{h-1} < B$ and $q_{h-1} < \cdots < q_1 < q_0 < k$,

$q_{h-1} = B\, 0 + a_h$, where $0 \le a_h < B$ and $0 < a_h = q_{h-1} < \cdots < q_1 < q_0 < k$.

Consequently, the number k can be written in the B-base (IV.2.1), that is,

$$k = (a_h\, a_{h-1}\, \ldots\, a_1\, a_0)_B.$$

In this way, we see that

$$(243)_{10} = (183)_{12}.$$

As usual, we shall write simply (in case of the decimal system) $a_h\, a_{h-1}\, \ldots\, a_1\, a_0$ instead of $(a_h\, a_{h-1}\, \ldots\, a_1\, a_0)_{10}$.

We now face the natural question of how to express a rational number $\frac{k}{n}$ in a B-base numeral system. Here we encounter a new phenomenon. Already for the decimal system, we see that we have to extend the form (IV.2.1) to a **periodic** infinite expansion, such as

$$\frac{1}{3} = 0.33333\ldots = 0.\overline{3} \text{ or } \frac{8}{7} = 1.\overline{142857},$$

and deal with the question of uniqueness of such expressions. Indeed, $1 = 0.\overline{9}$. As a matter of fact, the latter expression represents the only cases when the representation is not unique, and the same situation prevails in the case of rational numbers. We will therefore not pay particular attention to this problem of uniqueness of expression.

The following theorem will provide a theoretical basis for the expansion of a fraction $\frac{k}{n}$ in a base B.

Theorem IV.2.1. *Let n and $B \neq 1$ be positive integers. Let n_1 be the largest divisor of n that is relatively prime to B. Then there exists a natural number t such that n is a divisor of the product*

$$B^t\, (B^{\varphi(n_1)} - 1),$$

where φ is Euler's totient function.

Proof. We express n as a product of not necessarily distinct primes

$$n = \prod_{1 \le i \le r} p_i.$$

Then $n = n_1 m$, where

$$m = \prod_{1 \leq i \leq r,\ p_i | B} p_i.$$

Since the greatest common divisor $d(n_1, B) = 1$, n_1 is a divisor of $B^{\varphi(n_1)} - 1$ by Euler's theorem. On the other hand, m is a factor of a suitable power B^t of B, and thus the theorem follows. □

Corollary IV.2.1. *Let n and $B \neq 1$ be positive integers. Then*

$$\frac{k}{n} = (a_h\, a_{h-1}\, \dots\, a_1\, a_0 .\, a_{-1}\, \dots\, a_{-j}\, \underbrace{a_{-(j+1)}\, \dots\, a_{-(j+d)}}\, \underbrace{a_{-(j+1)}\, \dots\, a_{-(j+d)}}\, \dots)_B$$

$$= (a_h\, a_{h-1}\, \dots\, a_1\, a_0 .\, a_{-1}\, \dots\, a_{-j}\, \overline{a_{-(j+1)}\, \dots\, a_{-(j+d)}})_B,$$

where the length d of the period is a divisor of $\varphi(n)$.

Proof. By Theorem IV.2.1, there is an integer c such that

$$c\, n \;=\; B^t (B^{\varphi(n_1)} - 1),$$

and thus

$$\frac{k}{n} = \frac{k\, c}{B^{t+\varphi(n_1)}}\, \frac{B^{\varphi(n_1)}}{B^{\varphi(n_1)} - 1} = \frac{k\, c}{B^{t+\varphi(n_1)}} \sum_{q=0}^{\infty} \left(\frac{1}{B^{\varphi(n_1)}} \right)^q .$$

Therefore it is sufficient to express kc in the base B and remark that the period of the B-expansion of the fraction $\frac{k}{n}$ is a divisor of $\varphi(n_1)$, and hence also a divisor of $\varphi(n)$. □

We give a few illustrations.

Example IV.2.1. Express $\frac{1}{7}$ in base 5. Here we have $d(7,5) = 1$, $\varphi(7) = 6$, and thus

$$5^{\varphi(7)} - 1 = 15624 = 7 \times 2232.$$

Now $2232 = (3\,2\,4\,1\,2)_5$, and so

$$\frac{1}{7} = (0\,.\,0\,3\,2\,4\,1\,2\,0\,3\,2\,4\,1\,2\,\dots)_5 = (0\,.\,\overline{0\,3\,2\,4\,1\,2})_5.$$

Example IV.2.2. If we try to use Corollary IV.2.1 to express $\frac{1}{7}$ in base 60, we encounter the difficulty of calculating with large numbers like $60^6 - 1$. It is therefore easier to use successive approximations

$$60 = 7 \times \mathbf{8} + 4;\ 4 \times 60 = 7 \times \mathbf{34} + 2;\ 2 \times 60 = 7 \times \mathbf{17} + 1;\ 1 \times 60 = 7 \times \mathbf{1} + 4; \dots .$$

The expression $\frac{1}{7} = \frac{8}{60} + \frac{34}{60^2} + \frac{17}{60^3} + \frac{1}{60^4 \times 7}$ indicates that the period of the fraction $\frac{1}{7}$ in the sexagesimal system is 3. We see easily that $60^3 - 1 = 7 \times 30\ 857$ and that $30\ 857 = (\widehat{8}\,\widehat{34}\,\widehat{17})_{60}$; hence

$$\frac{1}{7} = (\widehat{0}\,.\,\underbrace{\widehat{8}\,\widehat{34}\,\widehat{17}}\,\underbrace{\widehat{8}\,\widehat{34}\,\widehat{17}}\ldots)_{60}.$$

In a similar fashion, we find that the period of the expression of $\frac{1}{11}$ in the sexagesimal system is 5 ($\varphi(11) = 10$) and the period of $\frac{1}{13}$ is 4 ($\varphi(13) = 12$). The expansion of $\frac{1}{17}$ in the sexagesimal system has period $\varphi(17) = 16$.

Example IV.2.3. We express $\frac{1}{5\ 120}$ in the duodecimal numeral system, that is, in the 12−base system. Here $5\ 120 = 5 \times 2^{10}$ and $12 = 2^2 \times 3$. Thus 5 120 is a divisor of the number $12^5\ (12^4 - 1)$, that is,

$$5\ 159\ 531\ 520 = 5\ 120 \times 1\ 007\ 721.$$

Thus

$$\frac{1}{5\ 120} = \frac{1\ 007\ 721}{12^9}\left(1 + \frac{1}{12^4} + \frac{1}{12^8} + \frac{1}{12^{12}} + \cdots\right)$$

and since $1\ 007\ 721 = (4\ 0\ 7\ 2\ 0\ 9)_{12}$, we obtain

$$\frac{1}{5\ 120} = (0.\ 0\ 0\ 0\ 4\ 0\ 7\ 2\ 0\ 9)_{12} \times (1.\overline{0\ \ 0\ 0\ 1})_{12} = (0.\ 0\ 0\ 0\ 4\ 0\ \overline{7\ 2\ 4\ 9})_{12}.$$

Of course, we can also get the result by multiplying

$$\frac{1}{5} \times \frac{243}{12^5} = (0.\ \overline{2\ 4\ 9\ 7}\ )_{12} \times (0.\ 0\ 0\ \ 1\ 8\ 3)_{12} = (0.\ 0\ 0\ 0\ 4\ 0\ \overline{7\ 2\ 4\ 9})_{12}.$$

Example IV.2.4. We can express in a similar way, by successive approximations

$$\sqrt{2} \approx (1.\ 0\ 1\ 1\ 0\ 1\ \ 0\ 1\ 0\ 0\ 0\ \ 0\ 0\ 1\ 0\ 0\ \ 1\ 1\ 1\ 1\ 0\ \ 0\ 1\ 1\ 0\ 0\ \ 1\ldots)_2$$

$$e \approx (1\ 0.\ 1\ 0\ 1\ 1\ 0\ \ 1\ 1\ 1\ 1\ 1\ \ 1\ 0\ 0\ 0\ 0\ \ 1\ 0\ 1\ 0\ 1\ \ 0\ 0\ 0\ 1\ 0\ \ 1\ldots)_2$$

and

$$\pi \approx (11.\ 0\ 0\ 1\ 0\ 0\ \ 1\ 0\ 0\ 0\ 0\ \ 1\ 1\ 1\ 1\ 1\ \ 1\ 0\ 1\ 1\ 0\ \ 1\ 0\ 1\ 0\ 1\ \ 0\ldots)_2$$

We mention that an approximate value of π expressed in the sexagesimal numeral system is

$$\pi \approx (\widehat{3}.\,\widehat{8}\,\widehat{29}\,\widehat{44}\,\widehat{0}\,\widehat{46}\ldots)_{60} \quad (\text{that is, } \approx 3.141592652).$$

The sexagesimal numeral system was used by the Babylonians. They used the value $\dfrac{25}{8} = 3.125 = (\widehat{3}.\,\widehat{7}\,\widehat{30})_{60}$ as an approximation of the number π. The Egyptians used the value $\frac{2^8}{3^4} = 3.160493827 = (\widehat{3}.\,\widehat{9}\,\widehat{37}\,\widehat{46}\,\widehat{40})_{60}$ as an approximation.

Exercise IV.2.1. Using the decimal expansion, show that every rational number u can be written in the form $u = \frac{n}{10^m(10^m-1)}$ for suitable integers n and $m \geq 0$. Conclude that, for any prime number $p > 5$, there is an exponent m such that p is a divisor of $10^m - 1$.

Exercise IV.2.2. Formulate, in general, the modified Euclidean algorithm used in Example IV.2.2, and prove that it either terminates or becomes periodic. Then use it to express the rational number $\frac{r}{s} = q + \frac{t}{s}, t < s$, in the B-base numeral system. The first three lines of the algorithm are

$$\begin{aligned} t\,B &= s\,a_{-1} + t_1, \quad 0 \le t_1 < s, \\ t_1\,B &= s\,a_{-2} + t_2, \quad 0 \le t_2 < s, \\ t_2\,B &= s\,a_{-3} + t_3, \quad 0 \le t_3 < s. \end{aligned}$$

Exercise IV.2.3. Express

(i) $\frac{4}{5}$ and $\frac{11}{5}$ in 7−base numeral system;

(ii) $\frac{20}{7}$ in the 5−base numeral system. To simplify the calculations, use the table in Exercise IV.2.4. In particular, note that $\dfrac{r}{s} + \dfrac{ts - r}{s} = t$.

Exercise IV.2.4. Verify the values in the following table, where T represents 10 in base 12.

Base B	1/2	1/3	1/4	1/5	1/6	1/7	1/8	1/9
$B=2$	0.1	$0.\overline{01}$	0.01	$0.\overline{0011}$	$0.0\overline{01}$	$0.0\overline{001}$	0.001	$0.\overline{000111}$
$B=3$	$0.\overline{1}$	0.1	$0.\overline{02}$	$0.\overline{0121}$	$0.0\overline{1}$	$0.\overline{010212}$	$0.\overline{01}$	0.01
$B=4$	0.2	$0.\overline{1}$	0.1	$0.\overline{03}$	$0.0\overline{2}$	$0.\overline{021}$	0.02	$0.0\overline{13}$
$B=5$	$0.\overline{2}$	$0.\overline{13}$	$0.\overline{1}$	0.1	$0.\overline{04}$	$0.\overline{032412}$	$0.\overline{03}$	$0.\overline{023421}$
$B=6$	0.3	0.2	0.13	$0.\overline{1}$	0.1	$0.\overline{05}$	0.043	0.04
$B=7$	$0.\overline{3}$	$0.\overline{2}$	$0.\overline{15}$	$0.\overline{1254}$	$0.\overline{1}$	0.1	$0.\overline{06}$	$0.\overline{053}$
$B=8$	0.4	$0.\overline{25}$	0.2	$0.\overline{1463}$	$0.1\overline{25}$	$0.\overline{1}$	0.1	$0.\overline{07}$
$B=9$	$0.\overline{4}$	0.3	$0.\overline{2}$	$0.\overline{17}$	$0.1\overline{4}$	$0.\overline{125}$	$0.\overline{1}$	0.1
$B=10$	0.5	$0.\overline{3}$	0.25	0.2	$0.1\overline{6}$	$0.\overline{142857}$	0.125	$0.\overline{1}$
$B=11$	$0.\overline{5}$	$0.\overline{37}$	$0.\overline{28}$	$0.\overline{2}$	$0.\overline{19}$	$0.\overline{163}$	$0.\overline{14}$	$0.\overline{124986}$
$B=12$	0.6	0.4	0.3	$0.\overline{2497}$	0.2	$0.\overline{186T35}$	0.16	0.14

Figure IV.2.1 Fractions in B−base numeral systems

IV.3 Fractional triangle à la Pascal

The Yang Hui-Pascal triangle (commonly known as Pascal's triangle and named after **Blaise Pascal (1623 - 1662)**) consists of a triangular arrangement of the **binomial coefficients** and was known to many mathematicians centuries before Pascal. Here are some of those historically documented: **Abu Bakr al-Karaji (953 - about 1029), Omar Khayaam (1048 - 1131), and Yang Hui (1238 - 1298)**. Pascal probably learned about the triangle from **Marin Mersenne (1588 - 1648)**.

We recall the definition.

Definition IV.3.1. Let n be a positive integer. Pascal's triangle is the array of positive integers

$$a_{n,k} = \frac{k}{n}\binom{n}{k} = \binom{n-1}{k-1} \quad \text{for } k \in \{1, \ldots, n\},$$

arranged in a triangle, where n indicates the row; thus

$$\begin{array}{ccccccccccccccccccc}
 & & & & & & & & & a_{1,1}{=}\mathbf{1} & & & & & & & & & \\
 & & & & & & & & a_{2,1}{=}\mathbf{1} & & \mathbf{1} & & & & & & & & \\
 & & & & & & & a_{3,1}{=}\mathbf{1} & & a_{3,2}{=}\mathbf{2} & & \mathbf{1} & & & & & & & \\
 & & & & & & a_{4,1}{=}\mathbf{1} & & \mathbf{3} & & \mathbf{3} & & \mathbf{1} & & & & & & \\
 & & & & & a_{5,1}{=}\mathbf{1} & & \mathbf{4} & & a_{5,3}{=}\mathbf{6} & & \mathbf{4} & & \mathbf{1} & & & & & \\
 & & & & a_{6,1}{=}\mathbf{1} & & \mathbf{5} & & \mathbf{10} & & \mathbf{10} & & \mathbf{5} & & \mathbf{1} & & & & \\
 & & & a_{7,1}{=}\mathbf{1} & & \mathbf{6} & & \mathbf{15} & & a_{7,4}{=}\mathbf{20} & & \mathbf{15} & & \mathbf{6} & & \mathbf{1} & & & \\
 & & a_{8,1}{=}\mathbf{1} & & \mathbf{7} & & \mathbf{21} & & \mathbf{35} & & \mathbf{35} & & \mathbf{21} & & \mathbf{7} & & \mathbf{1} & & \\
 & a_{9,1}{=}\mathbf{1} & & \mathbf{8} & & \mathbf{28} & & \mathbf{56} & & a_{9,5}{=}\mathbf{70} & & \mathbf{56} & & \mathbf{28} & & \mathbf{8} & & \mathbf{1} & \\
\cdots & & \cdots & & \cdots & & \cdots & & \cdots & & \cdots & & \cdots & & \cdots & & \cdots & & \cdots
\end{array}$$

Figure IV.3.1 Pascal's triangle

With reference to Chapter II, we can formulate the basic properties of this triangle as in Theorem IV.3.1. The reader is invited to recheck these.

Theorem IV.3.1. *Let n be a positive integer. The following equalities hold for $k \in \{1, 2, \ldots, n\}$:*

$$a_{n,1} = a_{n,n} = 1 \ \text{ and } \ a_{n,k} = a_{n,n-k+1},$$

$$a_{n+1,k+1} = a_{n,k} + a_{n,k+1},$$

$$\sum_{k=1}^{n} a_{n,k} = 2^{n-1},$$

$$\sum_{k=1}^{n} a_{n,k}^2 = a_{2n-1,n},$$

$$\sum_{t=0}^{s} a_{r+t,t+1} = a_{r+s+1,s+1} \text{ for integers } r \geq 1 \text{ and } s \geq 0.$$

The sequence $\mathbf{a}_k = \{a_{n,k} \mid n = k, k+1, k+2, \ldots\}$ *is an arithmetic progression of* $(k-1)$*-th order.*

Exercise IV.3.1. Define $b_{n,k} = (-1)^{k+1} a_{n,k} = (-1)^{k+1}\binom{n}{k}$ to obtain a "marked" Pascal's triangle. Prove that for $k, n \in \mathbb{N}$:

$$b_{n,1} = 1,\ b_{n,k} = (-1)^{n+1} b_{n,n-k+1} \text{ and } b_{n,n} = (-1)^{n+1},$$

$$b_{n+1,k+1} = b_{n,k+1} - b_{n,k},$$

$$\sum_{k=1}^{n} b_{n,k} = 0,$$

$$\sum_{k=1}^{n} b_{n,k}^2 = (-1)^{n+1} b_{2n-1,n},$$

$$\sum_{t=0}^{s} b_{r+t,r} = -b_{r+s+1,r+1} \text{ for } r \in \mathbb{N} \text{ and } s \in \mathbb{N}_0.$$

Exercise IV.3.2. Let $s \in \mathbb{N}_0$. Prove that

$$\sum_{t=0}^{3s+1} b_{2+3s+t,2+3s-t} = \sum_{n+k=4+6s} b_{n,k} = 0,$$

and

$$\sum_{t=0}^{3s+2} b_{4+3s+t,3+3s-t} = \sum_{n+k=7+6s} b_{n,k} = 0.$$

Determine the sums

$$\sum_{t=0}^{3s+2} b_{3+3s+t,2+3s-t} = \sum_{n+k=5+6s} b_{n,k},$$

$$\sum_{t=0}^{3s+2} b_{3+3s+t,3+3s-t} = \sum_{n+k=6+6s} b_{n,k},$$

$$\sum_{t=0}^{3s+2} b_{4+3s+t,4+3s-t} = \sum_{n+k=8+6s} b_{n,k},$$

and

$$\sum_{t=0}^{3s+2} b_{5+3s+t,4+3s-t} = \sum_{n+k=9+6s} b_{n,k}.$$

Exercise IV.3.3. Replace all the quantities $b_{x,y}$ by the quantities $a_{x,y}$ in the sums of Exercise IV.3.2, that is, these sums become sums of binomial coefficients. Show that they can be evaluated in terms of Fibonacci numbers.

In analogy to Pascal's triangle one can define a **fractional triangle**, sometimes referred to as Leibniz's triangle (**Gottfried Wilhelm Leibniz (1646 - 1716)**).

Definition IV.3.2. Let $n \in \mathbb{N}$. Leibniz's triangle is the array of positive rational numbers

$$f_{n,k} = \left\{ {n \atop k} \right\} = \frac{1}{k\binom{n}{k}} = \frac{1}{n\binom{n-1}{k-1}} \quad \text{for } k \in \{1, 2, \ldots, n\},$$

arranged in a triangle, where n indicates the row; thus

$$\begin{array}{c}
f_{1,1}=\mathbf{1} \\
f_{2,1}=\frac{1}{2} \quad \frac{1}{2} \\
f_{3,1}=\frac{1}{3} \quad f_{3,2}=\frac{1}{6} \quad \frac{1}{3} \\
f_{4,1}=\frac{1}{4} \quad \frac{1}{12} \quad \frac{1}{12} \quad \frac{1}{4} \\
f_{5,1}=\frac{1}{5} \quad \frac{1}{20} \quad f_{5,3}=\frac{1}{30} \quad \frac{1}{20} \quad \frac{1}{5} \\
f_{6,1}=\frac{1}{6} \quad \frac{1}{30} \quad \frac{1}{60} \quad \frac{1}{60} \quad \frac{1}{30} \quad \frac{1}{6} \\
f_{7,1}=\frac{1}{7} \quad \frac{1}{42} \quad \frac{1}{105} \quad f_{7,4}=\frac{1}{140} \quad \frac{1}{105} \quad \frac{1}{42} \quad \frac{1}{7} \\
f_{8,1}=\frac{1}{8} \quad \frac{1}{56} \quad \frac{1}{168} \quad \frac{1}{280} \quad \frac{1}{280} \quad \frac{1}{168} \quad \frac{1}{56} \quad \frac{1}{8} \\
f_{9,1}=\frac{1}{9} \quad \frac{1}{72} \quad \frac{1}{252} \quad \frac{1}{504} \quad f_{9,5}=\frac{1}{630} \quad \frac{1}{504} \quad \frac{1}{252} \quad \frac{1}{72} \quad \frac{1}{9} \\
\cdots \quad \cdots \quad \cdots \quad \cdots \quad \cdots \quad \cdots \quad \cdots \quad \cdots \quad \cdots \quad \cdots
\end{array}$$

Figure IV.3.2 Leibniz's triangle

Again, in analogy to Theorem IV.3.1, we formulate a series of statements the proofs of which are left as an exercise.

Theorem IV.3.2. *Let $n \in \mathbb{N}$. Then for $k \in \{1, 2, \ldots, n\}$ we have*

$$\left\{ {n \atop 1} \right\} = \left\{ {n \atop n} \right\} = \frac{1}{n} \quad \textit{and} \quad \left\{ {n \atop k} \right\} = \left\{ {n \atop n-k+1} \right\};$$

$$\left\{ {n \atop k+1} \right\} = \left\{ {n-1 \atop k} \right\} - \left\{ {n \atop k} \right\}.$$

Also

$$\sum_{k=2}^{n} \left\{ {k \atop 2} \right\} = \frac{n-1}{n}.$$

The sequence

$$S(n) = \sum_{k=1}^{n} \left\{ {n \atop k} \right\} = \frac{1}{n} \sum_{k=0}^{n-1} \frac{1}{\binom{n-1}{k}}$$

is a recursive sequence defined by

$$S(1) = 1 \ \textit{and} \ S(n) = \frac{1}{2} S(n-1) + \frac{1}{n} \ \textit{for} \ n = 2, 3, 4, \ldots.$$

Exercise IV.3.4. Let $n \in \mathbb{N}$. Show that, for $n > 4$,

(i) $\frac{2}{n-1} < S(n) < \frac{2}{n-2}$,

(ii) for $n = 2r + 1$ (odd), $S(n+1) = \sum_{k=0}^{r} \frac{1}{(2k+1)\binom{n}{2k+1}}$,

(iii) for $n = 2r$ (even), $S(n+1) = \sum_{k=0}^{r-1} \frac{1}{(2k+1)\binom{n}{2k+1}} + \frac{1}{n+1}$.

A consequence of the statements in Exercise IV.3.4 is the following theorem.

Theorem IV.3.3. *The sequence* $\{S(n) \mid n = 1, 2, 3, \ldots\}$ *is decreasing and*

$$\lim_{n \to \infty} S(n) = 0.$$

Remark IV.3.1. In fact, one can show that the sequence $\{nS(n) \mid n = 5, 6, 7, \ldots\}$ is also decreasing and satisfies $nS(n) > 2$. Hence, $\lim_{n \to \infty} nS(n)$ exists. If we denote this limit by L, then from the relation

$$nS(n) = \frac{1}{2}(n-1)S(n-1) + \frac{1}{2}S(n-1) + 1,$$

we obtain $L = \frac{1}{2}L + 1$, and thus

$$L = \lim_{n \to \infty} nS(n) = 2.$$

IV.4 Real numbers

The first meaningful acquaintance with real numbers occurs usually in a course in analysis. Their comforting representation as the points of a "real line" provides a feeling of familiarity and camouflages their complicated mathematical nature. It may be useful to point out that the history of the development of real numbers, as indeed of the negative numbers (and the complex numbers that we shall study in the next chapter), is rather involved. It was one of the greatest accomplishments of mathematics about 150 years ago to establish solid foundations for the real numbers, be it in terms of **Dedekind's cuts** or **Weierstrass' fundamental sequences (Karl Weierstrass (1815 - 1897))**. Indeed, the construction of the field $(\mathbb{R}, +, .)$ of real numbers requires a "limit", an analytical, non-algebraic process.

The field $(\mathbb{R}, +, .)$ extends the field of rational numbers $(\mathbb{Q}, +, .)$ in such a way that it also extends its linear order in a way that is **compatible** with the algebraic operations of addition and multiplication. Thus, for $a, b, c \in \mathbb{R}$, $a < b$ implies $a + c < b + c$ and $0 < a, 0 < b$ implies $0 < ab$. The linear order of the real numbers is, of course, emphasized by their representation as an "infinite, continuous" line, from $-\infty$ to $+\infty$ with zero at the middle.

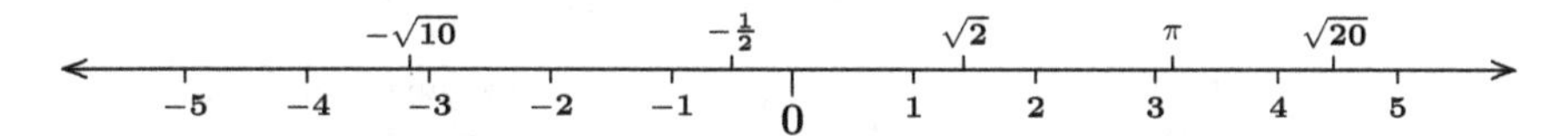

Figure IV.4.1 Real (continuous) line

The fundamental property of the field of real numbers, which distinguishes it profoundly from the field of rational numbers, is the fact that it is **complete**. This "analytical" property can be simply described in the following way (**existence of supremum**):

Every non-empty set of real numbers that is bounded from above has (in $\mathbb{R}$) a least upper bound (supremum).

As we have already mentioned, the field of rational numbers is not complete. For example the subset $S = \{x \in \mathbb{Q} \mid 0 \leq x^2 < 2\}$ of the set $\mathbb{Q}$ is bounded from above, but does not possess a least upper bound in $\mathbb{Q}$. On the other hand, the same set S has a supremum, namely the real number that we denote by $\sqrt{2}$ in $\mathbb{R}$. It is easy to show (see below) that the real number $\sqrt{2}$ is not rational. Any real number that is not rational we call **irrational.** We shall see that there is an abundance of irrational numbers. In fact the set of all irrational numbers is not countable! For instance, all positive real numbers r such that $r^n = a$ for some natural number a (which we denote by $r = \sqrt[n]{a}$) are irrational unless $a = b^n$ for a natural number b. This is a consequence of the following theorem, sometimes called the rational root theorem.

Theorem IV.4.1. *Suppose that the rational number u satisfies the equality*

$$u^n + a_{n-1}u^{n-1} + \cdots + a_2u^2 + a_1u + a_0 = 0,$$

where $n \in \mathbb{N}$ and $a_t \in \mathbb{Z}$ $(t = 0, 1, \ldots, n-1)$. Then $u \in \mathbb{Z}$.

Proof. We give an indirect proof. Suppose that u is not an integer. Then, as $u \in \mathbb{Q}$, we have $u = \frac{p}{q}$, where $p \in \mathbb{Z}$, $q \in \mathbb{N}$, $q > 1$ and $d(p, q) = 1$. From the given equality, we deduce

$$\frac{p^n}{q^n} + a_{n-1}\frac{p^{n-1}}{q^{n-1}} + \cdots + a_2\frac{p^2}{q^2} + a_1\frac{p}{q} + a_0 = 0.$$

Multiplying this equality by q^n, we obtain

$$p^n + a_{n-1}p^{n-1}q + \cdots + a_2p^2q^{n-2} + a_1pq^{n-1} + a_0q^n = 0.$$

Thus q is a divisor of p^n. As $q > 1$ there is a prime r dividing q. Thus r divides p^n and so by Remark III.5.2 r divides p. Hence p and q have r as a common factor contradicting that p and q are coprime. This contradicts our assumption and so u is an integer. $\square$

As a consequence, the numbers $\sqrt{2}$, $\sqrt{3}$, $\sqrt{5}, \ldots, \sqrt[3]{2}$, $\sqrt[3]{3}$, $\sqrt[3]{5}, \ldots, \sqrt[4]{2}$, $\sqrt[5]{2}, \ldots$, $\sqrt[10]{2012}$, $\sqrt[77]{1932}$, etc. are all irrational numbers. The irrational numbers that satisfy polynomial equations as in Theorem IV.4.1 (those that are roots of integral polynomials) are **algebraic numbers**. By definition, any number x that satisfies an algebraic equation, that is, an equation of the form

$$x^n + a_{n-1}x^{n-1} + \cdots + a_2x^2 + a_1x + a_0 = 0$$

with integral coefficients $a_t \in \mathbb{Z}$, $t \in \{0, 1, \ldots, n-1\}$ is called algebraic. Thus all rational numbers are algebraic. We will see in Chapter V that any such equation is satisfied by at most n different numbers (called roots). Use this fact in the following exercise.

Exercise IV.4.1. Show that the set of all algebraic equations with integral coefficients is countable. Hence, deduce that the set of all real algebraic numbers, and in particular all irrational algebraic numbers, is countable.

The irrational real numbers that are not algebraic, for example the number π, are called **transcendental numbers.** We will see in the next section that the set of all transcendental numbers (and therefore the set of all real numbers) is not countable.

IV.5 Continued fractions revisited

The earlier interpretation of the Euclidean division (and the resulting algorithm) for rational numbers allows us to formulate a similar procedure for real numbers.

Theorem IV.5.1. *Every real number a has a unique expression*

$$a = \lfloor a \rfloor + r, \; 0 \le r < 1,$$

where the symbol $\lfloor a \rfloor$ denotes the integer c satisfying $a \le c < a + 1$. Thus

$$a = q_0 + \frac{1}{s_1}, \text{ where } q_0 = \lfloor a \rfloor \text{ and } s_1 = \frac{1}{r} > 1.$$

Continuing recurrently, we obtain

$$a = [q_0; q_1, q_2, \ldots, q_k, s_{k+1}] = q_0 + \cfrac{1}{q_1 + \cfrac{1}{q_2 + \cfrac{1}{\cdots + \cfrac{\cdots}{\cdots + \cfrac{\cdots}{\cdots + \cfrac{1}{q_k + \cfrac{1}{s_{k+1}}}}}}}},$$

where

$$1 < s_t = \lfloor s_t \rfloor + r_t = q_t + \frac{1}{s_{t+1}}, \; s_{t+1} = \frac{1}{r_t} > 1 \textit{ for } t = 1, 2, \ldots, k.$$

We know from the previous section that the above process terminates in a finite number of steps (that is, $s_{k+1} = 1$ for some k) if and only if the original number a is rational. If the number a is irrational, the process continues unrestricted and the respective convergents $\frac{a_k}{b_k} = [q_0; q_1, q_2, \ldots, q_k]$ satisfy the inequalities

$$\frac{a_0}{b_0} < \frac{a_2}{b_2} < \cdots < \frac{a_{2n}}{b_{2n}} < \cdots < \frac{a_{2n+1}}{b_{2n+1}} < \cdots < \frac{a_3}{b_3} < \frac{a_1}{b_1} \quad \text{for } n = 0, 1, 2, \ldots.$$

Hence given an **infinite** simple continued fraction $[q_0; q_1, q_2, \ldots, q_n, \ldots]$, we may use the equalities stated in Theorem IV.1.4. to deduce

$$\lim_{n\to\infty} \left(\frac{a_{2n+1}}{b_{2n+1}} - \frac{a_{2n}}{b_{2n}} \right) = \lim_{n\to\infty} \frac{(-1)^{2n}}{b_{2n} b_{2n+1}} = 0,$$

and thus

$$\lim_{n\to\infty} \frac{a_{2n+1}}{b_{2n+1}} = \lim_{n\to\infty} \frac{a_{2n}}{b_{2n}} = a,$$

where a is an irrational number. Now suppose $a = [q_0; q_1, q_2, \ldots] = [q_0'; q_1', q_2', \ldots]$. Then $\lfloor a \rfloor = q_0 = q_0'$. Further

$$a = q_0 + \frac{1}{[q_1; q_2, q_3, \ldots]} = q_0' + \frac{1}{[q_1'; q_2', q_3', \ldots]}$$

so $[q_1; q_2, q_3, \ldots] = [q_1'; q_2', q_3', \ldots]$. Repeating this argument gives $q_1 = q_1'$, $q_2 = q_2'$, ... and so by mathematical induction $q_n = q_n'$ for all $n \in \mathbb{N}_0$. Thus distinct infinite simple continued fractions converge to different irrational numbers. Hence, we can formulate the following theorem

Theorem IV.5.2. *Every infinite simple continued fraction converges to an irrational number. On the other hand, every irrational number can be expanded into an infinite simple continued fraction, and this expansion is unique.*

Remark IV.5.1. The previous theorem provides a simple proof of the fact that the set $\mathbb{R}$ of all real numbers is not countable. The cardinality of $\mathbb{R}$ is said to be **continuum.** Here is an (indirect) proof.

Denote by $\mathbb{R}_I$ the subset of the field $\mathbb{R}$ of real numbers consisting of all irrational numbers. Assume that the set $\mathbb{R}_I$ is countable and present **all** its members as a sequence $\mathbf{r} = (r_0, r_1, \ldots, r_k, \ldots)$ of their continued fraction representations:

$$\begin{aligned} r_0 &= [q_{00}, q_{01}, \ldots, q_{0k}, \ldots] \\ r_1 &= [q_{10}, q_{11}, \ldots, q_{1k}, \ldots] \\ &\cdots\cdots\cdots\cdots\cdots\cdots \\ r_k &= [q_{k0}, q_{k1}, \ldots, q_{kk}, \ldots] \\ &\cdots\cdots\cdots\cdots\cdots\cdots \end{aligned}$$

The (diagonal) number $r = [q_{00} + 1, q_{11} + 1, \ldots, q_{kk} + 1, \ldots]$ belongs to $\mathbb{R}_I$. However r is not a member of the sequence $\mathbf{r}$. This contradicts our assumption. Hence the set $\mathbb{R}_I$, and therefore also the set of all real numbers $\mathbb{R}$ is **not countable**.

Example IV.5.1. The golden ratio is the irrational number $\frac{1+\sqrt{5}}{2} = 1.6180339887\ldots$. It is known to be equal to

$$\lim_{n\to\infty} \frac{F_{n+1}}{F_n},$$

where F_n is the n-th Fibonacci number. The continued fraction expansion of the golden ratio is

$$\frac{1+\sqrt{5}}{2} = [1; 1, 1, \ldots, 1, \ldots]$$

as

$$\frac{1+\sqrt{5}}{2} = 1 + \frac{1}{\frac{2}{\sqrt{5}-1}} = 1 + \frac{1}{1+\frac{1+\sqrt{5}}{2}}.$$

Similarly, since

$$\sqrt{7} = 2 + (\sqrt{7}-2); \quad \frac{1}{\sqrt{7}-2} = 1 + \frac{\sqrt{7}-1}{3}; \quad \frac{3}{\sqrt{7}-1} = 1 + \frac{\sqrt{7}-1}{2};$$

and

$$\frac{2}{\sqrt{7}-1} = 1 + \frac{\sqrt{7}-2}{3} \quad \text{and} \quad \frac{3}{\sqrt{7}-2} = 4 + (\sqrt{7}-2),$$

we obtain $\sqrt{7} = [2; 1, 1, 1, 4, 1, 1, 1, 4, 1, 1, 1, 4, \ldots] = [2; \overline{1, 1, 1, 4}]$.

Example IV.5.2. In order to find simple effective approximations to the number π, we start with the approximation $\pi = 3.14159$. We have the following successive divisions:

$$\frac{314159}{100000} = \mathbf{3} + \frac{1}{\frac{100000}{14159}}, \quad \frac{100000}{14159} = \mathbf{7} + \frac{1}{\frac{14159}{887}}, \quad \frac{14159}{887} = \mathbf{15} + \frac{1}{\frac{887}{854}},$$

$$\frac{887}{854} = \mathbf{1} + \frac{1}{\frac{854}{33}}, \quad \frac{854}{33} = \mathbf{25} + \frac{1}{\frac{33}{29}}, \quad \frac{33}{29} = \mathbf{1} + \frac{1}{\frac{29}{4}}, \quad \frac{29}{4} = \mathbf{7} + \frac{1}{\frac{4}{1}}, \quad \frac{4}{1} = \mathbf{4}.$$

They result in the following continued fraction representation

$$3.14159 = [3; 7, 15, 1, 25, 1, 7, 4].$$

Thus the respective convergents are

$$\frac{\mathbf{3}}{\mathbf{1}}, \frac{\mathbf{22}}{\mathbf{7}}, \frac{\mathbf{333}}{\mathbf{106}}, \frac{\mathbf{355}}{\mathbf{113}}, \frac{9208}{2931}, \frac{9563}{3044}, \frac{76149}{24239}, \frac{314159}{100000}.$$

Already the fourth convergent provides a close approximation: $\frac{355}{113} = 3.14159292$.

Exercise IV.5.1. Determine the continued fraction representations of $\sqrt{13}$ and $\sqrt{14}$.

Exercise IV.5.2. Find the best successive approximations to Euler's number $e = 2.71828\ldots$ using the expansion in the form of a continued fraction $[q_0; q_1, q_2, \ldots]$, so that $q_0 = 2$. Determine the integers q_1, q_2, q_3, q_4 and q_5 and the respective convergents.

Exercise IV.5.3. Let $\{\frac{a_k}{b_k} \mid k = 0, 1, 2, \dots\}$ be the sequence of convergents of the irrational number $\lim_{k\to\infty} \frac{a_k}{b_k} = a = [q_0; q_1, q_2, \dots, q_{k-1}, q_k, \dots]$. Show that, for every $k \in \mathbb{N}$,

$$\frac{a_k}{a_{k-1}} = [q_k; q_{k-1}, \dots, q_1, q_0] \text{ and } \frac{b_k}{b_{k-1}} = [q_k; q_{k-1}, \dots, q_1].$$

The sequence of convergents $\frac{a_k}{b_k}$ of the continued fraction representation of a real number $a = [q_0; q_1, q_2, \dots]$ gives valuable rational approximations of a. We have seen this already in some examples. In fact, this is one of the very important applications of the theory of continued fractions. Here, we present two results, without proofs, that underline the importance of this application by pointing out that the approximations are the best possible ones.

Theorem IV.5.3. *Let $a = [q_0; q_1, q_2, \dots, q_k, \dots]$. Let $\{\frac{a_k}{b_k} \mid k = 0, 1, 2, \dots\}$ be the sequence of convergents of a so that $\lim_{k\to\infty} \frac{a_k}{b_k} = a$. Then*

$$\left| a - \frac{a_k}{b_k} \right| < \left| a - \frac{a_{k-1}}{b_{k-1}} \right| \text{ for all } k \in \mathbb{N}.$$

In fact, the stronger inequality $| ab_k - a_k | < | ab_{k-1} - a_{k-1} |$ holds.

Theorem IV.5.4. *Let $a = [q_0; q_1, q_2, \dots, q_k, \dots]$ and let $\{\frac{a_k}{b_k} \mid k = 0, 1, 2, \dots\}$ be its sequence of convergents. If*

$$\left| a - \frac{r}{s} \right| < \left| a - \frac{a_k}{b_k} \right| \text{ for some integers } r, s \text{ with } s \in \mathbb{N},$$

then $s > b_k$.

The last theorem states that for all rational numbers $\frac{r}{s}$, where r and s are integers with $0 < s \leq b_k$, the difference $| a - \frac{r}{s} |$ cannot be smaller than $| a - \frac{a_k}{b_k} |$.

Example IV.5.3. If we want to realize a transmission 277 to 1000 by a gear train containing a reasonable number of cogs, we express the number $\frac{277}{1000}$ in the form of a continued fraction

$$\frac{277}{1000} = [0; 3, 1, 1, 1, 1, 3, 2, 1, 4],$$

and thus the best approximations are given by the convergents

$$0, \frac{1}{3}, \frac{1}{4}, \frac{2}{7}, \frac{3}{11}, \frac{5}{18}, \frac{18}{65}, \frac{41}{148}, \frac{59}{213}, \frac{277}{1000}.$$

The number of cogs of a required gear train can thus be selected using these fractions. Figure IV.5.1 illustrates the gear train approximating this transmission by the realistic transmission 5 to 18. We note that $\frac{5}{18} = 0.277777\dots$ can appear among the convergents of other fractions. For instance, the number $\frac{37}{133} = [0; 3, 1, 1, 2, 7]$ leads to the convergents

$$0, \frac{1}{3}, \frac{1}{4}, \frac{2}{7}, \frac{5}{18}, \frac{37}{133}.$$

We remark that **Christiaan Huygens (1629 - 1695)** used such approximations by continued fractions when constructing astronomical clocks.

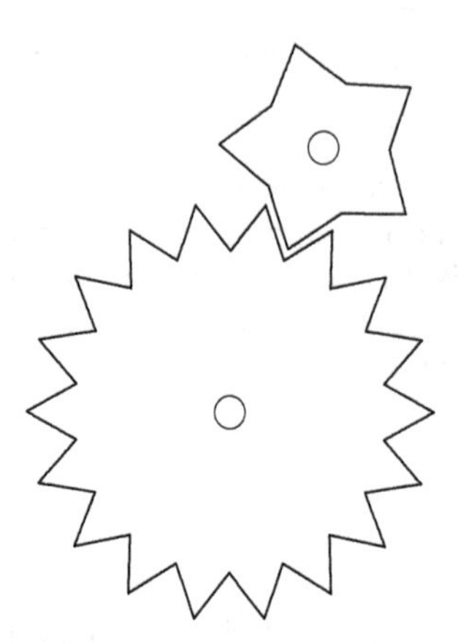

Figure IV.5.1 Gear train approximating transmission 277 to 1000

There is an important subset of infinite continued fractions that is in a bijection with the set of all irrational real numbers that satisfy quadratic equations with integral coefficients. The continued fractions in question are the **periodic continued fractions**.

Definition IV.5.1. An infinite simple continued fraction

$$[\underbrace{q_0; q_1, \dots, q_h,} \underbrace{q_{h+1}, \dots, q_{h+d},} \underbrace{q_{h+d+1} \dots, q_{h+2d},} \dots]$$

having a **non-repeating block** $\{q_0, q_1, \dots, q_h\}$ of length $h \geq 0$ and a **repeating block** $\{q_{h+1}, q_{h+2}, \dots, q_{h+d}\}$ of length $d \geq 1$, that is, satisfying

$$q_{h+r+sd} \quad = \quad q_{h+r} \text{ for all } r \in \{1, 2, \dots, d\} \text{ and all } s \in \mathbb{N},$$

is said to be a periodic continued fraction. This fact will be denoted by

$$[q_0; q_1, q_2, \dots, q_h, \overline{q_{h+1}, q_{h+2}, \dots, q_{h+d}}].$$

As mentioned earlier, there is a surprising relation between periodic continued fractions and real irrational numbers of the form

$$\frac{a + \sqrt{b}}{c}, \text{ where } a, b \text{ and } c \text{ are integers, } b > 1 \text{ and square-free,}$$

the so-called **quadratic irrationals** or **quadratic surds**. The relation was established by **Joseph-Louis Lagrange** and we are going to state it here without proof.

Theorem IV.5.5. *The simple continued fraction* $a = [q_0; q_1, q_2, \dots]$ *is periodic if and only if* a *is a quadratic irrational.*

Example IV.5.4. For any natural number a, we have

$$\sqrt{a^2+1} = [a; \overline{2a}\,].$$

Thus for $a = 1$ we have the following convergents of $\sqrt{2}$:

q_k	a_k	b_k
$q_0 = 1$	1	1
$q_1 = 2$	3	2
$q_2 = 2$	$2 \cdot 3 + 1 = 7$	$2 \cdot 2 + 1 = 5$
$q_3 = 2$	$2 \cdot 7 + 3 = 17$	$2 \cdot 5 + 2 = 12$
$q_4 = 2$	$2 \cdot 17 + 7 = 41$	$2 \cdot 12 + 5 = 29$
etc.		

Here are the first 15 convergents of the number $\sqrt{2}$:

$$\frac{1}{1}, \frac{3}{2}, \frac{7}{5}, \frac{17}{12}, \frac{41}{29}, \frac{99}{70}, \frac{239}{169}, \frac{577}{408}, \frac{1393}{985}, \frac{3363}{2378}, \frac{8119}{5741}, \frac{19601}{13860}, \frac{47321}{33461}, \frac{114243}{80782}, \frac{275807}{195025}, \frac{665857}{470832}, \ldots$$

Compare

$$\frac{665857}{470832} = \mathbf{1.}\underbrace{\mathbf{41421356237}}\mathbf{46}\ldots \text{ to } \sqrt{2} = \mathbf{1.}\underbrace{\mathbf{41421356237}}\mathbf{30}\ldots$$

Exercise IV.5.4. (i) Show that, for every natural number a,

$$\sqrt{a^2+2} = [a; \overline{a, 2a}\,] \text{ and } \frac{1}{2}\,(a + \sqrt{a^2+4}\,) = [a; \overline{a}\,].$$

(ii) Write the continued fraction expansion of $\sqrt{31} = [q_0; \overline{q_1, \ldots, q_{d-1}, q_d}\,]$. Notice that the $q_d = 2q_0$ and that $\{q_1, \ldots, q_{d-1}\}$ is a palindrome. In fact, this pattern is typical for all irrational numbers $\sqrt{c}$.

Exercise IV.5.5. Let $a \in \mathbb{N}$ satisfy $a > 1$. Show that the k-th convergent in the continued fraction representation of $\frac{1}{a}$ is the reciprocal of the $(k-1)$-th convergent of the continued fraction representation of a.

IV.6 ⋆ The Pell-Brouncker equation

Any study of continued fractions would be incomplete without a reference, no matter how brief, to Pell's equation (**John Pell (1611 - 1685)**), or more correctly the Pell-Brouncker (**William Brouncker (1620 - 1684)**) equation

$$x^2 = ny^2 + 1, \text{ where the natural number } n \text{ is not a perfect square.} \tag{IV.6.1}$$

A close connection between solutions of the Pell-Brouncker equation and the behavior of continued fractions was established by **Joseph Louis Lagrange** in the late 18-th century. However, this equation was studied by Indian mathematicians as early as 1000 years before Pell's and Brouncker's time. In fact the first contributions towards the solution of these equations should be attributed to **Brahmagupta (598 - 668)** and **Bhaskara II (1114 - 1185)**. Brahmaguptas' **Brahma Sphuta Siddhanta (628)**, which contained his **chakravala method** to solve the Pell-Brouncker equation, was translated into Arabic in 773 and was subsequently translated into Latin in 1126. In this way the equation reached Europe and excited broader interest mainly due to the fact that **Pierre de Fermat** became interested in the question in 1657. The solution

$$x = 32\ 188\ 120\ 829\ 134\ 849,\ y = 1\ 819\ 380\ 158\ 564\ 160$$

of the equation

$$x^2 = 313y^2 + 1$$

by Brouncker, as well as the solution

$$y = 12\ 055\ 735\ 790\ 331\ 359\ 447\ 442\ 538\ 767\ \approx 1.2 \times 10^{29}$$

of the equation

$$x^2 = 991y^2 + 1$$

by **Waclaw Sierpinski (1882 - 1969)** are well-known. They are the **least (fundamental) solutions** of the respective equations; the other solutions are obtained by a method described already by Brahmagupta:

If $(x, y) = (a, b)$ *and* $(x, y) = (c, d)$ *are solutions of the equation* (IV.6.1), *then*

$$(ac + nbd, ad + bc) \text{ and } (ac - nbd, ad - bc)$$

are also solutions of this equation. In particular (*taking* $(a, b) = (c, d)$)

$$(a^2 + nb^2, 2ab) \text{ and } (a^2 - nb^2, 0) = (1, 0)$$

are solutions of (IV.6.1).

Joseph Louis Lagrange described in 1771 the solutions of the Pell-Brouncker equation (IV.6.1) in the following way:

The Pell-Brouncker equation (IV.6.1) *has an infinite number of solutions for any square-free* n. *The solutions depend on the continued fraction of the number* $\sqrt{n}$ *that has the following form*

$$[q_0 = c; \overline{q_1 = c_1, q_2 = c_2, \ldots, q_{p-2} = c_2, q_{p-1} = c_1, q_p = 2c}]$$

with period of length p. *The* $(p-1)$*-th convergent* $\frac{a_{p-1}}{b_{p-1}}$ *satisfies the equality*

$$a_{p-1}{}^2 - nb_{p-1}{}^2 = (-1)^p.$$

Hence

$$(x, y) = (a_{p-1}, b_{p-1})$$

is the least (fundamental) solution of (IV.6.1) *if p is even. If p is odd, then the fundamental solution of* (IV.6.1) *is given by the* $(2p-1)$*-th convergent*

$$(x, y) = (a_{2p-1}, b_{2p-1}).$$

In both cases we have

$$a_{2p-1} + b_{2p-1}\sqrt{n} = (a_{p-1} + b_{p-1}\sqrt{n})^2.$$

Example IV.6.1. Lagrange has considered the equation

$$x^2 = 19y^2 + 1.$$

Here we have

$$\sqrt{19} = [4; \overline{2, 1, 3, 1, 2, 8}]$$

and the length $p = 6$ is even. The sequence of the convergents $\frac{a_k}{b_k}$ is

$$\frac{4}{1}\ \frac{9}{2}\ \frac{13}{3}\ \frac{48}{11}\ \frac{61}{14}\ \frac{\mathbf{170}}{\mathbf{39}}\ \frac{1421}{326}\ \frac{3012}{691}\ \frac{4433}{1017}\ \frac{16311}{3742}\ \frac{20744}{4759}\ \frac{\mathbf{57799}}{\mathbf{13260}}\ \frac{483136}{110839}\ \cdots$$

with the $(p-1=5)$-th and the $(2p-1=11)$-th convergents marked in bold. They provide the first solutions

$$(x, y) = (170, 39) \text{ and } (x, y) = (57799, 13260).$$

Thus $(x, y) = (170, 39)$ is the least (fundamental) solution, and $57799 + 13260\sqrt{19} = (170 + 39\sqrt{19})^2$. Similarly,

$$(170 + 39\sqrt{19})^3 = 19651490 + 4508361\sqrt{19} = a_{17} + b_{17}\sqrt{19}$$

yields the solution

$$(x, y) = (19651490, 4508361),$$

and we can continue in the same manner to obtain the solutions

$$(x, y) = (6681448801, 1532829480),$$

$$(x, y) = (2271672940850, 521157514839),$$

$$(x, y) = (772362118440199, 177192022215780),$$

$$(x, y) = (262600848596726810, 60244766395850361),$$

$$(x, y) = (89283516160768675201, 20483043382566906960), \text{ etc.}$$

Exercise IV.6.1. Solve the Pell-Brouncker equation

$$x^2 = 31y^2 + 1.$$

(Hint: $\sqrt{31} = [5; \overline{1, 1, 3, 5, 3, 1, 1, 10}]$.)

Example IV.6.2. Bhaskara II has solved the Pell-Brouncker equation for $n = 61$ by using the **Chakravala** circular method in his book **Bijaganita** (1150).

Here, $\sqrt{61} = [7; \overline{1, 4, 3, 1, 2, 2, 1, 3, 4, 1, 14}]$ has a period of odd length $p = 11$. The sequence of the first twelve convergents is easy to write down:

$$\frac{a_0}{b_0} = \frac{7}{1}; \frac{8}{1} \frac{39}{5} \frac{125}{16} \frac{164}{21} \frac{458}{58} \frac{1070}{137} \frac{1523}{195} \frac{5639}{722} \frac{24079}{3083}; \frac{a_{10}}{b_{10}} = \frac{\mathbf{29718}}{\mathbf{3805}}; \frac{a_{11}}{b_{11}} = \frac{440131}{56353}.$$

Now, the 10-th convergents yields the equality

$$29718^2 = 61 \times 3805^2 - 1,$$

and it is only the 21-st convergent

$$\frac{a_{21}}{b_{21}} = \frac{\mathbf{1766319049}}{\mathbf{226153980}}$$

that leads to the least (fundamental) solution $(x, y) = (1766319049, 226153980)$ obtained by Bhaskara II. Of course, these values can be easily obtained from the equality

$$(29718 + 3805\sqrt{61})^2 = 1766319049 + 226153980\sqrt{61}.$$

Exercise IV.6.2. Solve the Pell-Brouncker equation

$$x^2 = 13y^2 + 1.$$

IV.7 Congruence relations in $\mathbb{R}$

In this short section, we stress the importance of an equivalence that is familiar to us from geometry dealing with the measurements of angles. Our standard measurement of angles in degrees, which was influenced by the Babylonian sexagesimal numeric system, does not differentiate between 20° and 380°, between 100° and 820°; it uses a "new equality" where $360° \equiv 0°$, similar to the **congruence** of integers. Here is the definition of a **congruence in** $\mathbb{R}$.

Definition IV.7.1. Let g be a fixed positive real number. Two real numbers a, b are said to be **congruent modulo** g :

$$a \equiv b \pmod{g}$$

if $a - b = kg$ for some integer k.

This relation $\equiv$ is reflexive, symmetric and transitive. Thus $\equiv$ is an equivalence relation. We denote by $\widehat{a}$ the equivalence (congruence) class of a. There is a compatible (that is, not depending on the choice of representatives) definition of **addition**: $\widehat{a} \oplus \widehat{b} = \widehat{a + b}$ and, with respect to this operation, the set of all equivalence classes, denoted by $\mathbb{R}/g\mathbb{Z}$, is an **abelian group**. The equivalence class

$$\widehat{a} = \{x \in \mathbb{R} \mid x \equiv a \pmod{g}\} = a + \widehat{0},$$

where $\widehat{0} = g\mathbb{Z} = \{gz \mid z \in \mathbb{Z}\}$ is a subgroup of the additive group of $\mathbb{R}$.

Theorem IV.7.1. *The map $F : (\mathbb{R}, +) \to (\mathbb{R}/g\mathbb{Z}, \oplus)$ defined by $F(a) = \widehat{a}$ is a (canonical) homomorphism. The $\equiv$-equivalence classes of $\mathbb{R}$ are just the inverse images of the elements of $\mathbb{R}/g\mathbb{Z}$. In particular, the kernel of this homomorphism is the subgroup $g\mathbb{Z}$ of $(\mathbb{R}, +)$.*

Example IV.7.1 (Important). For $g = 2\pi$ or $g = 360$, we obtain the additive group of all **angles**, measured in radians or degrees, respectively.

Remark IV.7.1. We remark that there is no possibility of defining a compatible operation of multiplication on $\mathbb{R}/2\pi\mathbb{Z}$ as $ab \not\equiv (a + 2k\pi)(b + 2l\pi) \pmod{2\pi}$ for $kl \neq 0$.

Theorem IV.7.2. *Let g and h be positive real numbers. Then*

$$\mathbb{R}/g\mathbb{Z} \simeq \mathbb{R}/h\mathbb{Z},$$

that is, all groups $\mathbb{R}/x\mathbb{Z}, x > 0, x \in \mathbb{R}$ are isomorphic.

Proof. The map

$$\Phi : \mathbb{R}/g\mathbb{Z} \to \mathbb{R}/h\mathbb{Z}$$

defined by $\Phi(\widehat{a}) = \widehat{hg^{-1}a}$ is an isomorphism. □

IV.8 ⋆ A few words about p-adic numbers

We have already pointed out that the extension of the rational numbers to the field of real numbers requires analytical tools (see Section 4 of this chapter). The classical construction of Dedekind consists in partitioning the rational numbers into particular pairs of subsets called **Dedekind cuts.** Equally classical is the construction based on the concept of the **Bolzano-Cauchy (fundamental) sequences**, that is, sequences $\{a_n \mid n = 1, 2, \dots\}$ of rational numbers that satisfy the following condition:

For any $\epsilon > 0$ there is N such that $|a_n - a_m| < \epsilon$ for all $n, m > N$.

These sequences are named after **Bernard Bolzano (1781 - 1848)** and **Augustin Louis Cauchy (1789 - 1857)**. These two constructions depend on the **norm** of the rational numbers r, that is, the absolute value $|r|$. We recall the basic properties of the absolute value, namely, for all rational numbers a and b, we have

$$\begin{aligned} |a| &\geq 0, \\ |a| &= 0 \text{ if and only if } a = 0, \\ |ab| &= |a||b|, \\ |a + b| &\leq |a| + |b|. \end{aligned}$$

Should there be another way to define a norm ν in the field of rational numbers, we could copy the construction of Bolzano-Cauchy sequences to obtain $\nu-$sequences and use them to extend the field $\mathbb{Q}$ to a new field (depending on the norm ν). Well, there exists, for every prime p, a norm $\| r \|_p$ of the rational numbers r called the **p-adic norm** that leads to the construction of the **field of p-adic numbers**. In this section, we are going to sketch the basic idea of this construction and illustrate it on several examples and an application.

Let p be a prime. Every rational number $r \neq 0$ can be expressed in a unique way in the form

$$r = p^e \, \frac{a}{b}, \quad \text{where } a \in \mathbb{Z},\; b \in \mathbb{N}, \text{ and } d(p,a) = d(p,b) = d(a,b) = 1. \tag{IV.8.1}$$

Using the representation of r given in (IV.8.1), we define the $p-$**adic norm** of r by

$$\| r \|_p = p^{-e} \text{ (for } r \neq 0) \text{ and } \| 0 \|_p = 0.$$

For example,

$$\| \frac{360}{7} \|_2 = \frac{1}{8}, \; \| \frac{360}{7} \|_3 = \frac{1}{9}, \; \| \frac{360}{7} \|_5 = \frac{1}{5}, \; \| \frac{360}{7} \|_7 = 7, \; \| \frac{360}{7} \|_{11} = 1.$$

Similarly,

$$\| \frac{208}{4725} \|_2 = \frac{1}{16}, \; \| \frac{208}{4725} \|_3 = 27, \; \| \frac{208}{4725} \|_5 = 25, \; \| \frac{208}{4725} \|_7 = 7,$$

$$\| \frac{208}{4725} \|_{11} = 1 \text{ and } \| \frac{208}{4725} \|_{13} = \frac{1}{13}.$$

Exercise IV.8.1. Prove that $\| r \|_p$ has properties similar to those of $| r |$. In particular,

(i) $\| r_1 r_2 \|_p = \| r_1 \|_p \| r_2 \|_p$.

(ii) If $\| r_1 \|_p = \| r_2 \|_p$, then $\| r_1 + r_2 \|_p \leq \| r_1 \|_p$.

(iii) If $\| r_1 \|_p < \| r_2 \|_p$, then $\| r_1 + r_2 \|_p = \| r_2 \|_p$.

(iv) $\| r_1 + r_2 \|_p \leq \max(\| r_1 \|_p, \| r_2 \|_p)$.

(v) $\| r_1 + r_2 \|_p < \| r_1 \|_p + \| r_2 \|_p$ for $r_1 \neq 0, r_2 \neq 0$.

Furthermore, we define the **p-adic distance** of two rational numbers r_1, r_2 by

$$d_p(r_1, r_2) = \| r_1 - r_2 \|_p$$

and point out that the set of all rational numbers is a **metric space** with the **metric** d_p.

We recall that a metric space is a pair (X, d), where X is a non-empty set and d is a metric on X, that is, d is a mapping from $X \times X$ to $\{r \in \mathbb{R} \mid r \geq 0\}$ such that for all $x, y, z \in X$ we have

$$\begin{aligned} d(x,y) &= 0 \text{ if and only if } x = y, \\ d(x,y) &= d(y,x), \\ d(x,y) &\leq d(x,z) + d(z,y). \end{aligned}$$

Exercise IV.8.2. Prove the following (rather surprising) statements:

(i) For any three arbitrarily chosen rational numbers r_1, r_2, r_3, at least two of the distances $d_p(r_1, r_2), d_p(r_1, r_3), d_p(r_2, r_3)$ must be the same. Hence, in the p-adic metric, every triangle is an isosceles triangle.

(ii) For any prime p, the integers $a \in \mathbb{Z}$ satisfy $\| a \|_p \leq 1$, that is, all integers lie in a **closed disk** whose center is 0 and radius 1. For every $a \in \mathbb{Z}$, $d_p(0, a) \leq 1$.

(iii) Every inner rational point r of the disk described in (ii) is a "center" of this disk, that is, r is the same distance from all points of the disk.

Exercise IV.8.3. Define a fundamental p-adic sequence, where the role of $\epsilon = \frac{1}{n}$ in the usual metric is played by $\epsilon = p^n$. Show that the sum and the product of two such p-adic sequences is again a p-adic sequence.

As in the case of Bolzano-Cauchy sequences, we can define an equivalence on the set of all p-adic sequences and then define the operations of addition and multiplication on the set of the respective equivalence classes. It is easy to show that these operations enjoy the same properties as the usual operations in the field of rational numbers. Moreover, it is possible to show that every equivalence class contains a unique sequence that can be expressed in the form

$$s \;=\; a_{-k}\, p^{-k} + a_{-k+1}\, p^{-k+1} + \cdots + a_0\, p^0 + a_1\, p^1 + \cdots + a_t\, p^t, \; t \geq 0,$$

where $a_{-k} \neq 0$ and $0 \leq a_j \leq p - 1$ for all $-k \leq j \leq t$.

Thus, there is a bijection between the set of all the equivalence classes and the expressions of the form

$$s \;=\; \sum_{t=-k}^{\infty} a_t\, p^t, \; 0 \leq a_t \leq p-1, a_{-k} \neq 0 \qquad \text{(IV.8.2)}$$

that are called the **p-adic numbers.** These numbers form a field $\mathbb{Q}_{(p)}$ that contains the field $\mathbb{Q}$ of rational numbers. Below, we will illustrate this embedding for the prime $p = 3$.

It is easy to see that the definition of the p-adic norm for the rational numbers can be extended to the p-adic numbers: The norm of the number s in (IV.8.2) is $\| s \|_p = p^k$.

We point out that the field $\mathbb{Q}_{(p)}$ contains a subdomain of **p-adic integers**, that is, the p-adic numbers of the form (IV.8.2) with $k = 0$. Hence, these are just all the p-adic numbers whose norm is at most 1, that is, all s such that $\| s \|_p \leq 1$.

It is important to realize that the "sums" of the form (IV.8.2) are in $\mathbb{Q}_{(p)}$ (with respect to the p-adic norm) convergent and that we have lost the notation for "negative numbers" by means of the sign " $-$ " (in contrast to the field $\mathbb{R}$ of real numbers). Indeed, there is no compatible linear (total) order in $\mathbb{Q}_{(p)}$, that is, there is no comparison of p-adic numbers s_1 and s_2 like "$s_1 \leq s_2$ or $s_2 \leq s_1$" compatible with addition and multiplication.

Exercise IV.8.4. Describe the addition and the multiplication of two p-adic numbers in the form (IV.8.2).

A p-adic number s in (IV.8.2) is usually written (from the right to the left) as follows:

$$\ldots a_t a_{t-1} a_{t-2} \ldots a_2 a_1 a_0 \cdot a_{-1} a_{-2} \ldots a_{-k} \,|_p \,.$$

In this notation the number ...000000 is the zero element and ...000001 the identity element of the field $\mathbb{Q}_{(p)}$. Every natural number a has a very simple (finite) notation that is determined by expressing the number a in the base p numeral system (see Section 8 of Chapter III). For instance,

$$277\,|_{10} = 1 + 2.3 + 1.3^3 + 1.3^5 = 101021\,|_3 \,.$$

A rational number $r = \frac{a}{p^k}$, where $k \geq 1$ and a is a natural number that is not a multiple of p, has also a simple (finite) form. Such a number can be written in the form

$$r = b + \frac{c}{p^k}, \text{ where } b \text{ and } c \text{ are integers with } b \geq 0 \text{ and } 0 < c < p^k.$$

If we express b and c in the form

$$b = \sum_{t=0}^{n} a_t\, p^t \text{ and } c = \sum_{t=0}^{k-1} a_{t-k}\, p^t,$$

we obtain

$$r = \sum_{t=-k}^{n} a_t\, p^t, \text{ and thus } r = a_n\, a_{n-1} \ldots a_1\, a_0 \cdot a_{-1}\, a_{-2} \ldots a_{-k} \,|_p \,.$$

Of course, here some (or even all) a_t for $t \geq 0$, as well as some a_t with negative indices may be zero, but $a_{-k} \neq 0$.

We can easily verify that

$$p^{2n} - 1 = (p^n + 1)(p^n - 1) = (p^n + 1)(p^{n-1} + p^{n-2} + \cdots + 1)(p - 1) = \sum_{j=0}^{2n-1} (p-1)\, p^j.$$

Since

$$\| \sum_{j=0}^{2n-1} (p-1)\, p^j - (-1) \|_p = p^{-2n},$$

we obtain

$$-1 = \sum_{j=0}^{\infty}(p-1)\,p^j = \ldots(p-1)(p-1)\ldots(p-1)\,|_p\,. \qquad \text{(IV.8.3)}$$

We note that we can also get this expression simply from the equality

$$\left(\sum_{j=0}^{\infty} x_j\,p^j\right) + 1 = 0.$$

One can proceed similarly in the case of finding the expression for the negative number $-a$. The expression is simply the "zero complement" of the number $a = \sum_{t=0}^{n} a_t p^t$. In this way, we can get for example

$$-277 = 2 + 2.3^2 + 1.3^3 + 2.3^4 + 1.3^5 + \sum_{j=6}^{\infty} 2.3^j = \ldots 222121202\,|_3\,.$$

Exercise IV.8.5. Provide an alternative justification that $a_t = p-1$ for all $t = 0, 1, 2, \ldots$ in (IV.8.3).

To express the fraction $\frac{1}{b}$ in the form of a p-adic number when the natural number b is not a multiple of the prime p, one can use Euler's theorem (see Theorem III.3.8):

$$p^{\varphi(b)} = t\,b - (-1) \text{ for a suitable integer } t = a_q\,a_{q-1}\ldots a_1\,a_0\,|_p, \text{ where } q < \varphi(b).$$

Therefore

$$\|\,t - (-\frac{1}{b})\,\|_p = p^{-\varphi(b)}.$$

Exercise IV.8.6. (i) Using induction prove that

$$\|\,(p^{(2^h-1)\varphi(b)} + p^{(2^h-2)\varphi(b)} + \cdots + 1)t - (-\frac{1}{b})\,\|_p = p^{-2^h\varphi(b)}.$$

(ii) Express the rational number $\frac{1}{b}$ in the form of a p-adic number.

Example IV.8.1. We present an illustration of the described process for $p = 3$ and $b = 5$. First,

$$-1 = \sum_{j=0}^{\infty} 2.3^j = \ldots 2222\,|_3 \text{ and thus } -\frac{1}{2} = \sum_{j=0}^{\infty} 3^j = \ldots 1111\,|_3\,.$$

Furthermore,

$$\frac{3^4-1}{5} = 16\,|_{10} = 121\,|_3, \quad \frac{3^8-1}{5} = 1312\,|_{10} = 1210121\,|_3 \quad \text{etc., therefore}$$

$$-\frac{1}{5} = 1 + 2.3 + 1.3^2 + 1.3^4 + 2.3^5 + 1.3^6 + \cdots = \ldots 01210121\,|_3\,.$$

From here, quite simply either by multiplication or addition,

$$-\frac{2}{5} = \ldots 10121012\,|_3\,,\; -\frac{3}{5} = \ldots 12101210\,|_3\,,$$

$$-\frac{4}{5} = \ldots 21012101\,|_3 \text{ and finally } \frac{1}{5} = \ldots 21012102\,|_3\,.$$

Exercise IV.8.7. In $\mathbb{Q}_{(2)}$, express the numbers $-\frac{1}{3}, -\frac{2}{3}, \frac{1}{3}$ and $\frac{2}{3}$.

Exercise IV.8.8. Suppose that the p-adic norm of a rational number r satisfies $\| r \|_p \leq 1$ for every prime p. Prove that r is an integer.

Exercise IV.8.9. Prove that $\sqrt{2}$, that is, a number whose square equals 2, exists in the field $\mathbb{Q}_{(7)}$ and determine several of its last terms. Show that there are two values of $\sqrt{2}$.

Exercise IV.8.10. (i) Prove that $\sqrt{2}$ does not exist in the field $\mathbb{Q}_{(5)}$.

(ii) In $\mathbb{Q}_{(5)}$, solve the equation $x^2 = \ldots 00002100\,|_5 = 1.5^2 + 2.5^3$.

Exercise IV.8.11. In $\mathbb{Q}_{(5)}$, solve the equation $x^2 = -1$, that is, determine several last terms of the 5−adic number $\sqrt{-1}$.

It should be mentioned that the field of p-adic numbers plays a very important role in the theory of numbers. We should also add that **Alexander Markowich Ostrowski (1893 - 1986)** proved in 1916 that every nontrivial norm of the field $\mathbb{Q}$ of rational numbers is "equivalent" either to a p-adic norm $\| \bullet \|_p$ for a suitable prime p or to the absolute value $| \bullet |$.

We are now going to present an application of 2−adic numbers by using them to prove a theorem due to **Paul Monsky (1936 -)**, see The American Mathematical Monthly 77 (1970), pp. 161–164, which concerns dividing a square into triangles of equal area. Our proof is based on the proof by **B. Bekker**, **S. Vostokov** and **Yu. Ionin** in the Russian journal Kvant in 1979.

Theorem IV.8.1 (Monsky's theorem). *A square cannot be subdivided into an odd number of triangles of the same area.*

Note that such a division is trivially possible for an even number of triangles. Here, we avoid extending the p-adic norm to the real numbers and will prove Theorem IV.8.1 for "rational" subdivisions of the square.

Without loss of generality, we may assume that the coordinates of the vertices of the square $OABC$ are $O = (0,0), A = (1,0), B = (1,1), C = (0,1)$. By a subdivision into triangles we shall understand a **triangulation** in which the vertices of the triangles have rational coordinates.

Divide the set

$$S = \{(x, y) \mid 0 \leq x, y \leq 1, x \in \mathbb{Q}, y \in \mathbb{Q}\} \subset \mathbb{R}^2$$

of all points of the square $OABC$ with rational coordinates into three disjoint subsets

$$S_1 = \{(x, y) \mid \| x \|_2 < 1, \| y \|_2 < 1\},$$

$$S_2 = \{(x, y) \mid \| x \|_2 \geq 1, \| x \|_2 \geq \| y \|_2\},$$

$$S_3 = \{(x, y) \mid \| x \|_2 < \| y \|_2, \| y \|_2 \geq 1\}.$$

In Figure IV.8.1, which displays the points with coordinates $\left(\frac{r}{60}, \frac{s}{60}\right)$, the points in S_1 are marked by $\bullet$, the points in S_2 by $*$, and the points in S_3 by the symbol $\circ$.

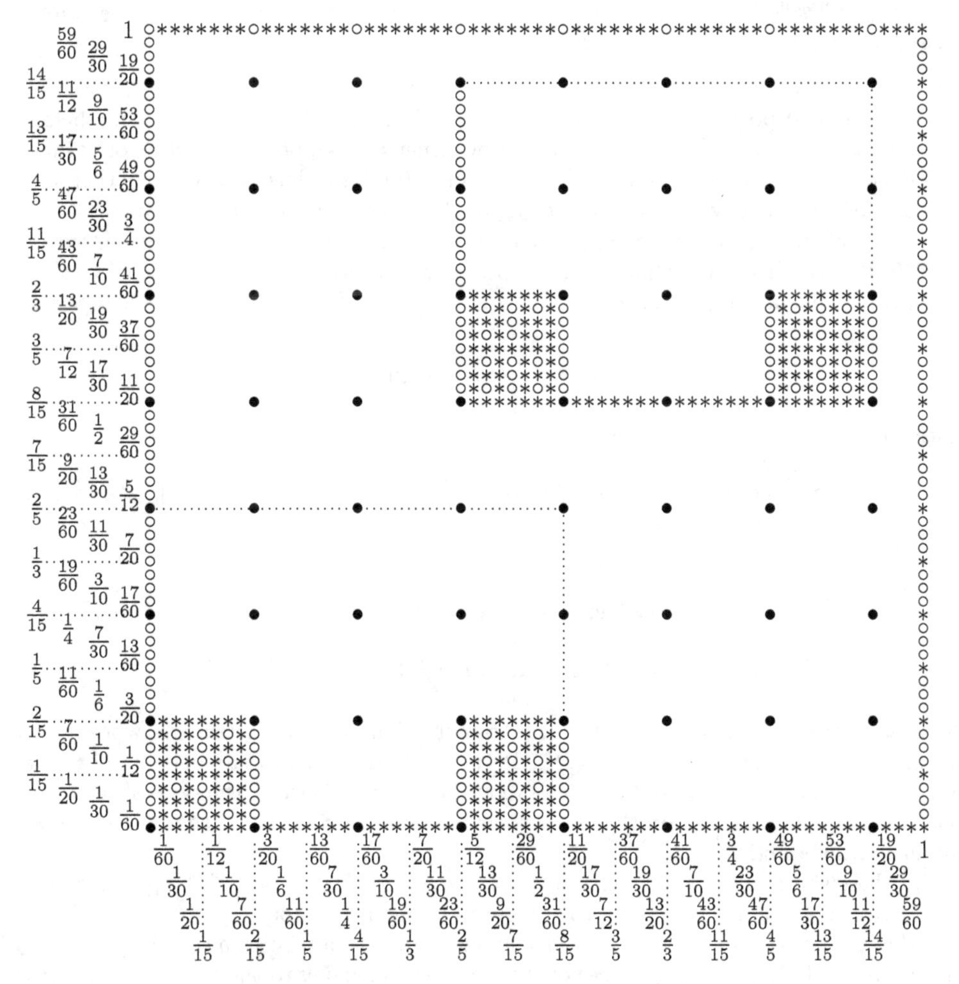

Figure IV.8.1 Points with coordinates $\left(\frac{r}{60}, \frac{s}{60}\right)$

The key element of our proof is the following theorem.

Theorem IV.8.2. *Let $\mathcal{T}$ be a (finite) rational triangulation of the square $OABC$. Then there is a triangle*

$$B_1B_2B_3 \in \mathcal{T} \text{ such that } B_t \in S_t,\ t = 1, 2, 3.$$

Proof. A proof of this statement will be given in three steps:
(1) If $(a, b) \in S_1$, then

$$(x, y) \in S_t \text{ if and only if } (a + x, b + y) \in S_t,\ t = 1, 2, 3.$$

This "translation" is well illustrated in Figure IV.8.1. The required equivalence follows from Exercise IV.8.1 (ii), (iii) and (iv): Let $\| a \|_2 < 1$ and $\| b \|_2 < 1$. If $(x, y) \in S_1$, then $\| a + x \|_2 < 1$ and $\| b + y \|_2 < 1$, and thus $(a + x, b + y) \in S_1$. If $(x, y) \in S_2$, then $\| a + x \|_2 = \| x \|_2 \geq 1$ and $\| b + y \|_2 \leq \| a + x \|_2$, and hence $(a + x, b + y) \in S_2$. Similarly $(x, y) \in S_3$ implies $(a + x, b + y) \in S_3$.

(2) The 2−norm of the area of a triangle whose vertices lie in the distinct subsets S_1, S_2 and S_3 is a (positive) power of 2. As a consequence, all points of every straight line belong to either S_1 and S_2 only (and are thus denoted by either $\bullet$ or $*$), or to S_1 and S_3 only (and are thus denoted by either $\bullet$ or $\circ$) or to S_2 and S_3 only (and are thus denoted by either $*$ or $\circ$). The area of a triangle whose vertices belong to the same line is of course zero.

In view of statement 1, we can assume that one of the vertices of the triangle is the origin $O = (0, 0) \in S_1$. If the coordinates of the other two vertices are (x_1, y_1) and (x_2, y_2), then the area of the triangle (see Theorem V.5.4) is

$$A = \frac{1}{2} \mid x_1y_2 - x_2y_1 \mid .$$

Without loss of generality, we may assume that

$$(x_1, y_1) \in S_2 \text{ and } (x_2, y_2) \in S_3, \text{ that is, } \| x_1 \|_2 \geq \| y_1 \|_2 \text{ and } \| x_2 \|_2 < \| y_2 \|_2 .$$

Therefore

$$\| x_1y_2 \|_2 = \| x_1 \|_2 \| y_2 \|_2 > \| x_2 \|_2 \| y_1 \|_2 = \| x_2y_1 \|_2, \text{ and thus}$$

$$\| A \|_2 = 2 \| x_1y_2 - x_2y_1 \|_2 = 2 \| x_1 \|_2 \| y_2 \|_2 \geq 2.$$

(3) Now it is sufficient to show that the triangulation contains a triangle whose vertices belong to the distinct subsets S_1, S_2, S_3. Recall that, in view of statement 2, all points of every line belong to just two of the three subsets S_1, S_2, S_3. Consider the sides of the triangles belonging to the triangulation that have their points from S_1 and S_2, that is, whose points are marked by $\bullet$ and $*$.

On these sides, consider all possible segments XY such that one of the points x, y (that is a vertex of a triangle belonging to the given triangulation) belongs to S_1 and the other one (that is also a vertex of a triangle belonging to the given triangulation) belongs to S_2 in such a way that there are no vertices of the triangles belonging to the triangulation between

X and Y. Call such segments **exceptional.** Since $O \in S_1$ and $A \in S_2$, the number l of all exceptional segments on the side OA of the square $OABC$ is odd. Equally true is the fact that there are no exceptional segments on the remaining sides of the square $OABC$. Denote the number of the exceptional segments inside the square $OABC$ by k. Then the number $l+2k$ is odd and therefore it is impossible that the number of the exceptional segments on all three sides of every triangle of the triangulation $\mathcal{T}$ be even. Consequently, there is a triangle $B_1B_2B_3 \in \mathcal{T}$ such that the number of all exceptional segmentation its sides is odd and, at the same time, its side B_1B_2 $(B_1 \in S_1, B_2 \in S_2)$ also contains an odd number of exceptional segments. But this implies that the remaining vertex must belong to S_3.

Now, to conclude the proof of Theorem IV.8.2, denote the number of the triangles in the triangulation $\mathcal{T}$ of the square $OABC$ by n. Then $A = \frac{1}{n}$ is their area and since $\| A \|_2 \geq 2$, n is necessarily an even number. □

IV.9 A few remarks

In this section we make a few remarks, together with some exercises, that relate to rational and real numbers. We start with an exercise about approximating a given real number by a rational number. It can be solved using a modified Euclidean algorithm and Bézout's theorem.

Exercise IV.9.1. Prove that for a given real number h and arbitrary $\epsilon > 0$, there exist integers a and $b > 0$ such that

$$\left|h - \frac{a}{b}\right| < \frac{\epsilon}{b}.$$

Our next exercise involves a recursively defined sequence of rational numbers $\{x_n \mid n = 1, 2, \ldots\}$, whose understanding originated in the study of the difference sequence $x_n - x_{n-1}$.

Exercise IV.9.2. Let a and b be natural numbers. Put $x_1 = 1, x_2 = \frac{a}{a+b}$ and for $n \geq 3$,

$$x_n = \frac{ax_{n-1} + bx_{n-2}}{a+b}.$$

Find a formula for x_n and determine $\lim_{n\to\infty} x_n$.

You have probably already seen the following strange historical expression for the golden section (golden ratio)

$$\sqrt{1+\sqrt{1+\sqrt{1+\sqrt{1+\sqrt{1+\cdots}}}}} = \frac{1+\sqrt{5}}{2}, \tag{IV.9.1}$$

which should have been written more properly as

$$\cdots\cdots\sqrt{1+\sqrt{1+\sqrt{\cdots\cdots\sqrt{1+\sqrt{1+\sqrt{1+\sqrt{1+\sqrt{1}}}}}}}} = \frac{1+\sqrt{5}}{2},$$

and mathematically formulated as follows:

If $a_0 = 1$ *and* $a_n = \sqrt{1 + a_{n-1}}$ *for* $n \in \mathbb{N}$, *so that*

$$a_0 = \sqrt{1},\ a_1 = \sqrt{1+\sqrt{1}},\ a_2 = \sqrt{1+\sqrt{1+\sqrt{1}}}, \ldots,$$

then

$$\lim_{n\to\infty} a_n = \frac{1+\sqrt{5}}{2}.$$

Exercise IV.9.3. Explain the following historical notation and prove that

$$\sqrt{2+\sqrt{2+\sqrt{2+\cdots}}} = 2.$$

In general, writing

$$a_n = \sqrt{k(k-1)+\sqrt{k(k-1)+\sqrt{\cdots+\sqrt{k(k-1)}}}} \text{ with } n \text{ symbols } \sqrt{},$$

show that

$$\lim_{n\to\infty} a_n \ = \ k \text{ for any real number } k \geq 1.$$

Now, returning to the expression (IV.9.1), prove the following general statement.

Exercise IV.9.4. Given arbitrary positive real numbers a and b, explain and prove the following expression:

$$\cdots\cdots a\sqrt{b+a\sqrt{b+\sqrt{\cdots\cdots a\sqrt{b+a\sqrt{b+a\sqrt{b+a\sqrt{b}}}}}}} \ = \ \frac{a+\sqrt{a^2+4b}}{2}.$$

Choose, in particular, $a = b = 1$ and $a = b = \frac{1}{2}$.

One has for an arbitrary positive real number r and arbitrary positive real number a with $a < r$, and $b = r(r-a)$ that

$$\cdots\cdots a\sqrt{b+a\sqrt{b+\sqrt{\cdots\cdots a\sqrt{b+a\sqrt{b+a\sqrt{b+a\sqrt{b+a\sqrt{b+a\sqrt{b}}}}}}}}} \ = \ r.$$

Further strange expressions, which are related (as we shall see in Chapter V) to Tartaglia's solution of cubic equations, are the equalities:

$$\sqrt[3]{\sqrt{108}+10} \ - \ \sqrt[3]{\sqrt{108}-10} \ = \ 2$$

and

$$\sqrt[3]{\sqrt{265226}+515} - \sqrt[3]{\sqrt{265226}-515} = 10.$$

We formulate a general form of these equalities in the following exercise.

Exercise IV.9.5. (i) Given an even natural number n and a natural number b, find a natural number a such that

$$n = \sqrt[3]{\sqrt{a^2+b^3}+a} - \sqrt[3]{\sqrt{a^2+b^3}-a}\,.$$

(ii) Given an odd natural number n and an odd natural number b, find a natural number a such that

$$n = \sqrt[3]{\sqrt{a^2+b^3}+a} - \sqrt[3]{\sqrt{a^2+b^3}-a}\,.$$

(iii) Find all possible pairs (a, b) of natural numbers so that

$$\sqrt[3]{\sqrt{a^2+b^3}+a} - \sqrt[3]{\sqrt{a^2+b^3}-a} = 1\,.$$

As a result, we can see that

$$\begin{aligned} 1 &= \sqrt[3]{\sqrt{5}+2} - \sqrt[3]{\sqrt{5}-2} = \sqrt[3]{\sqrt{52}+5} - \sqrt[3]{\sqrt{52}-5} \\ &= \sqrt[3]{\sqrt{189}+8} - \sqrt[3]{\sqrt{189}-8} = \cdots \end{aligned}$$

(iv) Using the solutions of (i) and (ii) derive Tartaglia's formula for a root of the equation $x^3 = px + q$ (see Section 1 of Chapter V).

CHAPTER V. COMPLEX NUMBERS AND PLANE GEOMETRY

V.1 A few historical remarks

Complex numbers have a long history. Nowadays, they are no longer regarded as being more "imaginary" or "complex" than real numbers. However, there has been a long arduous path to achieve this level of acceptance. The comforting presentation of real numbers as a line is conveniently extended to a picture of a plane for complex numbers. In this process, many algebraic properties of the real numbers are retained. One property of real numbers that is lost to complex numbers is linear order. On the other hand, some of the properties of real numbers are better understood when treated as elements of the field of complex numbers.

Contemporary understanding of number systems is that all of them are abstract constructions. The extensions of integers to rational numbers and real numbers to complex numbers are purely algebraic and rather unsophisticated. It is the extension of rational numbers to real numbers that represents a major non-trivial construction. This construction is beyond the scope of algebra and lies within the domain of analysis. The reader interested in learning about this construction might wish to consult Chapter 1 of *An Introduction to Analysis and Integration Theory* by Esther R. Phillips.

The history of complex numbers originates during the "Rinascimento" (Italian Renaissance) with **Italian algebraists** in competitions to solve algebraic equations, in particular those of degree 3 such as, for example, the equation

$$x^3 = 15x + 4. \tag{V.1.1}$$

Girolamo Cardano (1501 - 1576) published in 1547 in his **Ars Magna** the following result:

The equation $x^3 = px + q$ has a solution

$$\sqrt[3]{\frac{q}{2} + \sqrt{\frac{q^2}{4} - \frac{p^3}{27}}} + \sqrt[3]{\frac{q}{2} - \sqrt{\frac{q^2}{4} - \frac{p^3}{27}}}. \tag{V.1.2}$$

[In fact, the solution was discovered independently by **Scipione del Ferro (1465 - 1526)** and **Nicolo Fontana,** called **Tartaglia (1500 - 1557)**. Cardano got the solution from Tartaglia upon giving a solemn pledge of secrecy.]

It might be appropriate to point out that solving equations of the form $x^3 = px + q$ enables us to solve any cubic equation $z^3 + az^2 + bz + c = 0$. Indeed, the simple transformation $z = x - \frac{a}{3}$ brings this equation to such a form with

$$p = \frac{1}{3}(a^2 - 3b) \text{ and } q = \frac{1}{27}(9ab - 2a^3 - 27c).$$

Remark V.1.1. The formula (V.1.2) is of very limited practical use. For example this formula applied to the equation

$$x^3 = 3x + 2$$

gives the solution $x = 2$. However, in the case of the equation

$$x^3 = 3x + 52$$

the formula gives for the solution $x = 4$, the rather mysterious expression

$$4 = \sqrt[3]{26 + 15\sqrt{3}} + \sqrt[3]{26 - 15\sqrt{3}}.$$

Of course, $26 + 15\sqrt{3} = (2 + \sqrt{3})^3$ and $26 - 15\sqrt{3} = (2 - \sqrt{3})^3$, vindicating the expression. This is an important step, since this observation was an ingenious way for Bombelli to justify the emergence of complex numbers.

Exercise V.1.1. Using the formula (V.1.2), solve the equation

$$x^3 = 15x + 4.$$

In the case of this equation, the formula (V.1.2) produced a "formally imaginary" solution

$$\alpha = \sqrt[3]{2 + \sqrt{-121}} + \sqrt[3]{2 - \sqrt{-121}} \qquad \text{(V.1.3)}$$

without any understanding. Only ingenious **Rafaello Bombelli (1526 - 1572)** was able to work in a systematic way with these **"imaginary"** numbers involving the symbol $\sqrt{-1}$ to show in his **Algebra (1572)** that $\alpha = 4$. Bombelli was brilliant in distinguishing the symbols $+\sqrt{-1}$ and $-\sqrt{-1}$ and working with these symbols as we presently do with the symbols $+i$ and $-i$ (that is, $i^2 = (-i)^2 = -1$, $(-i)i = 1$ etc.), introduced later by **Leonhard Euler**. In this way he avoided such meaningless reasoning as

$$-1 = \sqrt{-1} \times \sqrt{-1} = \sqrt{(-1)(-1)} = \sqrt{1} = 1,$$

which one still meets occasionally in the "literature". Bombelli showed in contemporary notation that

$$(2 + i)^3 = 2 + 11\,i,\ (2 - i)^3 = 2 - 11\,i \text{ and } (11\,i)^2 = -121,$$

and in this way explained the value of one of the solutions to be $x_1 = 4$. The remaining two solutions are the roots of the quadratic equation $\frac{x^3-15x-4}{x-4} = x^2 + 4x + 1 = 0$, namely $x_2 = -2+\sqrt{3}$ and $x_3 = -2-\sqrt{3}$. It is important to be aware of the fact that the expression (V.1.3) describes **all three roots** of the equation (V.1.1). Indeed, there are three cube roots of the number $2 + 11\,i$. In addition to Bombelli's number $2 + i$,

$$\frac{1}{2}\,[(-2+\sqrt{3}) - (1+2\sqrt{3})\,i] \;\text{ and }\; \frac{1}{2}\,[(-2-\sqrt{3}) - (1-2\sqrt{3})\,i]$$

are also such roots. Equally, in addition to Bombelli's number $2 - i$,

$$\frac{1}{2}\,[(-2+\sqrt{3}) + (1+2\sqrt{3})\,i] \;\text{ and }\; \frac{1}{2}\,[(-2-\sqrt{3}) + (1-2\sqrt{3})\,i]$$

are also cube roots of $2 + 11\,i$. Then the sum of the respective roots as indicated in (V.1.3) results in the values of x_2 and x_3.

We remark that these two imaginary numbers were not accepted as a solution. The resistance not only to imaginary but also to negative numbers often lead as far as to academic hatred. For example Paolo Bonasoni in his book **Algebra Geometrica (1575)** accused Bombelli of having hallucinations when asserting that subtraction is possible in all cases. According to Bonasoni, the equation $(10-x)^2 = 9x$ has only one solution $x = 4$, for $x = 25$ cannot be a solution because it is impossible to subtract 25 from 10. Even **Thomas Harriot (1560 - 1621)** was still resisting negative and complex roots of equations. It was only **Albert Girard (1595 - 1632)** who routinely calculated with negative and complex numbers as we use them today.

Remark V.1.2. Later in the book, we shall see that the importance of the domain (field) of the complex numbers is due to the so-called **Fundamental theorem of algebra** often called **D'Alembert Theorem**:

Every complex non-constant polynomial has a complex root.

Starting with **Jean Le Rond d'Alembert (1717 - 1783)**, this theorem boasts of many (very often incomplete) proofs. Today, there are numerous indisputable proofs; the first one belongs to **Jean Robert Argand (1768 - 1822)**.

Remark V.1.3. Here is another glance into history. **Gottfried Wilhelm Leibniz** mentions in a letter of 1674 to **Christiaan Huygens (1629 - 1695)** a surprising relation

$$\sqrt{1+\sqrt{-3}} + \sqrt{1-\sqrt{-3}} = \sqrt{6}.$$

We point out that $\sqrt{6}$ is one of the possible values belonging to this algebraic relation since

$$\sqrt{1+\sqrt{3}\,i} = \pm\frac{\sqrt{2}}{2}(\sqrt{3}+i) \;\text{ and }\; \sqrt{1-\sqrt{3}\,i} = \pm\frac{\sqrt{2}}{2}(\sqrt{3}-i).$$

However, we agree that in this remark, we will follow the theory of real functions to assign to $\sqrt{x}$ for $x \geq 0$, the unique real positive value.

We may wonder whether Leibniz knew that his expression is a particular case of the expression

$$\sqrt{a+\sqrt{-b}}+\sqrt{a-\sqrt{-b}}=\sqrt{2(a+\sqrt{a^2+b})},$$

where $a \geq 0$ and $b \geq 0$ are real numbers.

Exercise V.1.2. Prove that for real numbers a and c satisfying $0 < a < c$,

$$\sqrt{a+\sqrt{a^2-c^2}}+\sqrt{a-\sqrt{a^2-c^2}}=\sqrt{2(a+c)}.$$

Thus $\sqrt{1+\sqrt{-8}}+\sqrt{1-\sqrt{-8}}=\sqrt{8},\ \sqrt{2+6\sqrt{-7}}+\sqrt{2-6\sqrt{-7}}=6,$ etc.

In this connection we formulate another similar exercise.

Exercise V.1.3. Prove that for a real number $a \geq 1$,

$$\sqrt{2a^2-1+2a\sqrt{a^2-1}}+\frac{1}{\sqrt{2a^2-1+2a\sqrt{a^2-1}}}$$

$$=\sqrt{2a^2+1+2a\sqrt{a^2+1}}-\frac{1}{\sqrt{2a^2+1+2a\sqrt{a^2+1}}}$$

$$=\sqrt{2a^2-1+2a\sqrt{a^2-1}}+\sqrt{2a^2-1-2a\sqrt{a^2-1}}$$

$$=\sqrt{2a^2+1+2a\sqrt{a^2+1}}-\sqrt{2a^2+1-2a\sqrt{a^2+1}}=2a.$$

Thus, for example, we have

$$\sqrt{49+20\sqrt{6}}+\frac{1}{\sqrt{49+20\sqrt{6}}}=10,\ \sqrt{7+4\sqrt{3}}+\frac{1}{\sqrt{7+4\sqrt{3}}}=4,\ \text{etc.}$$

The proof of the previous exercise indicates the way the equalities can be formulated to hold for an arbitrary complex number a; for instance the right hand side will appear as $2\sqrt{a^2}$.

V.2 Field of complex numbers

First encounters with complex numbers usually present a complex number z in the form

$$z = a + bi,$$

where a and b are real numbers and i is a symbol (that was introduced in the 18-th century by Euler) subject to Bombelli's multiplication table

mult	1	i	-1	$-i$
1	1	i	-1	$-i$
i	i	-1	$-i$	1
-1	-1	$-i$	1	i
$-i$	$-i$	1	i	-1

Remark V.2.1. In view of the fact that $i^2 = -1$, the symbol i is often, under the influence of the historical developments, introduced and confused with the square root of -1. Since $-i$ is also a square root of -1, it is an occasional misrepresentation of these two numbers that lead to dubious conclusions (see the previous section).

Recall that we deal with addition and multiplication, which are commutative, associative and satisfy the distributive law. If $z = a + b\,i$ and $w = c + d\,i$, then

$$z + w = (a + b\,i) + (c + d\,i) \;=\; (a + c) + (b + d)\,i$$

and

$$z \cdot w = (a + b\,i) \cdot (c + d\,i) \;=\; ac + ad\,i + bc\,i + bd\,i^2 = (ac - bd) + (ad + bc)\,i.$$

Moreover, the equation

$$z + x = w$$

has the solution $x = (c - a) + (d - b)\,i$, and if $a^2 + b^2 \neq 0$ the equation

$$z \cdot x = w$$

has the solution

$$x = \frac{ac + bd}{a^2 + b^2} + \frac{ad - bc}{a^2 + b^2}\,i.$$

Thus $(\mathbb{C}, +)$ is an additive abelian group and $(\mathbb{C} \setminus \{0 + 0\,i\}, \cdot)$ is a multiplicative abelian group. We say that $(\mathbb{C}, +, \cdot)$, where $\mathbb{C} = \{a + b\,i \mid a, b \in \mathbb{R}\}$ is the **field of complex numbers**. The field $\mathbb{R}$ of real numbers is then identified with the subfield of $\mathbb{C}$ of all complex numbers of the form $a + 0\,i$ with $a \in \mathbb{R}$. Indeed the mapping

$$F : \mathbb{R} \to \mathbb{C} \text{ defined by } F(a) = a + 0\,i$$

is an injective homomorphism, that is, an embedding of $\mathbb{R}$ into $\mathbb{C}$.

Remark V.2.2. We give an isomorphic description of the field $\mathbb{C}$ of complex numbers, that is, we construct a field, which is isomorphic to $\mathbb{C}$. We start with a cyclic group $G = \langle g \rangle = \{g, g^2, g^3, g^4 = e\}$ of order 4 and the respective real **group ring**

$$\mathbb{R}G \;=\; \{a_0 \cdot e + a_1 \cdot g + a_2 \cdot g^2 + a_3 \cdot g^3 \mid a_0, a_1, a_2, a_3 \in \mathbb{R}\},$$

that is, the real 4-dimensional vector space with G as its basis and with multiplication that is distributive defined by

$$(\sum_{r=0}^{3} a_r \cdot g^r) \cdot (\sum_{s=0}^{3} b_s \cdot g^s) = \sum_{k=0}^{3} (\sum_{\substack{r,\, s\, =\, 0 \\ r+s \equiv k \pmod 4}}^{3} a_r b_s) \cdot g^k.$$

Consider $\mathbb{K} = \{x \cdot e + y \cdot g \mid x, y \in \mathbb{R}\}$ as a subspace of the real vector space $\mathbb{R}G$. Define the following (new) multiplication $\circ$ on $\mathbb{K}$:

$$e \circ e = e \cdot e = e, \; e \circ g = e \cdot g = g, \; g \circ e = g \cdot e = g, \; g \circ g = -e.$$

Consequently, the linear transformation $F : \mathbb{R}G \to \mathbb{K}$ of the vector spaces defined by

$$F(a_0 \cdot e + a_1 \cdot g + a_2 \cdot g^2 + a_3 \cdot g^3) = (a_0 - a_2) \cdot e + (a_1 - a_3) \cdot g$$

is a mapping satisfying for $u = \sum_{k=0}^{3} a_k . \, g^k$ and $v = \sum_{k=0}^{3} b_k . \, g^k$,

$$F(u+v) = F(u) + F(v) \; \text{ and } \; F(u.v) = F(u) \circ F(v).$$

This means that F is a homomorphism of the ring $\mathbb{R}G$ onto $\mathbb{K}$.

The **kernel** Ker$F$ of F, that is, the set of all elements $w \in \mathbb{R}G$ satisfying $F(w) = 0 \cdot e + 0 \cdot g$ is the (zero) equivalence class of the equivalence $\equiv$ defined on $\mathbb{R}G$ by F (see Chapter I). Hence

$$\text{Ker}F = \{r \cdot (e + g^2) + s \cdot (g + g^3) \mid r, s \in \mathbb{R}\}$$

and the set of all $\equiv$ –equivalence classes is in a bijection with $\mathbb{K}$. Thus, $\mathbb{K}$ is a ring with the following addition and multiplication:

$$(a \cdot e + b \cdot g) + (c \cdot e + d \cdot g) = (a + c) \cdot e + (b + d) \cdot g$$

and

$$(a \cdot e + b \cdot g) \circ (c \cdot e + d \cdot g) = F((a \cdot e + b \cdot g)(c \cdot e + d \cdot g)$$
$$= F(ac \cdot e + (ad + bc) \cdot g + bd \cdot g^2) = (ac - bd) \cdot e + (ad + bc) \cdot g.$$

It is now easy to define an isomorphism between the field of complex numbers $\mathbb{C}$ and the ring $\mathbb{K}$ such that the image of i is the generator g of the group G.

It is the previous remark that clarifies the strictly formal introduction of the complex numbers introduced by **William Rowan Hamilton (1805 - 1865)**, namely as the pairs (a, b) of real numbers $a, b \in \mathbb{R}$:

$$\mathbb{C} = \{(a, b) \mid a, b \in \mathbb{R}\}$$

with the operations

$$(a, b) + (c, d) = (a + c, b + d) \; \text{ and } \; (a, b) \cdot (c, d) = (ac - bd, ad + bc).$$

As we have seen, the field $\mathbb{C}$ contains the field of real numbers $\mathbb{R} = \{(r, 0) \mid r \in \mathbb{R}\}$ and as such, $\mathbb{C}$ can be viewed as a two-dimensional real vector space:

$$(r, 0) \cdot (a, b) = (ra, rb).$$

We shall learn in Chapter IX that Ker$F$ is the ideal of the ring $\mathbb{R}G$ that defines the factor ring $\mathbb{K}$.

Now, the multiplication by the number $z = (a, b)$ maps the number (vector) (x, y) into the number (vector) $(ax - by, ay + bx)$, and hence yields a linear transformation of the vector space $\mathbb{C}_{\mathbb{R}}$ that is with respect to the basis $\{1 = (1, 0), i = (0, 1)\}$ represented by the matrix

$$\begin{pmatrix} a & b \\ -b & a \end{pmatrix} \tag{V.2.1}$$

This provides yet another description of the field of the complex numbers which we present here as an exercise.

Exercise V.2.1. Consider the set $\mathbf{M}_1$ of all real matrices of the form (V.2.1) together with their matrix addition and multiplication. Show that the map

$$F : \mathbb{C} \to \mathbf{M}_1 \text{ defined by } F(a + bi) = \begin{pmatrix} a & b \\ -b & a \end{pmatrix}$$

is an isomorphism (that is, a bijection compatible with the algebraic operations). Note that every real number r is represented by the matrix $\begin{pmatrix} r & 0 \\ 0 & r \end{pmatrix}$ and the number i by $\begin{pmatrix} 0 & 1 \\ -1 & 0 \end{pmatrix}$.

Exercise V.2.2. Similar to the previous exercise, consider the set $\mathbf{M}_2$ of all real matrices of the form

$$\begin{pmatrix} a + b & b \\ -2b & a - b \end{pmatrix}$$

together with their matrix addition and multiplication. Show that $\mathbf{M}_2$ is isomorphic to the field of complex numbers.

Remark V.2.3. In Chapter VII, we shall present an algebraic construction of the field of complex numbers as a **splitting field** of the polynomial $x^2 + 1$ over the field $\mathbb{R}$. This is a natural algebraic construction based on the concept of a polynomial congruence and first described by **Augustin Louis Cauchy**.

It is sometimes useful to speak about the **real part** $\Re(z)$ and the **imaginary part** $\Im(z)$ of the complex number $z = a + bi$:

$$\Re(z) = a \text{ and } \Im(z) = b.$$

Thus z is real if and only if $\Im(z) = 0$ and is **purely imaginary** if and only if $\Re(z) = 0$. An important concept is that of the **complex conjugate** of a complex number. The term "conjugate" was introduced by Augustin-Louis Cauchy in his **Cours d'analyse**.

Definition V.2.1. Let $z = a + bi$ be a complex number. Then its conjugate is the number $\overline{z} = a - bi$.

Thus

$$\Re(z) = \frac{1}{2}\,(z+\overline{z}) \text{ and } \Im(z) = \frac{1}{2i}\,(z-\overline{z}).$$

In the next section dealing with the geometry of the plane, we shall see the importance of the fact that $z\overline{z}$ is a real number ≥ 0; in fact, it is equal to 0 if and only if $z=0$. Here, we concentrate on the important map

$$\Phi : \mathbb{C} \to \mathbb{C} \text{ defined by } \Phi(z) = \overline{z}.$$

Exercise V.2.3. Show that the mapping Φ is an automorphism of the field $\mathbb{C}$, that is, for any $z, w \in \mathbb{C}$,

$$\overline{z+w} = \overline{z} + \overline{w} \text{ and } \overline{zw} = \overline{z}\,\overline{w}.$$

Theorem V.2.1. *Given two complex numbers z and w, we have*

(i) $z = \overline{z}$ *if and only if z is real, or if and only if z^2 is a non-negative real number;*

(ii) $z = -\overline{z}$ *if and only if z is purely imaginary, or if and only if z^2 is a non-positive real number;*

(iii) $\overline{z}w$ *is real if and only if $z=0$ or $w=rz$ with a real number r;*

(iv) *If $|z| = |w|$ and $|z+r| = |w+r|$ for a non-zero real number r, then $w=z$ or $w=\overline{z}$.*

Proof. (i) and (ii) are straightforward. To prove (iii), let $z = a+bi$ and $w = c+d\,i$. Then, $(a-b\,i)(c+d\,i) = ac+bd+(ad-bc)\,i$. Now, if $\overline{z}w$ is real, then $ad=bc$, that is, $dz=bw$. Assume $z \neq 0$. If $b \neq 0$, take $r = \frac{d}{b}$. Otherwise, both z and w are real. We leave the reader to check the converse. To prove (iv), again let $z=a+bi$ and $w=c+d\,i$. The assumptions yield $a^2+b^2=c^2+d^2$ and $a^2+2ar+r^2+b^2=c^2+2cr+r^2+d^2$. Thus $a=c$ and $b=\pm d$, that is, $w=z$ or $w=\overline{z}$. □

Exercise V.2.4. (i) Show that the (multiplicative) inverse of $z=a+bi \neq 0$ is

$$\frac{\overline{z}}{z\overline{z}} = \frac{a}{a^2+b^2} - \frac{b}{a^2+b^2}i.$$

(ii) Express $\sqrt{16+30i}$ in the form $a+bi$, where $a, b \in \mathbb{R}$.

Exercise V.2.5. If the complex numbers z_1 and z_2 satisfy $|z_1| = |z_2| = 1$ and $z_1z_2 \neq -1$, prove that the number $\dfrac{z_1+z_2}{1+z_1z_2}$ is real.

Exercise V.2.6. Determine all complex numbers z such that the number $(z-2)(\overline{z}+i)$ is real.

Exercise V.2.7. Let $n \in \mathbb{N}$. Suppose that $a_t \in \mathbb{R}$ for all $t \in \{0, 1, \ldots, n\}$ with $a_n \neq 0$. Let

$$P(z) = a_0 + a_1 z + a_2 z^2 + \cdots + a_n z^n. \tag{V.2.2}$$

If $z \in \mathbb{C}$ satisfies $P(z) = 0$ prove that we also have $P(\overline{z}) = 0$. In the next chapter, we shall prove that there are n complex solutions of (V.2.2). Assuming this fact, deduce that, in the case when n is odd, there must be a real number r such that $P(r) = 0$.

Exercise V.2.8. Consider the set $M = \mathbb{C} \times \mathbb{C} = \{[z_1, z_2] \mid z_1, z_2 \in \mathbb{C}\}$ of the ordered pairs of complex numbers together with the operations $\oplus$ (addition) and $\odot$ (multiplication) defined by

$$[z_1, z_2] \oplus [w_1, w_2] = [z_1 + w_1, z_2 + w_2], \quad [z_1, z_2] \odot [w_1, w_2] = [z_1 w_1 + z_2 \overline{w_2}, z_1 w_2 + z_2 \overline{w_1}].$$

Defining the mapping

$$\Phi : M(\oplus, \odot) \longrightarrow Mat_{2\times 2}(\mathbb{R})$$

by

$$\Phi([a + bi, c + di]) = \begin{pmatrix} a + d & c + b \\ c - b & a - d \end{pmatrix},$$

show that $M(\oplus, \odot)$ is isomorphic to the ring of all real 2×2 real matrices.

Exercise V.2.9. Using the notation of Exercise V.2.8, prove that

$$\Phi^{-1}\begin{pmatrix} 1 & 0 \\ 0 & 0 \end{pmatrix} = [\frac{1}{2}, \frac{1}{2}i],$$

and determine

$$\Phi^{-1}\begin{pmatrix} 0 & 1 \\ 0 & 0 \end{pmatrix}, \ \Phi^{-1}\begin{pmatrix} 0 & 0 \\ 1 & 0 \end{pmatrix} \text{ and } \Phi^{-1}\begin{pmatrix} 0 & 0 \\ 0 & 1 \end{pmatrix}.$$

In addition, prove that

$$[0,1] \odot [z,0] = [0,\bar{z}], \ [0,1] \odot [0,z] = [\bar{z},0], \ [z,0] \odot [0,1] = [0,z], \ [0,z] \odot [0,1] = [z,0],$$

and thus

$$[0,1] \odot [z,0] \odot [0,1] = [\bar{z},0] \quad \text{and} \quad [0,1] \odot [0,z] \odot [0,1] = [0,\bar{z}].$$

V.3 Geometric representation of complex numbers

To properly understand the structure and meaning of complex numbers it is necessary to associate them with the real plane and its geometry. Nowadays, after Hamilton's presentation of complex numbers as real pairs, this seems the natural way to proceed. A geometric interpretation of the numbers together with their algebraic operations of addition and multiplication is very comforting. Historically, the first vague notions on a representation of complex numbers by points of a plane can be traced to the work of **John Wallis (1616 - 1703)**. A substantive contribution was made by **Caspar Wessel (1745 - 1818)** who proposed considering vectors as complex numbers. He described this correspondence in terms of a real axis and an imaginary axis in detail. A thorough description of the representation was given by **Jean Robert Argand**, who interpreted multiplication by $\sqrt{-1}$ as a rotation about the origin by a right angle. It is probable that **Leonhard Euler** already considered both negative and complex numbers as points in the plane. He introduced the symbol i and described a complex number in the form $z = r(cos\ \alpha + i\ sin\ \alpha)$. The general acceptance of the representation of complex numbers as points of a plane was greatly strengthened by the authority of **Carl Friedrich Gauss**.

The complex numbers $z = a + b\,i$ are identified with the points $P(z) = (a, b)$ of a real plane with the orthonormal basis $\{(1,0),(0,1)\}$. The real numbers correspond to the points of the x-axis and the purely imaginary numbers to the points of the y-axis. Of course, the point $P(z)$ can also be interpreted as the **vector** $\vec{z} = (a, b)$. We shall move between these two descriptions quite freely.

The addition of complex numbers corresponds to the addition of the respective vectors (points), as indicated in Figure V.3.1. The product of complex numbers is more subtle (and thus, more interesting). It is natural to describe the product of two complex numbers in terms of their **polar coordinates**. This was already done by Argand. Every non-zero complex number $z = a + b\,i$ can be expressed (uniquely) in the **trigonometric form**

$$z = r(\cos\alpha + i\cos\alpha), \quad \text{where}$$

$$r = \mid z \mid = \sqrt{z\,\overline{z}} = \sqrt{a^2 + b^2}$$

is the **modulus**, or **absolute value** (the distance of the point $P(z)$ from the origin $(0,0)$) and $\alpha = \arg z$, the **argument** of z, is the unique angle $0 \leq \alpha < 2\pi$ such that

$$\cos\alpha = \frac{a}{r} \quad \text{and} \quad \sin\alpha = \frac{b}{r}.$$

We emphasize that the argument $\arg z$ satisfies $0 \leq \arg z < 2\pi$ and that the sum of $\arg z$ and $\arg w$ is performed modulo 2π and is denoted by $(\arg z + \arg w)[2\pi]$.

Remark V.3.1. The trigonometric form of a complex number defines the functions $\cos\alpha$ and $\sin\alpha$ for an angle α satisfying $0 \leq \alpha < 2\pi$; for a general value of the angle α, these functions are extended by means of the 2π-equivalence defined at the end of Chapter IV:

$$\alpha_1 \equiv \alpha_2 \pmod{2\pi} \text{ implies } \cos\alpha_1 = \cos\alpha_2 \text{ and } \sin\alpha_1 = \sin\alpha_2.$$

In what follows, we shall make use of the basic formulas

$$\cos(\alpha_1 \pm \alpha_2) = \cos\alpha_1\cos\alpha_2 \mp \sin\alpha_1\sin\alpha_2$$

and

$$\sin(\alpha_1 \pm \alpha_2) = \sin\alpha_1\cos\alpha_2 \pm \cos\alpha_1\sin\alpha_2.$$

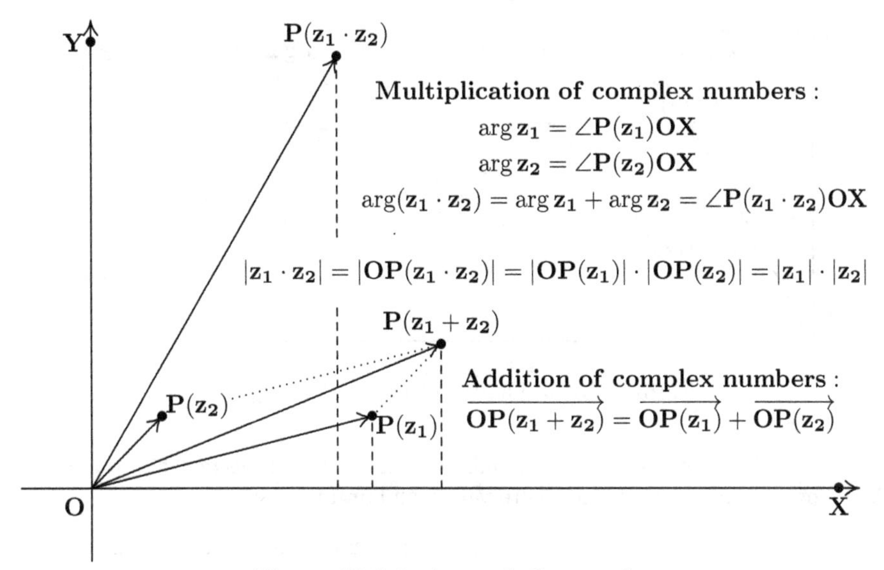

Figure V.3.1 Argand-Gauss plane

We reiterate that there are simple transformations between algebraic (cartesian) and polar coordinates:

$(0,0) \neq (a,b) \quad \longleftrightarrow \quad (r,\alpha)$ such that
$r = \sqrt{a^2+b^2} \neq 0$, $\cos\alpha = \frac{a}{r}$ and $\sin\alpha = \frac{b}{r}$

$r \neq 0, (r,\alpha) \quad \longleftrightarrow \quad (a,b) \neq (0,0)$ such that
$a = r\cos\alpha$ and $b = r\sin\alpha$.

It is important to realize that in dealing with angles between two vectors, we have to distinguish the **orientation** of those angles. Given two complex numbers z_1 and z_2, the angle between the respective vectors $\overrightarrow{OP(z_1)}$ and $\overrightarrow{OP(z_2)}$ has two orientations. We agree to denote by the symbol $\angle(z_1,z_2)$ the **oriented angle** $\arg z_2 - \arg z_1 = \arg \frac{z_2}{z_1}$. This is the angle by which we are required to rotate $\overrightarrow{OP(z_1)}$ to become a positive multiple of the vector $\overrightarrow{OP(z_2)}$. We note that

$$\angle(z_1,z_2) \;=\; \angle(rz_1,sz_2) \text{ for any positive real numbers } r \text{ and } s$$

and that in the trigonometric form $z = r(\cos\alpha + i\,\sin\alpha)$, we have

$$\alpha = \angle(z,1), \quad \cos\alpha = \frac{z+\bar{z}}{2r} \text{ and } \sin\alpha = \frac{z-\bar{z}}{2r\,i}.$$

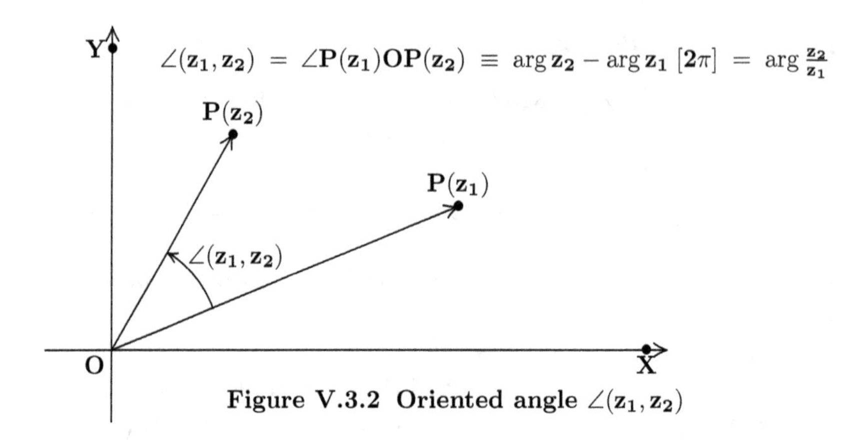

Figure V.3.2 Oriented angle $\angle(\mathbf{z_1}, \mathbf{z_2})$

Exercise V.3.1. For any two complex numbers z and w (when needed $z \neq 0$ and $w \neq 0$) prove the following statements for absolute values and arguments:

PRODUCT	$\lvert z \cdot w \rvert = \lvert z \rvert \cdot \lvert w \rvert$	$\arg(z \cdot w) \equiv (\arg z + \arg w)[2\pi]$
POWER	$\lvert z^n \rvert = \lvert z \rvert^n, n \in \mathbb{N}$	$\arg(z^n) \equiv n \arg z[2\pi]$
INVERSE	$\lvert \frac{1}{z} \rvert = \frac{1}{\lvert z \rvert}$	$\arg(\frac{1}{z}) \equiv -\arg z[2\pi]$
QUOTIENT	$\lvert \frac{z}{w} \rvert = \frac{\lvert z \rvert}{\lvert w \rvert}$	$\arg(\frac{z}{w}) \equiv (\arg z - \arg w)[2\pi]$
CONJUGATE	$\lvert \overline{z} \rvert = \lvert z \rvert$	$\arg(\overline{z}) \equiv -\arg z[2\pi]$
OPPOSITE	$\lvert -z \rvert = \lvert z \rvert$	$\arg(-z) \equiv \pi + \arg z[2\pi]$

Rewrite each line of the table in full. For instance the first line reads: Let $z = (r, \alpha)$ and $w = (s, \beta)$ be polar coordinates of the non-zero complex numbers z and w. Then (rs, γ) where $\gamma = (\alpha + \beta)\,[2\pi]$ are the polar coordinates of the product $z \cdot w$.

Exercise V.3.2. For any nonzero $z_1, z_2 \in \mathbb{C}$, prove that

$$\cos \angle(z_1, z_2) = \frac{\overline{z_1} z_2 + z_1 \overline{z_2}}{2 \mid z_1 \mid \mid z_2 \mid} \text{ and } \sin \angle(z_1, z_2) = \frac{\overline{z_1} z_2 - z_1 \overline{z_2}}{2i \mid z_1 \mid \mid z_2 \mid}.$$

Example V.3.1. The trigonometric form of $z = -1 + i\sqrt{3}$ is $2(\cos \frac{2\pi}{3} + i \sin \frac{2\pi}{3})$, but also for example $2(\cos \frac{8\pi}{3} + i \sin \frac{8\pi}{3})$. However $\arg z = \frac{2\pi}{3}$! Note that $\sqrt{5}(\frac{2}{\sqrt{5}} + i\frac{1}{\sqrt{5}})$ is the trigonometric form of $2 + i$. The trigonometric form of $-2(\cos \alpha + i \sin \alpha)$ is $2((\cos(\alpha + \pi) + i \sin(\alpha + \pi))$. If $z = r(\cos \alpha + i \sin \alpha)$, then the trigonometric form of $\frac{1}{z}$ is $\frac{1}{z} = \frac{1}{r}(\cos(-\alpha) + i \sin(-\alpha))$.

Example V.3.2. Calculate $(1 + i\sqrt{3})^5$. The trigonometric form of the complex number $z = 1 + i\sqrt{3}$ is $z = 2(\frac{1}{2} + i\frac{\sqrt{3}}{2})$, and thus $\arg z = \frac{\pi}{3}$. The modulus of z^5 is 32 and the argument $\frac{5\pi}{3}$. Hence,

$$(1 + i\sqrt{3})^5 = 32(\frac{1}{2} - i\frac{\sqrt{3}}{2}) = 16(1 - i\sqrt{3}).$$

We now present three simple results concerning the elementary geometry of triangles.

Theorem V.3.1 (Triangle inequality). *Given two complex numbers* z_1 *and* z_2, $|z_1+z_2| \leq |z_1| + |z_2|$; *the equality sign holds if and only if* $\overline{z_1}z_2$ *is real.*

Proof. By Theorem V.2.1(ii), $\overline{z_1}z_2 - z_1\overline{z_2}$ is purely imaginary and $(\overline{z_1}z_2 - z_1\overline{z_2})^2 \leq 0$. Expanding the square, we obtain $(\overline{z_1}z_2)^2 + (z_1\overline{z_2})^2 \leq 2z_1z_2\overline{z_1}\overline{z_2}$. Hence we have $(\overline{z_1}z_2 + z_1\overline{z_2})^2 \leq 4z_1z_2\overline{z_1}\overline{z_2}$. This implies

$$z_1\overline{z_1} + z_2\overline{z_2} + \overline{z_1}z_2 + z_1\overline{z_2} \leq z_1\overline{z_1} + z_2\overline{z_2} + 2\sqrt{z_1z_2\overline{z_1}\overline{z_2}}$$

and thus

$$\sqrt{(z_1+z_2)\overline{(z_1+z_2)}} \leq \sqrt{z_1\overline{z_1}} + \sqrt{z_2\overline{z_2}},$$

as required. □

We introduce two consequences of the preceding statements. The first one is a general formulation of the Pythagorean theorem attributed to the Persian mathematician **Ghiyath al-Din al-Kashi (1390 - 1450)**.

Theorem V.3.2 (Law of cosines, Theorem of al-Kashi). *Given* $z_1 \in \mathbb{C}$ *and* $z_2 \in \mathbb{C}$, *we have*

$$|z_1 + z_2|^2 = |z_1|^2 + |z_2|^2 + 2\Re(z_1\overline{z_2}),$$

where $2\Re(z_1\overline{z_2}) = \overline{z_1}z_2 + z_1\overline{z_2} = 2|z_1|\,|z_2|\cos\angle(z_1, z_2)$.

Proof. The proof is straightforward (see Exercise V.3.2). □

Theorem V.3.3 (Two squares theorem). *Given* $a, b, c, d \in \mathbb{R}$, *we have*

$$(a^2 + b^2)\,(c^2 + d^2) = (ac - bd)^2 + (ad + bc)^2.$$

Proof. Let $z = a + b\,i$ and $w = c + d\,i$. The two squares identity follows from the equality $|z|^2\,|w|^2 = |z \cdot w|^2$ as $|z|^2 = a^2 + b^2$, $|w|^2 = c^2 + d^2$, and $zw = (ac - bd) + (ad + bc)i$. □

The concept of the absolute value of a complex number is closely related to the following concept of the **scalar product** in $\mathbb{C}$.

Definition V.3.1. For $z_1 = a_1 + b_1 i \in \mathbb{C}$ and $z_2 = a_2 + b_2 i \in \mathbb{C}$ we define the scalar product $\langle z_1, z_2\rangle$ of z_1 and z_2 by

$$\langle z_1, z_2\rangle \;=\; \Re(\overline{z_1}\;z_2) = \frac{1}{2}\;(\overline{z_1}\;z_2 + z_1\;\overline{z_2}) = a_1a_2 + b_1b_2.$$

Hence we have $\mid z \mid \;=\; \sqrt{\langle z, z\rangle}$. Also $\langle z_1, z_2\rangle = \frac{1}{2}(\overline{z_1}\;z_2 + z_1\;\overline{z_2})$ and thus, if $z_1z_2 \neq 0$, $\langle z_1, z_2\rangle = 0$ if and only if $\cos\angle(z_1, z_2) = 0$, that is, if and only if z_1 and z_2 are **orthogonal (perpendicular)**.

Exercise V.3.3. Prove that the mapping $\mathbb{C} \times \mathbb{C} \to \mathbb{R}$ defined by $(z_1, z_2) \mapsto \langle z_1, z_2 \rangle$ is

(i) **$\mathbb{R}$-bilinear**: for all $z_1, z_2, z_3 \in \mathbb{C}$, $\langle z_1 + z_3, z_2 \rangle = \langle z_1, z_2 \rangle + \langle z_3, z_2 \rangle$ and $\langle r z_1, z_2 \rangle = r\langle z_1, z_2 \rangle$ for $r \in \mathbb{R}$;

(ii) **symmetric**: for all $z_1, z_2 \in \mathbb{C}$, $\langle z_1, z_2 \rangle = \langle z_2, z_1 \rangle$ and

(iii) **positive definite**, that is, $\langle z_1, z_1 \rangle > 0$ whenever $z_1 \neq 0$.

Exercise V.3.4. Show that the vectors described by the complex numbers $z, wz \in \mathbb{C}$ are orthogonal if and only if w is purely imaginary. Using this result, prove that the altitudes of a triangle $A = P(0), B = P(b), C = P(a + ci)$ meet in a common point $H = P(a + \frac{a(b-a)}{c}\, i)$, which is called the **orthocenter** of the triangle ABC.

Exercise V.3.5. Let $z \in \mathbb{C}$ with $|z| = 1$. Prove that for a suitable $w \in \{1, i, -1, -i\}$ the inequality

$$|z - w| \leq \sqrt{2 - \sqrt{2}} \tag{V.3.1}$$

holds.

The multiplicative group $(\mathbb{C}^\times, \cdot)$ of all non-zero complex numbers contains two important subgroups: The multiplicative group $(\mathbb{R}^<, \cdot)$ of all positive real numbers and the **circle group** $S^1 = \{z \in \mathbb{C} \mid |z| = 1\}$ of all complex numbers of modulus equal to 1 with respect to multiplication of complex numbers. These groups are closely related as described in the following theorem.

Theorem V.3.4. *The mapping $\mathbb{C}^\times \to \mathbb{R}^< \times S^1$ defined by $z \mapsto (|z|, \frac{z}{|z|})$ is an isomorphism of the group $\mathbb{C}^\times$ and the cartesian (direct) product of the group $\mathbb{R}^<$ and S^1 (with component wise multiplication).*

Proof. For $t = 1, 2$, write

$$z_t = |z_t|(\cos \alpha_t + i \sin \alpha_t) = |z_t| \cdot \frac{z_t}{|z_t|}.$$

The main step in the proof is the observation that

$$z_1 \cdot z_2 = (|z_1|\, |z_2|)(\cos(\alpha_1 + \alpha_2) + i \sin(\alpha_1 + \alpha_2)) = |z_1|\, |z_2| \cdot \frac{z_1}{|z_1|} \frac{z_2}{|z_2|}.$$

We leave the remaining details to the reader. □

The following statement indicates the important property of the field of complex numbers expressed in the "Fundamental theorem of algebra" on the existence of solutions of algebraic equations (see Chapter VII). In the next theorem (Theorem V.3.5) we show the existence of square roots of complex numbers. It suffices to construct the square roots of a complex number $z = a + bi$ with $b \neq 0$ as in the case $b = 0$ $(z = a)$ the square roots are either real $(a \geq 0)$ or purely imaginary $(a < 0)$. From Theorem V.3.5 we deduce the existence of solutions of quadratic equations with complex coefficients.

Theorem V.3.5 (Existence of square roots). *Any complex number $z = a+bi$ with $b \neq 0$ has exactly two distinct square roots, namely $\sqrt{a+bi} = \pm w$, where*

$$w = \sqrt{\frac{a+\sqrt{a^2+b^2}}{2}} + i\frac{b}{|b|}\sqrt{\frac{-a+\sqrt{a^2+b^2}}{2}}.$$

Proof. As $b \neq 0$ the quantity $\frac{b}{|b|}$ is defined and is equal to 1 if $b > 0$ and to -1 if $b < 0$. We let

$$x = \sqrt{\frac{a+\sqrt{a^2+b^2}}{2}} \quad \text{and} \quad y = \frac{b}{|b|}\sqrt{\frac{-a+\sqrt{a^2+b^2}}{2}},$$

so that $w = x + yi$. As $\sqrt{a^2+b^2} > |a|$, x and y are nonzero real numbers, which satisfy

$$x^2 - y^2 = \left(\frac{a+\sqrt{a^2+b^2}}{2}\right) - \left(\frac{-a+\sqrt{a^2+b^2}}{2}\right) = a$$

and

$$2xy = 2\frac{b}{|b|}\sqrt{\left(\frac{a+\sqrt{a^2+b^2}}{2}\right)\left(\frac{-a+\sqrt{a^2+b^2}}{2}\right)} = 2\frac{b}{|b|}\sqrt{\frac{b^2}{4}} = 2\frac{b}{|b|}\frac{|b|}{2} = b.$$

As $x, y \neq 0$ we deduce that $w \neq 0$. Further $(\pm w)^2 = w^2 = (x+yi)^2 = (x^2-y^2)+(2xy)i = a+bi$. Hence we have shown the existence of two complex numbers, which are square roots of $a+bi$, namely, w and $-w$. They are distinct as $w \neq 0$. Finally, suppose that $v \in \mathbb{C}$ is a square root of z. Then $v^2 = z = w^2$ so that $(v-w)(v+w) = 0$. As $v \pm w \in \mathbb{C}$ and $\mathbb{C}$ has no proper divisors of 0, we deduce that $v = w$ or $v = -w$, proving that z has exactly two complex square roots. □

If the complex number z is expressed in trigonometric form $z = |z|(\cos\alpha + i\sin\alpha)$ then $\sqrt{z} = |z|^{\frac{1}{2}}(\cos\frac{\alpha}{2} + i\sin\frac{\alpha}{2})$.

Now we show that a quadratic equation with complex coefficients has solutions which are complex numbers. We may take the equation as

$$z^2 + uz + v = 0, \quad \text{where } u, v \in \mathbb{C}.$$

Then, using the old Babylonian technique of completing the square, we obtain

$$(z + \frac{u}{2})^2 = \frac{u^2-4v}{4}.$$

From this, we have the two complex solutions (not necessarily distinct)

$$z_1 = \frac{-u+\sqrt{u^2-4v}}{4} \quad \text{and} \quad z_2 = \frac{-u-\sqrt{u^2-4v}}{4}. \tag{V.3.2}$$

Here, of course, in the formulas for both z_1 and z_2, $\sqrt{u^2-4v}$ denotes the same square root.

The corresponding linear factorization $z^2+uz+v = (z-z_1)(z-z_2)$ provides the formulas of **François Viète (1540 - 1603)**

$$z_1 + z_2 = -u \quad \text{and} \quad z_1 z_2 = v.$$

Example V.3.3. (i) Calculate $x + yi = \sqrt{5 - 12i}$. Here $x^2 - y^2 = 5$ and $2xy = -12$. Thus $x^2 + y^2 = \sqrt{169} = 13$. Hence $x^2 = 9$, that is, $x = \pm 3$ and $y = \pm 2$. Since $xy = -6 < 0$, the roots are $3 - 2i$ and $-3 + 2i$.

(ii) Find the roots of the equation $x^2 - 4ix - (9 - 12i) = 0$. We have $(x - 2i)^2 = -4 + 9 - 12i = 5 - 12i$; thus $x - 2i = \pm(3 - 2i)$, and the roots are $x_1 = 3$ and $x_2 = -3 + 4i$.

Exercise V.3.6. Solve the following equations, and represent the solutions as points in the Argand-Gauss plane:

(i) $(2 + i)z^2 - (9 + 2i)z + 5(3 - i) = 0$.

(ii) $z^2 + (1 - i)z + 4 + 7i = 0$.

(iii) $z^4 - z^2 + 1 = 0$.

(iv) $z^8 + z^4 + 1 = 0$.

(v) $z^2 - 2\,\overline{z} + 1 = 0$.

Exercise V.3.7. Let z_1 and z_2 be the roots of the quadratic equation $z^2 + uz + v = 0$ with $u, v \in \mathbb{C}, v \neq 0$. If $|z_1| = |z_2|$, prove that $\frac{u}{v}$ is a real number.

Exercise V.3.8. Suppose that one of the roots of the quadratic equation $z^2 + uz + v = 0$ with $|u| = |v| = 1$ has absolute value 1. Prove that $u^2 = v$.

Exercise V.3.9. Give a necessary and sufficient condition for both roots of $z^2 + uz + v = 0$ $(u, v \in \mathbb{C})$ to be real.

Exercise V.3.10. Give a necessary and sufficient condition for at least one of the roots of $z^2 + uz + v = 0$ $(u, v \in \mathbb{C})$ to be real.

V.4 Euler's exponential form, de Moivre's theorem

Definition V.4.1. *For $\alpha \in \mathbb{R}$, we define*

$$e^{i\alpha} = \cos\alpha + i\sin\alpha.$$

This definition allows us to write the trigonometric form of the complex number z, namely $z = r(\cos\alpha + i\sin\alpha)$, in the exponential form $z = re^{i\alpha}$, where $|z| = r$ and $\alpha \equiv \arg z \pmod{2\pi}$.

If $\alpha, \beta \in \mathbb{R}$ we have

$$\begin{aligned} e^{i\alpha} \cdot e^{i\beta} &= (\cos\alpha + i\sin\alpha)(\cos\beta + i\sin\beta) \\ &= (\cos\alpha\cos\beta - \sin\alpha\sin\beta) + i(\sin\alpha\cos\beta + \cos\alpha\sin\beta) \\ &= \cos(\alpha+\beta) + i\sin(\alpha+\beta), \end{aligned}$$

so that $e^{i\alpha} \cdot e^{i\beta} = e^{i(\alpha+\beta)}$, and by induction for $\alpha_1, \alpha_2, \ldots, \alpha_n \in \mathbb{R}$ we have

$$e^{i(\alpha_1+\alpha_2+\cdots+\alpha_n)} = e^{i\alpha_1}e^{i\alpha_2}\cdots e^{i\alpha_n}.$$

Choosing $\alpha_1 = \alpha_2 = \cdots = \alpha_n = \alpha$ we deduce

$$e^{in\alpha} = (e^{i\alpha})^n, \quad n \in \mathbb{N}.$$

Choosing $\beta = -\alpha$ we deduce

$$e^{i\alpha} \cdot e^{i(-\alpha)} = e^{i(\alpha+(-\alpha))} = e^{i0} = \cos 0 + i\sin 0 = 1, \text{ so } (e^{i\alpha})^{-1} = e^{-i\alpha}.$$

Example V.4.1. Taking $\alpha = 2\pi, \pi, \frac{\pi}{2}, \frac{\pi}{3}, \frac{\pi}{4}$ in Definition V.4.1, we obtain

$$e^{2i\pi} = 1; \; e^{i\pi} = -1; \; e^{i\frac{\pi}{2}} = i; \; e^{i\frac{\pi}{3}} = \frac{1}{2} + i\frac{\sqrt{3}}{2}; \; e^{i\frac{\pi}{4}} = \frac{1+i}{\sqrt{2}}.$$

Thus for example we have

$$(1-i)^8 = (\sqrt{2}e^{-i\frac{\pi}{4}})^8 = (\sqrt{2})^8 e^{-8i\frac{\pi}{4}} = 16.$$

Remark V.4.1. We now give a brief justification of the definition of the exponential form of a complex number by means of a brief digression into the theory of complex functions. The real exponential function e^x ($x \in \mathbb{R}$), as well as the functions $\cos x$ and $\sin x$, can be extended to complex functions $e^z, \cos z$ and $\sin z$ by making use of the Taylor series expansions

$$e^z = 1 + \frac{1}{1!}z + \frac{1}{2!}z^2 + \frac{1}{3!}z^3 + \frac{1}{4!}z^4 + \frac{1}{5!}z^5 + \frac{1}{6!}z^6 + \frac{1}{7!}z^7 + \cdots$$

$$\sin z = \frac{1}{1!}z - \frac{1}{3!}z^3 + \frac{1}{5!}z^5 - \frac{1}{7!}z^7 + \cdots,$$

$$\cos z = 1 - \frac{1}{2!}z^2 + \frac{1}{4!}z^4 - \frac{1}{6!}z^6 + \frac{1}{8!}z^8 + \cdots$$

Replacing z by ix ($x \in \mathbb{R}$), we deduce

$$e^{ix} = 1 + \frac{1}{1!}ix - \frac{1}{2!}x^2 - \frac{1}{3!}ix^3 + \frac{1}{4!}x^4 + \frac{1}{5!}ix^5 - \frac{1}{6!}x^6 - \frac{1}{7!}ix^7 + \frac{1}{8!}x^8 + \cdots = \cos x + i\sin x.$$

We now prove the well-known theorem of **Abraham de Moivre**.

Theorem V.4.1 (De Moivre's Theorem). *For $\alpha \in \mathbb{R}$ and $n \in \mathbb{Z}$, we have*

$$(\cos\alpha + i\,\sin\alpha)^n = \cos(n\alpha) + i\,\sin(n\alpha) \quad \textit{and} \quad (\cos\alpha - i\,\sin\alpha)^n = \cos(n\alpha) - i\,\sin(n\alpha).$$

Proof. We make use of the previously proved relations

$$(e^{i\alpha})^{-1} = e^{-i\alpha}, \quad (e^{i\alpha})^n = e^{in\alpha} \ (n \in \mathbb{N}).$$

Let $n \in \mathbb{Z}$. If $n > 0$ we have

$$(\cos\alpha + i\sin\alpha)^n = (e^{i\alpha})^n = e^{in\alpha} = \cos n\alpha + i\sin n\alpha.$$

If $n < 0$ then $n = -m$ with $m \in \mathbb{N}$ and

$$(\cos\alpha + i\sin\alpha)^{-1} = (e^{i\alpha})^{-1} = e^{-i\alpha}$$

so

$$(\cos\alpha + i\sin\alpha)^{-m} = (e^{-i\alpha})^m = e^{-im\alpha} = e^{in\alpha} = \cos n\alpha + i\sin n\alpha.$$

For any nonzero complex number w, w^0 is defined to be 1 so as to be consistent with the law of indices $w^k w^l = w^{k+l}$ $(k, l \in \mathbb{Z})$ when $l = 0$. Now $\cos\alpha + i\sin\alpha \neq 0$ as its modulus is 1. Thus, for $n = 0$ we have $(\cos\alpha + i\sin\alpha)^n = (\cos\alpha + i\sin\alpha)^0 = 1 = \cos 0 + i\sin 0 = \cos n\alpha + i\sin n\alpha$.

This completes the proof of the first equality. We note that the second equality follows from the first on replacing α by $-\alpha$. □

The exponential form of complex numbers allows one to express the values $\cos\alpha$ and $\sin\alpha$ in the exponential (the so-called Euler's) form:

$$\cos\alpha = \frac{e^{i\,\alpha} + e^{-i\,\alpha}}{2} \quad \text{and} \quad \sin\alpha = \frac{e^{i\,\alpha} - e^{-i\,\alpha}}{2\,i}.$$

Example V.4.2. To calculate $\left(\sqrt{3} - i\right)^7$, we proceed as follows:

$$(\sqrt{3} - i)^7 = [2\,(\frac{\sqrt{3}}{2} + (-\frac{1}{2})\,i)]^7 = 2^7\,(\cos\frac{11\pi}{6} + i\,\sin\frac{11\pi}{6})^7 = 2^7\,(\cos\frac{77\pi}{6} + i\,\sin\frac{77\pi}{6})$$
$$= 128\,(\cos\frac{5\pi}{6} + i\,\sin\frac{5\pi}{6}) = 64\,(-\sqrt{3} + i).$$

In 1706, **John Machin (1680 - 1752)** used the relation

$$\frac{(5+i)^4}{239+i} = 2 + 2i$$

to calculate the number π. The equality yields

$$\frac{\pi}{4} = 4\arctan\frac{1}{5} - \arctan\frac{1}{239}$$

and making use of this equality, Machin determined π to 100 decimal places. Indeed, he made use of the trigonometric form of complex numbers and applied the relation

$$\arg(x + yi) = \arctan\frac{y}{x}.$$

Exercise V.4.1. Justify the procedure of Machin. In a similar vein, following **John Dahse**, who in 1844 used the formula of Strassnicky

$$\frac{\pi}{4} = \arctan\frac{1}{2} + \arctan\frac{1}{5} + \arctan\frac{1}{8}$$

for his 2-month long calculation of π to 205 correct places, consider the following "8 squares display"

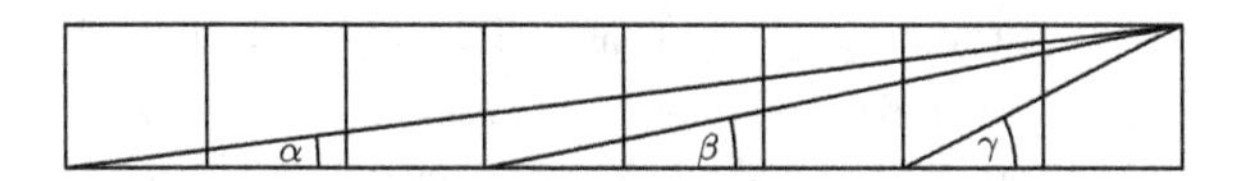

and justify Strassnicky's formula by showing that $\alpha + \beta + \gamma = \frac{\pi}{4}$.

Remark V.4.2. The technique of Machin was explored by many mathematicians. Thus Euler has his formula

$$\frac{\pi}{4} = 4\arctan\frac{1}{5} - \arctan\frac{1}{70} + \arctan\frac{1}{99},$$

whilst Gauss' formula reads

$$\frac{\pi}{4} = 12\arctan\frac{1}{18} + \arctan\frac{1}{57} - 5\arctan\frac{1}{239}.$$

Exercise V.4.2. Prove the above formula of Euler. You may also like the challenge of proving a remarkable formula of Takano:

$$\frac{\pi}{4} = 12\arctan\frac{1}{49} + 32\arctan\frac{1}{57} - 5\arctan\frac{1}{239} + 12\arctan\frac{1}{110\,443}.$$

The reader will find some very general arctangent formulas for π due to **Marc Chamberland** and **Eugene A. Herman** in their article in The American Mathematical Monthly 126 (2019), pp. 646–650.

De Moivre's theorem provides a direct way of determining the n-th roots of unity, that is, the roots of the polynomial $z^n - 1$. We leave the proof of the following theorem to the reader.

Theorem V.4.2. *For any natural number n, the equation $z^n = 1$ has n distinct roots (the so-called n-th roots of unity)*

$$z_k = e^{i\frac{2k\pi}{n}} \quad \textit{for } k = 0, 1, 2, \ldots, n-1.$$

More generally, the equation $z^n = re^{i\alpha}$ has n distinct roots

$$z_k = \sqrt[n]{r}\, e^{i\frac{\alpha+2k\pi}{n}} \quad \textit{for } k = 0, 1, 2, \ldots, n-1.$$

The respective points $P(z_k)$ in the Argand-Gauss plane form a regular n–polygon.

Theorem V.4.3. *The set G of n-th roots of unity for all $n = 1, 2, 3, \ldots$ forms a (commutative) multiplicative group. For each $n \in \mathbb{N}$, the set G_n of all n-th roots of unity forms a multiplicative group of n elements, a subgroup of G. The mapping $f : \mathbb{Z}_n \to G_n$ by $f([k]) = z_k$, $t = 0, 1, \ldots, n-1$ is a bijection between $\mathbb{Z}_n$ and G_n that satisfies $f([k_1] + [k_2]) = z_{k_1} \cdot z_{k_2}$ and hence is an* **isomorphism** *of the (additive) group $\mathbb{Z}_n$ and the (multiplicative) group G_n.*

The subgroups $G_n, n = 1, 2, 3, \ldots$ are the only **finite** *subgroups of G; they form a lattice isomorphic to $(\mathbb{N}, \preceq)$ of Section 4 of Chapter III.*

Proof. If $z_1^n = 1$ and $z_2^n = 1$, then $(z_1 z_2)^n = 1$, as well as $\left(\frac{z_1}{z_2}\right)^n = 1$, and thus G_n is a group, a subgroup of G. Associate with every element $g \in G$ the least natural number n_g such that g is a n_g-th root of unity. Such a number must exist. Indeed, if $g^{k_1} = g^{k_2} = 1$, then, in view of Bézout's theorem, $g^k = 1$, where k is the greatest common divisor of the numbers k_1 and k_2.

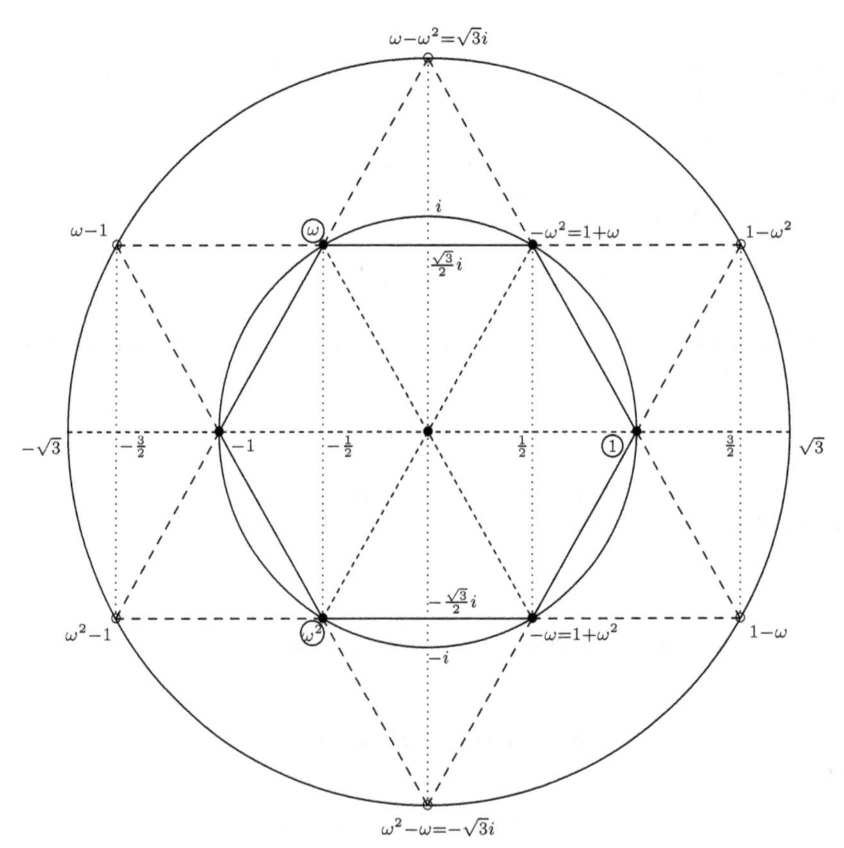

Figure V.4.1 Cube roots of unity in the Argand-Gauss plane

Now, if H is a finite subgroup of G (that is, a finite subset H of G such that $h_1, h_2 \in H$ implies that both $h_1 h_2$ and $\frac{h_1}{h_2}$ belong to H), there is $h_1 \in H$ such that n_{h_1} is the largest among all $n_h, h \in H$. We claim that H is the cyclic group $\langle h_1 \rangle = \{h_1, h_1^2, \ldots, h_1^{n_{h_1}} = 1\}$

generated by h_1. For, if $g \in H, g \notin \langle h_1 \rangle$, then $n_g < n_{h_1}$. But then the element $gh_1 \in H$ satisfies $(gh_1)^n = 1$, where $n > h_1$ is the least common multiple of n_g and n_{h_1} and $n = n_{gh_1}$, in contradiction to the choice of h_1.

Hence G_n is the **unique** subgroup of n elements of G and that $G_{n_1} \subseteq G_{n_2}$ if and only if n_1 is a divisor of n_2. The statement concerning the lattice of **finite** subgroups of G follows. □

Remark V.4.3. We point out that the group G of all n-th roots of unity has many proper **infinite** subgroups. For instance, for each prime p, G contains the subgroup of all p^k-th roots of unity for $k = 1, 2, \ldots$. This is a subgroup that is a union of the chain of subgroups G_{p^k}, the so-called Prüfer p–group, named after **Ernst Paul Heinz Prüfer (1896 - 1934)**.

Example V.4.3. G_3 consists of the elements $\omega_0 = 1, \omega_1 = e^{i\frac{2\pi}{3}} = -\frac{1}{2} + \frac{\sqrt{3}}{2}i$ and $\omega_2 = e^{i\frac{4\pi}{3}} = -\frac{1}{2} - \frac{\sqrt{3}}{2}i$. Thus, the cube roots of $z = re^{i\alpha}$ are $\sqrt[3]{r}e^{i\frac{\alpha}{3}}$, $\sqrt[3]{r}e^{i\frac{\alpha+2\pi}{3}}$ and $\sqrt[3]{r}e^{i\frac{\alpha+4\pi}{3}}$. Figure V.4.1 illustrates rather lucidly the place of the cube roots of unity in the Argand-Gauss plane.

Exercise V.4.3. Draw the n-th roots of unity for $1 \leq n \leq 10$ in the Argand-Gauss plane.

Example V.4.4. To find the solutions of the equation

$$(z + i)^n + (z - i)^n = 0$$

use the substitution $w = \frac{z+i}{z-i}$ and de Moivre's theorem:

$$z = \frac{s}{1-c} = \frac{1+c}{s}, \quad \text{where } s = \sin\frac{(2k+1)\pi}{n} \quad \text{and } c = \cos\frac{(2k+1)\pi}{n}.$$

Describe their representation in the Argand-Gauss plane. From the expression for z, we see that the largest absolute values of the solutions tend monotonically to infinity. For example, the solutions of the largest absolute value for $n = 8$ are $z = \pm 5,0273... = \pm(1 + \sqrt{2}(1 + \sqrt{2 + \sqrt{2}}))$, for $n = 1.000$, $z = \pm 636,6202...$ and for $n = 100.000$, $z = \pm 65.449,8469...$

Exercise V.4.4. Show that

$$z = \frac{(\alpha - \beta)si}{2(c-1)} - \frac{\alpha + \beta}{2},$$

where s and c are defined as in Example V.4.4, and α and β are arbitrary complex numbers, are solutions of the equation

$$(z + \alpha)^n + (z + \beta)^n = 0.$$

Describe the solutions of this equation in terms of the relation between the numbers α and β.

V.5 Geometry of complex numbers

The representation of complex numbers z as point $P(z)$ of the cartesian plane $\mathbb{R} \times \mathbb{R}$ described in the previous section provides a natural and efficient route to describe some phenomena of the geometry of the plane. This is mainly due to the fact that the operations of addition and multiplication have a geometrical meaning. Indeed, addition of two numbers

z_1 and z_2 corresponds to the addition of the respective vectors $\overrightarrow{z_1}$ and $\overrightarrow{z_2}$. And multiplication of z by i corresponds, as already Argand pointed out, to the rotation of $P(z)$ around the origin by 90^0 (in the positive direction, that is, counterclockwise). Thus, vectors $\overrightarrow{z}$ and $\overrightarrow{w}$ are orthogonal if and only if $w = r\,.\,zi$, where $r \in \mathbb{R}, r \neq 0$.

To understand the full value of this representation, it is important to realize that such a representation of the field $\mathbb{C}$ of complex numbers is one, albeit the "best one", of infinitely many other possibilities. To illustrate this situation, we consider the following example.

Example V.5.1. Consider the set $\mathbb{C}_\omega = \{a \oplus b\,\omega \mid a, b \in \mathbb{R}\}$ and define the operation $\oplus$ of addition and $\odot$ of multiplication by

$$(a \oplus b\,\omega) \oplus (c \oplus d\,\omega) = (a + c) \oplus (b + d)\,\omega \quad \text{and}$$

$$(a \oplus b\,\omega) \odot (c \oplus d\,\omega) = (ac - bd) \oplus (ad + bc + bd)\,\omega.$$

Notice that for real numbers a, c, that is, for elements $a \oplus 0\,\omega, c \oplus 0\,\omega$, of $\mathbb{C}_\omega$, the operation $\oplus$ and $+$ coincide and in that case we shall use both notations quite freely; in any case, this way we shall avoid using $\ominus$ for subtraction.

Now, define the mapping $F : (\mathbb{C}_\omega, \oplus, \odot) \to \mathbb{C}$ by

$$F(a \oplus b\,\omega) \;=\; (a + \frac{1}{2}\,b) + \frac{\sqrt{3}}{2}\,b\,i\,.$$

F is an injective mapping satisfying

$$F((a \oplus b\,\omega) \oplus (c \oplus d\,\omega)) = F(a \oplus b\,\omega) + F(c \oplus d\,\omega).$$

Moreover,

$$F((a \oplus b\,\omega) \odot (c \oplus d\,\omega)) = F((ac - bd) \oplus (ad + bc + bd)\,\omega)$$

$$= [ac + \frac{1}{2}(ad + bc - bd)] + [\frac{\sqrt{3}}{2}(ad + bc + bd)]\,i = \left[(a + \frac{1}{2}\,b) + \frac{\sqrt{3}}{2}\,b\,i\right]\left[(c + \frac{1}{2}\,d) + \frac{\sqrt{3}}{2}\,d\,i\right]$$

$$= F(a \oplus b\,\omega)F(c \oplus d\,\omega).$$

Furthermore

$$F^{-1}(a + b\,i) = (a - \frac{\sqrt{3}}{3}\,b) \oplus \frac{2\sqrt{3}}{3}\,b\,\omega.$$

Hence F is an isomorphism of $(\mathbb{C}_\omega, \oplus, \odot)$ and the field $\mathbb{C}$ of complex numbers. Here, $F(a) = F(a \oplus 0\,\omega) = a$ for all real a and

$$F(-\frac{\sqrt{3}}{3} \oplus \frac{2\sqrt{3}}{3}\,\omega) = i.$$

Note that

$$\omega^2 = -1 \oplus \omega,\ \omega^3 = -1,\ \omega^4 = -\omega,\ \omega^5 = 1 \oplus (-\omega) \ \text{ and } \omega^6 = 1.$$

The number ω therefore generates the cyclic group of all 6-th roots of unity. For a given $z = a \oplus b\,\omega$, the number

$$z \odot (-1 \oplus 2\,\omega) = (a \oplus b\,\omega) \odot (-1 \oplus 2\,\omega) = (-a - 2b) \oplus (2a + b)\,\omega$$

is "perpendicular" to z. One can define the "conjugate" $\overline{a \oplus b\,\omega}$ of $a \oplus b\,\omega$ by

$$\overline{a \oplus b\,\omega} = (a+b) \oplus (-b)\,\omega.$$

This is an involution, that is, $\overline{\overline{z}} = z$. Moreover, the absolute value of $z = a \oplus b\,\omega$, $|a \oplus b\,\omega| = \sqrt{z \odot \overline{z}} = \sqrt{a^2 + ab + b^2}$.

In conclusion, we point out that the described representation of the complex numbers relates to a choice of the coordinates of the real plane. While in the Argand-Gauss plane we deal with an orthonormal system of coordinates, in this example we deal with a system in which the axes make a 60^0 angle.

Exercise V.5.1. (i) Show that the set $\mathbb{C}_\kappa = \{a \oplus b\,\kappa \mid a, b \in \mathbb{R}\}$ with the operation $\oplus$ of addition and $\odot$ of multiplication defined by

$$(a \oplus b\,\kappa) \oplus (c + d\,\kappa) = (a+c) \oplus (b+d)\,\kappa \text{ and}$$

$$(a \oplus b\,\kappa) \odot (c \oplus d\,\kappa) = (ac - bd) \oplus (ad + bc + \sqrt{2}\,bd)\,\kappa$$

is isomorphic with the field of complex numbers.

(ii) Show that the set $\mathbb{C}_\lambda = \{a \oplus b\,\lambda \mid a, b \in \mathbb{R}\}$ with the operation $\oplus$ of addition and $\odot$ of multiplication defined by

$$(a \oplus b\,\lambda) \oplus (c + d\,\lambda) = (a+c) \oplus (b+d)\,\lambda \text{ and}$$

$$(a \oplus b\,\lambda) \odot (c \oplus d\,\lambda) = (ac - bd) \oplus (ad + bc + \frac{6}{5}\,bd)\,\lambda$$

is isomorphic with the field of complex numbers.

(iii) Based on the Example V.5.1 and parts (i) and (ii) of Exercise V.5.1, describe a general representation of the field of complex numbers as pairs of real numbers.

We start our brief excursion into the geometry of the real plane by considering equations of a straight line.

Theorem V.5.1 (Parametric equation of the straight line). *Let $P_1 = P(z_1)$ and $P_2 = P(z_2)$, where $z_1 \neq z_2$. Then all points on the line $P_1\,P_2$ correspond to the numbers*

$$z = z_1 + t(z_2 - z_1), t \in \mathbb{R}.$$

The point P of the segment P_1, P_2 such that $|P_1, P| : |P, P_2| = 1 : r$ corresponds to the number

$$z = z_1 + \frac{1}{r+1}(z_2 - z_1); \qquad \text{(V.5.1)}$$

in particular, the midpoint of P_1, P_2 corresponds to $z = \frac{z_1+z_2}{2}$.

The **centroid (barycenter)** *of a triangle P_1, P_2 and $P_3 = P(z_3)$, that is, the intersection of its medians, is the point*

$$P(\frac{z_1 + z_2 + z_3}{3}).$$

Remark V.5.1. We explicitly mention that the **gradient** of the straight line defined by the points $P_1 = P(z_1)$ and $P_2 = P(z_2)$, $z_1 \neq z_2$, is given by the vector $\overrightarrow{w}$, where $w = z_2 - z_1$, and hence is equal to

$$\frac{\Im(w)}{\Re(w)} = -i\,\frac{w - \overline{w}}{w + \overline{w}}.$$

The division of a line segment P_1, P_2 in the ratio $m : n$ $(m > 0, n > 0)$ is given by the point $P(z)$, where

$$z = \frac{mz_1 + nz_2}{m + n},$$

as indicated in Figure V.5.1.

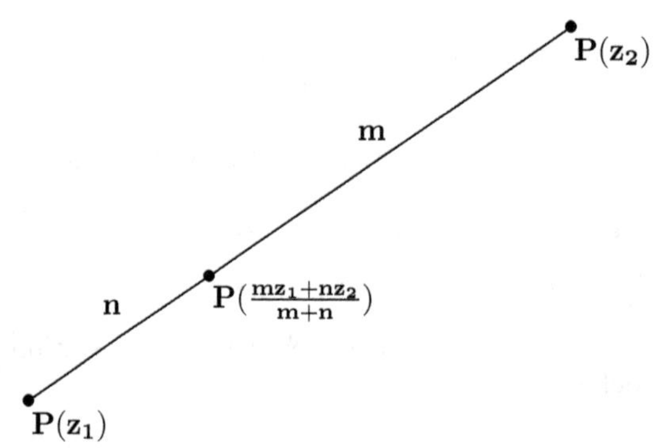

Figure V.5.1 Division of a line segment

Theorem V.5.1 leads straightforwardly to the **barycentric coordinates**, which were introduced in Example I.3.3 and Exercise I.3.1 of Chapter I. The equation (V.5.1) also allows us to describe a straight line by a complex equation. For, the parameter $t = \frac{z - z_1}{z_2 - z_1}$ is real if and only if

$$\frac{z - z_1}{z_2 - z_1} = \overline{\left(\frac{z - z_1}{z_2 - z_1}\right)}.$$

Observe that this relation means that the determinant

$$\begin{vmatrix} z & \overline{z} & 1 \\ z_1 & \overline{z_1} & 1 \\ z_2 & \overline{z_2} & 1 \end{vmatrix} = 0. \tag{V.5.2}$$

Hence the points $P(z)$ of the straight line $P_1\,P_2$ satisfy

$$\frac{i}{z_2 - z_1}\,z - \frac{i}{\overline{z_2} - \overline{z_1}}\,\overline{z} + \frac{\overline{z_1}\,i}{\overline{z_2} - \overline{z_1}} - \frac{z_1\,i}{z_2 - z_1} = 0,$$

and thus

$$wz + \overline{wz} = r \text{ with } w = \frac{i}{z_2 - z_1} \text{ and the real number } r = \frac{(z_1\,\overline{z_2} - \overline{z_1}\,z_2)\,i}{|z_2 - z_1|^2}.$$

Consequently, we can formulate the following theorem whose proof is left as an exercise.

Theorem V.5.2. *Every straight line of the Argand-Gauss plane can be described uniquely, up to a nonzero real multiple, by the equation*

$$wz + \overline{wz} = r, \quad \textit{where } 0 \neq w \in \mathbb{C} \textit{ and } r \in \mathbb{R}. \tag{V.5.3}$$

The vector $\overrightarrow{\overline{w}i}$ *defines the gradient of this line.*

If w *is a real number, then the line is parallel to the imaginary* y*-axis and intersects the real* x*-axis at the point* $P(\frac{r}{2w})$. *The equation of the imaginary axis is thus* $z + \overline{z} = 0$.

If w *is not a real number, then* $w - \overline{w} \neq 0$, *the straight line intersects the (imaginary)* y*-axis at the point* $P(\frac{r\,i}{w-\overline{w}})$ *and its gradient is* $\frac{w+\overline{w}}{w-\overline{w}}\,i$. *The line is parallel to the real* x*-axis if and only if* w *is purely imaginary. The equation of the real axis is* $z - \overline{z} = 0$.

Exercise V.5.2 (Normed equation of a straight line). Prove that every equation of the form (V.5.1) can be put in the form

$$-\frac{i}{2}\,z + \frac{i}{2}\,\overline{z} = r$$

or in the form

$$wz + \overline{wz} = r, \quad \text{where } w + \overline{w} = 1.$$

In the first case, the line is parallel to the x-axis and the point $P(ri)$ is the point of intersection with the y-axis. In the second ("general") case, $P(r)$ is a point on the line which is parallel to the y-axis if w is real. If $w \neq \overline{w}$, then the gradient of the straight line equals

$$\frac{w + \overline{w}}{w - \overline{w}}\,i.$$

Exercise V.5.3. Show that the "complex" equation (V.5.3) of a straight line is equivalent to the "real" equation

$$(w + \overline{w})\,x + (w - \overline{w})\,i\,y - r = 0,$$

and that, on the other hand, the equation $ax + by + c = 0$, where $a, b, c \in \mathbb{R}$, is equivalent to the equation

$$\frac{a - bi}{2}\,z + \frac{a + bi}{2}\,\overline{z} = -c.$$

Exercise V.5.4. Show that the distance of a given point $P(z_0)$ of the Argand-Gauss plane from the line given by the equation (V.5.1) is

$$\left|\frac{wz_0 + \overline{wz_0} - r}{2w}\right|.$$

Exercise V.5.5. Consider two lines, one given by the equation (V.5.3) and the other by the equation $vz + \overline{vz} = s$, where $v \neq 0$. Prove the following two assertions.

(i) The lines are parallel if and only if the number $\frac{w}{v}$ is real, that is, if $w\overline{v} = \overline{w}v$.

(ii) The lines are perpendicular if and only if the number $\frac{w}{v}$ is purely imaginary, that is, if $w\overline{v} + \overline{w}v = 0$.

We now turn our attention to circles in the Argand-Gauss plane. It is evident that a circle with center $P(z_0)$ and radius $r > o$ can be described by the relation

$$z = z_0 + re^{i\varphi}, \text{ where } 0 \leq \varphi < 2\pi,$$

or, equivalently, by $|z - z_0| = r$, that is, by $(z - z_0)(\overline{z} - \overline{z_0}) = r^2$. This can be rewritten as the following statement.

Theorem V.5.3. *The relation*

$$z\overline{z} + wz + \overline{wz} = s, \text{ where } s \in \mathbb{R} \text{ and } s + w\overline{w} > 0,$$

describes the circle whose center is the point $P(-\overline{w})$ and whose radius $r = \sqrt{s + w\overline{w}}$.

Exercise V.5.6. Let D be the set of all points z in the Argand-Gauss plane that satisfy the condition

$$\frac{|z - w_1|}{|z - w_2|} = r, \text{ where } w_1, w_2 \in \mathbb{C}, r \in \mathbb{R}.$$

Show that

(i) for $r = 1$, D is the line given by the points $P(w_1)$ and $P(w_2)$;

(ii) for $r \neq 1$, D is the circle whose center is the point $P(\frac{r^2 w_2 - w_1}{r^2 - 1})$ and whose radius is equal to $\frac{r}{|r^2 - 1|} |w_1 - w_2|$.

Exercise V.5.7. Justify the following statements. Every straight line perpendicular to the line given by the equation (V.5.1) is given by the equation $i\,wz + \overline{i\,wz} = s$, where $s \in \mathbb{R}$. Thus the equation of the line passing through the point $P(z_0)$ and perpendicular to the line given by (V.5.1) is the line whose equation is $i\,wz + \overline{i\,wz} = -2\Im(wz_0)$, that is, $wz - \overline{wz} = wz_0 - \overline{wz_0}$. The intersection point of those two perpendicular lines is the point

$$P(\frac{1}{2w}[w(w_0 + z_0) + \overline{w}(\overline{w_0} - \overline{z_0})]).$$

Hence the foot of the altitude from the point $P(z_3)$ in a triangle $\Delta = P(z_1)P(z_2)P(z_3)$ is the point

$$P(w_3), \text{ where } w_3 = \frac{1}{2(\overline{z_2} - \overline{z_1})}[(z_3 + z_1)(\overline{z_2} - \overline{z_1}) + (\overline{z_3} - \overline{z_1})(z_2 - z_1)].$$

Exercise V.5.8. Show that the equation of the tangent to the circle given by the equation $|z - z_0| = |w_0 - z_0|$ (whose center is the point $P(z_0)$ and whose radius is $r = |w_0 - z_0|$) at the point $P(w_0)$ is

$$(z - w_0)\,(\overline{z_0} - \overline{w_0}) + (\overline{z} - \overline{w_0})\,(z_0 - w_0) = 0.$$

An application of complex numbers allows us to easily calculate the area of a given triangle.

Theorem V.5.4. *Let $\Delta = P(z_1)P(z_2)P(z_3)$ be a triangle oriented counterclockwise. Then its area $A(\Delta)$ is given by the formula*

$$A(\Delta) \;=\; \frac{1}{2}\,\Im(\overline{z_1}\,z_2 + \overline{z_2}\,z_3 + \overline{z_3}\,z_1). \tag{V.5.4}$$

Hence the area of an unoriented triangle Δ is simply $|A(\Delta)|$.

Proof. We are going to give a proof in three steps. We will exploit the fact that a **translation** of the plane in the direction $\overrightarrow{w}$, that is, the mapping $T_w : z \mapsto z + w$ and the **rotation** of the plane around the origin $P(0)$ by the angle φ (counterclockwise), that is, the mapping $R_{0,\varphi} : z \mapsto ze^{i\varphi} = z\,(\cos\varphi + i\,\sin\varphi)$ preserve the distance between any two points, that is, they are **isometries** of the plane.

First step. The number $z = \overline{z_1}\,z_2 + \overline{z_2}\,z_3 + \overline{z_3}\,z_1$ does not change if we rotate the plane around the origin $P(0)$ (that is, z is an invariant of the rotation around the origin). We have only to observe that

$$\begin{aligned}
&\overline{R_{(0,\varphi)}(z_1)}R_{(0,\varphi)}(z_2) + \overline{R_{(0,\varphi)}(z_2)}R_{(0,\varphi)}(z_3) + \overline{R_{(0,\varphi)}(z_3)}\,R_{(0,\varphi)}(z_1)\\
&= \overline{z_1}\,e^{-i\varphi}z_2\,e^{i\varphi} + \overline{z_2}\,e^{-i\varphi}z_3\,e^{i\varphi} + \overline{z_3}\,e^{-i\varphi}z_1\,e^{i\varphi} \;=\; \overline{z_1}\,z_2 + \overline{z_2}\,z_3 + \overline{z_3}\,z_1 = z.
\end{aligned}$$

Second step. The real number $\Im z$ does not change if we arbitrarily translate the plane, that is, $\Im z$ is an invariant of any translation T_w. This is a consequence of the following calculation:

$$\begin{aligned}
&\overline{T_w(z_1)}\,T_w(z_2) + \overline{T_w(z_2)}\,T_w(z_3) + \overline{T_w(z_3)}\,T_w(z_1)\\
&= \overline{(z_1+w)}\,(z_2+w) + \overline{(z_2+w)}\,(z_3+w) + \overline{(z_3+w)}\,(z_1+w)\\
&= \overline{z_1}\,z_2 + \overline{z_2}\,z_3 + \overline{z_3}\,z_1 \;+\; r,
\end{aligned}$$

where $r = \overline{z_1}\,w \,+\, z_1\,\overline{w} \,+\, \overline{z_2}\,w \,+\, z_2\,\overline{w} \,+\, \overline{z_3}\,w \,+\, z_3\,\overline{w} \,+\, 3w\,\overline{w}$. We have $r = \overline{r}$, and thus r is a real number and therefore the real number $\Im(z)$ has been preserved.

Third step. By a suitable translation and rotation around the origin, the given triangle is mapped into the position when $z_1 = 0, z_2 = a$ and $z_3 = b + ci$, with real numbers $a > 0, b$ and $c > 0$. The area of such a triangle is $\frac{1}{2}ac$. Also

$$\frac{1}{2}\Im(\overline{z_1}\,z_2 + \overline{z_2}\,z_3 + \overline{z_3}\,z_1) \;=\frac{1}{2}\Im(ab + aci) \;=\; \frac{1}{2}ac.$$

This shows that the formula (V.5.4) for the area of the (counterclockwise) oriented triangle Δ is valid.

Now, if we consider the clockwise orientation of the same triangle we would get instead of the number $z = \overline{z_1}\,z_2 + \overline{z_2}\,z_3 + \overline{z_3}\,z_1$ the number $z' = \overline{z_1}\,z_3 + \overline{z_3}\,z_2 + \overline{z_2}\,z_1$. We readily see that $z' = \overline{z}$, that is, the numbers z' and z are complex conjugates. Thus the area of an unoriented triangle Δ is simply $|A(\Delta)|$. □

Remark V.5.2. We remark that the area of a counterclockwise oriented triangle Δ is often presented in the form of a determinant:

$$A(\Delta) \;=\; \frac{1}{4i}\begin{vmatrix} z_1 & \overline{z}_1 & 1\\ z_2 & \overline{z}_2 & 1\\ z_3 & \overline{z}_3 & 1 \end{vmatrix}.$$

Thus we have

$$\begin{vmatrix} z_1 & \overline{z_1} & 1 \\ z_2 & \overline{z_2} & 1 \\ z_3 & \overline{z_3} & 1 \end{vmatrix} = 0$$

if and only if the points $P(z_1), P(z_2)$ and $P(z_3)$ lie on a straight line, that is, if they are collinear.

We can apply Theorem V.5.4 to the computation of the area of an arbitrary plane polygon. By a polygon, we mean a part of the plane bounded by a closed finite piecewise linear boundary, that is, bounded by a finite number (greater than or equal to 3) of line segments that do not intersect. These segments are the **sides** and their end points the **vertices** of the polygon. If the number of vertices is n, we speak about a (general) $n-$gon.

Theorem V.5.5. *Let*

$$\Pi = P(z_1)\, P(z_2) \,\cdots\, P(z_n), \;\; n \geq 3,$$

be an counterclockwise oriented $n-$gon. Then its area $A(\Pi)$ is given by the following formula

$$A(\Pi) = \frac{1}{2}\,\Im(\overline{z_1}\, z_2 + \overline{z_2}\, z_3 + \cdots + \overline{z_n}\, z_1) = \frac{1}{2}\,\Im(\sum_{t=1}^{n} \overline{z_t}\, z_{t+1}), \;\; \textit{where } z_{n+1} = z_1.$$

Proof. To prove this theorem, we proceed inductively, making use of a triangulation of the polygon Π. There exists a vertex $P(z_t)$ such that the area of Π is the sum of the area of a triangle

$$\Delta \; = \; P(z_{t-1})\, P(z_t)\, P(z_{t+1})$$

and the area of the $(n-1)-$gon

$$\Pi' \; = \; P(z_1)\, P(z_2) \,\ldots\, P(z_{t-1})\, P(z_{t+1}) \,\ldots\, P(z_n),$$

both oriented counterclockwise. Thus $A(\Pi) \; = \; A(\Delta) + A(\Pi')$. By induction,

$$\begin{aligned} A(\Pi) &= \frac{1}{2}\,\Im(\overline{z_{t-1}}\, z_t + \overline{z_t}\, z_{t+1} + \overline{z_{t+1}}\, z_{t-1}) \\ &+ \frac{1}{2}\,\Im(\overline{z_1}\, z_2 + \overline{z_2}\, z_3 + \cdots + \overline{z_{t-2}}\, z_{t-1} + \overline{z_{t-1}}\, z_{t+1} + \overline{z_{t+1}}\, z_{t+2} + \cdots + \overline{z_n}\, z_1) \\ &= \frac{1}{2}\,\Im(\overline{z_1}\, z_2 + \overline{z_2}\, z_3 + \cdots + \overline{z_n}\, z_1), \end{aligned}$$

because the number $\overline{z_{t+1}}\, z_{t-1} + \overline{z_{t-1}}\, z_{t+1}$ is real. This completes the proof. □

Remark V.5.3. As in the case of a triangle, the area of an unoriented triangle is simply the absolute value of $A(\Pi)$. You may like to use the formula for calculating the area of a regular $n-$gon inscribed in a circle of radius r. It is not difficult to see that the answer is $\frac{nr^2}{2}\,\sin\frac{2\pi}{n}$ in the case that the center of the circle is at the origin $P(0)$. This provides a way of deriving the formula for the area of a circle

$$\lim_{n\to\infty} \frac{nr^2}{2} \sin\frac{2\pi}{n} = \pi r^2.$$

Example V.5.2. We find the area of the 10−gon Π described in Figure V.5.2.

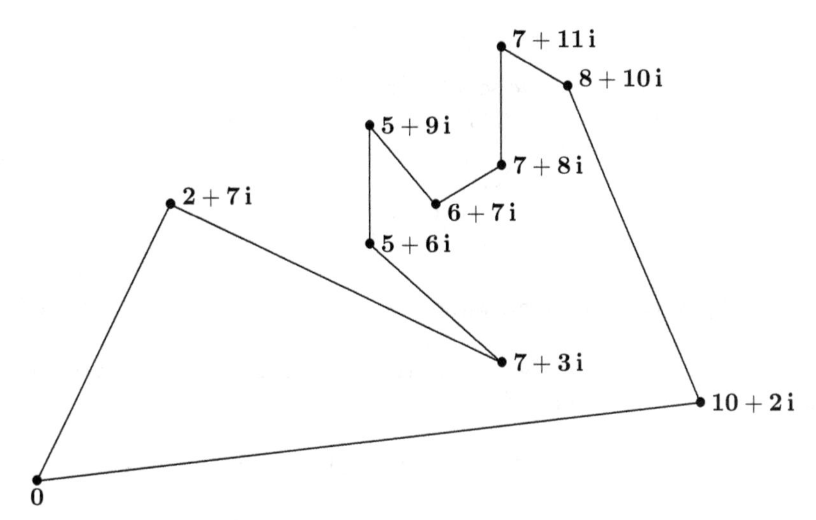

Figure V.5.2 The 10−gon Π

Here we have

$$\begin{aligned} A(\Pi) &= \frac{1}{2}\Im[(10-2\,i)(8+10\,i)+(8-10\,i)(7+11\,i)+(7-11\,i)(7+8\,i)+(7-8\,i)(6+7\,i) \\ &+ (6-7\,i)(5+9\,i)+(5-9\,i)(5+6\,i)+(5-6\,i)(7+3\,i)+(7-3\,i)(2+7\,i)] \\ &= \frac{1}{2}(100-16+88-70+56-77+49-48+54-35+30-45+15-42+49-6) \\ &= 51. \end{aligned}$$

Exercise V.5.9. Find the area of the $2n$−gon defined by the vertices $P(z_1)\,P(z_2)\,\cdots\,P(z_{2n})$, where

$$P(z_{2t-1}) = r_1\,e^{i\,\frac{2(t-1)\pi}{n}} \quad \text{and} \quad P(z_{2t}) = r_2\,e^{i\,\frac{(2t-1)\pi}{n}} \quad \text{for } 1 \le t \le n,\ 0 < r_1 < r_2.$$

Hence deduce the area of the region bounded by the line segments

$$P(r_1)\,P(r_2\,e^{i\,\frac{\pi}{n}}),\ \ P(r_1\,e^{i\,\frac{2\pi}{n}})\,P(r_2\,e^{i\,\frac{\pi}{n}})$$

and the arc of the circle with center at the origin and radius r_1.

An important use of complex numbers is to characterize in simple terms particular geometric objects. One of the most basic of such objects is an equilateral triangle. Here is a characterization.

Theorem V.5.6 (Equilateral triangle). *A triangle* $\Delta = P(z_1)\,P(z_2)\,P(z_3)$ *is equilateral if and only is*

$$\omega_1 z_1 + \omega_2 z_2 + \omega_3 z_3 = 0, \tag{V.5.5}$$

where $\{\omega_1, \omega_2, \omega_3\} = \{1, \omega, \omega^2\}$ *is the set of cube roots of unity. This condition is equivalent to the equality* $z_1^2 + z_2^2 + z_3^2 - z_1 z_2 - z_1 z_3 - z_2 z_3 = 0$, *that is, to*

$$\begin{vmatrix} z_1 & z_2 & 1 \\ z_2 & z_3 & 1 \\ z_3 & z_1 & 1 \end{vmatrix} = 0.$$

Proof. First of all, we point out that, without loss of generality, we may assume in (V.5.5) that $\omega_1 = 1$.

Applying the rotation around the vertex $P(z_2)$ by the angle $\frac{\pi}{3} = 60°$, we can see that the triangle Δ is equilateral if and only if

$$z_1 = z_2 + (z_3 - z_2)e^{i\frac{\pi}{3}} = z_2 + (z_3 - z_2)(-\omega^2) = -\omega z_2 - \omega^2 z_3,$$

that is, if and only if $z_1 + \omega z_2 + \omega^2 z_3 \;=\; 0$ (here, Figure V.4.1 may help).

In a similar way, we obtain for the triangle $P(z_1)\,P(z_3)\,P(z_2)$ (that is, considering the opposite orientation of Δ) that the necessary and sufficient condition is $z_1 + \omega^2 z_2 + \omega z_3 \;=\; 0$.

Finally, $(z_1 + \omega z_2 + \omega^2 z_3)\,(z_1 + \omega^2 z_2 + \omega z_3) \;=\; z_1^2 + z_2^2 + z_3^2 - z_1 z_2 - z_1 z_3 - z_2 z_3$. □

Exercise V.5.10. Two triangles $\Delta_1 = P(z_1)\,P(z_2)\,P(z_3)$ and $\Delta_2 = P(w_1)\,P(w_2)\,P(w_3)$ are said to be **similar** if every angle of one triangle is equal to the corresponding angle in the other triangle. Show that Δ_1 and Δ_2 are similar if and only if

$$\frac{z_2 - z_1}{z_3 - z_1} = \frac{w_2 - w_1}{w_3 - w_1}.$$

Hence, prove that Δ_1 and Δ_2 are similar if and only if

$$\begin{vmatrix} z_1 & w_1 & 1 \\ z_2 & w_2 & 1 \\ z_3 & w_3 & 1 \end{vmatrix} = 0.$$

Using the fact that $P(1)\,P(\omega)\,P(\omega^2)$ is an equilateral triangle, provide another proof of Theorem V.5.6.

Exercise V.5.11. Denote by $A(\Delta)$ the area of the equilateral triangle $\Delta = A\,B\,C$ such that $A = P(0)$ and B and C are points of the lines $x = -a$ and $x = b$, where $a > 0, b > 0$, which are both parallel to the y axis. Determine the values of $c, d \in \mathbb{R}$ as indicated in Figure V.5.3 and show that

$$A(\Delta) \;=\; \frac{a^2 + ab + b^2}{3}\,\sqrt{3}.$$

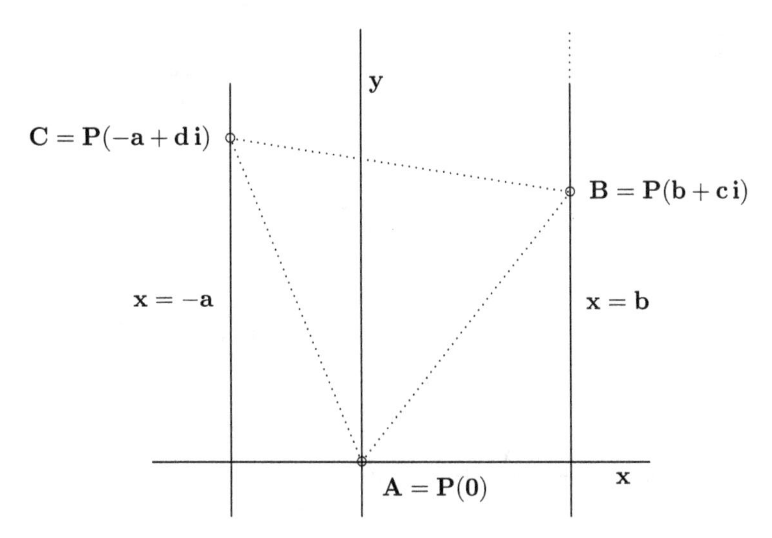

Figure V.5.3 Diagram for Exercise V.5.11

Theorem V.5.7 (Napoleon's triangle and the Fermat-Torricelli point). *Let $\Delta = Z_1Z_2Z_3$ be an* **arbitrary** *triangle. Erect on its sides externally the equilateral triangles $\Delta_1 = Z_1U_3Z_2$, $\Delta_2 = Z_2U_1Z_3$ and $\Delta_3 = Z_3U_2Z_1$. Denote by X_3, X_1 and X_2 the centroids of these triangles. Then the triangle $\Delta_{NAP} = X_3X_1X_2$ is equilateral and its centroid G coincides with the centroid of the original triangle Δ. Moreover, the centroid of the triangle $\Delta = U_3U_1U_2$ also coincides with G.*

If none of the angles of the triangle $Z_1Z_2Z_3$ exceeds $120°$, then the line segments Z_1U_1, Z_2U_2 and Z_3U_3 intersect at the same point F, which is called the Fermat-Torricelli point and has the following property: The sum $|Z_1F|+|Z_2F|+|Z_3F|$ of the distances of this point from the vertices of the original triangle is **minimal** *with respect of the sum of these distances from any other point, and all angles $\angle Z_1FU_3, \angle U_3FZ_2, \angle Z_2FU_1, \angle U_1FZ_3, \angle Z_3FU_2$ and $\angle U_2FZ_1$ are equal to $60°$ (see Figure V.5.4).*

Remark V.5.4. Although **Napoleon Bonaparte (1769 - 1821)** is considered to be the author of the first part of Theorem V.5.7, there are some historians who have some doubts about his authorship. However, it is a fact that Napoleon Bonaparte surrounded himself in his life with scientists, and mathematicians in particular, and that he was interested in mathematics. Therefore, to claim that he discovered this theorem may be more justified than to claim that Pythagoras discovered the so-called Pythagoras theorem, or Fibonacci the so-called Fibonacci sequence, or Pascal the so-called Pascal triangle. While the statement concerning Napoleon's triangle is true for an arbitrary triangle, the statement concerning the point described by **Pierre de Fermat** and **Evangelista Torricelli (1608 - 1647)** requires an additional assumption that no angle of the triangle is greater than $120°$. We add that the point F is also a common intersection point of the circumcircles of the triangles Δ_1, Δ_2 and Δ_3.

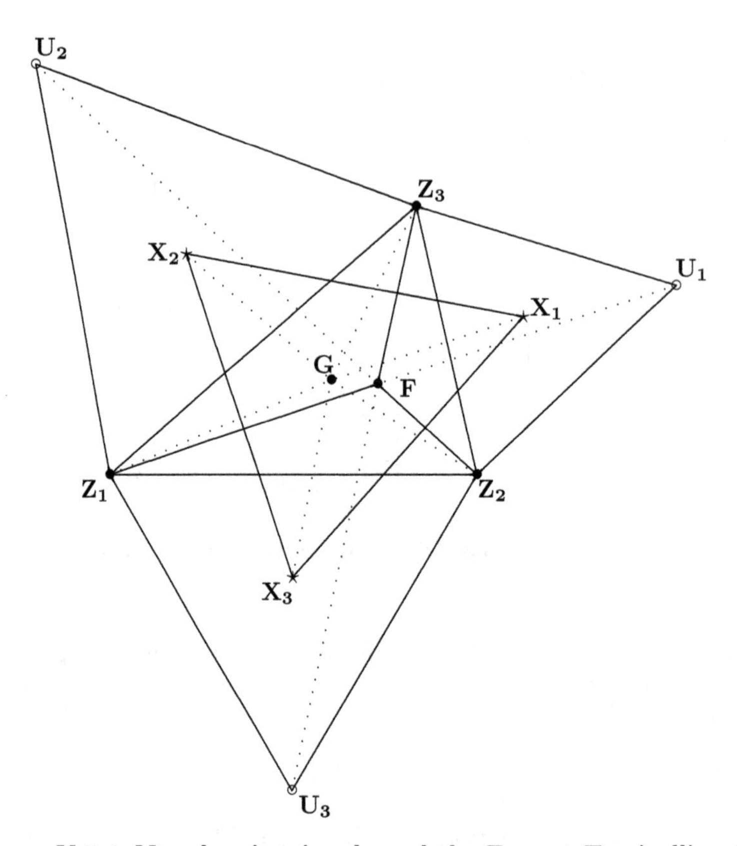

Figure V.5.4 Napoleon's triangle and the Fermat-Torricelli point

Proof of Theorem V.5.7. Write $Q = P(q)$ for $Q = Z_t, X_t, U_t,\ \ (t = 1, 2, 3)$. The affix g of the centroid G of the original triangle satisfies $g = \frac{1}{3}(z_1 + z_2 + z_3)$. By Theorem V.5.6 characterizing the equilateral triangles, we have

$$u_3 + \omega\, z_2 + \omega^2\, z_1 = u_1 + \omega\, z_3 + \omega^2\, z_2 = u_2 + \omega\, z_1 + \omega^2\, z_3 = 0.$$

Hence, adding these equalities, we get, in view of $\omega + \omega^2 = -1$, $u_1 + u_2 + u_3 = z_1 + z_2 + z_3$, and therefore the centroid of the triangle $\Delta_0 = U_3U_1U_2$ coincides with the centroid G of the original triangle Δ.

Now, $3\,x_3 = z_1 + u_3 + z_2$, $3\,x_1 = z_2 + u_1 + z_3$ and $3\,x_2 = z_3 + u_2 + z_1$. Thus $x_1 + x_2 + x_3 = z_1 + z_2 + z_3$, that is, the centroid of Δ_{NAP} coincides with G. Moreover,

$$\begin{aligned} 3(x_1 + \omega\, x_2 + \omega^2\, x_3) &= (z_2 + u_1 + z_3) + \omega\,(z_3 + u_2 + z_1) + \omega^2\,(z_1 + u_3 + z_2) \\ &= (u_1 + \omega\, z_3 + \omega^2\, z_2) + \omega\,(u_2 + \omega\, z_1 + \omega^2\, z_3) + \omega^2\,(u_3 + \omega\, z_2 + \omega^2\, z_1) = 0, \end{aligned}$$

and hence the triangle Δ_{NAP} is equilateral. The first part of the theorem is proved.

A proof of the second part of the theorem is illustrated in Figure V.5.5. Here, $Z_1U_3Z_2$, and $Z_3U_2Z_1$ are again the equilateral triangles erected on the sides of the triangle $\Delta = Z_1Z_2Z_3$, where P is an arbitrary point inside the triangle Δ and F is the intersection point of the segments Z_3U_3 and Z_2U_2. The rotation around the point Z_1 by 60° moves the triangle $\Delta_{Z_3} = Z_1FZ_3$ to $\Delta_{Z_3^\star} = Z_1F^\star Z_3^\star$, the segment U_3Z_3 to $Z_2Z_3^\star$ and the points P to $P^\star$ and F to $F^\star$, creating in this way two equilateral triangles $Z_1PP^\star$ and $Z_1FF^\star$.

Now, the Fermat-Torricelli sum equals

$$\sum(F) = |Z_1F| + |Z_2F| + |Z_3F| = |FF^\star| + |Z_2F| + |Z_3^\star F^\star| = |Z_2U_2|.$$

The respective sum corresponding to the point P is

$$\sum(P) = |Z_1P| + |Z_2P| + |Z_3P| = |PP^\star| + |Z_2P| + |Z_3^\star P^\star|,$$

that is, $\sum(P)$ is the length of the polygonal chain (piecewise linear curve) $Z_2PP^\star Z_3^\star$ and thus $\sum(F) \le \sum(P)$.

The statement concerning the angles can easily be deduced from Figure V.5.5. In view of the rotation by 60° the angle $\angle Z_3FU_2$ is 60°. Moreover, the angle $\angle U_2FZ_1$ is, as an inner angle of the equilateral triangle $Z_1FF^\star$ also equal to 60°. Finally, replacing the point U_2 by the point U_3 or by U_1, we obtain by the same reasoning the statements for the remaining angles. □

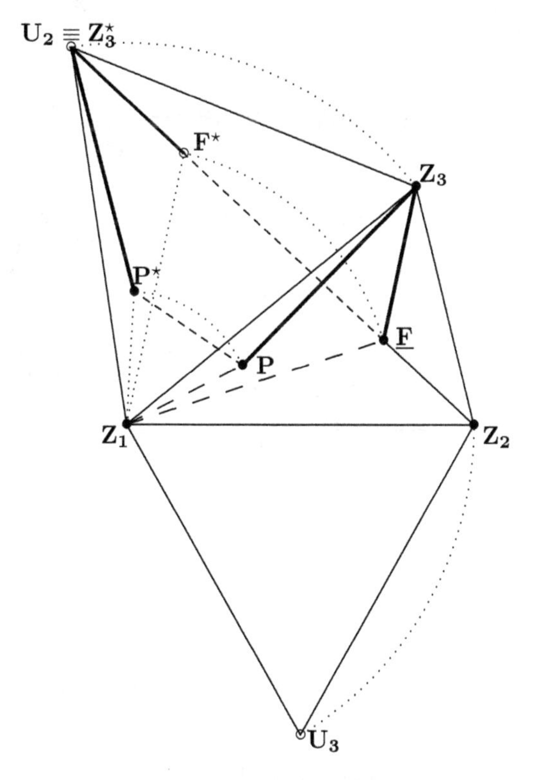

Figure V.5.5 The Fermat-Torricelli point

The existence of the Euler line and the Feuerbach-Euler circle (often called the nine point circle) can also be proved rather easily in the Argand-Gauss plane. Feuerbach refers to the mathematician **Karl Wilhelm Feuerbach (1800 - 1834)**.

Theorem V.5.8 (Euler line and the Feuerbach-Euler circle). *Let $\Delta = ABC$ be a triangle. Then there exists a circle, the Feuerbach-Euler circle, passing through the following* **nine** *points: the foot A_1 of the altitude from the point A, the foot B_1 of the altitude from the point B, the foot C_1 of the altitude from the point C, midpoints A_2, B_2, C_2 of the sides of the triangle and the midpoints A_3, B_3, C_3 of the segments $|AH|$, $|BH|$, $|CH|$, where H is the orthocenter of the triangle Δ (see Figure V.5.6).*

The center E of this circle lies on the line (the Euler line), connecting the point H and the center S of the circumscribed circle of the triangle Δ. This line contains also the centroid G of the triangle Δ whereby

$$| SG |= \frac{1}{3} \ | SH | \quad and \quad | SE |= \frac{1}{2} \ | SH | \, .$$

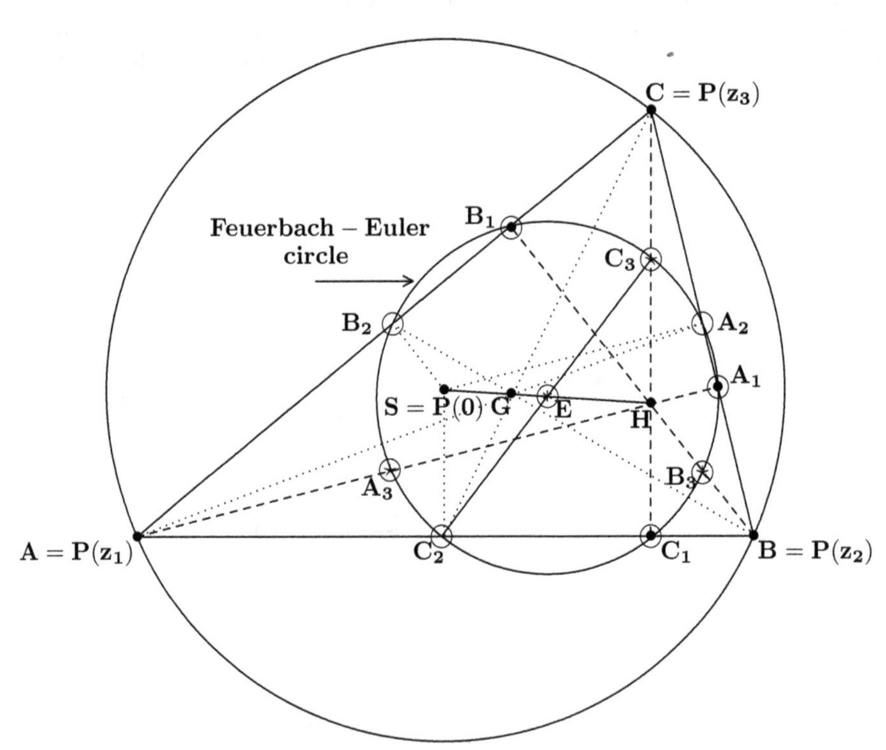

Figure V.5.6 Euler line and Feuerbach-Euler circle

Proof. We choose the center of the circumscribed circle as the origin of the Argand-Gauss plane and let $A = P(z_1)$, $B = P(z_2)$ and $C = P(z_3)$ (see Figure V.5.6). Central to the proof is the fact that $H = P(z_1 + z_2 + z_3)$. This is, of course, just an expression that the lines defined by the segments AH and BC are perpendicular. Indeed, since $|z_2| = |z_3|$, the complex number

$$\frac{z_1 + z_2 + z_3 - z_1}{z_2 - z_3} = \frac{z_2 + z_3}{z_2 - z_3} = \frac{(z_2 + z_3)(\overline{z_2} - \overline{z_3})}{(z_2 - z_3)(\overline{z_2} - \overline{z_3})} = \frac{\overline{z_2}z_3 - z_2\overline{z_3}}{|z_2 - z_3|^2},$$

is purely imaginary (recall Exercise V.5.5). Of course

$$G = P\left(\frac{z_1 + z_2 + z_3}{3}\right), \; A_2 = P\left(\frac{z_2 + z_3}{2}\right), \; B_2 = P\left(\frac{z_1 + z_3}{2}\right), \; C_2 = P\left(\frac{z_1 + z_2}{2}\right)$$

and

$$A_3 = P\left(z_1 + \frac{z_2 + z_3}{2}\right), \; B_3 = P\left(z_2 + \frac{z_1 + z_3}{2}\right), \; C_3 = P\left(z_3 + \frac{z_1 + z_2}{2}\right).$$

Now define

$$E = P\left(\frac{z_1 + z_2 + z_3}{2}\right).$$

To show that all nine points $A_1, B_1, \ldots, C_3$ belong to the circle whose center is the point E and radius $\frac{1}{2}|z_1|$ is a routine calculation. Similarly it is straightforward to justify the statements about the points

$$S = P(0), \; G = P\left(\frac{z_1 + z_2 + z_3}{3}\right), \; E = P\left(\frac{z_1 + z_2 + z_3}{2}\right) \quad \text{and} \quad H = P\left(z_1 + z_2 + z_3\right)$$

on the Euler line. The proof is now complete. □

Example V.5.3 (Construction of Henri van Aubel (1830 - 1906)). Given an arbitrary quadrilateral $ABCD$, we erect externally on its sides squares whose centers are successively denoted by M, N, P and Q; see Figure V.5.7. We are going to show that the segments MP and NQ have the same length and are perpendicular. To this end, we write

$$A = P(z_1), \; B = P(z_2), \; C = P(z_3) \;\text{ and }\; D = P(z_4)$$

and making simple calculations, we find that the midpoints U, V, W and Z of the sides of the quadrilateral are given by

$$U = P(\frac{z_1 + z_2}{2}), \; V = P(\frac{z_2 + z_3}{2}), \; W = P(\frac{z_3 + z_4}{2}), \; Z = P(\frac{z_4 + z_1}{2}).$$

Hence the centers of the squares satisfy

$$M = P(\frac{z_1 + z_2}{2} + \frac{z_1 - z_2}{2}\,i), \; N = P(\frac{z_2 + z_3}{2} + \frac{z_2 - z_3}{2}\,i),$$

$$P = P(\frac{z_3 + z_4}{2} + \frac{z_3 - z_4}{2}\,i) \;\text{ and }\; Q = P(\frac{z_4 + z_1}{2} + \frac{z_4 - z_1}{2}\,i).$$

Thus the vector $\mathbf{v} = \overrightarrow{MP}$ is described by

$$u = \frac{z_3 + z_4 - z_1 - z_2}{2} + \frac{z_3 - z_4 - z_1 + z_2}{2} i$$

and the vector $\mathbf{v} = \overrightarrow{NQ}$ by

$$\mathbf{v} = \frac{z_4 + z_1 - z_2 - z_3}{2} + \frac{z_4 - z_1 - z_2 + z_3}{2} i.$$

Since $u\,i = \mathbf{v}$, the segments MP and NQ are perpendicular and have the same length.

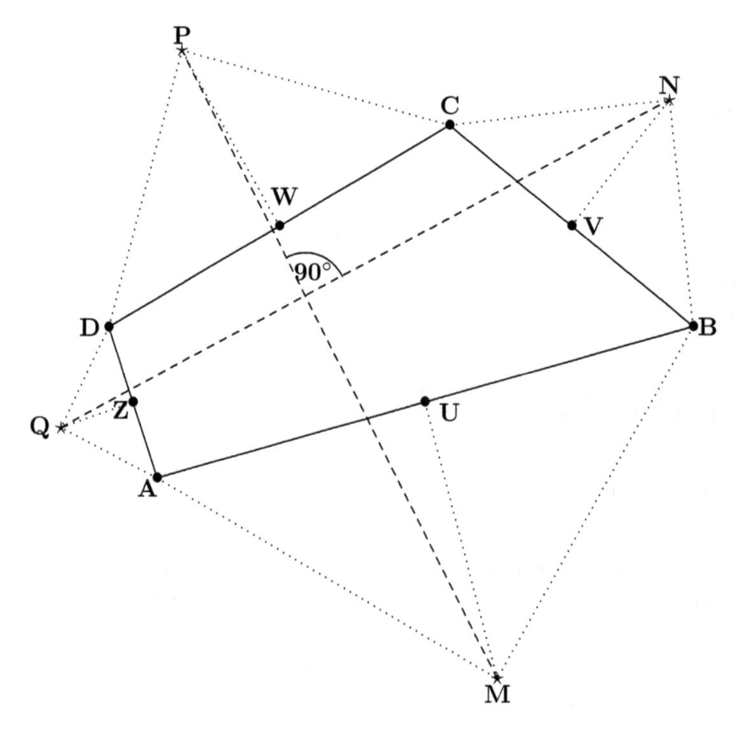

Figure V.5.7 Construction of Henri van Aubel

Exercise V.5.12. Referring to Example V.5.3, prove that the quadrilateral $ABCD$ is a parallelogram if and only if $MNPQ$ is also a parallelogram.

Moreover, prove that if $ABCD$ is not a parallelogram, then the quadrilateral $GHKL$, where G, H, K and L are the midpoints of the segments AC, MP, BD and NQ, respectively, is a square. What happens to the points G, H, K and L if $ABCD$ is a parallelogram?

Exercise V.5.13. Let

$$\Delta_1 = P(0)P(z_1)P(z_2), \ \Delta_2 = P(0)P(z_3)P(z_4) \ \text{and} \ \Delta_3 = P(0)P(z_5)P(z_6)$$

be three equilateral triangles with a common vertex $P(0)$ at the origin. Let A, B and C be the midpoints of the segments

$$P(z_2)P(z_3),\ P(z_4)P(z_5),\ \text{ and }\ P(z_6)P(z_1),\ \text{ respectively.}$$

Prove that the triangle $\Delta = ABC$ is equilateral.

Exercise V.5.14. Two squares $S_1 = ABCD$ and $S_2 = AEFG$ have a common vertex A. Complete EAD to a parallelogram $P_1 = EADH$ and BAG to a parallelogram $P_2 = BAGK$. Prove that the centers of S_1, S_2, P_1 and P_2 form a square.

Exercise V.5.15. Let $P_1, P_2, \ldots, P_n$ be the vertices of a regular n–gon inscribed in a circle of center C and radius r. Prove that for any point P of the plane

$$\sum_{t=1}^{n} |PP_t|^2 = n\,(|PC|^2 + r^2).$$

Note that this sum equals $2nr^2$ for points P on the circumcircle of the n–gon.

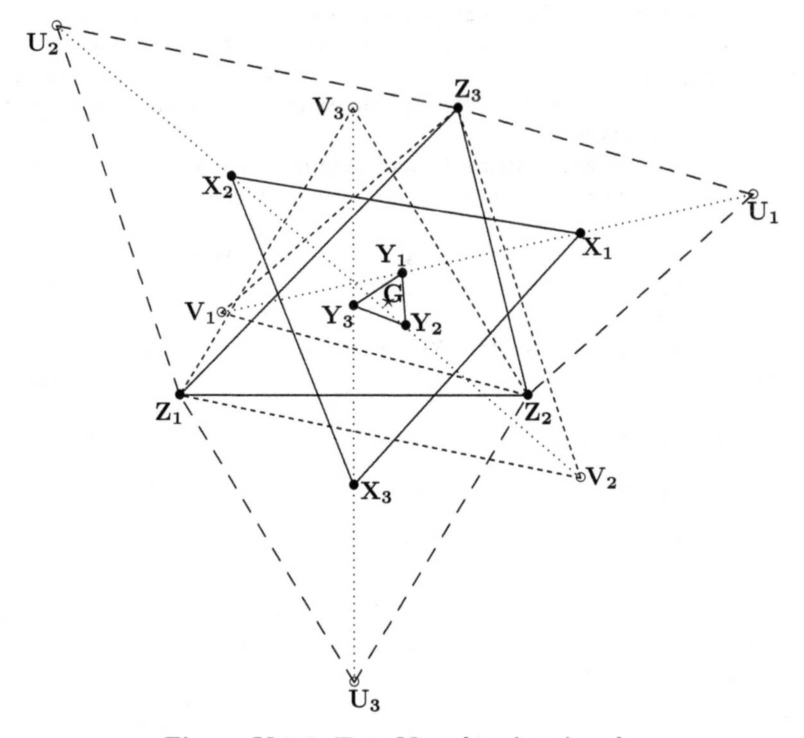

Figure V.5.8 Two Napoleon's triangles

V.6 ⋆ Petr's polygons

Exercise V.6.1 (revisiting Napoleon's theorem). Referring to Theorem V.5.7, erect on the sides of the given triangle $\Delta = Z_1Z_2Z_3$, (**external**) equilateral triangles $U_1Z_3Z_2$, $U_2Z_1Z_3$ and $U_3Z_2Z_1$ with centroids X_1, X_2 and X_3, as well as (**internal**) equilateral triangles $V_1Z_2Z_3$, $V_3Z_1Z_2$, $V_2Z_3Z_1$ with centroids Y_1, Y_3 and Y_2.

Thus we have two Napoleon's triangles: $\Delta_{NAP} = X_1X_2X_3$ and $\Delta_{nap} = Y_1Y_2Y_3$. As before G is the centroid of the given triangle Δ (see Figure V.5.8).

(i) Show that the triangle Δ_{nap} is equilateral and that its centroid coincides with the centroid of Δ.

(ii) Prove that the area of the triangle Δ equals the difference of the areas of the triangle Δ_{NAP} and the triangle Δ_{nap}.

Exercise V.6.2. Erect on the sides of a given triangle $\Delta = Z_1Z_2Z_3$ regular $n-$gons and denote their centers by X_1, X_2 and X_3. Prove that the triangle $\Delta_0 = X_1X_2X_3$ is equilateral if and only if $n = 3$.

A deeper understanding of Napoleon's theorem and van Aubel's construction is provided by an interesting construction of **Karel Petr (1868 - 1950)** that assigns to any $n-$gon (that is, to any ordered set of n points of the plane connected by n linear segments determined by that order) a regular convex $n-$gon. Here, we will use his construction to emphasize the role of complex numbers in elementary plane geometry as well as pointing out a relationship with finite recursive sequences.

To this end, we first derive a statement concerning recursive sequences that will describe Petr's construction. Consider the following $n \times n$ scheme of complex numbers $\mathbf{v}_{r,s}$

$$\mathbb{M} = \begin{matrix} \mathbf{v}_{0,1} & \mathbf{v}_{0,2} & \mathbf{v}_{0,3} & \cdots & \mathbf{v}_{0,s} & \mathbf{v}_{0,s+1} & \mathbf{v}_{0,s+2} & \cdots & \mathbf{v}_{0,n} \\ \mathbf{v}_{1,1} & \mathbf{v}_{1,2} & \mathbf{v}_{1,3} & \cdots & \mathbf{v}_{1,s} & \mathbf{v}_{1,s+1} & \mathbf{v}_{1,s+2} & \cdots & \mathbf{v}_{1,n} \\ \cdots & \cdots & \cdots & \cdots & \cdots & \cdots & \cdots & \cdots & \cdots \\ \mathbf{v}_{r-1,1} & \mathbf{v}_{r-1,2} & \mathbf{v}_{r-1,3} & \cdots & \mathbf{v}_{r-1,s} & \mathbf{v}_{r-1,s+1} & \mathbf{v}_{r-1,s+2} & \cdots & \mathbf{v}_{r-1,n} \\ \mathbf{v}_{r,1} & \mathbf{v}_{r,2} & \mathbf{v}_{r,3} & \cdots & \mathbf{v}_{r,s} & \mathbf{v}_{r,s+1} & \mathbf{v}_{r,s+2} & \cdots & \mathbf{v}_{r,n} \\ \cdots & \cdots & \cdots & \cdots & \cdots & \cdots & \cdots & \cdots & \cdots \\ \mathbf{v}_{n-1,1} & \mathbf{v}_{n-1,2} & \mathbf{v}_{n-1,3} & \cdots & \mathbf{v}_{n-1,s} & \mathbf{v}_{n-1,s+1} & \mathbf{v}_{n-1,s+2} & \cdots & \mathbf{v}_{n-1,n} \end{matrix},$$

that satisfies the condition

$$\sum_{s=1}^{n} \mathbf{v}_{0,s} = 0. \tag{V.6.1}$$

Remark V.6.1. Here, the r-th row ($r \in \{0, 1, \ldots, n-1\}$) of the scheme $\mathbb{M}$ represents an $(n+1)-$gon $A_{r,1}A_{r,2}\ldots A_{r,s}\ldots A_{r,n+1}$ in the Argand-Gauss plane with a fixed vertex $A_{r,1} = P(a_{r,1})$ defined as follows:

$$A_{r,s} = P(a_{r,s}), \text{ where } a_{r,s} = a_{r,1} + \sum_{t=1}^{s-1} \mathbf{v}_{r,t} \text{ for } s \in \{2, 3, \ldots, n+1\}.$$

In fact, due to the property (V.6.1), $A_{0,n+1} = A_{0,1}$, and thus the first row represents a closed $n-$gon. Theorem V.6.1 will provide conditions under which **all** the rows will represent (possibly degenerate) closed $n-$gons.

Let $\{1, \omega_1, \omega_2, \ldots, \omega_n\}$ be the set of all n-th roots of unity and $k_1, k_2, \ldots, k_{n-1}$ a sequence of nonzero complex numbers, that is, numbers satisfying the condition

$$\prod_{r=1}^{n-1} k_r \neq 0. \tag{V.6.2}$$

Recall that

$$\begin{aligned}
\omega_1 + \omega_2 + \cdots + \omega_{n-1} &= -1, \\
\omega_1\omega_2 + \omega_1\omega_3 + \cdots + \omega_2\omega_3 + \cdots + \omega_{n-2}\omega_{n-1} &= 1, \\
\omega_1\omega_2\omega_3 + \omega_1\omega_2\omega_4 + \cdots + \omega_{n-3}\omega_{n-2}\omega_{n-1} &= -1, \\
\omega_1\omega_2\omega_3\omega_4 + \omega_1\omega_2\omega_3\omega_5 + \cdots + \omega_{n-4}\omega_{n-3}\omega_{n-2}\omega_{n-1} &= 1, \\
\cdots\cdots\cdots\cdots\cdots\cdots\cdots\cdots\cdots\cdots &, \\
\omega_1\omega_2\ldots\omega_{n-1} &= (-1)^{n-1}.
\end{aligned}$$

These equalities reflect Viète's formulas since the numbers ω_r $(r \in \{1, 2, \ldots, n-1\})$ are the roots of the equation

$$\frac{x^n - 1}{x - 1} = x^{n-1} + x^{n-2} + \cdots + x + 1 = 0.$$

The following statement describes the essence of Petr's construction.

Theorem V.6.1. *Suppose that the scheme* $\mathbb{M}$ *satisfies the condition* (V.6.1). *Let*

$$\mathbf{v}_{r,s} = k_r\,(\omega_r\,\mathbf{v}_{r-1,s} - \mathbf{v}_{r-1,s+1}) \text{ for all } r \in \{1, 2, \ldots, n-1\},\ s \in \{1, 2, \ldots, n\}, \tag{V.6.3}$$

so that the conditions (V.6.2) *and* $\mathbf{v}_{r,n+1} = \mathbf{v}_{r,1}$ $(r \in \{1, 2, \ldots, n-1\})$ *hold. Then*

$$\mathbf{v}_{n-1,s} = 0 \text{ for all } s \in \{1, 2, \ldots, n\}.$$

Proof. We are going to prove that $\mathbf{v}_{n-1,1} = 0$. The equalities $\mathbf{v}_{n-1,s} = 0$ for the remaining indices s can then be obtained by a mere shift of the indices. In view of (V.6.1), we have

$$\begin{aligned}
\mathbf{v}_{n-1,1} &= k_{n-1}\,(\omega_{n-1}\,\mathbf{v}_{n-3,1} - \mathbf{v}_{n-3,2}) \\
&= k_{n-1}\,[\omega_{n-1}k_{n-2}(\omega_{n-2}\,\mathbf{v}_{n-3,1} - \mathbf{v}_{n-3,2}) - k_{n-2}(\omega_{n-2}\mathbf{v}_{n-3,2} - \mathbf{v}_{n-3,3})] \\
&= k_{n-1}k_{n-2}\,[\omega_{n-1}\omega_{n-2}\,\mathbf{v}_{n-3,1} - (\omega_{n-1} + \omega_{n-2})\,\mathbf{v}_{n-3,2} + \mathbf{v}_{n-3,3})].
\end{aligned}$$

The successive step results in

$$\begin{aligned}
\mathbf{v}_{n-1,1} = k_{n-1}\,k_{n-2}\,k_{n-3}\,[\,&\omega_{n-1}\omega_{n-2}\omega_{n-3}\,\mathbf{v}_{n-4,1} \\
&- (\omega_{n-1}\omega_{n-2} + \omega_{n-1}\omega_{n-3} + \omega_{n-2}\omega_{n-3})\,\mathbf{v}_{n-4,2} \\
&+ (\omega_{n-1} + \omega_{n-2} + \omega_{n-3})\,\mathbf{v}_{n-4,3} - \mathbf{v}_{n-4,4}\,].
\end{aligned}$$

Now we can see that proceeding by induction we obtain in $n-4$ steps

$$\begin{aligned}\mathbf{v}_{n-1,1} &= k_{n-1}\,k_{n-2}\ldots k_1\,[\,\omega_{n-1}\omega_{n-2}\ldots\omega_1\,\mathbf{v}_{0,1}\\ &- (\omega_{n-1}\omega_{n-2}+\omega_{n-1}\omega_{n-3}+\cdots+\omega_2\omega_1)\,\mathbf{v}_{0,2}\\ &+ (\omega_{n-1}\omega_{n-2}\omega_{n-3}+\omega_{n-1}\omega_{n-2}\omega_{n-4}+\cdots+\omega_3\omega_2\omega_1)\,\mathbf{v}_{0,3}-\cdots\\ &\cdots\cdots\cdots\cdots\cdots\cdots\cdots\cdots\cdots\cdots\cdots\cdots\\ &+ (-1)^{n-2}(\omega_{n-1}\omega_{n-2}\ldots\omega_2+\cdots+\omega_{n-2}\omega_{n-3}\ldots\omega_1)\,\mathbf{v}_{0,n-1}+(-1)^{n-1}\mathbf{v}_{0,n}]\\ &= (-1)^{n-1}\,k_{n-1}k_{n-2}\ldots k_1\,(\mathbf{v}_{0,1}+\mathbf{v}_{0,2}+\mathbf{v}_{0,3}+\cdots+\mathbf{v}_{0,n-1}+\mathbf{v}_{0,n})\;=\;0.\end{aligned}$$

This concludes the proof of the theorem. □

Remark V.6.2. It is also possible to show that the absolute values of all numbers $\mathbf{v}_{n-2,s}$ are equal, that is, that

$$|\mathbf{v}_{n-2,0}| \;=\; |\mathbf{v}_{n-2,1}| \;=\; \cdots \;=\; |\mathbf{v}_{n-2,n}|.$$

We shall sidestep the proof since these equalities as well as further valuable information will be obtained by the geometric interpretation of the preceding theorem in what follows.

The complex numbers $\mathbf{v}_{r,s} = x_{r,s}+y_{r,s}i$ of the scheme $\mathbb{M}$ will be interpreted as the plane vectors $(x_{r,s}, y_{r,s})$ and the number sequence in each of the rows of the scheme $\mathbb{M}$ as a polygon whose edges are defined by the numbers $\mathbf{v}_{r,s}$. Such an edge is defined by the initial point $A_{r,s}$ and the terminal point $A_{r,s+1}$:

$$\mathbf{v}_{r,s} \;=\; A_{r,s+1} \;-\; A_{r,s}.$$

We observe that, in view of the condition (V.6.1) the connected system of these segments is closed. As a matter of fact, this is true for the remaining rows, as well. Indeed, in view of (V.6.3),

$$\sum_{s=1}^{n}\mathbf{v}_{r,s} \;=\; k_r\,(\omega_r \;-\; 1)\sum_{s=1}^{n}\mathbf{v}_{r-1,s},$$

and thus, we obtain by induction

$$\sum_{s=1}^{n}\mathbf{v}_{r,s} \;=\; 0 \text{ for all } r\in\{0,1,\ldots,n-1\}.$$

The r-th polygon with the successive vertices

$$A_{r,1}, A_{r,2}, A_{r,3},\ldots,A_{r,s},A_{r,s+1},\ldots,A_{r,n},$$

satisfies

$$\mathbf{v}_{r,1}=A_{r,2}-A_{r,1},\mathbf{v}_{r,2}=A_{r,3}-A_{r,2},\ldots,\mathbf{v}_{r,s}=A_{r,s+1}-A_{r,s},\ldots,\mathbf{v}_{r,n}=A_{r,1}-A_{r,n}.$$

The following statement answers the basic question which asks for a geometric construction that corresponds to the recurrence relation (V.6.3).

Theorem V.6.2 (Petr's construction). *Let $A_0B_0C_0$ be a triangle whose sides A_0B_0 and B_0C_0 are the bases of similar isosceles triangles $A_0A_1B_0$ and $B_0B_1C_0$; thus, $\angle B_0A_1A_0 = \angle C_0B_1B_0$ (see* Figure V.6.1*). Then*

$$B_1 - A_1 \;=\; k\,[\,\omega(B_0 - A_0) \;-\; (C_0 - B_0)\,], \quad \textit{where } \omega = e^{-i\alpha} \textit{ and } k \neq 0.$$

In fact, $k = \dfrac{1}{\omega - 1}$.

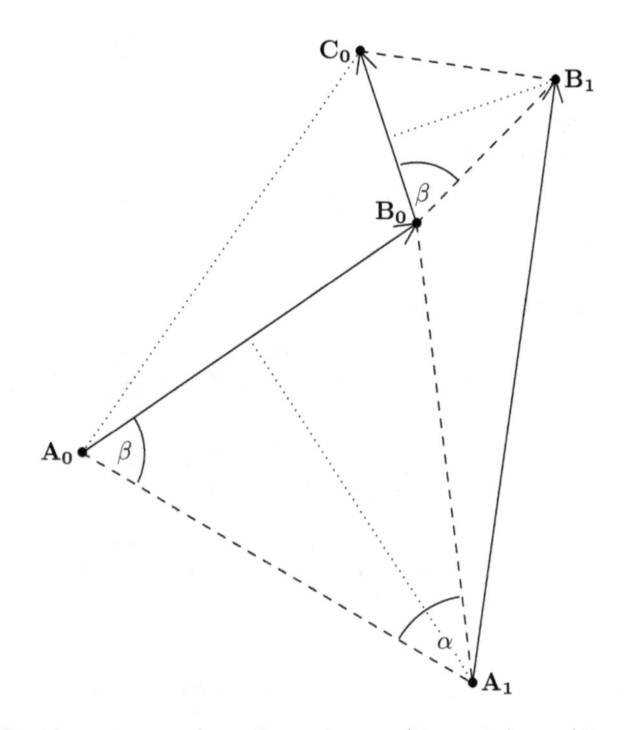

Figure V.6.1 Petr's construction: $\mathbf{B_1 - A_1 = x\,(B_0 - A_0) + y\,(C_0 - B_0)}$

Proof. We denote the ratio of the lengths of the vectors $A_1 - A_0$ and $A_0 - B_0$ by h:

$$h = \frac{|A_0A_1|}{|B_0A_0|},$$

and describe the rotation of the plane by the angle $\beta = \angle A_1A_0B_0$ clockwise and counterclockwise by the angle $\beta = \angle A_1A_0B_0$ about the points A_0 and B_0 counterclockwise and clockwise, respectively:

$$(A_1 - A_0)e^{i\beta} \;=\; h(B_0 - A_0) \text{ and } (A_1 - B_0)e^{-i\beta} \;=\; h(A_0 - B_0).$$

From here

$$(A_1 - A_0) = he^{-i\beta}(B_0 - A_0) \text{ and } (B_0 - A_1) = he^{i\beta}(B_0 - A_0);$$

and by adding these two equalities we obtain

$$h(e^{i\beta} + e^{-i\beta}) = 1. \tag{V.6.4}$$

Moreover, define

$$\omega = \frac{B_0 - A_1}{A_0 - A_1} = \frac{he^{i\beta}(B_0 - A_0)}{-he^{-i\beta}(B_0 - A_0)} = -e^{i\,2\beta} = -e^{i(\pi-\alpha)} = e^{-i\alpha}.$$

We have $(A_1 - A_0) = he^{-i\beta}(B_0 - A_0)$, and similarly $(B_1 - B_0) = he^{-i\beta}(C_0 - B_0)$. Using these equalities we can write the expression

$$B_1 - A_1 = (B_0 - A_1) + (B_1 - B_0) = he^{i\beta}(B_0 - A_0) + he^{-i\beta}(C_0 - B_0)$$
$$= k\,[\,\omega\,(B_0 - A_0) - (C_0 - B_0)\,], \text{ where } \omega = -e^{i\,2\beta} = e^{-i\alpha} \text{ a } k = -he^{-i\beta}.$$

Finally, using the equality (V.6.1). we deduce

$$k = -he^{-i\beta} = -\frac{e^{-i\beta}}{e^{i\beta} + e^{-i\beta}} = -\frac{1}{e^{i\,2\beta} + 1} = \frac{1}{\omega - 1}.$$

□

From here, we get the following important consequence.

Theorem V.6.3. *Let $\omega_1, \omega_2, \ldots, \omega_{n-1}$ be a sequence of all the n-th roots of unity excluding 1. In other words, these are the numbers $e^{i\,\frac{2s\pi}{n}}$ for $s \in \{1, 2, \ldots, n-1\}$, in an arbitrarily chosen order; write*

$$\omega_r = e^{i\,\frac{2s_r\pi}{n}}.$$

For each $r \in \{1, 2, \ldots, n-2\}$, perform Petr's construction successively on the polygon

$$A_{r,1}A_{r,2}A_{r,3}\ldots A_{r,s}A_{r,s+1}\ldots A_{r,n},$$

that is, construct over every edge $A_{r,s}A_{r,s+1}$ the isosceles triangle

$$A_{r,s}A_{r+1,s}A_{r,s+1},$$

whose angle at the vertex $A_{r+1,s}$ is $\dfrac{2s_r\pi}{n}$.

Then, for an **arbitrary polygon** $A_{0,1}A_{0,2}A_{0,3}\ldots A_{0,s}A_{0,s+1}\ldots A_{0,n}$,

$$A_{n-1,1} = A_{n-1,2} = A_{n-1,3} = \cdots = A_{n-1,s} = A_{n-1,s+1} = \cdots = A_{n-1,n},$$

so that

$$A_{n-2,1}A_{n-2,2}A_{n-2,3}\ldots A_{n-2,s}A_{n-2,s+1}\ldots A_{n-2,n}$$

is a **regular polygon**.

Remark V.6.3. Karel Petr published his results in Czech in 1905 (Časopis pro pěst. mat. a fyz. 34 (1905), pp. 166–172) and later in 1908 republished them in German (Archiv Math. Physik 13 (1908), pp. 29–31). His algorithm was later rediscovered by **Jesse Douglas (1897 - 1965)** in 1940 (J. Math. and Phys. 19 (1940), pp. 93–130) and by **Bernhard Neumann (1909 - 2002)** in 1941 (Journal of the London Mathematical Society 16 (1941), pp. 230–245.

Exercise V.6.3. Prove that the sum of the sequence $k_1, k_2, \ldots, k_{n-1}$ of complex numbers that together with $\omega_1, \omega_2, \ldots, \omega_{n-1}$ describe the construction of Theorem V.6.3, equals $\frac{1}{n}$.

Exercise V.6.4. The **centroid** C_r of a polygon $A_{r,1}A_{r,2}A_{r,3}\ldots A_{r,s}A_{r,s+1}\ldots A_{r,n}$, where $A_{r,s} = P(a_{r,s})$, is defined by

$$C_r = P(\frac{1}{n}\sum_{s=1}^{n} a_{r,s}).$$

Show that the centroids of all polygons obtained successively by Petr's construction coincide.

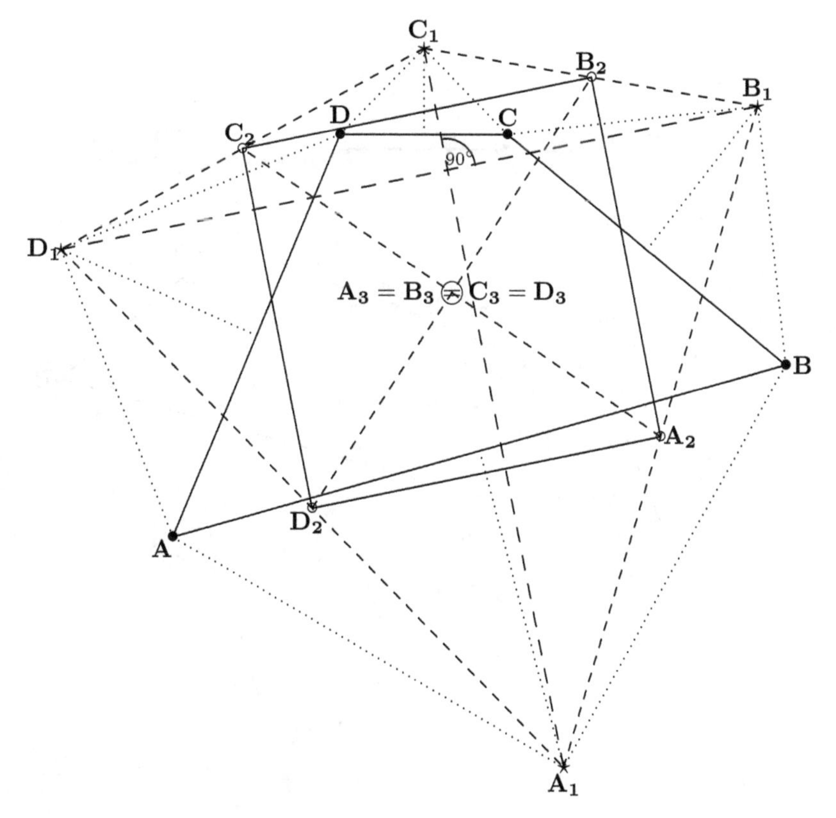

Figure V.6.2 Van Aubel's construction and Petr's square $A_2B_2C_2D_2$

Remark V.6.4. It is easy to see that Napoleon's theorem and van Aubel's construction of the previous section are Petr's constructions for $n = 3$ and $n = 4$. Napoleon's theorem is formulated in terms of the centroids of the equilateral triangles erected on the sides of a given triangle, and hence we deal with Petr's construction defined by the angle 120° (and 240°). Van Aubel's theorem is formulated in terms of the diagonals of the square whose vertices are the vertices of the right-angled isosceles triangles erected on the sides of a given quadrilateral. They are perpendicular and are of the same length. This is equivalent to saying that the mid points of such a quadrilateral define a square. Such a construction is therefore Petr's construction defined by the sequence of angles $90°, 180°$ (and 270°) (see Figure V.6.2).

Exercise V.6.5. The vertices of a given quadrilateral are defined by complex numbers $z_{0,1}, z_{0,2}, z_{0,3}$ and $z_{0,4}$. Determine the complex numbers $z_{2,1}, z_{2,2}, z_{2,3}$ and $z_{2,4}$ describing Petr's square constructed using the sequences

(i) $90°, 180°, 270°$;

(ii) $270°, 180°, 90°$.

Show that the position of the vertex $P(z_{2,1})$ of Petr's square depends only on the position of the vertices $P(z_{0,1}), P(z_{0,2})$ and $P(z_{0,3})$.

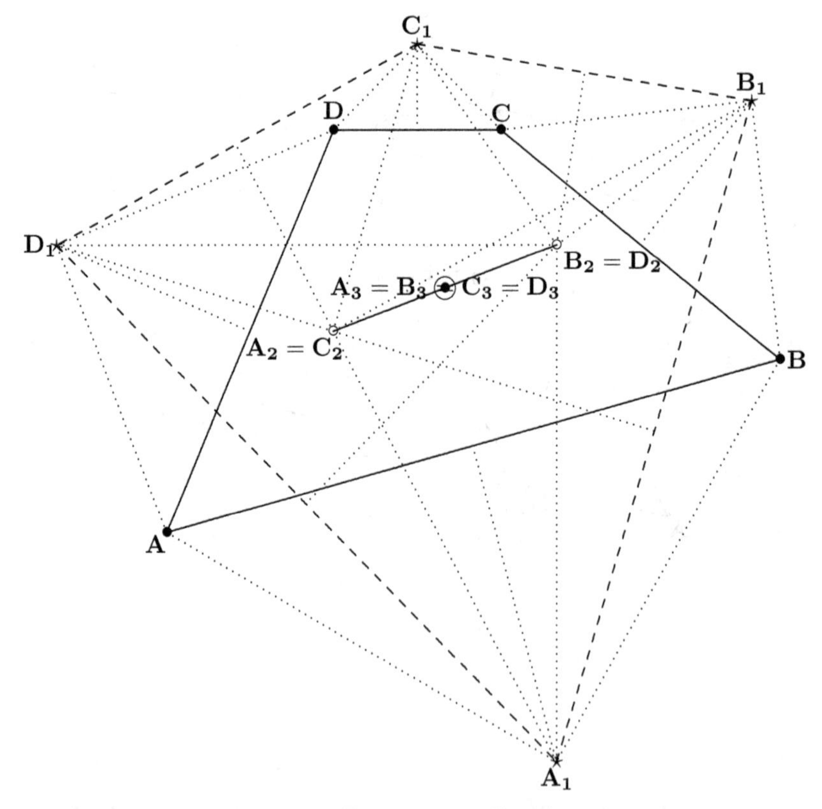

Figure V.6.3 Degenerate Petr's square

Exercise V.6.6. Show that the mid points of the diagonals AC and BD together with the mid points of the diagonals A_1C_1 and B_1D_1 of Figure V.6.2 form a square.

Remark V.6.5. It is important to be aware of the fact that Petr's construction is entirely general. It concerns general polygons that may become degenerate. Thus, for example, considering the initial quadrilateral $ABCD$ of Remark V.6.4 (Figure V.6.2) and performing the construction defined by the sequence of angles $90°, 270°, 180°$, we obtain Petr's square degenerated to a segment $A_2B_2C_2D_2$ ($A_2 = C_2,\ B_2 = D_2$). The midpoint of this square is the point $A_3 = B_3 = C_3 = D_3$ (see Figure V.6.3).

Example V.6.1. Of course, the initial polygon may already be degenerate. Let $ABCD$ be a quadrilateral such that $A = C$. Figure V.6.4 illustrates Petr's construction for the sequence of angles 90°, 180°, 270° (the square $A_2B_2C_2D_2$) and for the sequence 90°, 270°, 180° (degenerate square $A_2'B_2'C_2'D_2'$, in which $A_2' = C_2'$ and $B_2' = D_2'$).

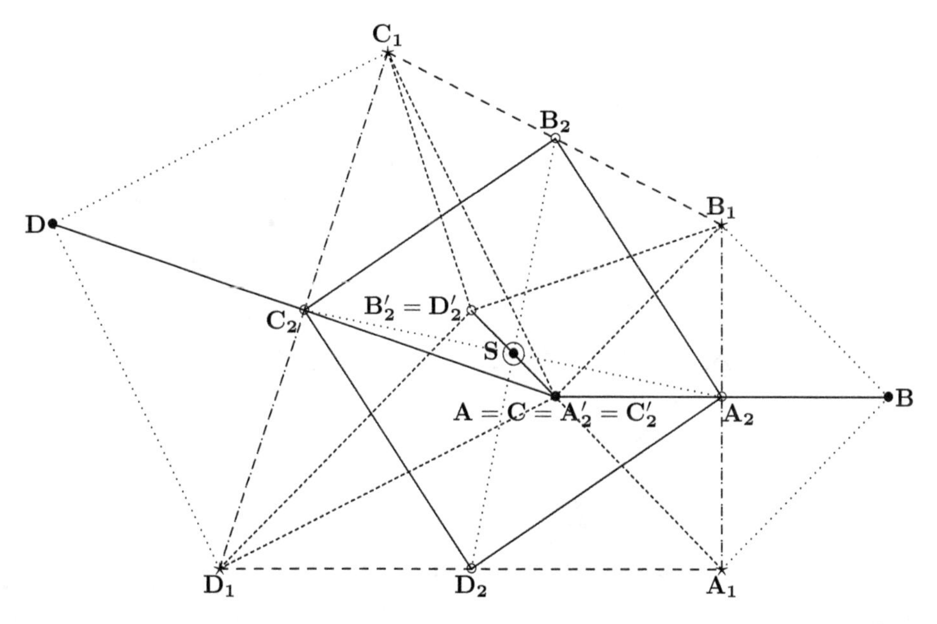

Figure V.6.4 Petr's square $A_2B_2C_2D_2$ and degenerate Petr's square $A_2'B_2'C_2'D_2'$

The construction of Example V.6.1 can be reformulated in the form of the following exercise.

Exercise V.6.7. Let $ABCD$ and $DEFG$ be two arbitrary squares with a common vertex D. Let P and R be their centers (that is, intersections of their diagonals). Let Q be the mid point of the segment CE and S the mid point of the segment AG. Show that the quadrilateral $PQRS$ is a square.

In fact, the very same exercise is often formulated as follows.

Exercise V.6.8. Let $ABCD$ and $DEFG$ be two arbitrary squares with a common vertex D. Let $CKED$ and $ADGH$ be the attached parallelograms as illustrated in Figure V.6.5. Show that the centers P, R of the squares and the centers Q, S of the parallelograms form a square.

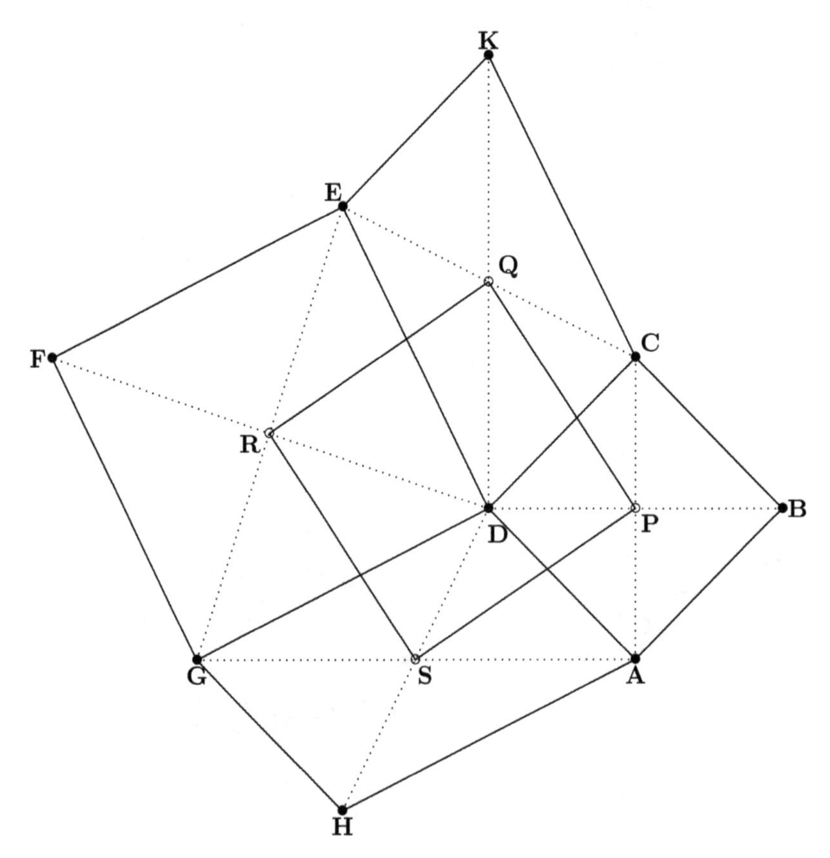

Figure V.6.5 Petr's square PQRS

We recall that the vertices of any parallelogram can be interpreted as the mid points of a suitable quadrilateral. As a consequence, it is easy to answer the following exercise.

Exercise V.6.9. Prove that the vertices of the right angled isosceles triangles erected on the sides of an arbitrary parallelogram form a square.

The next exercise concerns a degenerate triangle.

Exercise V.6.10. Let C be an arbitrary point of a given segment AB. Erect on the segments AB, AC and CB equally oriented isosceles triangles ABK, ALC and CMB with an equal angle α at the vertices K, L and M. Prove that the triangle ΔKLM is equilateral if and only if $\alpha = 120°$.

Example V.6.2. We conclude this section with two of Petr's constructions for $n = 6$. The first one corresponds to the sequence $60°, 120°, 180°, 240°, 300°$ (Figure V.6.6), the second, degenerate, to the sequence $60°, 120°, 180°, 300°, 240°$ (Figure V.6.7).

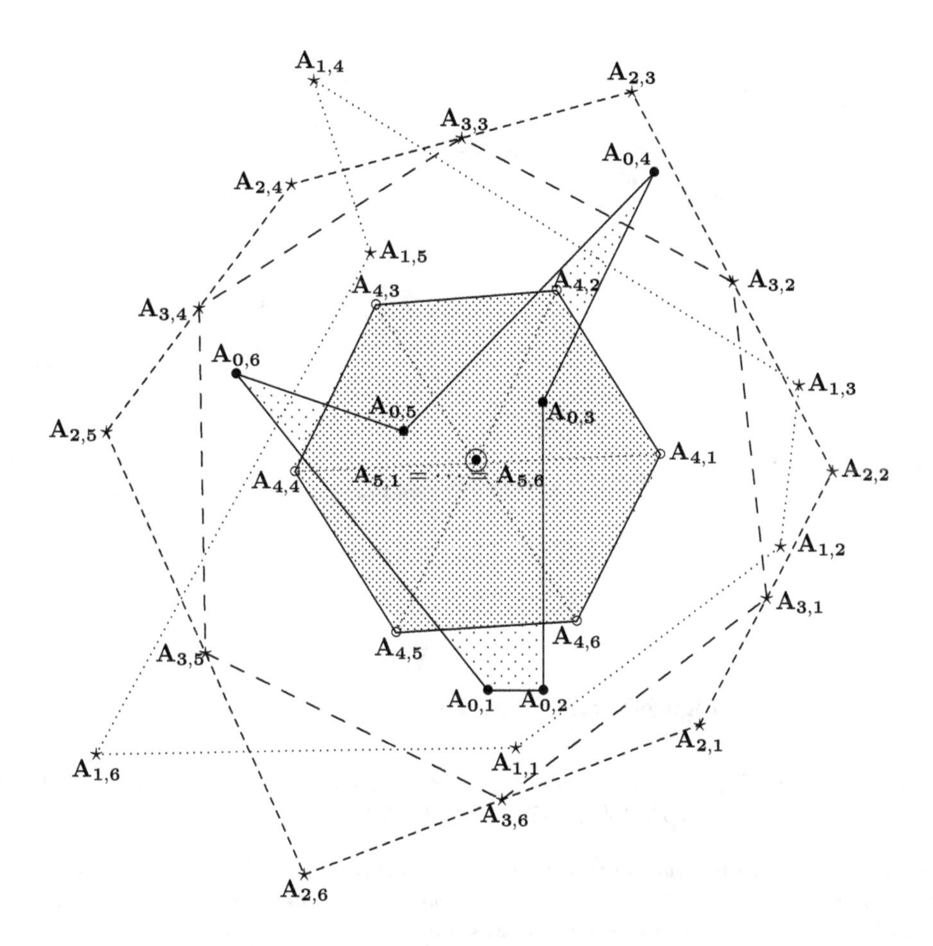

Figure V.6.6 Petr's hexagon using sequence 60°, 120°, 180°, 240°, 300°

In both Figures V.6.6 and Figure V.6.7, all hexagons are successively labeled, starting with an initial hexagon $A_{01}A_{02}A_{03}A_{04}A_{05}A_{06}$, continuing with $A_{11}A_{12}A_{13}A_{14}A_{15}A_{16}$, $A_{21}A_{22}A_{23}A_{24}A_{25}A_{26}$, $A_{31}A_{32}A_{33}A_{34}A_{35}A_{36}$, to arrive at the final regular hexagon $A_{41}A_{42}A_{43}A_{44}A_{45}A_{46}$ (that is in the second case degenerated: $A_{41} = A_{44}, A_{42} = A_{45}$ and $A_{43} = A_{46}$).

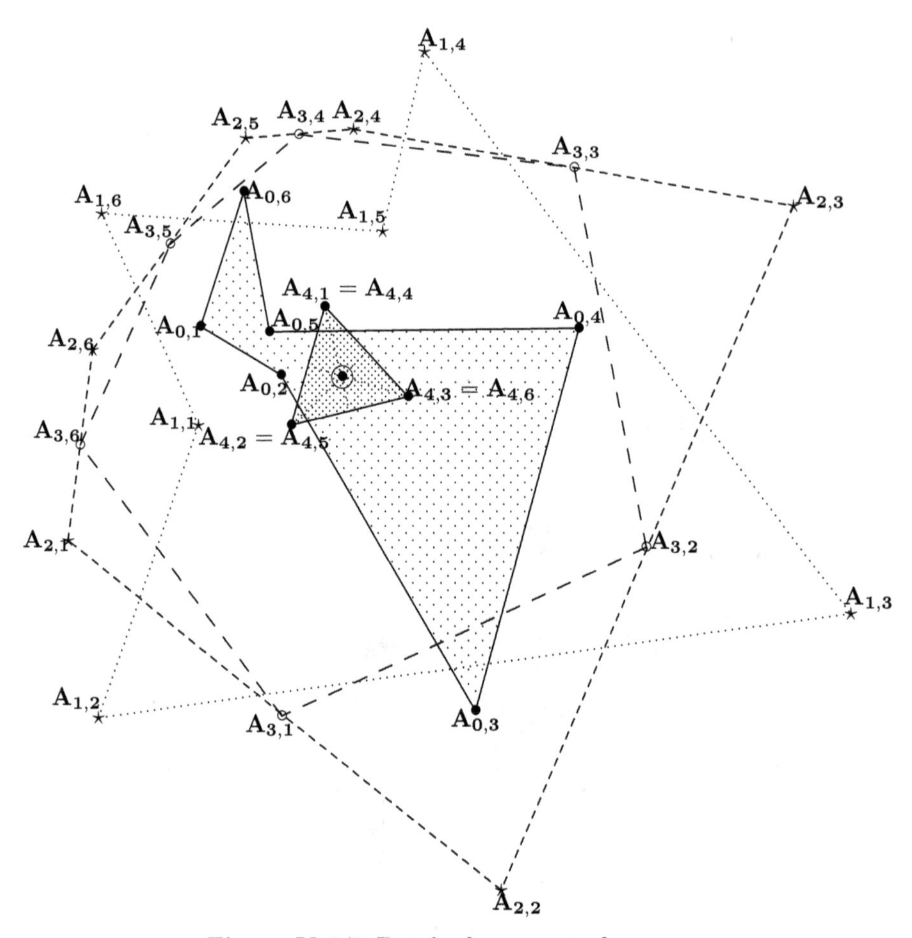

Figure V.6.7 Petr's degenerate hexagon

Exercise V.6.11. Given an arbitrary octagon, construct Petr's octagon using the following sequence of angles: 45°, 135°, 180°, 235°, 270°, 315°, 90°.

Remark V.6.6. We mention in passing that Petr's theorem can be generalized as follows: *Let M (model) and S (start) be two arbitrary n-gons. Then there exists a general algorithm that will carry S in $n-1$ steps into an n-gon similar to M. Petr's theorem is just the special case when M is regular. Such a generalization was published by Stephen B. Gray in 2003 in The American Mathematical Monthly 110 (2003), pp. 210–227.*

V.7 ⋆ Isometries of the plane

The concept of a plane isometry has already been used in the proof of Theorem V.5.4. We start with a formal definition.

Definition V.7.1. A mapping $F : \mathbb{C} \to \mathbb{C}$ is said to be an **isometry (distance preserving mapping)** of the Argand-Gauss plane if, for any $z_1, z_2 \in \mathbb{C}$,

$$| F(z_1) - F(z_2) | = | z_1 - z_2 | .$$

Surprisingly, such a mapping has a very limited form.

Theorem V.7.1. *Let F be an isometry of $\mathbb{C}$. Then*

$$F(z) = F(0) + wz \ \ \text{or} \ \ F(z) = F(0) + w\bar{z} \ \ \text{with} \ w \in \mathbb{C}, |w| = 1. \tag{V.7.1}$$

Proof. Note that $F(1) - F(0) = w$ with $|w| = 1$. Thus the mapping $G : \mathbb{C} \to \mathbb{C}$ defined by $G(z) = w^{-1}(F(z) - F(0))$ is also an isometry. Now, $G(0) = 0$ and $G(1) = 1$. Thus

$$|G(z)| = |G(z) - G(0)| = |z| \ \text{and} \ |G(z) - 1| = |G(z) - G(1)| = |z - 1|.$$

Therefore, according to Theorem V.2.1(iv), $G(z) = z$ or $G(z) = \bar{z}$. Hence (V.7.1) follows. □

Observe that, as a simple consequence of the previous theorem, every isometry F satisfying $F(0) = 0$ is a $\mathbb{R}$−linear mapping and thus a linear transformation of the real vector space $\mathbb{C} = \mathbb{R} \oplus \mathbb{R}$.

Any isometry F of the plane is a bijection and thus an inverse isometry F^{-1} always exists. A composition (product) $F \circ G$ of two isometries F and G is again an isometry. Hence the set of all isometries of the plane forms with respect to this product a **group**. We shall describe the structure of this group in Chapter VIII; now we shall only make a few calculations and remarks. But first, we introduce a few types of isometry and show that they exhaust all possibilities.

Definition V.7.2. *Given a (fixed) complex number c, the mapping $T : \mathbb{C} \to \mathbb{C}$ such that*

$$T(z) = c + z$$

is called a **translation** *of the plane $\mathbb{C}$ through c. We shall denote it by T_c (see Figure V.7.1).*

Given a (fixed) real angle φ, the mapping $R : \mathbb{C} \to \mathbb{C}$ such that

$$R(z) = e^{i\varphi} z$$

is called a **rotation** *of the plane around the origin $P(0)$ by the angle φ. More generally, $R(z) = c + e^{i\varphi}(z - c)$ is a rotation around the point $P(c)$ by the angle φ. We shall denote it by $R_{(c,\varphi)}$ (see Figure V.7.2).*

The mapping $F : \mathbb{C} \to \mathbb{C}$ defined by

$$F(z) = \bar{z}$$

is called a **reflection** *in the x-axis. More generally,*

$$F(z) = c + e^{i\varphi}(\bar{z} - \bar{c})$$

is called a reflection about the line $z = c + te^{i\frac{\varphi}{2}}$. We shall denote it by $F_{(c,\frac{\varphi}{2})}$ (see Figure V.7.3).

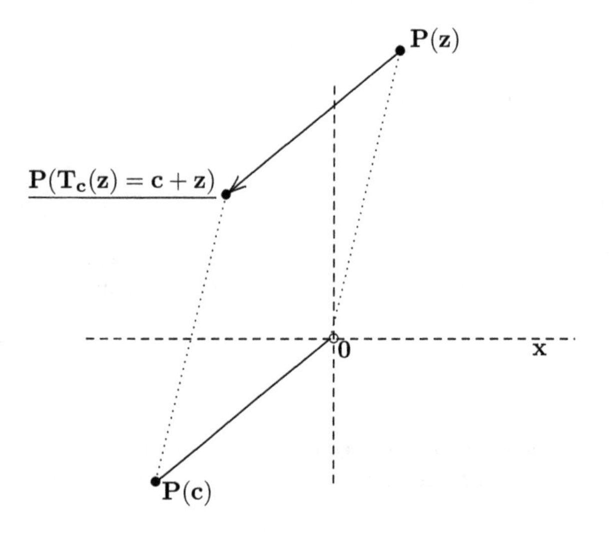

Figure V.7.1 Translation through c

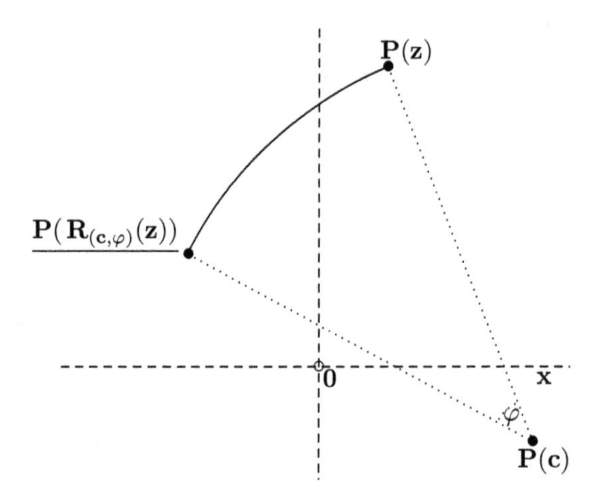

Figure V.7.2 Rotation around P(c) by angle φ

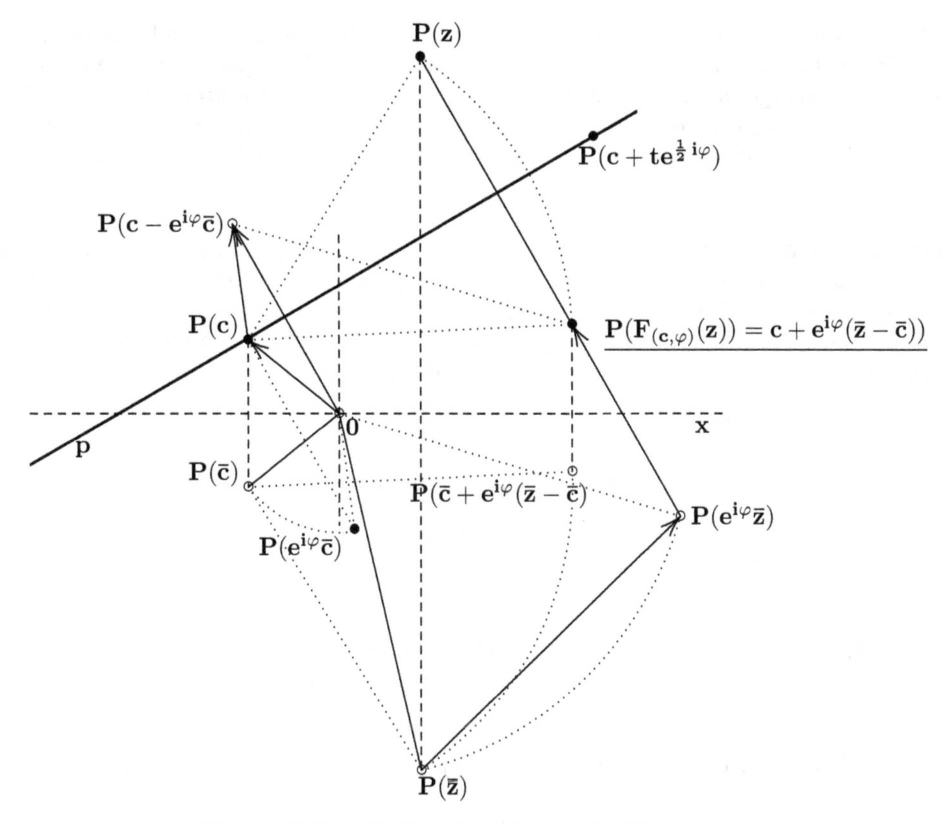

Figure V.7.3 Reflection about the line p

Remark V.7.1. The two families of isometries described in Theorem V.7.1 are often referred to as **isometries of the first class**, or **direct isometries**

$$F_{(c,\varphi,+)}(z) = c + e^{i\varphi}z,$$

and **isometries of the second class**, or **indirect isometries** since they change the orientation of the objects

$$F_{(c,\varphi,-)}(z) = c + e^{i\varphi}\bar{z}.$$

We point out all consequences of Theorem V.7.1. Observe that if $c = F(0) = 0$, we deal with a rotation or $R_{(0,\varphi)}$ or a reflection $F_{(0,\frac{\varphi}{2})}$. For $\varphi = 0$ we deal with the identical mapping $I = F_{(0,0,+)}(z) = z$ or the reflection $F_{(0,0,-)}(z) = \bar{z}$. Now, for $F(0) \neq 0$ and $\varphi = 0$ we get either a translation or a mirror translation. Finally, for $c = F(0) \neq 0$ and $\varphi \neq 0$, that is, $w \neq 1$, we get (for $w = e^{i\varphi}$) either

$$R_{(a,\varphi)} \text{ with } a = \frac{F(0)}{1-w} \quad \text{or} \quad T_d \circ F_{(b,\varphi)} \text{ with } b = \frac{\overline{F(0)}}{1-\bar{w}} \text{ and } d = \frac{F(0)}{1-w} - \frac{\overline{F(0)}}{1-\bar{w}}$$

(the case sometimes called a **glide reflection**). Thus, we can see that the isometries defined in Definition V.7.2 exhaust the entire set (group) of isometries of the plane.

A composition of two translations is again a translation. And, naturally an inverse of a translation T_c is a translation T_{-c}. Thus all translations of the plane form a group G_T, a subgroup of the group G_{S_+} of all direct isometries. Also a composition of two rotations around a point $P(c)$ is a rotation around the point $P(c)$ and the inverse mapping of such a rotation $R_{(c,\varphi)}$ is the rotation $R_{(c,-\varphi)}$. Thus these rotations also form a subgroup G_{R_c} of G_{S_+}.

Theorem V.7.2. *The bijective mapping Φ from the group G_T of all translations of the plane to $\mathbb{C}$ defined by $\Phi(T_c) = c$ is an isomorphism of the group G_T with the additive group $(\mathbf{C}, +)$ of complex numbers. The correspondence Ψ, mapping $R_{(c,\varphi)}$ to φ is an isomorphism of the group G_{R_c} of all rotations of the plane around the point $P(c)$ and the additive groups $(\mathbb{R}/2\pi\mathbb{Z}, +)$ of the real numbers modulo 2π. Furthermore,*

$$R_{(c,0)} = T_0 = I \text{ and } R_{(c,\varphi)} = T_{c_*} \circ R_{(0,\varphi)}, \text{ where } c_* = c - e^{i\varphi}c,$$

$$R_{(c_2,\varphi_2)} \circ R_{(c_1,\varphi_1)} = T_c \circ R_{(c_1,\varphi_1+\varphi_2)}, \text{ where } c = c_2 - c_1 + e^{i\varphi_2}(c_1 - c_2),$$

$$R_{(c_2,\varphi_2)} \circ R_{(c_1,\varphi_1)} = T_{c_0} \circ R_{(0,\varphi_1+\varphi_2)}, \text{ where } c_0 = c_2 + e^{i\varphi_2}(c_1 - c_2) - e^{i(\varphi_1+\varphi_2)}c_1$$

and

$$R_{(c_2,\varphi)} \circ T_{c_1} = T_{c_1} \circ R_{(c_2-c_1,\varphi)}.$$

In particular, $R_{(c,\varphi)} = T_c \circ R_{(0,\varphi)} \circ T_{-c}$. Thus, the group G_{S_+} of all (direct) isometries generated by all rotations and all translations of the plane satisfies $G_{S_+} = G_T \circ G_{R_0}$.

Exercise V.7.1. Prove Theorem V.7.2.

Exercise V.7.2. Prove that

$$R_{(\frac{c}{2},\pi)} \circ R_{(0,\pi)} = T_c$$

and, in general, for an arbitrary $c \in \mathbb{C}$ and an arbitrary angle φ, that

$$R_{(\frac{c}{1-e^{i\varphi}},\varphi)} \circ R_{(0,-\varphi)} = T_c.$$

Exercise V.7.3. Prove the following relations:

(i) $R_{(c,\varphi)} \circ T_{c_0} \circ R_{(c,-\varphi)} = T_{e^{i\varphi}c_0}$;

(ii) $F_{(c,\varphi)} \circ T_{c_0} \circ F_{(c,\varphi)} = T_{e^{2i\varphi}\bar{c_0}}$;

(iii) $T_{c_2-c_1} \circ R_{(c_1,\varphi)} \circ T_{-c_2+c_1} = R_{(c_2,\varphi)}$.

Exercise V.7.4. Show that

$$T_{h_2} \circ R_{(c_2,\varphi_2)} \circ T_{h_1} \circ R_{(c_1,\varphi_1)} = T_{c_0} \circ R_{(c_1,\varphi_1+\varphi_2)},$$

where $c_0 = c_2 - c_1 + h_2 + e^{i\varphi_2}(c_1 - c_2 + h_1)$.

Remark V.7.2. In conclusion, we record a few useful facts. First,

$$F_{(c,\varphi,-)} = T_c \circ R_{(0,\varphi)} \circ F_{(0,0,-)} \text{ (see Figure V.7.3).}$$

In particular, $F_{(0,0,-)}(z) = \bar{z}$. Furthermore

$$R_{(c,\varphi)} = T_{c(1-e^{i\varphi})} \circ R_{(0,\varphi)} \text{ and } T_c = R_{(\frac{c}{1-e^{i\varphi}},\varphi)} \circ R_{(0,-\varphi)}.$$

Hence, e.g. $T_c = R_{(\frac{c}{2},\pi)} \circ R_{(0,\pi)}$. We also point out that the group G_{s_+} is **not commutative**. We have for example,

$$R_{(c,\varphi)} \circ T_h = T_{e^{i\varphi}h} \circ R_{(c,\varphi)}.$$

Also,

$$R_{(c_2,\varphi_2)} \circ F_{(c_1,\varphi_1,-)} = T_{c_0} \circ F_{(c_1,\varphi_1+\varphi_2,-)},$$

where $c_0 = (1 - e^{i\varphi_2})(c_2 - c_1)$ and

$$F_{(c_1,\varphi_1,-)} \circ R_{(c_2,\varphi_2)} = T_{c_\star} \circ F_{(c_2,\frac{\varphi_1-\varphi_2}{2})},$$

where $c_\star = c_1 - c_2 + e^{i\varphi_1}\overline{c_2}$.

V.8 Quaternions

In this short section, we shall include a few remarks concerning the quaternions, mainly as a reaction to the following question: Can we imitate the process of generating the field of complex numbers as a 2–dimensional real vector space with a particular operation of multiplication to higher dimensional real vector spaces? These were questions that many mathematicians, and in particular **William Rowan Hamilton (1805 - 1865)**, asked and tried to resolve. After unsuccessfully trying to construct a field over the 3–dimensional real vector space (that is, trying to find a proper multiplication of real triplets), he succeeded in 1843 in constructing a **non-commutative division ring** $\mathbb{H}$ over the 4–dimensional real vector space.

In fact, it is quite easy to see that there is no $\mathbb{R}$–linear multiplication of the real triplets $(a, b, c) \in \mathbb{R} \times \mathbb{R} \times \mathbb{R} = \mathbb{R}^3$ that will extend the multiplication defined on the pairs $(a, b) \in \mathbb{C} = \mathbb{R} \times \mathbb{R} = \mathbb{R}^2 \subset \mathbb{R}^3$. For, denoting by

$$e = e_1 = (1, 0, 0), \ e_2 = (0, 1, 0) \text{ and } e_3 = (0, 0, 1)$$

the elements of the canonical basis of $\mathbb{R}^3$, then

$$e_2e_3 = r_1e_1 + r_2e_2 + r_3e_3 \text{ with } r_1, r_2, r_3 \in \mathbb{R}.$$

Assuming that $e_2e_1 = e_2, e_1e_3 = e_3$ and $e_2^2 = -e_1$, we have

$$e_2(e_2e_3) = (e_2e_2)e_3 = -e_1e_3 = -e_3 \text{ and}$$

$$-e_3 = e_2(r_1e_1 + r_2e_2 + r_3e_3) = r_1e_2 - r_2e_1 + r_1r_3e_1 + r_2r_3e_2 + r_3^2e_3 = s_1e_1 + s_2e_3 + r_3^2e_3$$

for suitable $s_1, s_2 \in \mathbb{R}$. From here (since e_1, e_2 and e_3 are independent) $r_3^2 = -1$, contradicting the fact that r_3 is a real number.

The construction of the division ring $\mathbb{H}$ of quaternions is as follows: Choose the canonical basis

$$\{e_0 = (1,0,0,0),\ e_1 = (0,1,0,0),\ e_2 = (0,0,1,0) \text{ and } e_3 = (0,0,0,1)\}$$

of the real vector space $\mathbb{R}^4$ and define Hamiltonian multiplication by the table

mult	e_0	e_1	e_2	e_3
e_0	e_0	e_1	e_2	e_3
e_1	e_1	$-e_0$	e_3	$-e_2$
e_2	e_2	$-e_3$	$-e_0$	e_1
e_3	e_3	e_2	$-e_1$	$-e_0$

Figure V.8.1 Hamiltonian multiplication

It is easy to check that this multiplication is associative, that e_0 is a neutral element (the identity) and that the distributive law holds; in fact, all will follow from Theorem V.8.1 identifying the elements of $\mathbb{H}$ with certain complex 2×2 matrices. Here, we only emphasize that the multiplication is not commutative: for example $e_1e_2 \neq e_2e_1$. Moreover, we mention that the elements of the basis e_0, e_1, e_2, e_3 are traditionally denoted by $1, i, j, k$. In this notation, $q = a_0 + a_1 i + a_2 j + a_3 k$ with $a_t \in \mathbb{R}\,(t = 0,1,2,3)$ and

$$i^2 = j^2 = k^2 = -1,\ ij = -ji = k,\ jk = -kj = i,\ ki = -ik = j.$$

Using the distributive law, we obtain the **product formula**

$$(a_0 + a_1 i + a_2 j + a_3 k)(b_0 e + b_1 i + b_2 j + b_3 k) = (a_0b_0 - a_1b_1 - a_2b_2 - a_3b_3)e$$
$$+ (a_0b_1 + a_1b_0 + a_2b_3 - a_3b_2)i + (a_0b_2 - a_1b_3 + a_2b_0 + a_3b_1)j + (a_0b_3 + a_1b_2 - a_2b_1 + a_3b_0)k.$$

We can also see that we may identify the field $\mathbb{C}$ with the subring whose elements are from the real vector space generated either by one of $\{1, i\}$, $\{1, j\}$ or $\{1, k\}$. In fact, there are infinitely many different embeddings of $\mathbb{C}$ into $\mathbb{H}$.

Exercise V.8.1. Show that every real vector subspace of $\mathbb{H}$ generated by 1 and $\omega = a_1e_1 + a_2e_2 + a_3e_3$ such that $a_1^2 + a_2^2 + a_3^2 = 1$ forms a subring isomorphic to the field $\mathbb{C}$.

Exercise V.8.1 also shows that a quadratic equation can have an infinite number of solutions! Indeed all elements ω from that exercise are solutions of the equation $h^2 + 1 = 0$.

We define the **conjugate** quaternion $\overline{h}$ to the quaternion $h = a_0 + a_1 i + a_2 j + a_3 k$ to be $\overline{h} = a_0 - a_1 i - a_2 j - a_3 k$. Thus, $h\overline{h} = a_0^2 + a_1^2 + a_2^2 + a_3^2$ is a real non-negative number that is zero if and only if $h = 0$. We call the square root $\sqrt{h\overline{h}} = |h|$ the **norm (modulus)** of h. For any $h \neq 0$, we have

$$h\,\frac{\overline{h}}{|h|^2} = 1,$$

and thus every non-zero quaternion h has a (unique) inverse $\frac{\overline{h}}{|h|^2}$. This is the reason why we speak of **H** being a **division ring**.

Here is the promised embedding of $\mathbb{H}$ into the ring of all 2×2 matrices with complex coefficients. The proof of the isomorphism is left as an exercise.

Theorem V.8.1. *The mapping* $\Phi : \mathbb{H} \to \mathrm{Mat}_{2\times 2}(\mathbb{C})$ *defined by*

$$\Phi(a_0 + a_1\,i + a_2\,j + a_3\,k) = \begin{pmatrix} a_0 + a_1\,i & -a_2 - a_3\,i \\ a_2 - a_3\,i & a_0 - a_1\,i \end{pmatrix}$$

is an isomorphism of $\mathbb{H}$ *with a subring of* $\mathrm{Mat}_{2\times 2}(\mathbb{C})$. *Thus*

$$\Phi(i) = \begin{pmatrix} i & 0 \\ 0 & -i \end{pmatrix},\ \Phi(j) = \begin{pmatrix} 0 & -1 \\ 1 & 0 \end{pmatrix},\ \Phi(k) = \begin{pmatrix} 0 & -i \\ -i & 0 \end{pmatrix}$$

and every quaternion $h = a_0 + a_1\,i + a_2\,j + a_3\,k$ *satisfies the real quadratic equation*

$$h^2 - 2a_0 h + |h|^2 = 0.$$

Exercise V.8.2. Using Theorem V.8.1, or otherwise, show that $\mathbb{H}$ is isomorphic to a subring of the ring of real 4×4 matrices.

Exercise V.8.3. Let a, b be nonzero real numbers. Show that the subring

$$\mathbb{R}\left[\frac{a\,i + b\,j}{\sqrt{a^2+b^2}}\right] = \left\{r + s\,\frac{a\,i + b\,j}{\sqrt{a^2+b^2}} \mid r, s \in \mathbb{R}\right\}$$

is isomorphic to the field of complex numbers.

Exercise V.8.4. Prove that the product

$$(a_1^2 + a_2^2 + a_3^2 + a_4^2)(b_1^2 + b_2^2 + b_3^2 + b_4^2),$$

where $a_1, a_2, \ldots, b_4 \in \mathbb{Z}$, can be expressed in the form

$$c_1^2 + c_2^2 + c_3^2 + c_4^2,$$

where $c_1, c_2, c_3, c_4 \in \mathbb{Z}$.

Exercise V.8.5. Let $\mathbb{V}$ be the ring defined over the complex vector space $\mathbb{V} = e_0\mathbb{C} + e_1\mathbb{C} + e_2\mathbb{C} + e_3\mathbb{C}$ by the Hamiltonian definition of multiplication. Thus the elements of $\mathbb{V}$ are $v = a_0\ e_0 + a_1\ e_1 + a_2\ e_2 + a_3\ e_3$, where a_0, a_1, a_2, a_3 are arbitrary complex numbers and multiplication is defined as in Figure V.8.1.

Defining the mapping $\Phi : \mathbb{V} \to \mathrm{Mat}_{2\times 2}(\mathbb{C})$ that satisfies

$$\begin{pmatrix} 1 & 0 \\ 0 & 0 \end{pmatrix} = \Phi(\frac{1}{2}(1 + ie_1)),\ \begin{pmatrix} 0 & 0 \\ 0 & 1 \end{pmatrix} = \Phi(\frac{1}{2}(1 - ie_1)),$$

$$\begin{pmatrix} 0 & 1 \\ 0 & 0 \end{pmatrix} = \Phi(\frac{1}{2}(ie_2 - e_3)) \text{ and } \begin{pmatrix} 0 & 0 \\ 1 & 0 \end{pmatrix} = \Phi(\frac{1}{2}(ie_2 + e_3)),$$

show that $\mathbb{V}$ is isomorphic to $\mathrm{Mat}_{2\times 2}(\mathbb{C})$.

Exercise V.8.6. Let the set $\mathbf{S} = \mathbb{C}\times\mathbb{C} = \{[z, w] \mid z, w \in \mathbb{C}\}$ of (ordered) pairs of complex numbers be endowed with the operations of addition and multiplication defined as follows

$$[z_1, w_1] \oplus [z_2, w_2] = [z_1 + z_2, w_1 + w_2]$$

and

$$[z_1, w_1] \odot [z_2, w_2] = [z_1 z_2 + w_1 \overline{w}_2, z_1 w_2 + w_1 \overline{z}_2].$$

Show that the mapping

$$\Phi : \mathbf{S} \longrightarrow \mathrm{Mat}_{2\times 2}(\mathbb{R})$$

of $(\mathbf{S}; \oplus, \odot)$ into the ring of all 2×2 real matrices defined by

$$\Phi([a + bi, c + di]) = \begin{pmatrix} a + d & c + b \\ c - b & a - d \end{pmatrix}.$$

is an isomorphism of $(\mathbf{S}; \oplus, \odot)$ and $\mathrm{Mat}_{2\times 2}(\mathbb{R})$. Show, in particular, that

$$\Phi^{-1}\begin{pmatrix} p & q \\ r & s \end{pmatrix} = \left[\frac{p+s}{2} + \frac{q-r}{2}\, i, \frac{q+r}{2} + \frac{p-s}{2}\, i\right].$$

Exercise V.8.7. Using the well-known fact that the mapping $F : \mathbb{C} \longrightarrow \mathrm{Mat}_{2\times 2}(\mathbb{R})$ defined by

$$F(a + b\,i) = \begin{pmatrix} a & b \\ -b & a \end{pmatrix}$$

is an embedding, show that every real matrix $A = \begin{pmatrix} p & q \\ r & s \end{pmatrix}$ can be written in the form

$$A = F(z) + F(w)\ J \text{ with suitable } z, w \in \mathbb{C} \text{ and } J = \begin{pmatrix} 0 & 1 \\ 1 & 0 \end{pmatrix}.$$

Determine z and w in terms of p, q, r and s.

Exercise V.8.8. Using the notation of Exercise V.8.7, show that

$$J^2 = \begin{pmatrix} 1 & 0 \\ 0 & 1 \end{pmatrix}, \quad J\,F(z)\,J = F(\overline{z})$$

and prove that the product of the matrices

$$A_t = F(z_t) + F(w_t)\,J, \; t = 1, 2,$$

is equal to

$$A_1\,A_2 = F(z_1 z_2 + w_1 \overline{w}_2) + F(z_1 w_2 + w_1 \overline{z}_2)\,J.$$

Compare this result with the definition of product in Exercise V.8.6.

Remark V.8.1. The complex numbers were presented earlier as the ordered pairs (a_0, a_1) of real numbers $a_0, a_1 \in \mathbb{R}$ with the operations of addition

$$(a_0, a_1) + (b_0, b_1) = (a_0 + b_0, a_1 + b_1), \quad a_0, a_1, b_0, b_1, \in \mathbb{R},$$

and multiplication

$$(a_0, a_1) \cdot (b_0, b_1) = (a_0 b_0 - a_1 b_1, a_0 b_1 + a_1 b_0). \tag{V.8.1}$$

The complex conjugate $\overline{(a_0, a_1)}$ of the pair (a_0, a_1) is the pair $(a_0, -a_1)$. We have seen that the conjugation yields a norm

$$|(a_0, a_1)| = \sqrt{(a_0, a_1) \cdot \overline{(a_0, a_1)}} = \sqrt{a_0^2 + a_1^2}$$

which provides a multiplicative inverse

$$\frac{\overline{(a_0, a_1)}}{|(a_0, a_1)|} \tag{V.8.2}$$

of any nonzero complex number (a_0, a_1).

In a similar fashion, the quaternions can be presented as pairs (α_0, α_1) of complex numbers $\alpha_0, \alpha_1 \in \mathbb{C}$ with the operations of addition

$$(\alpha_0, \alpha_1) + (\beta_0, \beta_1) = (\alpha_0 + \beta_0, \alpha_1 + \beta_1), \quad \alpha_0, \alpha_1, \beta_0, \beta_1 \in \mathbb{C},$$

and multiplication

$$(\alpha_0, \alpha_1) \cdot (\beta_0, \beta_1) = (\alpha_0 \beta_0 - \alpha_1 \overline{\beta_1}, \alpha_0 \beta_1 + \alpha_1 \overline{\beta_0}), \tag{V.8.3}$$

that is commutative, but no longer associative. The quaternion conjugate $\overline{(\alpha_0, \alpha_1)}$ of (α_0, α_1) is $(\overline{\alpha_0}, -\alpha_1)$. Again, it allows to express in a simple way a norm

$$|(\alpha_0, \alpha_1)| = \sqrt{a_0^2 + a_1^2 + a_2^2 + a_3^2}$$

and a multiplicative inverse of any nonzero quaternion similar to (V.8.2).

All this follows immediately from the fact that the quaternion $a_0 + a_1 i + a_2 j + a_3 k$ can be written in the form $(a_0 + a_1 i) + (a_2 + a_3 i)j = \alpha_0 + \alpha_1$.

Also the well-known bijection between the complex numbers and the 2×2 real matrices

$$a_0 + a_1 i \longleftrightarrow \begin{pmatrix} a_0 & -a_1 \\ a_1 & a_0 \end{pmatrix}$$

can be extended to the bijection between the quaternions and the 2×2 complex matrices

$$\alpha_0 + \alpha_1 j \longleftrightarrow \begin{pmatrix} \alpha_0 & -\alpha_1 \\ \overline{\alpha_1} & \overline{\alpha_0} \end{pmatrix}.$$

The last bijection can be rewritten further to the following bijection between the quaternions $a_0 + a_1 i + a_2 j + a_3 k$ and the 4×4 real matrices

$$a_0 + a_1 i + a_2 j + a_3 k \longleftrightarrow \begin{pmatrix} a_0 & -a_1 & -a_2 & -a_3 \\ a_1 & a_0 & a_3 & -a_2 \\ a_2 & -a_3 & a_0 & a_1 \\ a_3 & a_2 & -a_1 & a_0 \end{pmatrix}.$$

The reader is adviced to check all these statements in detail.

We can see that the division rings $\mathbb{R}, \mathbb{C}$ and $\mathbf{H}$ contain a copy of the field $\mathbb{R}$:

$$\mathbb{R} = \{(r,0) | \, r \in \mathbb{R}\} \subset \mathbb{C} = \{(\alpha,0) | \, \alpha \in \mathbb{C}\} \subset \mathbb{H}$$

and that $r \cdot h = h \cdot r$ for all $r \in \mathbb{R}$ and $h \in \mathbf{H}$. In fact, $\mathbb{R}, \mathbb{C}$ and $\mathbf{H}$ are the only finite dimensional associative division rings over the real numbers. This is the content of the **Frobenius theorem** proven by **Ferdinand Georg Frobenius (1849 - 1917)** in 1877 (for a proof see for instance **Richard Sheldon Palais (1931 -)**, The American Mathematical Monthly 75 (1968), pp. 366–368).

The constructions of Remark V.8.1 can be extended to pairs of quaternions. This way we obtain the **octonions (Cayley numbers)** discovered by **John T. Graves (1806 - 1870)** in 1843. Their properties are presented in the following theorem.

Theorem V.8.2. *Denote by $\mathbb{O}$ the set of all ordered pairs (h_0, h_1) of quaternions $h_0, h_1 \in \mathbf{H}$, the 8-dimensional vector space over the field $\mathbb{R} = \{(r,0) | \, r \in \mathbb{R}\}$. The set $\mathbb{O}$ admits the addition*

$$(h_0, h_1) + (g_0, g_1) = (h_0 + g_0, h_1 + g_1), \; h_0, h_1, g_0, g_1 \in \mathbf{H},$$

and, in the spirit of (V.8.1) *and* (V.8.3), *the non-commutative multiplication*

$$(h_0, h_1) \cdot (g_0, g_1) = (h_0 \cdot g_0 - \overline{g_1} \cdot h_1, \; g_1 \cdot h_0 + h_1 \cdot \overline{g_0}), \tag{V.8.4}$$

which is no longer associative. Every octonion (h_0, h_1) yields a conjugate $\overline{(h_0, h_1)} = (\overline{h_0}, -h_1)$ that allows to define a norm $|(h_0, h_1)| = \sqrt{(h_0, h_1) \cdot \overline{(h_0, h_1)}}$ and, as in (V.8.2), *a multiplicative inverse of any nonzero octonion.*

Proof. The proof is left for the reader as an exercise. □

Exercise V.8.9. Display three octonions $\alpha, \beta, \gamma \in \mathbb{O}$ such that

$$(\alpha \cdot \beta) \cdot \gamma \neq \alpha \cdot (\beta \cdot \gamma).$$

Here we should point out that, due to the fact that multiplication of quaternions is not commutative, the order of the factors in the formula (V.8.4) is important. Moreover, due to the fact that multiplication of octonions is not associative, octonions have no matrix representation (such as complex numbers and quaternions have).

CHAPTER VI. ALGEBRAIC STRUCTURES - AN OVERVIEW

In preceding chapters, we have met a series of **algebraic structures**, that is, sets with **algebraic operations**; these were almost exclusively additions or multiplications. To classify these structures, we have used terms like **semigroups, groups, integral domains** or **fields**. We have now assembled a sufficient body of concrete examples of these structures to define them formally. Some of them will be studied in more detail in subsequent chapters.

VI.1 Semigroups and groups

We start with a simple concept of an operation on a set.

Definition VI.1.1. *A* **binary operation** $\star$ *on a set* S *is a mapping* $\star : S \times S \to S$.

Hence, a binary operation assigns to a every (ordered) pair (a, b) of elements of the set S in a unique way a new element $c \in S$; we usually write $a \star b = c$.

In the previous chapters, we have seen a number of such operations with particular properties like commutativity, associativity, etc. Typical examples of such operations are addition and multiplication in number domains.

Remark VI.1.1. In a similar way, one can define a **tertiary algebraic operation** on a set S as a mapping $S \times S \times S \to S$, or indeed an n**-ary algebraic operation** as a mapping $S^n \to S$. **Unary algebraic operations** on a set S are just transformations $S \to S$. For example, the transformation $a \mapsto -a$ is a unary operation on the sets $\mathbb{Z}$, $\mathbb{Q}$, $\mathbb{R}$, $\mathbb{C}$ or $\mathbb{H}$, while $a \mapsto a^{-1}$ is a unary operation on $\mathbb{Q} \setminus \{0\}, \mathbb{R} \setminus \{0\}, \mathbb{C} \setminus \{0\}, \mathbb{H} \setminus \{0\}$ and $\{a \in \mathbb{R} \mid a > 0\}$, but not on $\mathbb{Q}$, $\mathbb{R}$, $\mathbb{C}$ or $\mathbb{H}$. The mapping $(a, b, c) \mapsto \sqrt{a^2 + b^2 + c^2}$ for $a, b, c \in \mathbb{R}$, is an example of a tertiary operation on $\mathbb{R}$.

Definition VI.1.2. *A* **semigroup** *is a set* S *with an* **associative** *binary operation* $\star$ *(usually called addition, multiplication, composition or the like), that is,*

$$a \star (b \star c) = (a \star b) \star c \text{ for all } a, b, c \in S.$$

If, in addition, there is a **neutral element** e *in* S*, that is, an element* e *satisfying*

$$a\star\ e\ =\ e\star\ a\ =\ a\ \textit{for every elment}\ a\in S$$

we say that $(S,\star)$ *is a* **monoid**.

It is straightforward that two neutral elements e_1 and e_2 satisfy $e_1 = e_1 \star\ e_2 = e_2$ and thus the neutral element is determined uniquely. If $\star$ is addition, we speak about **zero** and denote it by 0; if it is multiplication, we call the neutral element an **identity element** and denote it by 1:

$$a+0=0+a=a \quad \text{and} \quad a\cdot 1 = 1\cdot a = a.$$

Similarly we speak about subsemigroups of a given semigroup and submonoids of a given monoid. We emphasize that we require that a submonoid have the same neutral element as the monoid. Thus both $S_1 = \{(a,b) \mid a,b\in\mathbb{Z}\}$ and $S_2 = \{(a,0) \mid a,b\in\mathbb{Z}\}$ are monoids with respect to coordinate multiplication and $S_2 \subset S_1$, but S_2 is not in this sense a submonoid of the monoid S_1. On the other hand, for any choice of the positive integers m and n, the sets $S_{m,n} = \{(m^k,n^l) \mid k,l\in\mathbb{N}_0\}$ are submonoids of the monoid S_1.

Definition VI.1.3. *A monoid* $(G,\star)$ *is said to be a* **group** *if every element* $g\in G$ *has an* **inverse**, *that is, for every* $g\in G$ *there is an element* $g^\bullet\in G$ *such that*

$$g\star\ g^\bullet\ =\ g^\bullet\star\ g\ =\ e.$$

Using associativity of the binary operation of a group, we see that an inverse of an element is determined uniquely. For, if $g_1^\bullet$ and $g_2^\bullet$ are two inverses of g, then

$$g_1^\bullet\ =\ g_1^\bullet\star e\ =\ g_1^\bullet\star(g\star\ g_2^\bullet)\ =\ (g_1^\bullet\star g)\star\ g_2^\bullet\ =\ e\star\ g_2^\bullet\ =\ g_2^\bullet.$$

Of course, in the case that $\star$ is an addition, we talk about an **opposite** element $-g$ of an element g; also in case of a multiplication, we denote the inverse element usually by g^{-1}. Thus

$$g+(-g)=(-g)+g=0 \quad \text{and} \quad g\cdot g^{-1}=g^{-1}\cdot g=1.$$

Here, we should like to emphasize the fact that the mapping $g\mapsto g^\bullet$ is bijective; in fact,

$$(g^\bullet)^\bullet\ =\ g,$$

and thus g and $g^\bullet$ are mutually inverse elements. Be aware of the fact that

$$(g\star\ h)^\bullet\ =\ h^\bullet\star\ g^\bullet\ !$$

We add that in the case that the group operation $\star$ is commutative, we call the group commutative or **abelian** (relating the concept to the Norwegian mathematician **Niels Henrik Abel (1802 - 1829)**).

There is again a concept of a subgroup of a given group : It is a subset of the group that is "closed" with respect to both binary operation $\star$ and the unary operation of taking the inverse of a given element. In Chapter VIII we study the lattice of subgroups of a group as well as homomorphisms of groups. This leads us to the important concepts of a normal subgroup and the internal structure of groups.

Here may be the proper place to recall two concepts that are mentioned freely almost in any textbook, without further consideration and explanation of their properties: the **binary operations of subtraction and division**. We should be aware that improper use and frequent misunderstandings of algebraic operations lead to banal questions like: What is the value of $98 : 14 : 7$? Is it 49 or 1? Of course, without any further specification, we should consider $a : b = a \cdot b^{-1}$ and thus calculate $98 : 14 : 7 = 98 \cdot \frac{1}{14} \cdot \frac{1}{7} = 1$. For details, refer to Exercise III.2.1.

VI.2 Rings and fields

We now turn our attention to structures with two operations.

Definition VI.2.1. *A set R with two binary operations, an addition denoted by $+$ and a multiplication denoted by $\cdot$, is said to be a* **ring** *if*

(i) $(R, +)$ *is an (additive) abelian group with the neutral element* 0,

(ii) *multiplication is associative and*

(iii) *both operations are related by the* **distributive laws**

$$(a+b)\cdot c = a\cdot c + b\cdot c \text{ and } a\cdot(b+c) = a\cdot b + a\cdot c \text{ for all } a,b,c \in R.$$

Here the two distributive laws reflect the fact that the multiplication may be non-commutative. In this compendium, unless otherwise stated, we shall always consider **rings with identity**, that is, those rings for which $(R, \cdot)$ is a monoid (that also has a zero element); the subrings will be required to contain the identity element (sometimes called the unity) of this monoid.

As in the case of any algebraic structure, one of the fundamental concepts is that of a **homomorphism** between two rings. This concept will come under closer scrutiny later in the text, however we mention here the related concept of an **ideal** of a ring that appears naturally as a kernel of a homomorphism:

Let $\varphi : R \to S$ be a ring homomorphism, that is,

$$\varphi(a+b) = \varphi(a) + \varphi(b) \text{ and } \varphi(a\cdot b) = \varphi(a)\cdot\varphi(b) \text{ for all } a,b \in R.$$

Then the **kernel** $\mathrm{Ker}\varphi = \{a \in R \mid \varphi(a) = 0\}$ is an ideal in the following sense.

Definition VI.2.2. *A subset I of a ring R is said to be an ideal of R if*

(i) $(I, +)$ *is a subgroup of the additive group* $(R, +)$ *and if*

(ii) $a \cdot r$ *and* $r \cdot a$ *belong to I for every $a \in I$ and every $r \in R$.*

Especially important in the case of a commutative ring R is the concept of a **principal ideal**. An ideal I of a commutative ring R is called a principal ideal if $I = \{ra \mid r \in R\}$ for some element a of I. In this case we say that the ideal I is generated by a, that a is a generator of I, and we write $I = < a >$. We have seen that every ideal of the integral domain of integers $\mathbb{Z}$ is principal and is of the form $n\mathbb{Z}$ for some $n \in \mathbb{N}$. Each of these ideals describes a **congruence** $(\bmod\, n)$. This congruence relation results in the concept of a **quotient ring** which can be extended to any ring and will be studied extensively in Chapters VII and IX.

Definition VI.2.3. *Let I be an ideal of a ring R. Then the relation $a \equiv b \pmod I$ defined by the fact that $a - b = a + (-b) \in I$ is a congruence relation and the respective equivalence classes form with respect to the induced addition and multiplication a ring called a quotient ring of r modulo I and denoted by R/I.*

As we have seen, an important example of a ring is the ring of all integers. In fact, this ring is an example of a commutative ring, which has the property concerning divisors of zero described in Definition VI.2.4.

Definition VI.2.4. *An* **integral domain** $(D; +, \cdot)$ *is a commutative ring (with identity $\neq 0$) that has no non-trivial divisors of zero, that is for all $a \neq 0$ and $b \neq 0$ we have $a \cdot b \neq 0$.*

An integral domain can be defined by an equivalent condition of right and left cancellation in the following sense:

$$a \cdot c = b \cdot c \text{ with } c \neq 0 \text{ implies } a = b \text{ and } c \cdot a = c \cdot b \text{ with } c \neq 0 \text{ implies } a = b.$$

These two conditions are equivalent in the case of a commutative ring.

The integral domain of the integers was a subject of Chapter IV. In Chapter VII, we shall see that the integral domain of polynomials shares a lot of common properties related to divisibility, in particular Euclidean division, with the integers. Another important integral domain is the following subring of the domain $\mathbb{C}$ of all complex numbers, namely, the **domain of Gaussian integers** $\mathbb{Z}[i] = \{a + bi \mid a, b \in \mathbb{Z}\}$. These three examples share a property that is expressed in the notion of an **Euclidean domain**.

Definition VI.2.5. *An integral domain D is said to be an* **Euclidean domain** *if there is a function $\nu : D\setminus\{0\} \to \mathbb{N} \cup \{0\}$ called a* **norm** *(or measure, or degree) satisfying the following properties:*

(i) *$\nu(a) \leq \nu(a \cdot b)$ for all nonzero elements a and b of D,*

(ii) *for any two elements a and $b \neq 0$ from D, there exist elements q and r in D such that*

$$a = b \cdot q + r, \quad \text{where either } r = 0 \text{ or } \nu(r) < \nu(b).$$

We have already seen that the **absolute value** of a number plays the role of the norm in the integral domain $\mathbb{Z}$. In the next chapter we will use the **degree** of a polynomial as a norm. The function

$$\nu(a + bi) = a^2 + b^2$$

plays the role of the norm in the integral domain of Gaussian integers.

The methods that we have applied in Chapter III can be extended to the study of divisibility in an arbitrary Euclidean domain. The **greatest common divisor** $d(a,b)$ of the elements a and b can be found using the Euclidean algorithm, and Bézout's theorem provides a presentation of $d(a,b)$ as a linear combination of a and b. It is important to keep in mind that all expressions are unique up to replacing any element by an associated element. We say that two elements are **associated** (one is an associate of the other) if each is a divisor of the other. "Being associated to" is an equivalence relation so we can talk about the classes of associated elements. The elements a and b of an integral domain D are associated if and only if $a = g \cdot b$, where g is invertible, that is, where g is a **unit** in the sense that $g^{-1} \in D$ exists such that $g \cdot g^{-1} = 1$. All the units of an integral domain D form an abelian group of units $\mathbf{U}(D)$. It follows that each equivalent class of associated elements of D has the same number of elements (the same cardinality) as the group $\mathbf{U}(D)$.

Exercise VI.2.1. Show that each class of associated elements in the domain $\mathbb{Z}[i]$ of Gaussian integers has four elements. If a is one of them, determine the others.

Exercise VI.2.2. Let $d > 1$ be an integer. Determine the group $\mathbf{U}(D)$ of units of the domain $D = \mathbb{Z}[\sqrt{d}\, i] = \{a + b\,\sqrt{d}\, i \mid a, b \in \mathbb{Z}\}$.

Exercise VI.2.3. Denote by $\mathbb{Z}[\frac{1}{2}]$ the integral domain of all rational numbers of the form $2^k\, a$, where a and k are integers. Determine the group $\mathbf{U}(\mathbb{Z}[\frac{1}{2}])$.

Exercise VI.2.4. Prove that the subset of divisors of zero of a domain D and the set of elements of the group $\mathbf{U}(D)$ are disjoint. Determine the group $\mathbf{U}(D_{[0,1]})$ of the domain $D_{[0,1]}$ of all real continuous functions on the interval $[0,1]$ and show that there exist elements in $D_{[0,1]}$ that are neither units nor divisors of zero. Here, for $f, g \in D_{[0,1]}$, we have $(f+g)(x) = f(x) + g(x)$ and $(fg)(x) = f(x)g(x)$ for all $x \in [0,1]$.

Example VI.2.1. In the integral domain $\mathbb{Z}[i]$ of the Gaussian integers, find the greatest common divisor d of the numbers $a = 17 + 9\,i$ and $b = 7 + 4i$, and express it as a linear combination of a and b. Using the Euclidean algorithm, we successively get

$$\begin{aligned}
17 + 9\,i &= (7 + 4\,i) \cdot 2 + (3 + i),\\
7 + 4\,i &= (3 + i) \cdot 2 + (1 + 2\,i),\\
3 + i &= (1 + 2\,i) \cdot (1 - i),
\end{aligned}$$

and thus $d = 1 + 2\,i$ and

$$d = (-2) \cdot a + 5 \cdot b. \tag{VI.2.1}$$

The algorithm is not unique; we may also proceed as follows:

$$\begin{aligned}
17 + 9\,i &= (7 + 4\,i) \cdot (2 - i) + (-1 + 8\,i),\\
7 + 4\,i &= (-1 + 8\,i) \cdot (-i) + (-1 + 3\,i),\\
-1 + 8\,i &= (-1 + 3\,i) \cdot 2 + (1 + 2\,i),\\
-1 + 3\,i &= (1 + 2\,i) \cdot (1 + i),
\end{aligned}$$

and obtain

$$d = (1 - 2\,i) \cdot a + (-2 + 5\,i) \cdot b. \tag{VI.2.2}$$

Exercise VI.2.5. By yet another choice of the algorithm, we can get

$$d = (-2 - i) \cdot a + (5 + 2i) \cdot b. \tag{VI.2.3}$$

Compare (VI.2.1), (VI.2.2) and (VI.2.3) and explain their relationship.

In the above calculations, we are using the fact that for every complex number $z \in \mathbb{C}$, there is a Gaussian integer $c \in \mathbb{Z}[i]$ such that the absolute value $|z - c|$, that is, the distance of the corresponding points in the Argand-Gauss plane, satisfies the inequality

$$|z - c|^2 \le \frac{1}{2}.$$

Check the inequality (a simple sketch provides a proof). Thus, in order to divide $a \in \mathbb{Z}[i]$ by $0 \neq b \in \mathbb{Z}[i]$, we put $z = \frac{a}{b} \in \mathbb{C}$ and write $a = b \cdot c + b \cdot (z - c)$. Therefore either $z = c$ or $\nu[b \cdot (z - c)] = \nu(b)\nu(z - c) < \nu(b)$. Recall that a nonzero element a of an integral domain D is said to be **irreducible** if a is not a unit and, whenever $a = b \cdot c$ with b and c from D, then either b or c is a unit.

Exercise VI.2.6. Show that in a Euclidean domain, an element a is irreducible if and only if it is **prime** in the sense that it is not a unit and whenever a divides a product $b \cdot c$, a divides either b or c.

Equally as in $\mathbb{Z}$, in a Euclidean domain, every element is a finite product of **irreducible elements**. This product is determined, up to order of the factors and their associates, uniquely. Thus, every Euclidean domain is a **unique factorization domain**. Not every integral domain is a unique factorization domain.

Exercise VI.2.7. Put $D = \mathbb{Z}[2i] = \{a + 2bi \mid a, b \in \mathbb{Z}\}$. Considering the product $4 = 2 \cdot 2 = (-2i) \cdot (2i)$ justify that D is an integral domain that is not a unique factorization domain. Observe that while 2 and $2i$ are not associates in D, they are associates in a bigger domain of the Gaussian integers.

On the other hand, we shall see that there are unique factorization domains that are not Euclidean; in the next chapter, we will see that the ring of polynomials with integral coefficients $\mathbb{Z}[x]$ is a simple example of such a domain.

Example VI.2.2. Consider, for every natural number n, the integral domain

$$\mathbb{Z}[\sqrt{n}\, i] = \{a + b\sqrt{n}\, i \mid a, b \in \mathbb{Z}\}.$$

We have already seen that this domain is Euclidean for $n = 1$ and thus is a unique factorization domain. Also $\mathbb{Z}[\sqrt{2}\, i]$ is a unique factorization domain. In $\mathbb{Z}[\sqrt{3}\, i]$ and $\mathbb{Z}[\sqrt{5}\, i]$ every element can be written as a product of irreducible elements, but as we can see, such expressions are not unique:

$$4 = 2 \cdot 2 = (1 + \sqrt{3}\, i) \cdot (1 - \sqrt{3}\, i) \text{ or } 6 = 2 \cdot 3 = (1 + \sqrt{5}\, i) \cdot (1 - \sqrt{5}\, i).$$

It is a simple important fact that every ideal I of a Euclidean domain is principal: I is "generated" by an element with minimal norm. Thus every Euclidean domain is a **principal ideal domain**, that is, a domain all of whose ideals are principal (see Theorem IX.3.4). We mention, without giving a proof, that every principal ideal domain is a unique factorization domain and that there are domains that are unique factorization domains that are not principal ideal domains. The following Figure VI.2.1 illustrates the relations among these related concepts.

We mention that the only unique factorization domains of the type $\mathbb{Z}[\sqrt{n}\ i]$, where $n \in \mathbb{N}$, are those for $n = 1$ and 2. In fact, they are Euclidean domains (see Exercise IX.3.7).

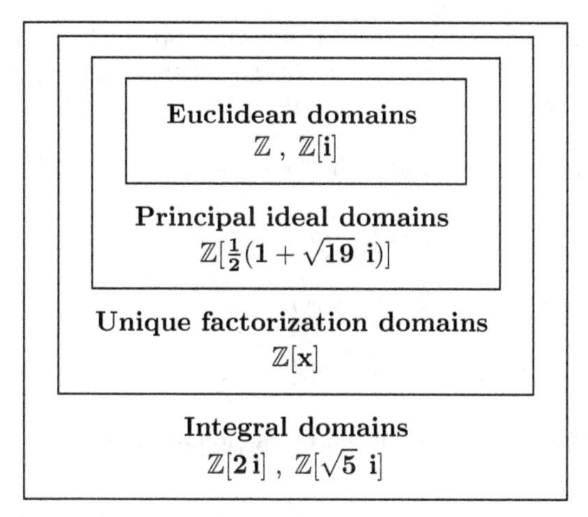

Figure VI.2.1 Hierarchy of integral domains

Exercise VI.2.8. Write down two different decompositions of a number in $\mathbb{Z}[\sqrt{13}\ i]$ and in $\mathbb{Z}[\sqrt{277}\ i]$.

Finally, we give a formal definition of a (skew) field.

Definition VI.2.6. *A ring $D \neq \{0\}$ is said to be a* **skew field** *(or a division ring) if every nonzero element of D is invertible, that is, if the group $\mathbf{U}(D)$ of units is equal to $D \setminus \{0\}$. A commutative division ring is called a* **field**.

Thus, $\mathbb{Q}, \mathbb{R}, \mathbb{C}$ and $\mathbb{Z}_p = \mathbb{Z}/p\mathbb{Z}$ (p prime) are examples of fields, while the quaternions $\mathbb{H}$ form a skew field. Similarly as in Example VI.2.2, we can define for every natural number n, the fields

$$\mathbb{Q}[\sqrt{n}] \;=\; \{a + b\sqrt{n} \mid a.b \in \mathbb{Q}\} \subseteq \mathbb{R} \text{ and } \mathbb{Q}[\sqrt{n}\ i] \;=\; \{a + b\sqrt{n}\ i \mid a.b \in \mathbb{Q}\} \subseteq \mathbb{C}.$$

These fields are examples of subfields of $\mathbb{R}$ and $\mathbb{C}$. We may also write $\mathbb{R}[i] \;=\; \mathbb{C}$.

Fields (and their subfields) will be a subject of study in Chapters VII and IX. Here, we describe the **prime fields** and present a construction of a finite field; a complete description of all finite fields will be given in Chapter VII.

An intersection of any number of subfields is again a subfield. Thus every skew field D contains a (unique) **prime field**, that is, a minimal subfield, a field that contains no proper subfield. Consider the homomorphism

$$\varphi : \mathbb{Z} \to D \text{ given by the rule } \varphi(z) = z \cdot e \text{ for } z \in \mathbb{Z} \text{ and } e \text{ the identity element of } D.$$

If the kernel $\text{Ker}\varphi \neq \{0\}$, then $\text{Ker}\varphi = p\mathbb{Z}$ and $\text{Im}\varphi \simeq \mathbb{Z}/p\mathbb{Z} = \mathbb{Z}_p$ is a prime field of the skew field D. Otherwise, that is, if $\text{Ker}\varphi = \{0\}$, the image $\text{Im}\varphi \simeq \mathbb{Z}$ and the prime subfield of D is isomorphic to the field $\mathbb{Q}$ of rational numbers, which is the field of fractions of $\mathbb{Z}$.

Exercise VI.2.9. Write down details of the preceding proof, that is, of the following statement: *Every prime field is isomorphic either to the countable field $\mathbb{Q}$ of rational numbers or the finite Galois field $GF(p) \simeq \mathbb{Z}_p$ of p elements, where p is a prime.*

Example VI.2.3. Consider the ideal I_n of the Gaussian domain $\mathbb{Z}[i]$ defined by

$$I_n = \{a + b\,i \mid a, b \in n\mathbb{Z}\}.$$

Consider the quotient ring $D_n = \mathbb{Z}[i]/I_n$. For $n = 3$,

$$D_3 = \{[0], [1], [2], [i], [1+i], [2+i], [2\,i], [1+2\,i], [2+2\,i]\}$$

has nine elements with multiplication given by the following table

mult	$[1]$	$[2]$	$[i]$	$[1+i]$	$[2+i]$	$[2\,i]$	$[1+2\,i]$	$[2+2\,i]$
$[1]$	$[1]$	$[2]$	$[i]$	$[1+i]$	$[2+i]$	$[2\,i]$	$[1+2\,i]$	$[2+2\,i]$
$[2]$	$[2]$	$[1]$	$[2\,i]$	$[2+2\,i]$	$[1+2\,i]$	$[i]$	$[2+i]$	$[1+i]$
$[i]$	$[i]$	$[2i]$	$[2]$	$[2+i]$	$[2+2\,i]$	$[1]$	$[1+i]$	$[1+2\,i]$
$[1+i]$	$[1+i]$	$[2+2\,i]$	$[2+i]$	$[2\,i]$	$[1]$	$[1+2\,i]$	$[2]$	$[i]$
$[2+i]$	$[2+i]$	$[1+2\,i]$	$[2+2\,i]$	$[1]$	$[i]$	$[1+i]$	$[2\,i]$	$[2]$
$[2\,i]$	$[2\,i]$	$[i]$	$[1]$	$[1+2\,i]$	$[1+i]$	$[2]$	$[2+2\,i]$	$[2+i]$
$[1+2\,i]$	$[1+2\,i]$	$[2+i]$	$[1+i]$	$[2]$	$[2\,i]$	$[2+2\,i]$	$[i]$	$[1]$
$[2+2\,i]$	$[2+2\,i]$	$[1+i]$	$[1+2\,i]$	$[i]$	$[2]$	$[2+i]$	$[1]$	$[2\,i]$

Thus D_3 is a finite field, the Galois field $GF(3^2)$ and its multiplicative group $U(D_3) = \langle [1+i] \rangle$ is isomorphic to the cyclic group of order 8.

Exercise VI.2.10. Describe, in a similar way, the rings D_2, D_5 and D_7. Show that D_2 is not a field, D_5 has non-trivial divisors of zero and D_7 is a field.

VI.3 Path algebras

There is yet another important construction that we like to recall here. It will be applied in the next chapter in order to introduce the concept of a (formal) polynomial over a field. It falls under the general concept of an F-algebra that we have already used in some examples earlier. We are going to define it formally now. Recall that a vector space A over a field F is an (additive) abelian group A subject to action by elements a, b of the field F so that, for any $x, y \in A$,

$$\begin{aligned} a(x+y) &= ax + ay, \\ (a+b)x &= ax + bx, \\ (ab)x &= a(bx), \\ 1x &= x. \end{aligned}$$

Definition VI.3.1. *An* **algebra** *A over a field F, or briefly an F-***algebra***, is a vector space A over the field F together with a bilinear associative multiplication, that is, a multiplication $A \times A \to A$ that satisfies the following axioms*

$$\begin{aligned} x(y+z) &= xy + xz, \\ (y+z)x &= yx + zx, \\ (ax)y &= x(ay) = a(xy), \\ (xy)z &= x(yz), \end{aligned}$$

where x, y, z are elements from A and a an element (scalar) from the field F.

An F-algebra A is said to be **finite dimensional** or **infinite dimensional** according to whether the space A is finite or infinite dimensional. We shall always assume that A has an element $e \in A$, the **identity** of the algebra satisfying

$$xe = ex = x \text{ for all } x \in A.$$

Observe that the identity e is unique: if e' is another identity, then $e = ee' = e'$.

The following exercise shows that the existence of the identity is a non-essential restriction.

Exercise VI.3.1. If A is an F-algebra without an identity, show that it is always possible to "adjoin" an identity by considering the algebra $\tilde{A}$ of all pairs (x, a), where $x \in A, a \in F$, with componentwise addition, scalar multiplication and the multiplication defined by

$$(x, a)\,(y, b) \;=\; (xy + ay + xb, ab).$$

Example VI.3.1. The following are examples of F-algebras.

(1) The set of all square matrices of order n with entries from a field F forms an algebra with respect to the ordinary operations on matrices. It is a finite dimensional algebra of dimension n^2.

(2) If A is a vector space over the field F, then the linear transformations of the space A form an F-algebra that is finite dimensional if and only if A is finite dimensional.

(3) Every field K that contains the field F as a subfield can be considered as an F-algebra.

(4) In Chapter V, we have considered a four-dimensional $\mathbb{R}$-algebra called the **quaternion algebra** $\mathbb{H}$. Historically, it is one of the first examples of a non-commutative algebra.

Exercise VI.3.2. Show that the set of all elements c of an F-algebra A which commute with all elements of the algebra, that is, such that $ca = ac$ for all $a \in A$, form a subalgebra $C(A)$ of A. The subalgebra $C(A)$ is called the **center** of the F-algebra A.

Exercise VI.3.3. Show that the set of all scalar multiples of the identity of an F-algebra A, that is, of the elements of the form ae with a from the field F, forms a subalgebra Fe isomorphic to F.

Exercise VI.3.4. Prove that a two-dimensional F-algebra A is either a field or it possesses a basis $\{e, x\}$ such that $x^2 = x$ or $x^2 = 0$.

Now, return to the concept of a **path algebra** that is an extension of the concept of path semigroup, as it was introduced in Chapter I. Given a finite oriented graph Γ as defined in I.8, denote by $\mathcal{S}(\Gamma)$ the corresponding path semigroup, as introduced there. Recall that the set $\mathcal{S}(\Gamma)$ consists of all (oriented) paths, including the "trivial" ones e_t corresponding to the vertices of the graph Γ and a formal zero 0.

Choose a field F and denote by $F(\Gamma)$ the vector space over F with $\mathcal{S}(\Gamma) \setminus \{0\}$ taken as its basis. Hence, since the elements of $F(\Gamma)$ are finite F-linear combinations of the elements of $\mathcal{S}(\Gamma)$ (that has a multiplicative structure), we can (applying the distributive law) extend the multiplication to all elements of $F(\Gamma)$. In particular, the element $e = \sum_{t \in \mathbf{V}} e_t$ satisfies the equalities $e \circ p = p \circ e = p$ for every path $p \in \mathcal{S}(\Gamma)$, that is, e is the identity element of the ring $F(\Gamma)$. This ring is at the same time an F-space with a compatible multiplication. In fact, we can identify the elements a of the field F with the elements $a \circ e$ of $F(\Gamma)$ and thus consider the field F to be a (central) subfield of the path algebra $F(\Gamma)$; in this sense, we speak about F-algebra $F(\Gamma)$.

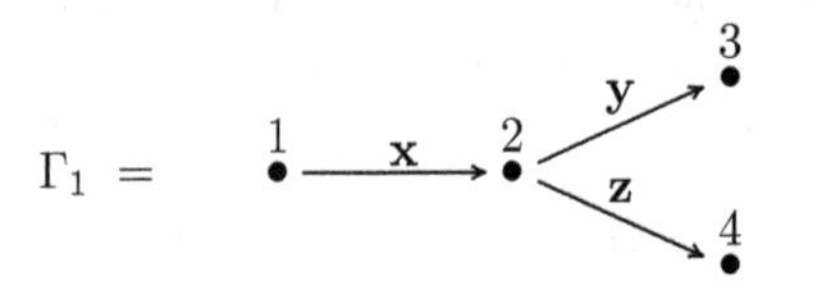

Figure VI.3.1 Example of a graph

Thus, for instance, the graph (of Exercise I.7.1) in Figure VI.3.1 defines the F-algebra $A_1 = F(\Gamma_1)$ with the following multiplication table

mult	e_1	e_2	e_3	e_4	x	y	z	xy	xz
e_1	e_1	0	0	0	x	0	0	xy	xz
e_2	0	e_2	0	0	0	y	z	0	0
e_3	0	0	e_3	0	0	0	0	0	0
e_4	0	0	0	e_4	0	0	0	0	0
x	0	x	0	0	0	xy	xz	0	0
y	0	0	y	0	0	0	0	0	0
z	0	0	0	z	0	0	0	0	0
xy	0	0	xy	0	0	0	0	0	0
xz	0	0	0	xz	0	0	0	0	0

As a result, A_1 is isomorphic to the algebra of all matrices of the form

$$\begin{pmatrix} a_1 & a_5 & a_8 & a_9 \\ 0 & a_2 & a_6 & a_7 \\ 0 & 0 & a_3 & 0 \\ 0 & 0 & 0 & a_4 \end{pmatrix}, \quad a_t \in F$$

(corresponding to the element $a_1 e_1 + a_2 e_2 + a_3 e_3 + a_4 e_4 + a_5 x + a_6 y + a_7 z + a_8 xy + a_9 xz$).

Example VI.3.2. The 4-dimensional F-algebra A_2, defined by the graph Γ_2 in Figure VI.3.2

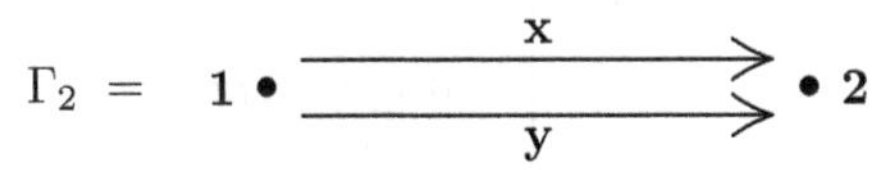

$$\Gamma_2 \;=\; \mathbf{1}\bullet \underset{\mathbf{y}}{\overset{\mathbf{x}}{\rightrightarrows}} \bullet\,\mathbf{2}$$

Figure VI.3.2 Example of a graph

is determined by the multiplication table

mult	e_1	e_2	x	y
e_1	e_1	0	x	y
e_2	0	e_2	0	0
x	0	x	0	0
y	0	y	0	0

and is isomorphic to the F-algebra of all matrices of the form

$$\begin{pmatrix} a_1 & 0 & a_3 & 0 \\ 0 & a_1 & 0 & a_4 \\ 0 & 0 & a_2 & 0 \\ 0 & 0 & 0 & a_2 \end{pmatrix}, \quad a_t \in F.$$

A_2 is a very important algebra, the **Kronecker algebra** that fully classifies pairs (M_1, M_2) of $r \times s$ matrices over the field F under the following "similarity" equivalence

$$(M_1, M_2) \simeq (N_1, N_2) \text{ if and only if } N_t\,P = Q\,M_t\ , t = 1, 2,$$

for some invertible matrices P and Q.

Example VI.3.3. The infinite dimensional F-algebra A_3, defined by the graph Γ_3 in Figure VI.3.3

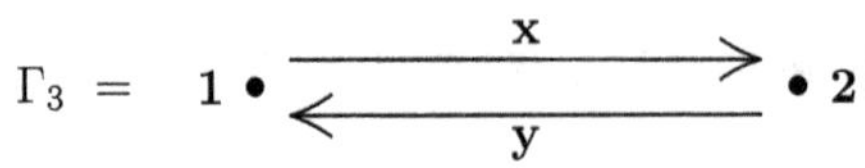

$$\Gamma_3 \ = \ \mathbf{1} \bullet \underset{\mathbf{y}}{\overset{\mathbf{x}}{\rightleftarrows}} \bullet \mathbf{2}$$

Figure VI.3.3 Example of a graph

is determined by the infinite multiplication table

$mult$	e_1	e_2	x	y	xy	yx	xyx	yxy	$\cdots\cdots$
e_1	e_1	0	x	0	xy	0	xyx	0	$\cdots\cdots$
e_2	0	e_2	0	y	0	yx	0	yxy	$\cdots\cdots$
x	0	x	o	xy	0	xyx	0	$xyxy$	$\cdots\cdots$
y	y	0	yx	0	yxy	0	$yxyx$	0	$\cdots\cdots$
xy	xy	0	xyx	0	$xyxy$	0	$xyxyx$	0	$\cdots\cdots$
yx	0	yx	0	yxy	0	$yxyx$	0	$yxyxy$	$\cdots\cdots$
xyx	0	xyx	0	$xyxy$	0	$xyxyx$	0	$xyxyxy$	$\cdots\cdots$
yxy	yxy	0	$yxyx$	0	$yxyxy$	0	$yxyxyx$	0	$\cdots\cdots$
$\vdots$	$\vdots$	$\vdots$	$\vdots$	$\vdots$	$\vdots$	$\vdots$	$\vdots$	$\vdots$	$\ddots$

The F-subalgebra $F(1 = e_1 + e_2, x, y) \subseteq A_3$ is isomorphic to the algebra of all polynomials in two noncommuting indeterminates x and y over the field F modulo the ideal generated by x^2 and y^2.

Exercise VI.3.5. Describe the path algebra of the graph in Figure VI.3.4.

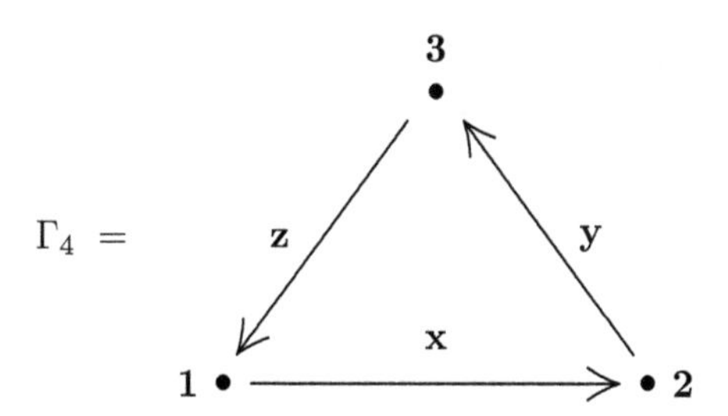

Figure VI.3.4 Example of a graph

The importance of the concept of a path algebra cannot be better documented than by the fact that, for **algebraically closed fields** F (such as for example the field $\mathbb{C}$ of complex numbers) *any F-algebra is isomorphic to a path algebra over an appropriate graph modulo a suitable ideal.*

CHAPTER VII. POLYNOMIALS

VII.1 Basic definition and properties

Simple operations with polynomials already appear nowadays as part of the high school curriculum. They are usually presented and studied as special mappings, as polynomial functions. Historically, polynomials were one of the basic concepts of algebra considered either as a formal construction or as a function. The resulting dichotomy will be clarified in this chapter.

The integral domain of the polynomials over a field is often introduced in textbooks on algebra by using an undefined concept of an "indeterminate", formally in terms of infinite sequences of elements of a given field with simple coordinate addition and a peculiar multiplication. We shall see, however, that this domain is a simple path algebra of the following graph Γ_1 with a single vertex **1** and a single loop **x** (see Figure VII.1.1) that shares many common properties with the integral domain $\mathbb{Z}$ of integers.

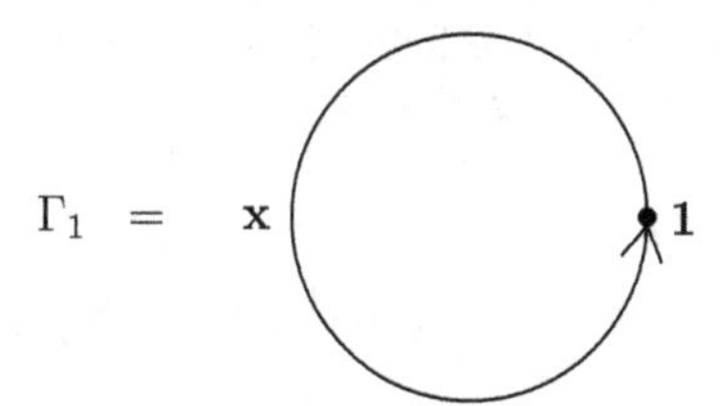

Figure VII.1.1 Graph defining polynomials

But, prior to clarifying this assertion, we recall the traditional construction of the integral domain $F[x]$ of the **(formal) polynomials over a field** F in the spirit of the methods of **James Joseph Sylvester, Arthur Cayley** and **William Rowan Hamilton**. You may think of the field F to be the field of rational numbers $\mathbb{Q}$, or the field of real numbers $\mathbb{R}$, or the field of complex numbers $\mathbb{C}$, or one of the (finite) fields $\mathbb{Z}_p$, where p is a prime.

Consider the vector space $\mathbb{V}_F$ of all infinite sequences

$$f = (a_0, a_1, a_2, \dots, a_t, \dots) = (a_t)_{t\geq 0} \text{ with } a_t \in F$$

such that **only finitely many** of their terms are different from the zero of F. Thus, for every non-zero $f \in \mathbb{V}_F$, there is $a_n \neq 0$ and $a_t = 0$ for all $t > n$. We will call n the **degree** of f and write $\deg f = n$.

The set $B = \{\mathbf{e}_n \mid n \geq 0\}$, where $\mathbf{e}_n = (a_t)$ with $a_n = 1$ and $a_t = 0$ for all $t \neq n$ is a basis of the vector space $\mathbb{V}_F$.

Now, define on $\mathbb{V}_F$ a multiplication compatible with the structure of the vector space by

$$a_r\,\mathbf{e}_r \;\times\; a_s\,\mathbf{e}_s = a_r a_s\,\mathbf{e}_{r+s}$$

and extend it to the entire space $\mathbb{V}_F$ by distributivity. This means the following:

$$\text{If } f = (a_r) = \sum_{r\geq 0} a_r \mathbf{e}_r \text{ and } g = (b_s) = \sum_{s\geq 0} b_s\,\mathbf{e}_s,$$

then

$$f \times g = (c_t) = \sum_{t\geq 0} c_t\,\mathbf{e}_t,$$

where $c_t = \sum_{q=0}^{t} a_q b_{t-q}$. Indeed,

$$f \times g = \sum_{r\geq 0}\sum_{s\geq 0} a_r b_s\,\mathbf{e}_{r+s} = \sum_{t\geq 0}\Big(\sum_{\substack{r\geq 0,\, s\geq 0\\ r+s=t}} a_r b_s\Big)\,\mathbf{e}_t.$$

This way, we have endowed $\mathbb{V}_F$ with a **ring** structure. The zero is the zero vector, the identity the vector $\mathbf{e}_0$; write simply $\mathbf{e}_0 = 1$. It is easy to show that the map $\Phi : F \to \mathbb{V}_F$ defined by $\Phi(a) = a\,\mathbf{e}_0$ is an **isomorphism** of the field F and the **subring** $\{a\,\mathbf{e}_0 \mid a \in F\}$ of all (a_t) such that $a_t = 0$ for all $t \geq 1$. We shall identify this subring with F. Hence, the field F is contained in the ring $\mathbb{V}_F$ which we shall denote, following the usual notation by $F[x]$ indicating that this algebra, that is, this ring with a compatible vector space structure, represents a transition from the field F to $F[x]$ known as the **adjunction** of an "indeterminate" x. Indeed, following this historical notation, we write $\mathbf{e}_1 = x$ and see that $\mathbf{e}_2 = \mathbf{e}_1 \times \mathbf{e}_1 = x^2, \mathbf{e}_3 = \mathbf{e}_2 \times \mathbf{e}_1 = x^3, \dots, \mathbf{e}_t = x^t, \dots$. Thus

$$(a_0, a_1, \dots, a_t, \dots) = a_0 + a_1 x + \cdots + a_t x^t + \cdots = \sum_{t=0}^{\infty} a_t x^t = f(x). \qquad \text{(VII.1.1)}$$

Observe that here we have identified the "addition" in the expression (VII.1.1) with the addition defined on $\mathbb{V}_F$;

$$(a_t) \;=\; a_0\,\mathbf{e}_0 + a_1\,\mathbf{e}_1 + \cdots + a_t\,\mathbf{e}_t + \cdots$$

These inconveniences, as well as a requirement for identification of the expressions of the form $0\,x^2$ and $0\,x^7$, or $5\,x^3 + 2\,x$ and $2\,x - (-5)\,x^3$ arise as a consequence of the fact that we have defined the polynomials in "normal form". The following definition of the algebra of polynomials over a field F eliminates these steps.

Definition VII.1.1. *The algebra $F[x]$ of (formal)* **polynomials** *over a field F is defined as a F-***path algebra** *of the graph Γ_1 in Figure VII.1.1.*

Referring to Chapter VI and to the illustrations VI.3.3 and VI.3.4, in particular, recall what is involved here: In Chapter I, we have defined the **path semigroup** $\mathcal{S}(\Gamma)$ of a given graph Γ. Here, we take such a semigroup $\mathcal{S}(\Gamma_1)$ of our graph Γ_1 for a basis of a vector space over a field F, extend the semigroup multiplication to the entire space by distributivity, identify the F-multiples of the trivial path ϵ_1 (that is, the one-dimensional subspace generated by ϵ_1) with the field F, and denote the resulting ring containing the field F by $F[x]$. Note that the field F lies in the **center** of $F[x]$, that is, $a\, f \;=\; f\, a$ for all elements $a \in F$ and $f \in F[x]$.

Every nonzero element $f = f(x)$ of $F[x]$ can be expressed in its **normal form** with $a_n \neq 0$

$$f(x) \;= a_n\, x^n + a_{n-1}\, x^{n-1} + \cdots + a_2\, x^2 + a_1\, x + a_0 = \sum_{t=0}^{n} a_{n-t}\, x^{n-t}\,. \qquad \text{(VII.1.2)}$$

Definition VII.1.2. *The term $a_n\, x^n$ in* (VII.1.2) *is called the* **leading term** *of f, the coefficient a_n the* **leading coefficient** *of f, a_0 the* **constant term** *of f and the number n is called the* **degree** *of the polynomial f and will be denoted by* $\deg f$.

Remark VII.1.1. Note that the polynomial 0 does not have a degree in this sense. Be aware that it is **not of degree** 0. The polynomials of degree zero are the nonzero elements of the field F, that is, the polynomials for which $a_0 \neq 0$ and $a_t = 0$ for all $t \geq 1$. Whenever a need for the degree of the zero polynomial appears (like in the formulation of the next theorem), we shall use the convention that it is $-\infty$. We conventionally agree that

$$-\infty \;+\; -\infty \;=\; -\infty, \quad -\infty\; + a \;=\; -\infty \;\text{ and }\; -\infty \;<\; a \;\text{ for every integer } a.$$

There will be no need for any other operation with $-\infty$ and thus no other operation is defined.

The proof of the following theorem is straightforward and is left as an exercise.

Theorem VII.1.1. *Let f and g be polynomials in $F[x]$. Then*

$$\deg(fg) = \deg f + \deg g \;\; \textit{and} \;\; \deg(f+g) \leq \max(\deg f, \deg g).$$

As a consequence, $F[x]$ has no divisors of zero and is thus an integral domain.

Remark VII.1.2. The following definition eliminates the restrictions that are required when one deals with the zero polynomial 0. In analogy to the degree, define for every integer $k > 1$ the $k-$**absolute value** $\|f\|_k$ of a polynomial $f \in F[x]$ as follows:

$$\|0\|_k = 0 \;\text{ and }\; \|f\|_k = k^n \;\text{ for }\; f \in F[x] \;\text{ of degree }\; n.$$

Trivially, $\|-f\|_k = \|f\|_k$ and $\|f+g\|_k \leq \|f\|_k + \|g\|_k$ for any $f, g \in F[x]$.

Exercise VII.1.1. Show that for any $f, g \in F[x]$,

(i) $\|fg\|_k = \|f\|_k \|g\|_k$ and

(ii) $\|f+g\|_k \leq \max(\|f\|_k, \|g\|_k)$. In fact, for $\|f\|_k \neq \|g\|_k$, we have $\|f+g\|_k = \max(\|f\|_k, \|g\|_k)$.

Exercise VII.1.2. Prove the following assertion: *Every nonempty set $S \subseteq F[x]$ contains a polynomial $f_\star$ such that $\|f_\star\|_k \leq \|f\|_k$ for all $f \in S$.* Of course, there may be many polynomials like $f_\star$ in S.

The following is an important remark.

Remark VII.1.3. In a similar manner as we have defined polynomials over a field, we can define polynomials over an arbitrary ring R (such as Z or $\mathbb{Z}_m$, where m is composite), and study the respective ring $R[x]$. In this case, instead of dealing with a vector space we deal with the abelian group of finite linear combinations of the paths with coefficients from R. We shall consider $\mathbb{Z}_m[x]$ in several examples and remarks to point out the importance of our assumption that F is a field. Indeed, we shall see that the theory of $\mathbb{Z}_m[x]$ with a composite m is dramatically different to that of $F[x]$, where F is a field. We already know that *every equation of the form $ax + b = 0$ with $ax + b \in \mathbb{Z}_m[x], a \neq 0$ has a solution in $\mathbb{Z}_m$ if and only if m is a prime number.* Such a solution is then unique. In contrast, for example $[6]\,x + [1] = [0]$ has no solution in $\mathbb{Z}_9[x]$ (verify this claim!).

It is also easy to see that a quadratic equation over $\mathbb{Z}_p$, where p is a prime, has **at most** two solutions. Thus, for instance, the equation $x^2+[3]x+[2] = 0$ has two solutions [3] and [4] in $\mathbb{Z}_5$, since $x^2+[3]x+[2] = (x+[1])(x+[2])$. On the other hand, the equation $x^2+x+[1] = 0$ has no solution in $\mathbb{Z}_5$. Moreover, quadratic equations over $\mathbb{Z}_m$ may have, in general, **more than two solutions**: For instance, the quadratic equation $x^2 + [3]x + [2] = [0]$ has four solutions [1], [2], [4] and [5] in $\mathbb{Z}_6[x]$. Indeed,

$$x^2 + [3]x + [2] \;=\; (x + [1])(x + [2]) \;=\; (x + [4])(x + [5])\;!$$

Exercise VII.1.3. Show that the ring $D[x]$ of all polynomials with coefficients from a ring D is an integral domain if and only if D is an integral domain.

We now turn our attention to two important mappings from the integral domain $F[x]$. One is the interpretation of the elements of $F[x]$ as functions defined on F, the other is the evaluation of elements of $F[x]$ at elements of F.

Assign to every polynomial $f \in F[x]$ of the form (VII.1.2) the **polynomial function over the field** F. The mapping $f_e : F \to F$ is defined by

$$f_e(c) \;=\; f(c) \;=\; \sum_{t=0}^{n} a_{n-t}\, c^{n-t} \;\text{ for every }\; c \in F.$$

The sum $f_e + g_e$ and the product $f_e g_e$ defined by $f_e g_e(c) = f_e(c) g_e(c)$ of two polynomial functions derived from the polynomials f and g of $F[x]$, are again polynomial functions over F. Hence the set of all polynomial functions forms a ring with respect to these operations; denote it by $P(F)$. Moreover,

$$f_e \;+\; g_e \;=\; (f+g)_e \;\text{ and }\; f_e g_e \;=\; (fg)_e,$$

and thus the mapping

$$\Phi : F[x] \to P(F) \;\text{ defined by }\; \Phi(f) \;=\; f_e$$

is a ring homomorphism of $F[x]$ onto $P(F)$, that is, an epimorphism. We shall describe the relationship between $F[x]$ and $P(F)$ fully later in this chapter. Here, we just point out that Φ is not, in general, an isomorphism. Indeed if the field F is finite, then $F[x]$ is (as for any field) infinite while $P(F)$ is finite.

Exercise VII.1.4. Show that if F has q elements, then the set $F[x]$ is countable and the set $P(F)$ has q^q elements. Give an example of two different polynomials f and g over the field $\mathbb{Z}_3$ such that $f_e = g_e$.

The other important mapping is the **evaluation mapping** $\Psi_c : F[x] \to F$ at an element c of the field F defined by

$$\Psi_c(f) \;=\; f(c) \;\text{ for all }\; f \in F[x].$$

We observe that given $f, g \in F[x]$ then $\Psi_c(f+g) = \Psi_c(f) + \Psi_c(g)$ and $\Psi_c(fg) = \Psi_c(f)\Psi_c(g)$. Thus for every $c \in F$, Ψ_c is a **homomorphism** from $F[x]$ onto F and the **kernel** of this homomorphism is the **ideal** $\{(x-c)h(x) \mid h(x) \in F[x]\}$. This follows from the fact that $\Psi_c(x-c) = 0$. We mention that this is a special case of congruence on $F[x]$ that we shall consider later in this chapter. The structure of the homomorphism Ψ_c is described by the congruence relation

$$f \equiv g \pmod{(x-c)}$$

expressing the fact that the polynomial $f - g$ is a multiple of the linear polynomial $x - c$. We shall deal with this concept in Section VII.3. The evaluation of a given polynomial $f(x) = a_n x^n + a_{n-1}x^{n-1} + \cdots + a_1 x + a_0$ at $c \in F$ can be expediently computed as

$$f(c) = [(((\ldots\{[(a_n c + a_{n-1})c + a_{n-2}]c + a_{n-3}\}\ldots)))c + a_1]c + a_0.$$

Remark VII.1.4. The homomorphism $\Psi_c : F[x] \to F$, where $c \in F$, can easily be extended to a homomorphism $\Psi_c : F[x] \to R$, where R is an arbitrary ring containing F, $c \in R$ and $as = sa$ for all $a \in F$ and all $s \in R$. Thus, for example, we can define an "evaluation" Ψ_T of polynomials from $F[x]$ at a given linear transformation T of a vector space over F (or at a matrix over F). The Cayley-Hamilton theorem of linear algebra asserts that the characteristic polynomial

$$f(x) = \operatorname{char}_M(x) = \det \begin{vmatrix} x - a_{11} & -a_{12} & \cdots & -a_{1n} \\ -a_{21} & x - a_{22} & \cdots & -a_{2n} \\ \vdots & \vdots & \ddots & \vdots \\ -a_{n1} & -a_{n2} & \cdots & x - a_{nn} \end{vmatrix}$$

of the matrix

$$M = \begin{pmatrix} a_{11} & a_{12} & \cdots & a_{1n} \\ a_{21} & a_{22} & \cdots & a_{2n} \\ \vdots & \vdots & \ddots & \vdots \\ a_{n1} & a_{n2} & \cdots & a_{nn} \end{pmatrix}$$

is in the kernel of Ψ_T.

Exercise VII.1.5. Determine elements c and d of the field $\mathbb{Z}_7$ such that the polynomial $f(x) = x^4+[3]\,x^3+[5]\,x^2+c\,x+d \in \mathbb{Z}_7[x]$ is the square of a quadratic polynomial $g(x) \in \mathbb{Z}_7[x]$, that is, $f(x) = g(x)^2$.

Exercise VII.1.6. Let a, b, c and d belong to a field F. Find necessary and sufficient conditions for the polynomial $f(x) = x^4 + 2a\,x^3 + b\,x^2 + c\,x + d \in F[x]$ to be the square of a quadratic polynomial in $F[x]$.

Exercise VII.1.7. Express the polynomial $f(x) = 2\,x^4 + 8\,x^3 + 21\,x^2 + 26\,x + 15 \in \mathbb{Q}[x]$ in the form of a quadratic polynomial $a\,y^2 + b\,y + c$ with $a, b, c \in \mathbb{Q}$ by choosing a new "indeterminate" y. Hence, find all roots $r \in \mathbb{Q}$ of the polynomial f, that is, all r for which $f(r) = 0$.

Exercise VII.1.8. Show that for any natural number n, the polynomial

$$n\,x^{n+2} - (n+2)\,x^{n+1} + (n+2)\,x - n,$$

is divisible by $(x-1)^3$ in $\mathbb{Z}[x]$.

VII.2 Euclidean division

We have seen that for any field F, $F[x]$ is an integral domain. In fact, $F[x]$ (like the integral domain $\mathbb{Z}$) a **Euclidean domain** in the sense of the following (fundamental) theorem (see Chapter VI).

Theorem VII.2.1 (Euclidean division). *Let f and $g \neq 0$ be polynomials from $F[x]$. Then there exist polynomials q and r in $F[x]$ such that*

$$f = g\,q + r, \quad \textit{with} \quad r = 0 \quad \textit{or} \quad \deg r < \deg g.$$

The polynomials q and r are uniquely determined by these conditions.

Proof. The statement is trivial in case that $f = 0$ or $\deg f < \deg g$. Thus, assume that

$$f(x) = a_n x^n + a_{n-1} x^{n-1} + \cdots + a_1 x + a_0, \quad \text{with } a_n \neq 0$$

and

$$g(x) = b_m x^m + b_{m-1} x^{m-1} + \cdots + b_1 x + b_0, \quad \text{with } b_m \neq 0,$$

where $n \geq m$. We proceed by induction on n. Write

$$\widehat{f}(x) = f(x) - a_n b_m^{-1} x^{n-m} g(x).$$

Then $\widehat{f} = 0$ or $\deg \widehat{f} < \deg f = n$ and thus, by the inductive hypothesis,

$$\widehat{f} = \widehat{q}\, g + r \text{ with } r = 0 \text{ or } \deg r < \deg g.$$

Hence

$$\begin{aligned} f(x) = a_n b_m^{-1} x^{n-m} g(x) + \widehat{f}(x) &= (a_n b_m^{-1} x^{n-m} + \widehat{q}(x))\, g(x) + r(x) \\ &= g(x) q(x) + r(x) \end{aligned}$$

expressing our polynomial f in the desired form.

To prove the uniqueness of this expression, suppose that

$$f = g\, q_1 + r_1 = g\, q_2 + r_2, \text{ with } \deg r_1 < \deg g \text{ and } \deg r_2 < \deg g.$$

Then $g\,(q_1 - q_2) = r_2 - r_1$, and thus, comparing the degrees, the only possibility is that both sides are equal to zero, as required. □

Exercise VII.2.1. Reformulate the Euclidean division of polynomials (Theorem VII.2.1) using as a norm the $k-$absolute value of polynomials as defined in Remark VII.1.2.

One consequence of the Euclidean division is the following corollary that concerns roots of polynomials. Recall that $c \in F$ is said to be a **root** of a polynomial $f \in F[x]$ in the field F if $f(c) = f_e(c) = 0$.

Corollary VII.2.1. *Let F be a field. Let f be a non-zero polynomial in $F[x]$. Then $c \in F$ is a root of f if and only if*

$$f(x) = (x - c)\, q(x), \text{ where } q(x) \in F[x] \text{ and } \deg q = \deg f - 1. \qquad \text{(VII.2.1)}$$

Thus, if the field F has the property that every non-constant polynomial in $F[x]$ has a root in F, that is, if F is **algebraically closed**, *then every $f \in F[x]$ of degree n has the form*

$$f(x) = a(x - c_1)(x - c_2)\dots(x - c_n) \text{ for some } a(\neq 0), c_1, \dots, c_n \in F.$$

This expression is, up to the order of the factors, unique.

Proof. Euclidean division by the linear polynomial $g(x) = x - c$ yields

$$f(x) = (x - c)\, q(x) + r,\ r \in F.$$

Thus $f(c) = 0$ if and only if $r = 0$.

The second assertion follows simply by using induction: Assuming that the polynomial q of degree $n - 1$ has the form

$$q(x) = a(x - c_1)(x - c_2)\dots(x - c_{n-1}),$$

$$f(x) = a(x - c_1)(x - c_2)\dots(x - c_{n-1})(x - c_n) \text{ with } c_n = c.$$

The fact that this expression is up to the order of the factors unique is evident. □

Remark VII.2.1. We express the coefficients of the polynomial $q(x)$ of (VII.2.1), namely,

$$q(x) = b_{n-1}\, x^{n-1} + b_{n-2}\, x^{n-2} + \cdots + b_1\, x + b_0$$

explicitly in terms of the coefficients of $f(x) = a_n\, x^n + a_{n-1}\, x^{n-1} + \cdots + a_1\, x + a_0$. We have

$$b_{n-1} = a_n,\ b_{n-2} = a_{n-1} + c\, b_{n-1}, \ldots, b_t = a_{t+1} + c\, b_{t+1}, \ldots, b_1 = a_2 + c\, b_2$$

and $b_0 = a_1 + c\, b_1$. Note that $a_0 + cb_0 = 0$.

In general, $r = f(c) = a_0 + cb_0$. An efficient way of finding the value $f(c)$ consists in **Horner's method (William George Horner (1786 - 1837))**, already used by Chinese mathematicians including **Qin Jiushao (1202 - 1261)** several hundred years earlier. The method is usually described by the following scheme:

Horner scheme

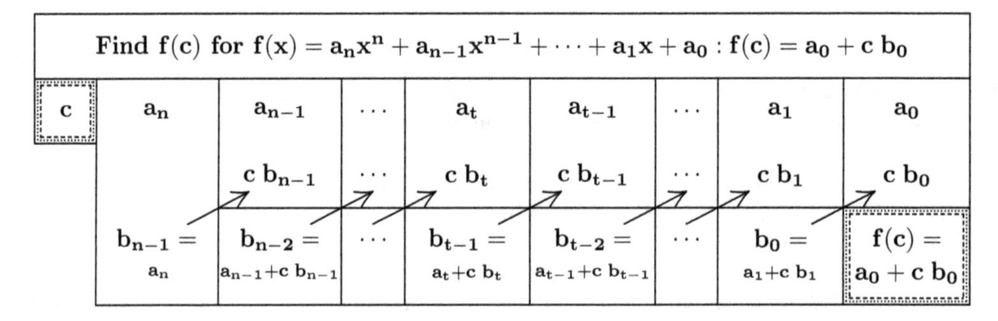

Thus, to find $f(-2), f(3)$ and $f(7)$ for $f(x) = x^5 - 3x^3 + 2x + 12$, we write

1	0	−3	0	2	12	1	0	−3	0	2	12
	−2	4	−2	4	−12		3	9	18	54	168
1	−2	1	−2	6	**0**	1	3	6	18	56	**180**

and

1	0	−3	0	2	12
	7	49	322	2254	15792
1	7	46	322	2256	**15804**

to find that $f(-2) = 0, f(3) = 180$ and $f(7) = 15804$.

Exercise VII.2.2. Show that the coefficient b_t of $q(x)$ in (VII.2.1) is equal to $\sum_{k=1}^{n-t} a_{t+k}\, c^{k-1}$, and thus derive that $f(c) = a_0 + c\, b_0$.

Remark VII.2.2. In Chapter IX, we sketch a proof that the field $\mathbb{C}$ of complex numbers is algebraically closed. Neither $\mathbb{Q}$ nor $\mathbb{R}$ is algebraically closed as the polynomial x^2+1 does not have a root in $\mathbb{R}$. We shall see that the fields $\mathbb{Z}_p$, for any prime p, are also not algebraically closed. Corollary VII.2.1 implies that a polynomial of degree n over an algebraically closed field cannot have more than n roots. The next theorem shows that this is true for any field F.

Theorem VII.2.2. *A non-zero polynomial $f \in F[x]$ of degree n has at most n roots. If*

$$f(x) = a_0 + a_1 x + \cdots + a_n x^n \ \text{ and } \ g(x) = b_0 + b_1 x + \cdots + b_n x^n$$

are two polynomials such that $f(c) = g(c)$ for all $c \in F$ and n is smaller than the number of elements in F, then

$$a_t = b_t \ \text{ for all } \ 0 \leq t \leq n.$$

Proof. This is a consequence of the last Corollary. We proceed by induction: If c is a root of f then $f(x) = (x - c)\ h(x)$ with $\deg h = n - 1$. Now, any root d of $f, d \neq c$, must be a root of h; for, $0 = f(d) = (d - c)\ h(d)$ and thus $h(d) = 0$. By the inductive hypothesis, h has at most $n - 1$ roots. Hence f has at most n roots.

Then apply this statement to the polynomial $f - g$ to obtain the second part of the theorem. □

The following two corollaries are of fundamental importance.

Corollary VII.2.2. *If the field F is infinite, then $f = g$ if and only if $f(c) = g(c)$ for all $c \in F$, that is, $f = g$ if and only if $f_e = g_e$. There is an isomorphism between $F[x]$ and $P(F)$.*

Corollary VII.2.3 (Universal property of polynomials over finite fields). *Let F be a finite field of q elements. Then there is a one-to-one correspondence between the set of all polynomials of $F[x]$ of degree $< q$ and the set of all (point-wise) mappings of F into itself, that is, for any mapping $\phi : F \to F$ there exists a* **unique** *polynomial $f \in F[x]$ of degree $< q$ such that $f(c) = \phi(c)$.*

Proof. If the degrees of f and g from $F[x]$ are smaller than q, then $f = g$ if and only if $f(c) = g(c)$ for all $c \in F$. For, the degree of the polynomial $f - g$ cannot be smaller than the number of roots, unless $f = g$. Since the number of polynomials of degree less than q is q^q which is the number of elements of the set of all mappings of F into F, the bijective correspondence follows. □

Remark VII.2.3. We observe that in the proof of Theorem VII.2.2, we have used the important fact that F has no non-trivial divisors of zero. Otherwise, as we have remarked earlier, a polynomial of degree n can have more than n roots : $x^2 + [3]\,x + [2]$ has 4 roots in $\mathbb{Z}_6$ and $x^2 + [28]$ has 8 roots ([2], [6], [10], [14], [18], [22], [26] and [30]) in $\mathbb{Z}_{32}$. We add that all elements of $\mathbb{Z}_6$ are roots of the (Fermat) polynomial $x^3 + [5]x = x\ (x - [1])\ (x - [5]) = (x - [1])\ (x - [2])\ (x - [3]) = (x - [3])\ (x - [4])\ (x - [5])$.

We formulate an expansion of polynomials that one can easily obtain by making repeated use of Theorem VII.2.1.

Theorem VII.2.3. *Let $f \in F[x]$ be a polynomial of degree n and let $c_0,\ c_1, \ldots, c_{n-1} \in F$. Then there are (unique) elements $a_0,\ a_1,\ a_2, \ldots, a_n \in F$ such that*

$$f(x) = a_0 + a_1(x - c_0) + a_2(x - c_0)(x - c_1) + \cdots + a_n(x - c_0) \ldots (x - c_{n-1}). \qquad \text{(VII.2.2)}$$

Proof. Successively applying the Euclidean divisions, we obtain

$$f(x) = (x - c_0)f_1(x) + a_0 \text{ with } \deg f_1 = n - 1$$

....

$$f_t(x) = (x - c_t)f_{t+1}(x) + a_t \text{ with } deg f_{t+1} = n - t - 1$$

....

$$f_{n-1}(x) = (x - c_{n-1})f_n(x) + a_{n-1} \text{ with } deg f_n = 0$$

$$f_n(x) = a_n.$$

□

Remark VII.2.4. The expression (VII.2.2) is called **Newton's interpolation formula** for the polynomial f **(Sir Isaac Newton (1643 - 1727))**. Given $n+1$ **distinct** elements $c_0, c_1, \ldots, c_n$ in F and $n+1$ **arbitrary** elements $d_0, d_1, \ldots, d_n$ in F, we can determine recursively the elements $a_0, a_1, a_2, \ldots, a_n \in F$ such that

$$f(x) = \sum_{t=0}^{n} a_t\, p_t(x), \text{ where } p_0(x) = 1 \text{ and } p_t(x) = \prod_{r=0}^{t-1}(x - c_r), \qquad \text{(VII.2.3)}$$

is a (unique) polynomial of degree at most n that satisfies $f(c_t) = d_t$ for all $0 \le t \le n$. To this end, we apply Theorem VII.2.3 and recursively calculate

$a_0 = d_0,$

$a_1 = [d_1 - a_0]\,[c_1 - c_0]^{-1},$

$a_2 = [d_2 - a_1(c_2 - c_0) - a_0]\,[(c_2 - c_0)(c_2 - c_1)]^{-1},$

....

$a_n = [d_n - a_{n-1}(c_n - c_0)\ldots(c_n - c_{n-2}) - \cdots - a_1(c_n - c_0) - a_0]\,[(c_n - c_0)\ldots(c_n - c_{n-1})]^{-1}.$

We add that the polynomials $p_t(x)$ are called **Newton's basis polynomials.** Newton's interpolation formula is closely related to **Lagrange's formula.** The expression (VII.2.3) can be rewritten in the form

$$f(x) = \sum_{t=0}^{n} b_t\, x^t,$$

which yields the following theorem.

Theorem VII.2.4 (Interpolation à la Joseph-Louis Lagrange). *Given $n+1$ distinct elements $c_0, c_1, \ldots, c_n$ and $n+1$ arbitrary elements $d_0, d_1, \ldots, d_n$ in F, there is a unique polynomial $f \in F[x]$ of degree $\le n$ such that $f(c_t) = d_t$ for every $0 \le t \le n$.*

Proof. We prove this result by constructing explicitly the required polynomial. We define **Lagrange's basic polynomials** q_t $(t = 0, 1, \ldots, n)$ by

$$q_t(x) = \prod_{\substack{0 \leq k \leq n \\ k \neq t}} \frac{(x - c_k)}{(c_t - c_k)}.$$

We have $q_t(c_k) = 0$ for $k \neq t$ and $q_t(c_t) = 1$. Hence the polynomial

$$f = \sum_{t=0}^{n} d_t \, q_t$$

satisfies all the properties required by the theorem. In order to show that f is unique, let g be another polynomial with these properties. Then all $n + 1$ elements c_t are roots of the polynomial $f - g$, and since the degree of $f - g$ cannot be smaller than the number of roots, necessarily $f - g = 0$. □

We now illustrate the previous procedures in the following example.

Example VII.2.1. Find the polynomial $f \in \mathbb{Q}[x]$ of degree 4 such that

$$f(-2) = 46, f(-1) = 12, f(0) = 8, f(1) = 10 \text{ and } f(2) = 42.$$

We first find such a polynomial using Theorem VII.2.3, as described in Remark VII.2.3. We look for this polynomial in the form

$$f(x) = a_0 + a_1(x+2) + a_2(x+2)(x+1) + a_3(x+2)(x+1)x + a_4(x+2)(x+1)x(x-1).$$

Substituting $x = -2, -1, 0, 1$ and 2, we obtain the equations

$$a_0 = 46,\ a_0 + a_1 = 12,\ a_0 + 2a_1 + 2a_2 = 8,\ a_0 + 3a_1 + 6a_2 + 6a_3 = 10,$$
$$a_0 + 4a_1 + 12a_2 + 24a_3 + 24a_4 = 42.$$

Solving these equations we find

$$a_0 = 46,\ a_1 = -34,\ a_2 = 15,\ a_3 = -4 \text{ and } a_4 = 2.$$

Hence

$$b_0 = a_0 + 2a_1 + 2a_2 = 8,\ b_1 = a_1 + 3a_2 + 2a_3 - 2a_4 = -1,$$

$$b_2 = a_2 + 3a_3 - a_4 = 1,\ b_3 = a_3 + 2a_4 = 0,\ b_4 = 2$$

and thus $f(x) = 2x^4 + x^2 - x + 8$ is the required polynomial.

Now, we use the method in the proof of Theorem VII.2.4 to find the polynomial f.

$$\begin{aligned}
q_0(x) &= \tfrac{1}{24}(x+1)x(x-1)(x-2): && q_0(-2)=1\\
& && q_0(-1)=q_0(0)=q_0(1)=q_0(2)=0;\\
q_1(x) &= -\tfrac{1}{6}(x+2)x(x-1)(x-2): && q_1(-1)=1\\
& && q_1(-2)=q_1(0)=q_1(1)=q_1(2)=0;\\
q_2(x) &= \tfrac{1}{4}(x+2)(x+1)(x-1)(x-2): && q_2(0)=1\\
& && q_2(-2)=q_2(-1)=q_2(1)=q_2(2)=0;\\
q_3(x) &= -\tfrac{1}{6}(x+2)(x+1)x(x-2): && q_3(1)=1\\
& && q_3(-2)=q_3(-1)=q_3(0)=q_3(2)=0;\\
q_4(x) &= \tfrac{1}{24}(x+2)(x+1)x(x-1): && q_4(2)=1\\
& && q_4(-2)=q_4(-1)=q_4(0)=q_4(1)=0.
\end{aligned}$$

Thus, $f(x) = 46q_0(x)+12q_1(x)+8q_2(x)+10q_3(x)+42q_4(x) = 2x^4+x^2-x+8$ is the required polynomial.

Alternatively, we may also apply elementary linear algebra. We evaluate the polynomial $f(x)$, written in the form $f(x) = ax^4+bx^3+cx^2+dx+e$, at the numbers $-2,-1,0,1$ and 2 to obtain a system of five linear equations in the five unknowns a,b,c,d and e, namely,

$$\begin{pmatrix} 16 & -8 & 4 & -2 & 1\\ 1 & -1 & 1 & -1 & 1\\ 0 & 0 & 0 & 0 & 1\\ 1 & 1 & 1 & 1 & 1\\ 16 & 8 & 4 & 2 & 1 \end{pmatrix}\begin{pmatrix} a\\ b\\ c\\ d\\ e\end{pmatrix} = \begin{pmatrix} 46\\ 12\\ 8\\ 10\\ 42\end{pmatrix},$$

In this particular situation, reduction of the system yields $e = 8$ and

$$\begin{pmatrix} 1 & 1\\ 4 & 1\end{pmatrix}\begin{pmatrix} a\\ c\end{pmatrix} = \begin{pmatrix} 3\\ 9\end{pmatrix} \text{ and } \begin{pmatrix} 1 & 1\\ 4 & 1\end{pmatrix}\begin{pmatrix} b\\ d\end{pmatrix} = \begin{pmatrix} -1\\ -1\end{pmatrix}.$$

Solving these equations, we find a,b,c and d.

Exercise VII.2.3. (i) Determine the polynomial $f(x) \in \mathbb{Z}_5[x]$, $\deg f \leq 4$ such that

$$f([0]) = [4],\ f([1]) = [4],\ f([2]) = [4],\ f([3]) = [0],\ \text{and}\ f([4]) = [2].$$

(ii) Construct all polynomials $g(x) \in \mathbb{Z}_5[x]$ of degree ≤ 4 such that

$$g([0]) = [0],\ g([1]) = [3],\ g([2]) = [2]\ \text{and}\ g([4]) = [0].$$

Exercise VII.2.4. Prove the following statement: *Let f be a polynomial from $F[x]$ of degree n and let $c \in F$. Then there are (unique) elements $a_0,\ a_1,\ a_2,\dots,a_n \in F$ such that*

$$f(x) = a_0 + a_1(x-c) + a_2(x-c)^2 + \cdots + a_n(x-c)^n = \sum_{t=0}^{n} a_t(x-c)^t.$$

Describe the conditions under which there exists $c \in F$ such that $a_{n-1} = 0$.

Exercise VII.2.5. (i) Find all $n \in \mathbb{N}$ such that $n^2 + n + 1$ is a multiple of 13.

(ii) Using the fact that in $\mathbb{Z}_4[x]$, $[2](x+[1]) = [2](x+[3])$, describe the set of all polynomials $f(x) \in \mathbb{Z}_4[x]$ of degree ≤ 3 such that $f(x) = [0]$ for all $x \in \mathbb{Z}_4$.

(iii) Show that there are an infinite number of polynomials $f \in \mathbb{Z}_4[x]$ such that $f^2 = 1$. [Hint: $f(x) = [2]x + [3]$ is such a polynomial.]

Exercise VII.2.6. Let d_t $(1 \leq t \leq n)$ be arbitrary elements of a field F. Let f be the polynomial of (7.5) satisfying $f(t) = d_t$. Determine the elements a_t.

Example VII.2.2. If $F = \mathbb{Z}_p$, then, by Fermat's Little Theorem, $c^p = c$ for every $c \in \mathbb{Z}_p$. This means that all elements of $\mathbb{Z}_p$ are roots of **Fermat's polynomial**

$$f^*(x) = x^p - x = x^p + [p-1]x \in \mathbb{Z}_p[x].$$

Hence, for every $f \in \mathbb{Z}_p[x], \deg f \geq p$, there is a polynomial $g \in \mathbb{Z}_p[x]$ with $\deg g < p$ such that $f_e = g_e$, that is, such that $f(c) = g(c)$ for all $c \in \mathbb{Z}_p$; just use the Euclidean division $f = f^*q + g$. The number of such polynomials is p^p. For small values of p, we can display easily all polynomials from $\mathbb{Z}_p[x]$ having given values at the elements of $\mathbb{Z}_p$. Thus, for $p = 2$, we have all polynomials divided into the following four families:

The polynomials f

from $\{(x^2+x)\,h(x) | h(x) \in \mathbb{Z}_2[x]\}$ satisfy $f([0]) = f([1]) = [0]$,
from $\{[1] + (x^2+x)\,h(x) | h(x) \in \mathbb{Z}_2[x]\}$ satisfy $f([0]) = f([1]) = [1]$,
from $\{x + (x^2+x)\,h(x) | h(x) \in \mathbb{Z}_2[x]\}$ satisfy $f([0]) = [0]$ and $f([1]) = [1]$ and
from $\{x + [1] + (x^2+x)\,h(x) | h(x) \in \mathbb{Z}_2[x]\}$ satisfy $f([0]) = [1]$ and $f([1]) = [0]$.

A list of **all** polynomials f from $\mathbb{Z}_3[x]$ as polynomial functions contains already 27 entries: Here, $F^*(x)$ is Fermat's polynomial $f^*(x) = x^3 + [2]\,x$ multiplied by an arbitrary polynomial $h(x) \in \mathbb{Z}_3[x]$ and the symbols

abc $f(x)$

denotes all functions such that $f([0]) = \mathbf{a}$, $f([1]) = \mathbf{b}$, $f([2]) = \mathbf{c}$.

000	$F^*(x) = (x^3 + [2]\,x)\,h(x)$	**001**	$[2]\,x^2 + x + F^*(x)$
002	$x^2 + [2]\,x + F^*(x)$		
100	$[2]\,x^2 + [1] + F^*(x)$	**101**	$x^2 + x + [1] + F^*(x)$
102	$[2]\,x + [1] + F^*(x)$		
200	$x^2 + [2] + F^*(x)$	**201**	$x + [2] + F^*(x)$
202	$[2]\,x^2 + [2]\,x + [2] + F^*(x)$		

010	$[2]\,x^2 + [2]\,x + F^*(x)$	**011**	$x^2 + F^*(x)$
012	$x + F^*(x)$		
110	$x^2 + [2]\,x + [1] + F^*(x)$	**111**	$[1] + F^*(x)$
112	$[2]\,x^2 + x + [1] + F^*(x)$		
210	$[2]\,x + [2] + F^*(x)$	**211**	$[2]\,x^2 + [2] + F^*(x)$
212	$x^2 + x + [2] + F^*(x)$		
020	$x^2 + x + F^*(x)$	**021**	$[2]\,x + F^*(x)$
022	$[2]\,x^2 + F^*(x)$		
120	$x + [1] + F^*(x)$	**121**	$[2]\,x^2 + [2]\,x + [1] + F^*(x)$
122	$x^2 + [1] + F^*(x)$		
220	$[2]\,x^2 + x + [2] + F^*(x)$	**221**	$x^2 + [2]\,x + [2] + F^*(x)$
222	$[2] + F^*(x)$		

The following theorem describes all polynomials of $\mathbb{Z}_p[x]$ as polynomial functions.

Theorem VII.2.5 (Polynomial functions over $\mathbb{Z}_p$). *As before, we denote Fermat's polynomial by f^*, that is, $f^*(x) = x^p - x$.*

(i) *The polynomials $f \in \mathbb{Z}_p[x]$ such that $f(c) = 0$ for all elements $c \in \mathbb{Z}_p$ (in other words, those polynomials f for which the polynomial function f_e is the zero function) form an* **ideal** *of $\mathbb{Z}_p[x]$, namely,*

$$I_0 = \{f^*h \mid h \in \mathbb{Z}_p[x]\} \subseteq \mathbb{Z}_p[x].$$

(ii) *Given $g \in \mathbb{Z}_p[x]$, all polynomials $f \in \mathbb{Z}_p[x]$ such that $f(c) = g(c)$ for all elements $c \in \mathbb{Z}_p$ (in other words, those polynomials f that satisfy $f_e = g_e$) form the* **coset**

$$I_{g^*} = \{g^* + f^*h \mid h \in \mathbb{Z}_p[x]\} \text{ of } I_0 \text{ in } \mathbb{Z}_p[x],$$

where g^* is the (unique) polynomial of degree $< p$ such that $g^*_e = g_e$.

Proof. By Fermat's little Theorem, $f^*(c) = 0$ for every $c \in \mathbb{Z}_p$. Since the number of roots of a non-zero polynomial cannot be larger than its degree, there is no non-zero polynomial r of smaller degree than p such that $r(c) = 0$ for all $c \in \mathbf{Z_p}$. Hence, if $f(c) = 0$ for all $c \in \mathbb{Z}_p$, then, by Euclidean division, $f = f^*h + r$ and r must be the zero polynomial, since $r(c) = 0$ for all $c \in \mathbb{Z}_p$. The statement (i) follows.

By an earlier result, there is a (unique) $g^* \in \mathbb{Z}_p[x]$ of degree $< p$ and such that ${g^*}_e = g_e$, that is, such that $g^*(c) = g(c)$ for all $c \in \mathbb{Z}_p$. Again, $f - g^* = f^*h$ and (ii) follows. □

Remark VII.2.5. As in the case of the linear polynomial $x - c$ earlier, here we deal with a congruence: f and g belong to the same coset I_{g^*} modulo I_0 means that $f \equiv g \pmod{g^*}$, that is, $f - g \in I_0$. In other words, $f - g$ is a multiple of g^*. We shall deal with this topic in Sections 3 and 4 of this chapter.

VII.3 Divisibility

This section will reveal a very similar behaviour of the rings $F[x]$ and $\mathbb{Z}$ as far as divisibility is concerned. This close relationship is not a result of mere fact that both are integral domains; we have seen in Chapter VI that the structure of some integral domains is quite distinct from that of $\mathbb{Z}$. The close relationship stems from the fact that both $F[x]$ and $\mathbb{Z}$ are Euclidean domains. Consequently, we will witness similarity of the upcoming definitions and statements with those of Chapter III.

Definition VII.3.1. *Let f and g be two polynomials. We say that g is a* **divisor**, *or* **factor of** *f and write $g|f$ if there is a polynomial h such that $f = gh$. This situation can also be expressed by saying that f is a* **multiple of** *g. If both $g|f$ and $f|g$, we call f and g* **associates**, *or that f and g are associated.*

Remark VII.3.1. Observe that "to be associated" is an equivalence relation and thus, one can speak about the classes of associates in $F[x]$. The set of all **units** of $F[x]$, that is, the set of all polynomials $f \in F[x]$ that possess a (multiplicative) inverse, is such a class; it consists of all nonzero elements of F. In view of the Euclidean division, one can see that f and g are associates if and only if $f = a\,g$, where a is a unit of the field F. In particular, such polynomials have the same degree. Thus, in every class of associated polynomials there is a unique **monic polynomial**, that is, a polynomial with the leading (highest) coefficient equal to 1. We shall therefore often restrict our considerations concerning the divisibility of polynomials to monic polynomials, without loss of generality.

Exercise VII.3.1. Considering polynomials over an integral domain D, prove that the set of all units of $D[x]$ equals to the set of all units of D.

Definition VII.3.2. *Given two polynomials $f, g \in F[x], g \neq 0$, a* **monic polynomial** *$d \in F[x]$ is said to be their* **greatest common divisor** *if d is a divisor of both f and g, and moreover, whenever d' divides both f and g, then d' is a divisor of d. We shall write $d = d(f, g)$.*

The existence of greatest common divisors is established by the following **Euclidean algorithm.**

Theorem VII.3.1. *Let f and $g \neq 0$ be two polynomials from $F[x]$. The following series of Euclidean divisions*

$$f = g\,q_0 + r_0,\ r_0 \neq 0,\ \deg r_0 < \deg g;$$

$$g = r_0\,q_1 + r_1,\ r_1 \neq 0,\ \deg r_1 < \deg r_0;$$

$$r_0 = r_1\,q_2 + r_2,\ r_2 \neq 0,\ \deg r_2 < \deg r_1;$$

....

$$r_{t-1} = r_t\,q_{t+1} + r_{t+1},\ r_{t+1} \neq 0,\ \deg r_{t+1} < \deg r_t;$$

....

is finite: it terminates when the remainder in the Euclidean division becomes zero.

The monic multiple of the **last non-zero remainder is the greatest common divisor** *$d = d(f, g)$ of f and g. If $r_0 = 0$, that is, if $g|f$, then $d(f, g)$ is the monic multiple of g.*

The proof of the Euclidean algorithm theorem follows line by line the proof of the similar theorem for the integral domain of integers. We remark that this theorem implies that the greatest common divisor of f and g is the (unique) monic divisor of both f and g of the highest degree.

Exercise VII.3.2. Define the **least common multiple** $m(f,g)$ of two polynomials f and g of $F[x]$, prove its existence and show that $d(f,g)\,m(f,g) \;=\; f\,g$.

Exercise VII.3.3. Let $f_t, 1 \le t \le n$ be polynomials from $F[x]$. Prove that

$$d(f_1, d(f_2, f_3)) \;=\; d(d(f_1, f_2), f_3) \text{ and } m(f_1, m(f_2, f_3)) \;=\; m(m(f_1, f_2), f_3).$$

Define the greatest common divisor $d(f_1, f_2, \ldots f_n)$ and the least common multiple $m(f_1, f_2, \ldots, f_n)$ of the polynomials $f_t, 1 \le t \le n$.

Example VII.3.1. (i) To find the greatest common divisor of $f(x) = 2x^4 + 5x^3 - x$ and $g(x) = 2x^3 - 3x^2 + 1$ from $\mathbb{Q}[x]$, we use the Euclidean algorithm (replacing in each step, whenever possible, a polynomial by its "simpler" associate):

$$2x^4 + 5x^3 - x = (2x^3 - 3x^2 + 1)(x + 4) + (12x^2 - 2x - 4);$$

$$2x^3 - 3x^2 + 1 = (6x^2 - x - 2)(\frac{1}{3}x - \frac{4}{9}) + (\frac{2}{9}x + \frac{1}{9});$$

$$6x^2 - x - 2 = (2x + 1)(3x - 2).$$

Thus $d(x) \;=\; x + \frac{1}{2}$ is the greatest common divisor of f and g. Tracing the above relations back, we can also write d as a (linear) combination of f and g:

$$x + \frac{1}{2} = (-\frac{3}{4}x + 1)(2x^4 + 5x^3 - x) + (\frac{3}{4}x^2 + 2x + \frac{1}{2})(2x^3 - 3x^2 + 1).$$

(ii) In a similar way, we find that $d(x) = x + [2]$ is the greatest common divisor of

$$f(x) = x^4 + x^3 + x \text{ and } g(x) = [2]x^3 + [1] \text{ in } \mathbb{Z}_3[x]:$$

$$x^4 + x^3 + x = (x^3 + [2])(x + [1]) + ([2]x + [1]), \;\; x^3 + [2] = (x + [2])(x^2 + x + [1]).$$

From there

$$x + [2] = [2]\,(x^4 + x^3 + x) + ([2]x + [2])([2]x^3 + [1]).$$

Remark VII.3.2. Note that, in the previous examples, we could have used (ii) to get a hint on the outcome of (i). The (canonical) homomorphism $\varphi : \mathbb{Z} \to \mathbb{Z}_3$ can be extended to a homomorphism $\varphi^* : \mathbb{Z}[x] \to \mathbb{Z}_3[x]$. We formulate this relationship in the form of an exercise.

Exercise VII.3.4. Consider the subring $\mathbb{Z}[x]$ of $\mathbb{Q}[x]$ of all polynomials from $\mathbb{Q}[x]$ whose coefficients are integers. Define the mapping

$$\varphi : \mathbb{Z}[x] \to \mathbb{Z}_n[x] \text{ by } \varphi(\sum_t a_t\, x^t) = \sum_t [a_t]\, x^t,$$

where $[a_t]$ is the congruence class mod n containing the integer a_t.

(i) Show that φ is a ring homomorphism and determine its kernel.

(ii) Prove that $f|g$ in $\mathbb{Z}[x]$ implies that $\varphi(f)|\varphi(g)$ in $\mathbb{Z}_n[x]$.

(iii) Provide an example when $\varphi(f)|\varphi(g)$ in $\mathbb{Z}_n[x]$ although f is not a divisor of g in $\mathbb{Z}[x]$.

Exercise VII.3.5. Let $f(x) = [6]\,x^5 + [4]\,x^3 + x^2 + [3]$ and $g(x) = x^7 + [3]\,x^5 + [5]\,x^3 + x + [2]$ be polynomials from $\mathbb{Z}_7$. Find their greatest common divisor $d(x)$ and polynomials $u(x)$ and $v(x)$ such that $f(x)\,u(x) + g(x)\,v(x) = d(x)$.

We have already seen in Chapters III and VI that the existence of the Euclidean division has important consequences. We refresh our earlier reasonings.

First, the set

$$I(f,g) = \{fu + gv \mid u, v \in F[x]\}$$

is an additive abelian group. In fact, since for any element $h \in F[x]$ and for any element $w = fu+gv \in I(f,g)$, also $hw = f(hu)+g(hv)$ belongs to $I(f,g)$, $I(f,g) \subseteq F[x]$ is an **ideal**.

Now, look at an arbitrary nonzero ideal I of $F[x]$. The set of the degrees of all polynomials in I has a least element, and thus there is a monic polynomial $d \in I$ of the least degree. Consequently, if f is any polynomial of I, then necessarily $f = d\,q$ for some $q \in F[x]$. This is a consequence of the fact that $f = d\,q+r$ with $r \neq 0, \deg r < \deg d$ (and thus $r = f - d\,q \in I$) would contradict the choice of d. Hence, $I = \{d\,h \mid h \in F[x]\}$ is generated by a single element, that is, every ideal of $F[x]$ is a **principal ideal**.

In the case of the ideal $I(f,g)$ we infer that $d = f\,u + g\,v$ for some $u, v \in F[x]$ and that d is a divisor of both f and g. In fact, one can see that d is the greatest common divisor of f and g. Indeed, if for some d', $f = d'\,q_1$ and $g = d'\,q_2$, then $d = f\,u + g\,v = d'\,(uq_1 + vq_2)$.

Before we formulate some consequences, we recall several concepts.

Definition VII.3.3. *A polynomial $p \in F[x]$ is said to be* **irreducible** *(over F) if it is of degree ≥ 1, and whenever $p = f\,g$ with $f, g \in F[x]$, one of the factors must be a unit in $F[x]$. A polynomial that is not irreducible is called* **reducible**. *Polynomials f and g are said to be* **relatively prime** *if their greatest common divisor is 1.*

Corollary VII.3.1 (Bézout's theorem). *Let f and g be two nonzero polynomials from $F[x]$ and $d = d(f,g)$. Then there are polynomials u and v in $F[x]$ such that*

$$d = fu + gv.$$

In particular, for relatively prime f and g, there are u and v in $F[x]$ such that

$$fu + gv = 1.$$

Corollary VII.3.2. *Let f, g and h be three non-zero polynomials.*

(i) *If h divides the product fg, and h and g are relatively prime, then h divides f.*

(ii) *An irreducible polynomial p is* **prime** *in the following sense: If p divides a product fg, then it divides either f or g. More generally, if an irreducible polynomial p divides a product $f_1 f_2 \dots f_m$, then it must divide one of the factors. If all these factors are irreducible, then up to a multiple by a unit, p must be one of the factors.*

Exercise VII.3.6. Using Corollary VII.3.1, give a proof of Corollary VII.3.2.

Corollary VII.3.3 (Diophantine equations). *Given polynomials $f, g, h \in F[x]$, with f and g non-zero, the equation*

$$fs + gr = h \tag{VII.3.1}$$

for polynomials s and r has a solution $s, r \in F[x]$ if and only if $d = d(f,g) \mid h$. If (s_0, r_0) is a particular solution, then

$$\{(s_0 + g_1 w, r_0 - f_1 w) \,|w \in F[x]\}, \text{ where } f = f_1 d \text{ and } g = g_1 d,$$

is the set of all solutions.

Proof. Every common divisor of f and g divides $fs + gr = h$. On the other hand, if $h = h_1 d$ with $h_1 \in F[x]$, then the set of all solutions of (VII.3.1) equals the set of all solutions of $f_1 s + g_1 r = h_1$. Since $d(f_1, g_1) = 1$, using Bézout's theorem, there are $u, v \in F[x]$ such that $f_1 u + g_1 v = 1$ and thus $(s_0 = u h_1, r_0 = v h_1)$ is a solution of (VII.3.1). Now, if (s, r) is a solution of (VII.3.1), then $f_1 (s - s_0) = g_1 (r_o - r)$, and thus, by Corollary VII.3.2(i), $r_0 - r = f_1 w$ for some $w \in F[x]$. From here, $r = r_0 - f_1 w$ and $s - s_0 = g_1 w$, as required. □

The following theorem is an analogue of the **Fundamental theorem of arithmetic** (replacing the domain $\mathbb{Z}$ by the algebra of polynomials $F[x]$).

Theorem VII.3.2. *Every polynomial f in $F[x]$ of degree ≥ 1 can be expressed as a product*

$$f \;=\; a\, p_1 p_2 \dots p_m$$

of irreducible monic polynomials with a representing the highest coefficient of f. In such a product, the polynomials $p_1, p_2, \dots, p_m$ are uniquely determined up to ordering.

Proof. We can restrict our proof to the case when f is monic. The existence of the factorization with irreducible factors follows by induction on the degree of polynomials. If f is irreducible, we are done. Otherwise $f = g\, h$ with both g and h monic polynomials of smaller degrees than degree of f. Thus both g and h are products of monic irreducible polynomials, and so is their product.

To prove uniqueness, we proceed inductively, applying Corollary VII.3.2(ii). Every nonzero polynomial can therefore be written in the form

$$f = a\, p_1^{k_1}\, p_2^{k_2} \dots p_r^{k_r}$$

with distinct irreducible monic polynomials p_t and natural numbers $k_t, 1 \leq t \leq r$. The integral domains having the property described in the previous theorem are called, as we have already mentioned in Chapter VI, **unique factorization domains**; thus all our polynomial domains $F[x]$ are unique factorization domains. The nature of the irreducible polynomials in these domains depends on the nature of the field F. Let us provide a brief overview of some cases. □

Corollary VII.3.4. *Let the field F be algebraically closed. Then every polynomial $f \in F[x]$ of degree $n \geq 1$ has a factorization*

$$f(x) = a(x - c_1)(x - c_2) \cdots (x - c_n) \text{ with } a \in F \text{ and } c_t \in F, 1 \leq t \leq n.$$

Thus all irreducible polynomials over an algebraically closed field are linear polynomials. This is the case when $F = \mathbb{C}$ is the field of complex numbers.

Exercise VII.3.7. Let $f \in \mathbb{C}[x]$ be a polynomial with **real** coefficients and c be a complex root of f. Show that the conjugate $\overline{c}$ is also a root of f. Hence show that the irreducible monic polynomials in $\mathbb{R}[x]$ are either linear polynomials $x - c$ with $c \in \mathbb{R}$ or quadratic polynomials $(x - a)^2 + b^2$ with $a, b \in \mathbb{R}$ and $b \neq 0$.

It can be shown that in $\mathbb{Q}[x]$, there are irreducible polynomials of arbitrarily high degree. The number of monic irreducible polynomials of a given degree d over the field $\mathbb{Z}_p$ can be calculated. Denote this number by $N_p(d)$. Then

$$N_p(d) = \frac{1}{d} \sum_{t|d} \mu(\frac{d}{t}) \cdot p^t, \qquad \text{(VII.3.2)}$$

where μ is the Möbius function: Recall that $\mu(m) = 0$ if m is divisible by square of prime, and $\mu(p_1 p_2 \dots p_k) = (-1)^k$ if $p_1, p_2, \dots, p_k$ are distinct primes. Thus,

$$N_2(2) = 1, N_2(3) = 2, N_2(5) = 6, N_2(7) = 18, N_2(11) = 186, N_2(13) = 630,$$

$$N_2(17) = 7710, \; N_2(19) = 27594, \dots, N_2(q) = 2q^{-1}(2^{q-1} - 1) \text{ for any prime } q.$$

The formula (VII.3.2) can be obtained from a result of **Carl Friedrich Gauss** in his Disquisitiones Arithmeticae, see p. 610.

Exercise VII.3.8. (i) Calculate $N_2(30)$ and $N_5(4)$.

(ii) List all irreducible polynomials of $\mathbb{Z}_3[x]$ of degree ≤ 3.

(iii) Find the greatest common divisor of $3x^5 + 3x^3 + x^2 - 6x + 2$ and $7x^4 - x^3 + 15x^2 - 2x + 2$ in $\mathbb{Q}[z]$.

[Answer: $x^2 + 2$.]

(iv) Find the greatest common divisor $d(x)$ of $f(x) = [3]x^5 + [3]x^3 + x^2 + [4]x + [2]$ and $g(x) = [2]x^4 + [4]x^3 + [3]x + [2]$ in $\mathbb{Z}_5[x]$ and find the polynomials $u, v \in \mathbb{Z}_5[x]$ such that $d = uf + vg$.

In the same way as we have constructed the field of rational numbers over the integral domain of integers, we can construct the **field of fractions** (quotient field), the field of **rational functions** $F(x)$ over the integral domain $F[x]$.

Theorem VII.3.3. *The elements of $F(x)$ are fractions $\frac{f}{g}, g \neq 0$. Hereby,*

$$\frac{f_1}{g_1} = \frac{f_2}{g_2} \text{ if and only if } f_1\, g_2 = f_2\, g_1,$$

$$\frac{f_1}{g_1} + \frac{f_2}{g_2} = \frac{f_1\, g_2 + f_2\, g_1}{g_1 g_2} \text{ and } \frac{f_1}{g_1}\frac{f_2}{g_2} = \frac{f_1\, f_2}{g_1 g_2}.$$

The polynomials $f \in F[x]$ are identified with the fractions $\frac{f}{1}$.

Remark VII.3.3. Every nonzero rational function $\frac{f}{g} \in F(x)$ can be expressed, after canceling their greatest common divisor, in the form $c_1\, \frac{f_1}{g_1}$, where f_1 and g_1 are relatively prime monic polynomials and $c_1 \in F$. Such an expression is unique, since $c_1\, \frac{f_1}{g_1} = c_2\, \frac{f_2}{g_2}$ implies that $c_1 f_1 g_2 = c_2 f_2 g_1$ and from the unique factorization of polynomials we conclude that $f_1 = f_2$ and $g_1 = g_2$. Finally, the Euclidean division $f_1 = g_1\, q + r$ means that every element of $F(x)$ has a **unique form**

$$q + c\, \frac{r}{s},$$

where q is a polynomial from $F[x]$, $r = 0$ or r and s are relatively prime monic polynomials, $\deg r < \deg s$ and $c \in F$. Factorizing both r and s into products of irreducible polynomials from $F[x]$, we obtain a **unique factorization** of the rational function.

Exercise VII.3.9. Let D be an integral domain, F its field of fractions and $\varphi : D \to F$ the canonical embedding of D into F. Let $\psi : D \to E$ be an embedding of D into some field E. Then there exists a unique extension $\omega : F \to E$ of the embedding ψ such that the diagram

$$\begin{array}{ccc} D & \xrightarrow{\varphi} & F \\ \| & & \downarrow \omega \\ D & \xrightarrow{\psi} & E \end{array}$$

commutes. Recall that commutativity of this diagram means that $(\omega \circ \varphi)(d) = \psi(d)$ for all $d \in D$.

In a number of applications we are required to decompose a rational function into a sum of rational functions whose denominators are powers of irreducible elements. Such a decomposition is called a **partial fraction decomposition**. The basis for such decompositions are the following two theorems.

Theorem VII.3.4. *Let f, g_1, g_2 be polynomials from $F[x]$ such that g_1 and g_2 are relatively prime and* $\deg f < \deg(g_1 g_2)$. *Then there is a* **unique** *decomposition*

$$\frac{f}{g_1 g_2} = \frac{f_1}{g_1} + \frac{f_2}{g_2} \quad \text{with } f_t \in F[x],\ f_t = 0 \text{ or } \deg f_t < \deg g_t, t = 1, 2.$$

Theorem VII.3.5. *Let $f, g \in F[x]$ $g \neq 0$. Then there is a decomposition*

$$\frac{f}{g^n} = \frac{h}{g^{n-1}} + \frac{f_n}{g^n} \quad \textit{with } h, f_n \in F[x] \textit{ and } f_n = 0 \textit{ or } \deg f_n < \deg g.$$

Hence, there is a **unique** *decomposition*

$$\frac{f}{g^n} = \frac{f_n}{g^n} + \frac{f_{n-1}}{g^{n-1}} + \cdots + \frac{f_2}{g^2} + \frac{f_1}{g} + f_0 \qquad \text{(VII.3.3)}$$

with $f_t \in F[x]$ and $f_n = 0$ or $\deg f_t < \deg g$ for all $1 \leq t \leq n$.

Proof. The first theorem is a simple consequence of Bézout's theorem: There are $u_1, u_2 \in F[x]$ such that

$$1 = u_2\, g_1 + u_1\, g_2\,, \quad \text{and thus} \quad \frac{f}{g_1 g_2} = \frac{f u_1}{g_1} + \frac{f u_2}{g_2} = q_1 + \frac{f_1}{g_1} + q_2 + \frac{f_2}{g_2},$$

where $f u_t = g_t\, q_t + f_t$, $f_t = 0$ or $\deg f_t < \deg g_t$ for all $1 \leq t \leq n$. Since $\deg f < \deg(g_1 g_2)$, $q_1 + q_2 = 0$ and the expression is unique, see Remark VII.3.3.

The first part of Theorem VII.3.5 is a consequence of the Euclidean division $f = g\, h + f_n$ and then the second part follows by induction. □

Corollary VII.3.5. *Every rational function in $F(x)$ can be written uniquely in the form*

$$\frac{f_1}{p_1^{k_1}} + \frac{f_2}{p_2^{k_2}} + \cdots + \frac{f_n}{p_n^{k_n}} + f_{n+1}, \qquad \text{(VII.3.4)}$$

where $p_1, p_2, \ldots, p_n$ are distinct irreducible monic polynomials, $k_1, k_2, \ldots, k_n$ are natural numbers, $f_1, f_2, \ldots, f_n, f_{n+1}$ are polynomials from $F[x]$ and $\deg f_t < \deg p_t^{k_t}$, $t = 1, 2, \ldots, n$. If we express the rational function (VII.3.4) *in the form $\frac{f}{g}$, then $g = p_1^{k_1}\, p_2^{k_2} \ldots p_n^{k_n}$ is the* (*unique*) *prime power factorization of g, and $f_{n+1} = 0$ if and only if $\deg f < \deg g$.*

Exercise VII.3.10. Give the details of a proof of Corollary VII.3.5.

Example VII.3.2. (i) If $c_1, \ldots, c_n$ are n distinct elements of a field F, then there are $a_1, \ldots, a_n \in F$ such that

$$\frac{1}{(x - c_1)(x - c_2) \ldots (x - c_n)} = \frac{a_1}{(x - c_1)} + \frac{a_2}{(x - c_2)} + \cdots + \frac{a_n}{(x - c_n)}.$$

(This is a particular case of Corollary VII.3.5.)

(ii) Let $f \in F[x]$ of degree n and $c \in F$. Then for every $k \geq 0$, there are uniquely determined $a_t \in F$, $k - n \leq t \leq k$, such that

$$\frac{f(x)}{(x - c)^k} = \frac{a_k}{(x - c)^k} + \frac{a_{k-1}}{(x - c)^{k-1}} + \cdots + \frac{a_1}{x - c}$$
$$+ \; a_o + a_{-1}(x - c) + a_{-2}(x - c)^2 + \cdots + a_{k-n}(x - c)^{n-k}.$$

Of course, for $n < k$, $a_t = 0$ for all $t \geq 0$. (This is a particular case of Theorem VII.3.5).

(iii) The coefficients of the polynomials f_t in the partial fraction decomposition (VII.3.3) of $\frac{f}{g}$ with $\deg f < \deg g$ are the solutions of a system of linear equations. The number of unknowns equals the number of equations, namely

$$\deg g = \sum_{t=1}^{n} k_t\, d_t, \quad \text{where } d_t = \deg p_t,\ 1 \le t \le n.$$

Corollary VII.3.5 guarantees that this system has a unique solution. Thus, to determine $a, b, c, d, e, f \in \mathbb{Z}_5$ that satisfy

$$\frac{[2]x^4 + x^3 + x^2 + x + [1]}{(x+[1])^3\,(x+[2])^2\,(x+[3])} = \frac{ax^2 + bx + c}{(x+[1])^3} + \frac{dx + e}{(x+[2])^2} + \frac{f}{x+[3]},$$

we solve the following system of linear equations modulo 5

$$\begin{pmatrix} [1] & [0] & [0] & [1] & [0] & [1] \\ [2] & [1] & [0] & [1] & [2] & [2] \\ [1] & [2] & [1] & [2] & [1] & [4] \\ [2] & [1] & [2] & [0] & [2] & [0] \\ [0] & [2] & [1] & [3] & [0] & [1] \\ [0] & [0] & [2] & [0] & [3] & [4] \end{pmatrix} \begin{pmatrix} [a] \\ [b] \\ [c] \\ [d] \\ [e] \\ [f] \end{pmatrix} = \begin{pmatrix} [0] \\ [2] \\ [1] \\ [1] \\ [1] \\ [1] \end{pmatrix}$$

to get $a = [3], b = [3], c = [1], d = [1], e = [0]$ and $f = [1]$.

(iv) The expansion (VII.3.3) can be treated in a similar way as the expression in (iii). Let $\deg f < \deg g$ and g be an irreducible polynomial of degree d. Then $f_0 = 0$ and each f_t has d coefficients. These dn coefficients are the solutions of a system of dn linear equations. Theorem VII.3.5 asserts that there is a unique solution of the system. For example, to determine $a, b, c \in \mathbb{Z}_5$ that satisfy

$$\frac{[3]x^2 + [3]x + [1]}{(x+1)^3} = \frac{a}{(x+1)^3} + \frac{b}{(x+1)^2} + \frac{c}{x+1},$$

we obtain $a = [1], b = [2], c = [3]$ as the unique solution of

$$\begin{pmatrix} [1] & [1] & [1] \\ [0] & [1] & [2] \\ [0] & [0] & [1] \end{pmatrix} \begin{pmatrix} a \\ b \\ c \end{pmatrix} = \begin{pmatrix} [1] \\ [3] \\ [3] \end{pmatrix}.$$

Have you found something particular about the matrix of the system this time? Formulate your finding.

Exercise VII.3.11. The $(x - c)$-adic expansion of a polynomial f in Exercise VII.2.4 can be extended to an expansion of f by any nonzero polynomial $p \in F[x]$:

$$f(x) = f_0 + f_1\,p + f_2\,p^2 + \cdots + f_n\,p^n \text{ with } f_t = 0 \text{ or } \deg f_t < \deg p \text{ for all } 0 \le t \le n. \tag{VII.3.5}$$

Prove the existence and uniqueness of this p-**adic expansion** (VII.3.5).

The partial fraction decompositions of rational functions depend on the field F. A polynomial that is irreducible over a field F does not have to be irreducible over a field extension E, that is, over a field E that contains the field F as its subfield. We are going to clarify such a situation in the following exercise.

Exercise VII.3.12. Decompose the rational function

$$r(x) = \frac{x}{x^8 - 1}$$

into the partial fractions over the field (i) $\mathbb{Q}$; (ii) $\mathbb{R}$ and (iii) $\mathbb{C}$.

Exercise VII.3.13. Decompose the rational function $r \in \mathbb{R}(x)$

$$r(x) = \frac{2x^6 + 8x^5 + 19x^4 + 30x^3 + 31x^2 + 20x + 10}{(x^2 + x + 2)^2\,(x + 1)^3}.$$

VII.4 Congruences

The concept of an equivalence that is compatible with the algebraic structure of the underlying objects (such as groups or rings) and its corresponding quotient structure (whose elements are the respective equivalence classes) has been extensively studied in case of the integral domain $\mathbb{Z}$ of integers in Chapter III. On several occasions we have touched upon the subject of equivalence later (see for example Definition VII.4.1 or Remark VII.2.5). Here, we shall provide the relevant formal definitions for polynomial algebras.

Definition VII.4.1. *Let I be a proper (that is, distinct from $\{0\}$ and $F[x]$) ideal of $F[x]$. Two polynomials f and g of $F[x]$ are said to be* **congruent modulo** *I if their difference $f - g$ belongs to I. We may also say that f is* **congruent to** *g* **modulo** *I and write*

$$f \equiv g \pmod{I}.$$

Recall that (similarly to the situation in the integral domain $\mathbb{Z}$) every ideal I of $F[x]$ is principal, that is, there is a polynomial $q \in F[x]$ such that $I = \{q\,h | h \in F[x]\}$. Thus $f \equiv g \pmod{I}$ means that $f - g = qh$ for some $h \in F[x]$ and so we often express this fact equivalently by saying that f is **congruent to** g **modulo** q and write

$$f \equiv g \pmod{q}.$$

Exercise VII.4.1. Prove that two polynomials $f, g \in F[x]$ are congruent modulo q if and only if the remainders in the Euclidean division by q are associates.

Exercise VII.4.2. Prove the following simple facts:

(i) $f \equiv f \pmod{q}, f \equiv g \pmod{q}$ implies $g \equiv f \pmod{q}$ and $f \equiv g \pmod{q}$ together with $g \equiv h \pmod{q}$ imply $f \equiv h \pmod{q}$, that is, the congruence relation is an equivalence.

(ii) If $f_1 \equiv g_1 \pmod{q}$ and $f_2 \equiv g_2 \pmod{q}$, then

$$f_1 + f_2 \equiv g_1 + g_2 \pmod{q} \text{ and } f_1\, f_2 \equiv g_1\, g_2 \pmod{q}.$$

Thus, in particular, $f_1 + h \equiv f_1 + h \pmod{q}$ and $f_1\, h \equiv g_1\, h \pmod{q}$ for any $h \in F[x]$. The congruence relation is compatible with respect to the operations of addition and multiplication.

Exercise VII.4.3 (Chinese remainder theorem for polynomials). Let $(f_1, f_2, \dots, f_n)$ be a sequence of arbitrary polynomials from $F[x]$ and $\{q_1, q_2, \dots, q_n\}$ a set of polynomials from $F[x]$ that are mutually relatively prime. Prove that there is a polynomial that satisfies $f \equiv f_t \pmod{q_t}$ for all $1 \le t \le n$. If g is another polynomial satisfying $g \equiv f_t \pmod{q_t}$ for all $1 \le t \le n$, then f is congruent to g modulo the product $q_1\, q_2 \dots q_n$.

Exercise VII.4.4. Solve the simultaneous congruences

$$f(x) \equiv 2\,x^2 \pmod{x + 1},$$
$$f(x) \equiv 7x + 5 \pmod{x^2 + 1}.$$

As a consequence of the statements in Exercise VII.4.2, there exists a (disjoint) partition of $F[x]$ into the **congruence classes**. Denote the congruence class containing a polynomial $f \in F[x]$ by

$$\bar{f} = \{h \in F[x] \mid h \equiv f \pmod{q}\}.$$

Exercise VII.4.5. Prove that the congruence class $\bar{0}$ containing 0 consists just of all elements of the ideal $I = \langle q \rangle = \{fq \mid f \in F[x]\}$ and that in every other congruence class $\bar{f}$ modulo q there is a unique polynomial of degree less than $\deg q$, that is, that there is a bijection between the set of all nonzero congruence classes modulo q and the set of all monic polynomials of degree less than $\deg q$.

We shall occasionally denote this unique representative of $\bar{f}$ by f^* in which case we should be aware that $f^* = g^*$ if and only if $f \equiv g \pmod{q}$. If $\deg q = 1$, say $q = x - c$, then the congruence classes are in one-to-one correspondence with the elements of the field F: $\bar{f}$ corresponds to the coefficient a_0 of (VII.2.2) in Theorem VII.2.3. For this reason, the definition of being congruent modulo q is often restricted to polynomials q of degree greater than 1. Denote the set of all congruence classes modulo q by

$$\overline{F[x]} = F[x]/\langle q \rangle.$$

Exercise VII.4.6. As in the case of congruence classes in $\mathbb{Z}$, define on $\overline{F[x]}$ the operations of addition and multiplication by

$$\bar{f} + \bar{g} = \overline{f + g} \text{ and } \bar{f} \cdot \bar{g} = \overline{fg},$$

show that these operations are well defined, that is, do not depend on the choice of representatives and that the set of all congruence classes is with respect to these operations a ring.

Definition VII.4.2. *The resulting ring* $(\overline{F[x]}, +, \cdot)$ *together with the canonical homomorphism*

$$\varphi : F[x] \to \overline{F[x]} = F[x]/\langle q \rangle$$

defined by $\varphi(f) = \bar{f}$, *is called the* **quotient ring** $F[x]$ **modulo** $\langle q \rangle$.

The following two theorems illustrate a sharp distinction in the behaviour of $F[x]/\langle q \rangle$ according to whether q is an irreducible or a reducible polynomial.

Theorem VII.4.1. $\overline{F[x]} = F[x]/\langle q \rangle$ *has non-trivial divisors of zero if and only if* q *is reducible.*

Proof. We observe that $q = fg, 1 \leq \deg f < \deg q$, in $F[x]$ implies $\bar{f}\bar{g} = \bar{0}$ with $\bar{f} \neq \bar{0}$ and $\bar{g} \neq \bar{0}$ in $\overline{F[x]}$. On the other hand, if q is irreducible and $\bar{f} \neq \bar{0}$, $\bar{g} \neq \bar{0}$ in $\overline{F[x]}$, then the greatest common divisors $d(f^*, q) = 1$ and $d(g^*, q) = 1$. Hence, also $d(f^*g^*, q) = 1$ and therefore $\bar{f}\bar{g} = \overline{fg} \neq \bar{0}$ in $\overline{F[x]}$. □

Theorem VII.4.2. *A polynomial* $p \in F[x]$ *is irreducible if and only if* $\overline{F[x]} = F[x]/\langle p \rangle$ *is a field.*

Proof. Theorem VII.4.1 shows that $\overline{F[x]}$ cannot be a field if p is reducible. Let p be irreducible and $\bar{f} \neq \bar{0}$ in $\overline{F[x]}$. Since f^* and p are relatively prime, there are, by Bézout's theorem, polynomials u and v such that $uf^* + vp = 1$. But then $\bar{f} \cdot \bar{u} = \bar{1}$ in $\overline{F[x]}$. Hence nonzero classes of $\overline{F[x]}$ form a multiplicative group and thus $\overline{F[x]}$ is a field. □

The construction of the field $F[x]/\langle p \rangle$ (when p is an irreducible polynomial over F) owns its importance to the following simple statement concerning solutions of algebraic equations. We shall illustrate this aspect of the construction with some examples below. We recall that an irreducible polynomial $p \in F[x]$ cannot have a root in F.

Theorem VII.4.3 (Kronecker's theorem). *Let* p *be an irreducible polynomial over a field* F. *Then the field* $E = \overline{F[x]} = F[x]/\langle p \rangle$ *is a (proper)* **algebraic field extension** *of the field* F *of degree* > 1, *that is,* $F \subset E$, *and every polynomial of* $F[x]$ *can be considered as a polynomial of* $E[x]$. *The class* $\bar{x} \in E = \overline{F[x]}$ *is a root of the polynomial* p *and thus, as a polynomial from* $E[x], p$ *is no longer irreducible.*

Proof. The statement follows from our preceding considerations. Here, we just clarify the embedding of $F[x]$ into $E[x]$: If $\varphi : F \to E$ is the canonical embedding of the field F into the extension field E, then the extension $\overline{\varphi} : F[x] \to E[x]$ of φ is defined by $\overline{\varphi}\ (a_n x^n + a_{n-1}x^{n-1} + \cdots + a_1 x + a_0) = \varphi(a_n)x^n + \varphi(a_{n-1})x^{n-1} + \cdots + \varphi(a_1)x + \varphi(a_0)$. Trivially, $\bar{\varphi}(p)(\bar{x}) = \overline{p(x)} = \bar{0}$ and thus $(x - \bar{x})$ is a factor of the polynomial $\overline{\varphi}(p) \in E[x]$. □

The construction of a field extension described in Kronecker's theorem can be continued until we reach the field over which the original irreducible polynomial p can be expressed as a product of linear factors, that is, until we reach a field containing all roots of the polynomial p. We call such a (minimal) field the **splitting field** of the polynomial p. We shall show in Chapter IX that, up to an isomorphism, the splitting field of a polynomial is uniquely determined (see Theorem IX.4.3).

Example VII.4.1. The field $E = \mathbb{Z}_2[x]/\langle x^4 + x + 1\rangle$ is the splitting field of the irreducible polynomial polynomial $p(x) = x^4 + x + [1] \in \mathbb{Z}_2[x]$. Denoting $\overline{x} = \alpha$, one can see that over the field E, $p(x) = (x+\alpha)\,(x+\alpha+[1])\,(x+\alpha^2)\,(x+\alpha^2+1)$. Write down the multiplication table of the field E and verify the decomposition of p. Also verify the existence of the chain of subfields

$$\begin{aligned}\mathbf{GF}_2 &= \{[0],[1]\} \subset \mathbf{GF}_{2^2} = \{[0],[1],\alpha^2+\alpha,\alpha^2+\alpha+[1]\} \subset \mathbf{GF}_{2^3}\\ &= \{[0],[1],\alpha^2+\alpha,\alpha^2+\alpha+[1],\alpha^3,\alpha^3+[1],\alpha^3+\alpha,\alpha^3+\alpha^2\} \subset E.\end{aligned}$$

We are now ready for some illustrations.

Example VII.4.2. The following examples illustrate various features of the extension constructions.

(i) Let $F = \mathbb{R}$ and $p(x) = x^2 + 1 \in \mathbb{R}[x]$. Put $E = \mathbb{R}[x]/\langle x^2+1\rangle$ and show that $E \simeq \mathbb{C}$. Writing $\bar{x} = i$, that is, $\overline{a+bx} = a + bi$, we have

$$(a+bi)+(c+di) = (a+c)+(b+d)i \;\text{ and }\; (a+bi)(c+di) = (ac-bd)+(ad+bc)i.$$

We note that

$$\mathbb{R}[x]/\langle x^2+r\rangle \simeq \mathbb{C} \;\text{ for any real }\; r > 0.$$

We have $\overline{a+bx} = a + b\bar{x}$ and

$$(a+b\bar{x})+(c+d\bar{x}) = (a+c)+(b+d)\bar{x} \;\text{ and}$$
$$(a+b\bar{x})(c+d\bar{x}) = (ac-bdr)+(ad+bc)\bar{x}.$$

Thus, defining the mapping

$$\psi : \mathbb{R}[x]/\langle x^2+r\rangle \to \mathbb{C} \;\text{ by }\; \psi(a+b\bar{x}) = a + b\sqrt{r}\,i,$$

we see that ψ is a one-to-one correspondence satisfying
$\psi[(a+b\bar{x})+(c+d\bar{x})] = \psi(a+b\bar{x})+\psi(c+d\bar{x})$ and
$\psi[(a+b\bar{x})(c+d\bar{x})] = \psi[(ac-bdr)+(ad+bc)\bar{x})] = (ac-bdr)+(ad+bc)\sqrt{r}\,i$
$= (a+b\sqrt{r}\,i)(c+d\sqrt{r}\,i) = \psi(a+b\bar{x})\,\psi(c+d\bar{x})$,
as required.

(ii) Let $F = \mathbb{Z}_2$ and $p(x) = x^2 + x + [1]$. Put $E = \mathbb{Z}_2[x]/\langle x^2+x+[1]\rangle \simeq \mathbf{GF}_{2^2}$. Here we have a field of four elements: the Galois field $\mathbf{GF}_{2^2}$ with obvious addition and the following multiplication table (in contrast to the ring $\mathbb{Z}_4$ of four elements that contains nontrivial divisors of zero).

mult	$\overline{1}$	$\overline{x}$	$\overline{1+x}$
$\overline{1}$	$\overline{1}$	$\overline{x}$	$\overline{1+x}$
$\overline{x}$	$\overline{x}$	$\overline{x+1}$	$\overline{1}$
$\overline{1+x}$	$\overline{1+x}$	$\overline{1}$	$\overline{x}$

We point out that here and in the examples that follow we omit for simplicity the proper notation of the elements of F in the symbols for the elements of the field extensions; thus instead of $\overline{[1]}$ we write simply $\overline{1}$, instead of $\overline{x+[1]}$ we write $\overline{x+1}$, etc.

(iii) Let $F = \mathbb{Z}_2$ and $p(x) = x^3 + x + [1]$. Put $E_1 = \mathbb{Z}_2[x]/\langle x^3 + x + [1]\rangle \simeq \mathbf{GF}_{2^3}$. This Galois field has $2^3 = 8$ elements. We observe that, as in the previous example (ii), it contains a (unique) copy of the (**prime**) field $\mathbf{GF}_2 = \mathbb{Z}_2$. Again, the addition table is easy to write down and the multiplication table is as follows.

mult	$\overline{1}$	$\overline{x}$	$\overline{x^2}$	$\overline{x+1}$	$\overline{x^2+x}$	$\overline{x^2+x+1}$	$\overline{x^2+1}$
$\overline{1}$	$\overline{1}$	$\overline{x}$	$\overline{x^2}$	$\overline{x+1}$	$\overline{x^2+x}$	$\overline{x^2+x+1}$	$\overline{x^2+1}$
$\overline{x}$	$\overline{x}$	$\overline{x^2}$	$\overline{x+1}$	$\overline{x^2+x}$	$\overline{x^2+x+1}$	$\overline{x^2+1}$	$\overline{1}$
$\overline{x^2}$	$\overline{x^2}$	$\overline{x+1}$	$\overline{x^2+x}$	$\overline{x^2+x+1}$	$\overline{x^2+1}$	$\overline{1}$	$\overline{x}$
$\overline{x+1}$	$\overline{x+1}$	$\overline{x^2+x}$	$\overline{x^2+x+1}$	$\overline{x^2+1}$	$\overline{1}$	$\overline{x}$	$\overline{x^2}$
$\overline{x^2+x}$	$\overline{x^2+x}$	$\overline{x^2+x+1}$	$\overline{x^2+1}$	$\overline{1}$	$\overline{x}$	$\overline{x^2}$	$\overline{x+1}$
$\overline{x^2+x+1}$	$\overline{x^2+x+1}$	$\overline{x^2+1}$	$\overline{1}$	$\overline{x}$	$\overline{x^2}$	$\overline{x+1}$	$\overline{x^2+x}$
$\overline{x^2+1}$	$\overline{x^2+1}$	$\overline{1}$	$\overline{x}$	$\overline{x^2}$	$\overline{x+1}$	$\overline{x^2+x}$	$\overline{x^2+x+1}$

(iv) Similarly, let $F = \mathbb{Z}_2$ and $p(x) = x^3 + x^2 + [1]$. Put $E_2 = \mathbb{Z}_2[x]/\langle x^3 + x^2 + [1]\rangle$ and see that the algebra E_1 of (iii) and the algebra E_2 are isomorphic. The multiplication table of E_2 is given below and thus it is easy to verify that there are three isomorphisms $\varphi : E_1 \to E_2$ defined by the image $\varphi(\bar{x})$. You may choose

$$\varphi(\overline{x}) = \overline{x+1} \in E_2 \text{ or } \varphi(\overline{x}) = \overline{x^2+1} \in E_2 \text{ or } \varphi(\overline{x}) = \overline{x^2+x} \in E_2.$$

We shall see in the next section that for any power p^n of a prime p there is up to an isomorphism just one field of p^n elements.

mult	$\overline{1}$	$\overline{x}$	$\overline{x^2}$	$\overline{x^2+1}$	$\overline{x^2+x+1}$	$\overline{x+1}$	$\overline{x^2+x}$
$\overline{1}$	$\overline{1}$	$\overline{x}$	$\overline{x^2}$	$\overline{x^2+1}$	$\overline{x^2+x+1}$	$\overline{x+1}$	$\overline{x^2+x}$
$\overline{x}$	$\overline{x}$	$\overline{x^2}$	$\overline{x^2+1}$	$\overline{x^2+x+1}$	$\overline{x+1}$	$\overline{x^2+x}$	$\overline{1}$
$\overline{x^2}$	$\overline{x^2}$	$\overline{x^2+1}$	$\overline{x^2+x+1}$	$\overline{x+1}$	$\overline{x^2+x}$	$\overline{1}$	$\overline{x}$
$\overline{x^2+1}$	$\overline{x^2+1}$	$\overline{x^2+x+1}$	$\overline{x+1}$	$\overline{x^2+x}$	$\overline{1}$	$\overline{x}$	$\overline{x^2}$
$\overline{x^2+x+1}$	$\overline{x^2+x+1}$	$\overline{x+1}$	$\overline{x^2+x}$	$\overline{1}$	$\overline{x}$	$\overline{x^2}$	$\overline{x^2+1}$
$\overline{x+1}$	$\overline{x+1}$	$\overline{x^2+x}$	$\overline{1}$	$\overline{x}$	$\overline{x^2}$	$\overline{x^2+1}$	$\overline{x^2+x+1}$
$\overline{x^2+x}$	$\overline{x^2+1}$	$\overline{1}$	$\overline{x}$	$\overline{x^2}$	$\overline{x^2+1}$	$\overline{x^2+x+1}$	$\overline{x+1}$

(v) In this example, we illustrate the situation when the polynomial defining the congruence is not irreducible. Let $F = \mathbb{Z}_2$ and $q(x) = x^2+x$. The multiplication table of the algebra $A = \mathbb{Z}_2[x]/\langle x^2 + x\rangle$ is as follows:

$mult$	$\overline{x}$	$\overline{x+1}$	$\overline{1}$
$\overline{x}$	$\overline{x}$	$\overline{0}$	$\overline{x}$
$\overline{x+1}$	$\overline{0}$	$\overline{x+1}$	$\overline{x+1}$
$\overline{1}$	$\overline{x}$	$\overline{x+1}$	$\overline{1}$

Here $A \simeq \mathbf{GF}_2 \times \mathbf{GF}_2$. This notation requires some explanation. The algebra A is a vector space over F and the notation $A \simeq F \times F$ refers to this fact. So every element has a (unique) form $a = (a_1, a_2)$ that represents the element $a = a_1 + a_2$ of the algebra A. Here the set $\{\overline{x},\ \overline{x+1}\}$ is a basis of the vector space A_F, $\overline{x}$ is the identity element of the first copy of $\mathbf{GF}_2$, $\overline{x+1}$ is the identity element of the second copy of $\mathbf{GF}_2$ and $\overline{1} = (\overline{x},\ \overline{x+1}) = \overline{x} + \overline{x+1}$.

(vi) Here is another example with a reducible polynomial. Choosing $F = \mathbb{Z}_3$ and $q(x) = x^2 + [2] = (x+[1])(x+[2])$, we see that the algebra $A = \mathbb{Z}_3[x]/\langle x^2 + [2]\rangle \simeq \mathbf{GF}_3 \times \mathbf{GF}_3$. The multiplication table is chosen so that it displays the decomposition:

$mult$	$\overline{x+2}$	$\overline{2x+1}$	$\overline{2x+2}$	$\overline{x+1}$	$\overline{1}$	$\overline{2}$	$\overline{x}$	$\overline{2x}$
$\overline{x+2}$	$\overline{x+2}$	$\overline{2x+1}$	$\overline{0}$	$\overline{0}$	$\overline{x+2}$	$\overline{2x+1}$	$\overline{2x+1}$	$\overline{x+2}$
$\overline{2x+1}$	$\overline{2x+1}$	$\overline{x+2}$	$\overline{0}$	$\overline{0}$	$\overline{2x+1}$	$\overline{x+2}$	$\overline{x+2}$	$\overline{2x+1}$
$\overline{2x+2}$	$\overline{0}$	$\overline{0}$	$\overline{2x+2}$	$\overline{x+1}$	$\overline{2x+2}$	$\overline{x+1}$	$\overline{2x+2}$	$\overline{x+1}$
$\overline{x+1}$	$\overline{0}$	$\overline{0}$	$\overline{x+1}$	$\overline{2x+2}$	$\overline{x+1}$	$\overline{2x+2}$	$\overline{x+1}$	$\overline{2x+2}$
$\overline{1}$	$\overline{x+2}$	$\overline{2x+1}$	$\overline{2x+2}$	$\overline{x+1}$	$\overline{1}$	$\overline{2}$	$\overline{x}$	$\overline{2x}$
$\overline{2}$	$\overline{2x+1}$	$\overline{x+2}$	$\overline{x+1}$	$\overline{2x+2}$	$\overline{2}$	$\overline{1}$	$\overline{2x}$	$\overline{x}$
$\overline{x}$	$\overline{2x+1}$	$\overline{x+2}$	$\overline{2x+2}$	$\overline{x+1}$	$\overline{x}$	$\overline{2x}$	$\overline{1}$	$\overline{2}$
$\overline{2x}$	$\overline{x+2}$	$\overline{2x+1}$	$\overline{x+1}$	$\overline{2x+2}$	$\overline{2x}$	$\overline{x}$	$\overline{2}$	$\overline{1}$

Notice that the element $\overline{x+2}$ is the identity of the first copy of $\mathbf{GF}_3$, and the element $\overline{2x+2}$ is the identity of the second copy of $\mathbf{GF}_3$. As in the previous example, we have for instance

$$\overline{1} = (\overline{x+2},\ \overline{2x+2}) \ = \ \overline{x+2} + \overline{2x+2},\ \ \overline{x} = (\overline{2x+1},\ \overline{2x+2}) \ = \ \overline{2x+1} + \overline{2x+2}, \text{ etc.}$$

(vii) Let $F = \mathbb{Z}_3$ and $q(x) = x^3 - x$. Observe that, as in (v), $q = f^*$ is the Fermat polynomial. Then $A = \mathbb{Z}_3[x]/\langle x^3 - x\rangle \simeq \mathbf{GF}_3 \times \mathbf{GF}_3 \times \mathbf{GF}_3$. The multiplication table (that is, the essential part of it) is as follows:

mult	$\overline{2x^2+1}$	$\overline{x^2+2}$	$\overline{2x^2+2x}$	$\overline{x^2+x}$	$\overline{2x^2+x}$	$\overline{x^2+2x}$	...
$\overline{2x^2+1}$	$\overline{2x^2+1}$	$\overline{x^2+2}$	$\overline{0}$	$\overline{0}$	$\overline{0}$	$\overline{0}$	...
$\overline{x^2+2}$	$\overline{x^2+2}$	$\overline{2x^2+1}$	$\overline{0}$	$\overline{0}$	$\overline{0}$	$\overline{0}$	...
$\overline{2x^2+2x}$	$\overline{0}$	$\overline{0}$	$\overline{2x^2+2x}$	$\overline{x^2+x}$	$\overline{0}$	$\overline{0}$	...
$\overline{x^2+x}$	$\overline{0}$	$\overline{0}$	$\overline{x^2+x}$	$\overline{2x^2+2x}$	$\overline{0}$	$\overline{0}$	...
$\overline{2x^2+x}$	$\overline{0}$	$\overline{0}$	$\overline{0}$	$\overline{0}$	$\overline{2x^2+x}$	$\overline{x^2+2x}$	...
$\overline{x^2+2x}$	$\overline{0}$	$\overline{0}$	$\overline{0}$	$\overline{0}$	$\overline{x^2+2x}$	$\overline{2x^2+x}$	...
...	...	...	...	...	...	...	...

Observe that the identities of the respective three copies of $\mathbf{GF}_3$, namely, $\alpha = \overline{[2]x^2 + [1]}$, $\beta = \overline{[2]x^2 + [2]x}$ and $\gamma = \overline{[2]x^2 + x}$ correspond to the polynomial functions that map the elements $\{[0], [1], [2]\}$ (in that order) onto

$$\{[1], [0], [0]\}, \ \{[0], [1], [0]\} \ \text{and} \ \{[0], [0], [1]\}, \ \text{respectively.}$$

The set $\{\alpha, \beta, \gamma\}$ is a basis of the vector space A_F such that $\alpha\beta = \alpha\gamma = \beta\gamma = 0$ and $\alpha + \beta + \gamma = 1$.

(viii) In general, the algebra $\mathbf{A}$ of all functions (mappings) $\varphi : \mathbb{Z}_p \to \mathbb{Z}_p$ satisfies

$$A \simeq \mathbb{Z}_p[x]/\langle x^p - x\rangle \simeq \underbrace{\mathbf{GF}_p \times \mathbf{GF}_p \times \cdots \times \mathbf{GF}_p}_{p \text{ copies}}.$$

To see this, one can use the idea of Lagrange interpolation as we have seen in Theorem VII.2.4. Indeed, choose the set of the images of the basic Lagrange polynomials q_t $(t = 1, 2, \ldots, p)$ as a basis of the vector space A_F, $F = \mathbf{GF}_p$

$$\{\alpha_t = \overline{q_t(x)} \mid t = 1, 2, \ldots, p\}.$$

We note here that in Chapter IX we will speak about **direct sums** of rings and with that notation we can also write A in the form

$$\underbrace{\mathbf{GF}_p \oplus \mathbf{GF}_p \oplus \cdots \oplus \mathbf{GF}_p}_{p \text{ copies}}.$$

Exercise VII.4.7. Show that the elements $\alpha_t \in A$ of Example VII.4.2(viii) are the identity elements of the p copies of $\mathbf{GF}_p$, that is, that $\alpha_{t_1}\alpha_{t_2} = 0$ for $t_1 \neq t_2$, ${\alpha_t}^2 = \alpha_t$ for all t and $\sum_{t=1}^{p} \alpha_t = 1$. Check that α, β and γ of Example VII.4.2(vii) are just the images of the basic Lagrange polynomials.

Exercise VII.4.8. Describe the structure of $\mathbb{Z}_5[x]/\langle x^5 - x\rangle$ as the (direct) sum of 5 copies of $\mathbf{GF}_5$ and write down the (unique) expression of $\overline{x^3 + 4x + 1}$ as a sum of elements of individual summands.

In the previous examples we have met a number of finite fields. We now describe their structure.

Theorem VII.4.4 (Finite fields). *For each prime number p and each natural number n, there is, up to an isomorphism, a* **unique** *field $\mathbf{GF_{p^n}}$ of $q = p^n$ elements. These elements are just the roots of the polynomial $f^*(x) = x^q - x \in \mathbf{GF_{p^n}}[x]$. All finite fields are of this form.*

We give an illustration of this theorem for $p = 3$ and $n = 2$. In this case, the polynomial $x^9 - x$ has the following (unique) factorization over $\mathbf{GF_3} \simeq \mathbb{Z}_3$:

$$x^9 - x = x(x + [1])(x + [2])(x^2 + [1])(x^2 + x + [2])(x^2 + [2]x + [2]).$$

The field $\mathbf{GF_{3^2}} \simeq \mathbb{Z}_3[x]/\langle x^2 + [1]\rangle (\simeq \mathbb{Z}_3[x]/\langle x^2 + x + [2]\rangle \simeq \mathbb{Z}_3[x]/\langle x^2 + [2]x + [2]\rangle)$ has the following multiplication table (making use of the presentation $\mathbb{Z}_3[x]/\langle x^2 + [1]\rangle$).

mult	$\overline{1}$	$\overline{2}$	$\overline{x}$	$\overline{2x}$	$\overline{x+1}$	$\overline{2x+1}$	$\overline{x+2}$	$\overline{2x+2}$
$\overline{1}$	$\overline{1}$	$\overline{2}$	$\overline{x}$	$\overline{2x}$	$\overline{x+1}$	$\overline{2x+1}$	$\overline{x+2}$	$\overline{2x+2}$
$\overline{2}$	$\overline{2}$	$\overline{1}$	$\overline{2x}$	$\overline{x}$	$\overline{2x+2}$	$\overline{x+2}$	$\overline{2x+1}$	$\overline{x+1}$
$\overline{x}$	$\overline{x}$	$\overline{2x}$	$\overline{2}$	$\overline{1}$	$\overline{x+2}$	$\overline{x+1}$	$\overline{2x+2}$	$\overline{2x+1}$
$\overline{2x}$	$\overline{2x}$	$\overline{x}$	$\overline{1}$	$\overline{2}$	$\overline{2x+1}$	$\overline{2x+2}$	$\overline{x+1}$	$\overline{x+2}$
$\overline{x+1}$	$\overline{x+1}$	$\overline{2x+2}$	$\overline{x+2}$	$\overline{2x+1}$	$\overline{2x}$	$\overline{2}$	$\overline{1}$	$\overline{x}$
$\overline{2x+1}$	$\overline{2x+1}$	$\overline{x+2}$	$\overline{x+1}$	$\overline{2x+2}$	$\overline{2}$	$\overline{x}$	$\overline{2x}$	$\overline{1}$
$\overline{x+2}$	$\overline{x+2}$	$\overline{2x+1}$	$\overline{2x+2}$	$\overline{x+1}$	$\overline{1}$	$\overline{2x}$	$\overline{x}$	$\overline{2}$
$\overline{2x+2}$	$\overline{2x+2}$	$\overline{x+1}$	$\overline{2x+1}$	$\overline{x+2}$	$\overline{x}$	$\overline{1}$	$\overline{2}$	$\overline{2x}$

Here, $\overline{0}$ is a root of the polynomial $x \in \mathbf{GF_{3^2}}[x]$, $\overline{[1]}$ is a root of the polynomial $x + [2]$ and $[2]$ of the polynomial $x + [1]$. Furthermore, $\overline{x}$ and $\overline{2x}$ are roots of the polynomial $x^2 + [1] = (x - \overline{x})(x - \overline{2x})$, $\overline{x+1}$ and $\overline{2x+1}$ of the polynomial $x^2 + x + [2] = (x - \overline{x+1})(x - \overline{2x+1})$ and $\overline{x+2}$ and $\overline{2x+2}$ are roots of the polynomial $x^2 + [2]x + [2] = (x - \overline{x+2})(x - \overline{2x+2}) \in \mathbf{GF_{3^2}}[x]$.

In the proof of Theorem VII.4.4 we need the following important statement concerning certain powers of elements in finite fields.

Theorem VII.4.5. *In every field F of* **characteristic** *p (that is, in a field containing $\mathbb{Z}_p$ as a prime field),*

$$(a \pm b)^{p^n} = a^{p^n} \pm b^{p^n} \text{ for all elements } a, b \in F \text{ and any natural number } n. \quad \text{(VII.4.1)}$$

Proof. The binomial theorem for numbers holds in any commutative ring:

$$(a+b)^p = \sum_{t=0}^{p} \binom{p}{t} a^{p-t} b^t. \tag{VII.4.2}$$

Now, for $t = 1, 2, \ldots, p$, we have

$$\binom{p}{t} = \frac{p(p-1)(p-2)\cdots(p-t+1)}{1 \cdot 2 \cdot 3 \cdots t} \equiv 0 \pmod{p},$$

since the factor p in the numerator cannot be canceled and thus (VII.4.2) follows. Setting $c + b = a$, we get $a^p = (c+b)^p = (a-b)^p + b^p$, therefore $(a-b)^p = a^p - b^p$.

The formula (VII.4.1) then follows by induction.

Proof of Theorem VII.4.4. Let F be a finite field. We have already seen in the previous chapter that F contains a (uniquely defined) prime field $\mathbb{Z}_p$ for some prime p. Consequently, the field is a finite dimensional vector space $F_{\mathbb{Z}_p}$ over the field $\mathbb{Z}_p$. Let $\dim F_{\mathbb{Z}_p} = n$ and $\{b_1, b_2, \ldots, b_n\} \subseteq F$ be a basis of the vector space $F_{\mathbb{Z}_p}$. Thus, every element $a \in F$ is a (unique) linear combination

$$a = a_1 b_1 + a_2 b_2 + \cdots + a_n b_n, \quad \text{where } a_1, a_2, \ldots, a_n \in \mathbb{Z}_p,$$

and therefore the number of elements of the field F is $q = p^n$.

Now, the set $S = F \setminus \{0\}$ of all nonzero elements of the field F forms a multiplicative group of the field F of $q - 1 = p^n - 1$ elements. Hence, for every nonzero element $a \in F$, the sequence

$$a^0 = 1, a, a^2, a^3, \ldots, a^{q-1}$$

must contain two equal elements: $a^k = a^l$ for $1 \le k < l \le q-1$ and therefore there exists $s = l - k > 0$ such that $a^s = 1$. Let d be the least such exponent that satisfies $a^d = 1$, that is, $C = \{a, a^2, \ldots, a^d = 1\}$ is a cyclic subgroup of S of order d (having d elements).
Define on S the following relation $\sim$ by

$$x \sim y \text{ if and only if } x = ya^s \text{ for some } 1 \le s \le d.$$

This relation is an equivalence relation. All equivalence classes

$$xC = \{xa^s \mid 1 \le s \le d\}$$

have the same number d of elements. The number t of these classes satisfies $q - 1 = dt$ and therefore

$$a^{q-1} = (a^d)^t = 1^t = 1 \text{ for every element } a \in S.$$

Every nonzero element of the group S is thus a root of the polynomial $x^{q-1} - 1 \in F[x]$, that is, **all elements** of the field F are roots of Fermat's polynomial

$$f^\star(x) = x^q - x \in F[x].$$

Therefore

$$f^\star(x) = \prod_{r=1}^{q} (x - a_r),$$

where the a_r $(r = 1, 2, \ldots, q)$ are all the elements of the field F. The field F is therefore, up to an isomorphism, the uniquely determined field of $q = p^n$ elements that is obtained by adjunction of the roots of the polynomial $f^\star(x) = x^q - x \in \mathbb{Z}_p[x]$ to $\mathbb{Z}_p$ (in the way, as a splitting field, that we have described earlier).

It remains to prove that for any prime p and any natural number n there **exists** a field of $q = p^n$ elements. We start with the field $\mathbb{Z}_p$. First we form an extension E of $\mathbb{Z}_p$ in which the polynomial $f^*(x) = x^q - x$ splits into linear factors. Consider the set G of all roots of f^* in the extension field E. We are going to show that G forms a subfield of E. In view of Theorem VII.4.4, a difference of two q-th roots is a q-th root. Moreover, for $b \neq 0$,

$$\left(\frac{a}{b}\right)^{p^n} = \frac{a^{p^n}}{b^{p^n}},$$

and thus also a quotient of two roots is again a root. In order to see that the number of elements in G is q, it remains to show that all the roots of f^* are simple, that is, that there are q distinct roots. This is usually done by using a formal derivative of a polynomial. Here we give an elementary proof: Assuming that a multiple root a of $x^q - x$ exists, we can see first that $a \neq 0$ and secondly that over $G[x]$

$$x^q - x = (x^2 - 2ax + a^2)(x^{q-2} + c_3x^{q-3} + c_4x^{q-4} + \cdots + c_{q-2}x^2 + c_{q-1}x + c_q).$$

Expanding the product and successively comparing the coefficients of x^{q-1}, x^{q-2}, ..., x^2, we obtain

$$c_3 = 2a,\ \ c_4 = 3a^2,\ \ c_5 = 4a^3, \ldots, c_q = (q-1)a^{q-2}.$$

Comparing constant terms, we see that $a^2c_q = 0$ so that $c_q = 0$, contradicting that

$$c_q = (q-1)a^{q-2} \neq 0.$$

□

Remark VII.4.1. Exploring properties of the groups of the roots of identity, it is possible to describe the structure of the field $\mathbf{GF}_{p^n}$ more closely: The multiplicative group of this field is always cyclic (generated by the **primitive** $(p^n - 1)$**-th root** of identity) and the lattice of all subfields of $\mathbf{GF}_{p^n}$ is isomorphic to the lattice of all divisors of the natural number n. In other words, for each divisor d of the number n, there is a unique subfield $\mathbf{GF}_{p^d}$ in the field $\mathbf{GF}_{p^n}$ and there are no other subfields of $\mathbf{GF}_{p^n}$.

Exercise VII.4.9. Show that every element a in a Galois field $\mathbf{GF}_{p^n}$ of characteristic p has precisely one p-th root $\sqrt[p]{a}$ in $\mathbf{GF}_{p^n}$.

Exercise VII.4.10. Prove that the mapping $\varphi : \mathbf{GF}_{p^n} \to \mathbf{GF}_{p^n}$ defined by $\varphi(a) = a^p$ is a field automorphism of the field $\mathbf{GF}_{p^n}$. Describe the group of automorphisms of $\mathbf{GF}_{p^n}$ generated by φ. The mapping $\varphi(a) = a^p$ is called the **Frobenius map** after **Ferdinand Georg Frobenius (1849 - 1917)**.

Exercise VII.4.11. Formulate the "generalization" of Fermat's theorem and of Wilson's theorem from $\mathbf{GF}_p$ to $\mathbf{GF}_{p^n}$.

VII.5 Polynomials over $\mathbb{Z}$ and $\mathbb{Q}$

In this section, we investigate the properties of the polynomials from the integral domain $\mathbb{Z}[x]$, that is, of polynomials over the integers. In this domain the set $\{2f + xg \mid f, g \in \mathbb{Z}[x]\}$ is an ideal which is not principal. However all ideals in a Euclidean domain are principal. Thus, unlike in the case of the polynomials from $F[x]$, where F is a field, the domain $\mathbb{Z}[x]$ is not a Euclidean domain.

Exercise VII.5.1. Show that there are $f, g \in \mathbb{Z}[x]$ for which there are no $q, r \in \mathbb{Z}[x]$ such that $f = g\,q + r$ with either $r = 0$ or $\deg r < \deg g$, that is, that there is in general no Euclidean division in $\mathbb{Z}[x]$ except when g is monic. However, show that for every $f, g \in \mathbb{Z}[x]$ there are $q, r \in \mathbb{Z}[x]$ and $c \in \mathbb{Z}$ such that $c\,f = g\,q + r$ with either $r = 0$ or $\deg r < \deg g$.

There are only two units, namely 1 and -1 in $\mathbb{Z}[x]$, that is, the group of units is the cyclic group of order 2. We still have the concept of an irreducible and a prime polynomial. Recall that these concepts depend substantially on the domain from which we consider the coefficients of the polynomials. For example, the polynomial $2x^2 + 8$ is irreducible in $\mathbb{Q}[x]$ and $\mathbb{R}[x]$, but reducible in $\mathbb{C}[x]$ and $\mathbb{Z}[x]$: $2x^2 + 8 = 2(x + 2\,i)\,(x - 2\,i) = 2\,(x^2 + 4)$.

We would like to answer two questions that were established for polynomials over fields using the Euclidean division:

Is every irreducible polynomial in $\mathbb{Z}[x]$ *prime?*

Is factorization of polynomials in $\mathbb{Z}[x]$ *unique?*

Both questions will be answered in the affirmative by using one of the following two reductions: Either the canonical embedding of $\mathbb{Z}[x]$ into $\mathbb{Q}[x]$ or the canonical surjective ring homomorphism (epimorphism) $\varphi : \mathbb{Z}[x] \to \mathbb{Z}_p[x]$ for some prime number p. Here φ is defined by

$$\varphi(a_0 + a_1x + a_2x^2 + \cdots + a_nx^n) = [a_0] + [a_1]x + [a_2]x^2 + \cdots + [a_n]x^n,$$

where $[a_t] \in \mathbb{Z}_p, t = 0, 1, \ldots, n$. In both instances, it is a reduction to the situation when polynomials have their coefficients in a field and thus answers to both these questions are already known.

There is an additional concept that will assist in the main proofs. We define the **content** of a nonzero polynomial $f(x) = a_0 + a_1x + a_2x^2 + \cdots + a_nx^n \in \mathbb{Z}[x]$ to be the greatest common divisor of the integers $\{a_0, a_1, a_2, \ldots, a_n\}$ and we denote it by $\mathrm{cont}f$. A polynomial is said to be **primitive** if $\mathrm{cont}f = 1$. This concept allows us to describe the close relationship between $\mathbb{Z}[x]$ and $\mathbb{Q}[x]$ mentioned above.

Theorem VII.5.1. *Each equivalence class of associates in* $\mathbb{Q}[x]$ *contains a unique monic polynomial* $f_m \in \mathbb{Q}[x]$ *and a unique primitive polynomial* $f_p \in \mathbb{Z}[x]$ *whose leading coefficient is positive. Thus there exists a bijection between the set of all monic polynomials in* $\mathbb{Q}[x]$ *and the set of all primitive polynomials with positive leading coefficients in* $\mathbb{Z}[x]$.

Exercise VII.5.2. Give a detailed proof of Theorem VII.5.1 showing that

(i) for every monic $f \in \mathbb{Q}[x]$ there is $c \in \mathbb{Z}, c \neq 0$, such that $c\,f \in \mathbb{Z}[x]$ is a primitive polynomial with a leading coefficient that is positive and

(ii) for every primitive $f \in \mathbb{Z}[x]$ there is $c \in \mathbb{Z}, c \neq 0$, such that $c^{-1} f \in \mathbb{Q}[x]$ is monic.

The basic fact concerning the content of polynomials is expressed in the following theorem.

Theorem VII.5.2. *Let $f, g \in \mathbb{Z}[x]$. Then*

$$\text{cont}\,(f\,g) \;=\; \text{cont}\, f \,\text{cont}\, g. \tag{VII.5.1}$$

In particular, the product of primitive polynomials is a primitive polynomial.

Proof. Let cont $f = r$ and cont $g = s$. Write $f = r\, f_1,\;\; g = s\, g_1$ and $h = f_1\, g_1$. Thus, both f_1 and g_1 are primitive polynomials and $f\, g = rs\, g_1\, g_1$. To prove (VII.5.1), it is enough to show that $h = f_1\, g_1$ is primitive. Assume that this is false and that there is a prime p dividing cont h. Use the homomorphism φ mentioned above reducing h to $\varphi(h) = \varphi(f_1)\,\varphi(g_1) = 0 \in \mathbb{Z}_p[x]$. Thus either $\varphi(f_1) = 0$ or $\varphi(g_1) = 0$ in the field $\mathbb{Z}_p$. Therefore all coefficients of f_1 or g_1 must be multiples of the prime p which contradicts the fact that they are both primitive. □

We formulate two important corollaries.

Corollary VII.5.1 (Gauss' Lemma). *Let $f = g\,h \in \mathbb{Z}[x]$, where g and h are monic polynomials from $\mathbb{Q}[x]$. Then both polynomials g and h belong to $\mathbb{Z}[x]$.*

Proof. This follows from the previous theorems and Exercise VII.5.2. □

Corollary VII.5.2 (Unique factorization). *A nonzero polynomial $f \in \mathbb{Z}[x]$ has, up to the order of the factors, a unique factorization*

$$f = c\, p_1\, p_2\, \ldots p_n,$$

where $c \in \mathbb{Z}$ and p_t $(t = 1, 2, \ldots, n)$, are irreducible primitive polynomials with positive leading coefficients.

Proof. Again, a proof follows from the above described relationship between $\mathbb{Z}[x]$ and $\mathbb{Q}[x]$ making use of Exercise VII.5.2. □

Remark VII.5.1. Gauss' lemma can be formulated more generally as follows: Let $f \in \mathbb{Z}[x]$ be a product $f = g\,h$ with $g, h \in \mathbb{Q}[x]$. Then there is a nonzero integer c and a nonzero rational number r such that

$$f \;=\; g_1\, h_1, \;\text{ where }\; g_1 = (c\,r)\, g \in \mathbb{Z}[x] \;\text{ and }\; h_1 = (r^{-1})\, h \in \mathbb{Z}[x].$$

As a consequence, if one of the factors in $f = g\,h$ is a primitive polynomial from $\mathbb{Z}[x]$, then the other factor is also from $\mathbb{Z}[x]$. Also, if $f \in \mathbb{Z}[x]$ is reducible in $\mathbb{Q}[x]$, then it is already reducible in $\mathbb{Z}[x]$. So, a primitive polynomial $f \in \mathbb{Z}[x]$ is irreducible in $\mathbb{Z}[x]$ if and only if it is irreducible in $\mathbb{Q}[x]$.

This remark, together with the fact that a polynomial of $\mathbb{Q}[x]$ is irreducible if and only if it is prime, provides a solution to the first part of the next exercise.

Exercise VII.5.3. (i) Prove that a polynomial $f \in \mathbb{Z}[x]$ is irreducible if and only it is prime.

(ii) Let cont $f = n$ for $f \in \mathbb{Z}[x]$. Prove that f is irreducible in $\mathbb{Q}[x]$ if and only if $n^{-1} f$ is irreducible in $\mathbb{Z}[x]$.

Factorization in $\mathbb{Q}[x]$ is difficult. To decide whether a polynomial is irreducible by testing whether it has or does not have a root in $\mathbb{Q}$ is feasible only for polynomials up to degree 3.

Exercise VII.5.4. Let F be a field, $f \in F[x]$ and $\deg f = 2$ or 3. Prove that f is irreducible if and only if f has no root in F.

Example VII.5.1 (Rational root theorem). Let $f(x) = a_0 + a_1x + \cdots + a_nx^n \in \mathbb{Z}[x]$, where $n \in \mathbb{N}$, $a_0 \neq 0$ and $a_n \neq 0$. Suppose that $r = \frac{b}{c}$, where $c \in \mathbb{N}$ and $d(b,c) = 1$, is a rational root of f. Then

$$f(x) \;=\; (x-r)\,g(x), \text{ that is, } c\,f(x) \;=\; (cx-b)\,g(x) \text{ in } \mathbb{Q}[x].$$

The polynomial $cx - b \in \mathbb{Z}[x]$ is primitive. Thus, by Gauss' lemma (see Remark VII.5.1), we have $g \in \mathbb{Z}[x]$. Hence

$$a_0 + a_1x + \cdots + a_nx^n \;=\; (-b+cx)\,(b_0 + b_1x + \cdots + b_{n-1}x^{n-1}), \text{ where } b_0, b_1, \ldots, b_{n-1} \in \mathbb{Z}.$$

Comparing the leading coefficients and the constant terms, we obtain

$$a_n = cb_{n-1} \text{ and } a_0 = -b\,b_0.$$

Thus c **is a divisor of** $a_n (\neq 0)$ and b **is a divisor of** $a_0 (\neq 0)$. Hence we have proved that there are only finitely many possibilities for the rational roots of f. Moreover these possibilities are the rational numbers $\frac{b}{c}$, where b is an integer dividing a_0, c is a positive integer dividing a_n, and $d(b,c) = 1$.

For example, for $f(x) = x^k - 2$, where k is a positive integer greater than or equal to 2, the only possible rational roots are $r = \pm 1, \pm 2$. It follows that $x^k - 2$ has no rational roots, that is, all the quantities $\sqrt[k]{2}$ are irrational.

If $f(x) = 5\,x^3 + x - 3$ the only possible rational roots of f are $r = \pm 1, \pm 3, \pm\frac{1}{5}, \pm\frac{3}{5}$. Since none of these numbers is a root of f, the polynomial f is irreducible in $\mathbb{Q}[x]$ (see Exercise VII.5.4) and thus also in $\mathbb{Z}[x]$.

We now prove a simple useful test, named after **Ferdinand Gotthold Eisenstein (1823 - 1852)**, for the irreducibility of polynomials in $\mathbb{Z}[x]$. It is a sufficient condition, but not a necessary condition, for irreducibility. It enables us to show, among other things, that there are irreducible polynomials in $\mathbb{Z}[x]$ of arbitrarily high degree.

Theorem VII.5.3 (Eisenstein's criterion). *Let*

$$f(x) = a_0 + a_1x + \cdots + a_nx^n \;\in\; \mathbb{Z}[x],\; n \geq 1$$

be a primitive polynomial. Suppose there is a prime number p such that

$$p \text{ is a divisor of all } a_t \;(t = 0, 1, \ldots, n-1), \text{ but } p^2 \text{ is not a divisor of } a_0. \tag{VII.5.2}$$

Then f is irreducible in $\mathbb{Z}[x]$ and hence in $\mathbb{Q}[x]$.

Proof. We are going to use the epimorphism $\varphi : \mathbb{Z}[x] \to \mathbb{Z}_p[x]$ mentioned earlier; it will make our indirect proof rather simple. Thus, assume that f is a primitive polynomial that has no irreducible factor of degree zero and that

$$f = g\,h \text{ with } g, h \in \mathbb{Z}[x] \text{ and } \deg g = k, \ \deg h = l, \ k + l = n.$$

Then $\varphi(f) = \varphi(g)\,\varphi(h)$, where $\varphi(f) = [a_n]x^n$ with $[a_n] \neq [0]$ in $\mathbb{Z}_p$ in accordance with the conditions (VII.5.2) and the fact that f is primitive. Making use of unique factorization in $\mathbb{Z}_p[x]$ and the fact that the polynomial $x \in \mathbb{Z}_p[x]$ is irreducible,

$$\varphi(g) = [c]\,x^k \text{ and } \varphi(h) = [d]\,x^l.$$

This yields that the constant terms b_0 of $g(x)$ and c_0 of $h(x)$ are multiples of the prime p and thus $a_0 = b_0\,c_0$ is a multiple of p^2, contrary to (VII.5.2). The theorem follows. □

A well-known application of Eisenstein's criterion provides a proof that the **cyclotomic** (circle dividing) **polynomial** $\Phi_p(x) = \frac{x^p - 1}{x - 1} = 1 + x + x^2 + \cdots + x^{p-1}$ for a prime number p is irreducible over $\mathbb{Z}[x]$.

Corollary VII.5.3. *$\Phi_p(x)$ is irreducible over $\mathbb{Z}[x]$ (and thus over $\mathbb{Q}[x]$) for every prime p.*

Proof. We shall use the ring automorphism $\alpha : \mathbb{Z}[x] \to \mathbb{Z}[x]$ defined by $\alpha(f) = g$, where $g(x) = f(x+1)$. It establishes a bijection between irreducible polynomials. In particular, to prove our theorem, it is sufficient to show that

$$\begin{aligned} g(x) = \alpha(\Phi_p(x)) &= \frac{(x+1)^p - 1}{(x+1) - 1} = \frac{(x+1)^p - 1}{x} \\ &= \binom{p}{1} + \binom{p}{2} x + \cdots + \binom{p}{p-1} x^{p-2} + x^{p-1} \end{aligned}$$

is irreducible over $\mathbb{Z}[x]$. The polynomial is primitive and all its coefficients $\binom{p}{t}$ $t = 1, 2, \ldots,$ $p - 1$ are multiples of p. At the same time, p^2 is not a divisor of $\binom{p}{1} = p$. Thus Eisenstein's criterion applies completing the proof of the theorem. □

Remark VII.5.2. Let $n \in \mathbb{N}$. The n-th roots of unity are the n complex numbers $e^{2\pi i \frac{t}{n}}$ $(t = 1, 2, \ldots, n)$. They are the n roots of the polynomial $x^n - 1$ so that

$$x^n - 1 = \prod_{t=1}^{n} (x - e^{2\pi i \frac{t}{n}}).$$

The primitive n-th roots of unity are the complex numbers $e^{2\pi i \frac{t}{n}}$ $(t = 1, \ldots, n,\ d(t, n) = 1)$. There are $\varphi(n)$ of them, where φ is Euler's totient function. They are the roots of the polynomial

$$\Phi_n(x) = \prod_{\substack{t = 1 \\ d(t,n) = 1}}^{n} (x - e^{2\pi i \frac{t}{n}}).$$

We have

$$x^n - 1 = \prod_{k|n} \prod_{\substack{t=1 \\ d(t,n)=k}}^{n} (x - e^{2\pi i \frac{t}{n}}) = \prod_{k|n} \prod_{\substack{u=1 \\ d(u,n/k)=1}}^{n/k} (x - e^{2\pi i \frac{u}{n/k}})$$

$$= \prod_{k|n} \Phi_{n/k}(x) = \prod_{d|n} \Phi_d(x).$$

This allows us to express $\Phi_n(x)$ recursively in terms of $\Phi_d(x)$ with $d < n$ and $d \mid n$, namely,

$$\Phi_n(x) = (x^n - 1) \prod_{\substack{d=1 \\ d \mid n}}^{n-1} \Phi_d(x)^{-1}.$$

Thus, for example, with $n = 6$ we have

$$\Phi_6(x) = \frac{x^6 - 1}{\Phi_1(x)\,\Phi_2(x)\,\Phi_3(x)} = \frac{x^6 - 1}{(x-1)\,(x+1)\,(x^2+x+1)} = x^2 - x + 1.$$

All cyclotomic polynomials can be shown to be irreducible (see for example *Elements of Algebra* by J. Stillwell or *Modern Algebra* I by B. L. van der Waerden).

Exercise VII.5.5. Prove the *back-to-front* Eisenstein's criterion:

Let $f(x) = a_0 + a_1 x + a_2 x^2 + \cdots + a_n x^n \in \mathbb{Z}[x]$ be a polynomial of degree $n \geq 1$ that satisfies, for a prime p,

$$p \mid a_n, p \mid a_{n-1}, \ldots, p \mid a_1 \text{ and } p \nmid a_0, p^2 \nmid a_n.$$

Then the polynomial f is irreducible in $\mathbb{Z}[x]$ and thus also in $\mathbb{Q}[x]$.

Exercise VII.5.6. Transform the polynomial $f(x) = x^4 - 4x^3 + 6x^2 + x + 1$ into a polynomial that satisfies Eisenstein's criterion and thus show that f is irreducible over $\mathbb{Q}$.

Exercise VII.5.7. Determine the polynomials $\Phi_{12}(x)$, $\Phi_{24}(x)$, $\Phi_{15}(x)$ and $\Phi_{30}(x)$.

Exercise VII.5.8. Let p be a prime number and let k be a positive integer. Prove that

$$\Phi_{p^k}(x) = \Phi_p(x^{p^{k-1}}).$$

Exercise VII.5.9. Let $\omega_1, \omega_2, \ldots, \omega_n$ be the n-th roots of unity in $\mathbb{C}$. Show that

$$\sum_{t=1}^{n} \omega_t^k = \begin{cases} n & \text{if } k \equiv 0 \pmod{n}, \\ 0 & \text{if } k \not\equiv 0 \pmod{n}. \end{cases}$$

VII.6 Polynomials in several indeterminates

We turn our attention to an important class of polynomials over a field F, namely to the class of **symmetric polynomials** (symmetric functions) over F. We have already underlined the importance of defining polynomials over an arbitrary integral domain, see Exercise VII.1.3. We shall now use this fact to introduce polynomials in n indeterminates over a field F. Starting with the integral domain (of one indeterminate) $F[x_1]$, define successively $F[x_1, x_2]$ as $(F[x_1])[x_2]$, $F[x_1, x_2, x_3]$ as $(F[x_1.x_2])[x_3]$ etc. up to the integral domain

$$F[\mathbf{x}] = F[x_1, x_2, \dots, x_n] = (\dots((F[x_1])[x_2])\dots)[x_n].$$

Example VII.6.1. We consider the polynomial f of three indeterminates x_1, x_2, x_3 given by

$$\begin{aligned} f(x_1, x_2, x_3) &= x_1^4\, x_3^8 + [(x_1^2 + x_1)\, x_2^4 + x_1^4\, x_2]\, x_3^7 + [(x_1^3 - 2)\, x_2^2 + x_1 + 3]\, x_3^3 \\ &\quad + x_1^5\, x_3 + (x_1^3 + x_1^2)\, x_2^4 + x_1^5\, x_2 \;=\; x_1^5 x_2 + x_1^5 x_3 + x_1^4 x_2 x_3^7 + x_1^4 x_3^8 + x_1^3 x_2^4 \\ &\quad + x_1^3 x_2^2 x_3^3 + x_1^2 x_2^4 x_3^7 + x_1^2 x_2^4 + x_1 x_2^4 x_3^7 + x_1 x_3^3 - 2 x_2^2 x_3^3 + 3 x_3^3. \end{aligned}$$

We have here two expressions for f the second one being in the **lexicographical order**: the term $a_{k_1 k_2 \dots k_n} x_1^{k_1}\, x_2^{k_2} \dots x_n^{k_n}$ precedes the term $b_{l_1 l_2 \dots l_n} x_1^{l_1}\, x_2^{l_2} \dots x_n^{l_n}$ if and only if

$$k_1 = l_1, k_2 = l_2, \dots, k_{t-1} = l_{t-1} \text{ and } k_t > l_t \text{ for some } 1 \le t \le n.$$

Denote by Σ_n the **symmetric group** of all permutations of the set $I_n = \{1, 2, \dots, n\}$: the (associative) multiplication is just the composition of the permutations (bijective mappings) of the set I_n whereby there is an identity element, namely the identity mapping, and to each of the $n!$ permutations there is an inverse permutation. For every $\sigma \in \Sigma_n$ and every polynomial $f \in F[\mathbf{x}] = F[x_1, x_2, \dots, x_n]$ define the polynomial $f^\sigma \in F[\mathbf{x}]$ by

$$f^\sigma(x_1, x_2, \dots, x_n) = f(x_{\sigma(1)}, x_{\sigma(2)}, \dots, x_{\sigma(n)}).$$

Observe that for $\sigma_1, \sigma_2 \in \Sigma_n$, $(f^{\sigma_1})^{\sigma_2} = f^{\sigma_1 \sigma_2}$, that is, that the group Σ_n operates on the integral domain $F[\mathbf{x}]$.

Definition VII.6.1. *A polynomial $f = f(x_1, x_2, \dots, x_n) \in F[\mathbf{x}]$ is a* **symmetric polynomial** *if $f = f^\sigma$ for every permutation $\sigma \in \Sigma_n$.*

Remark VII.6.1. Denote by $\sigma_{r,s}$ the permutation (called a **transposition**) of I_n that maps r to s, s to r and satisfies $\sigma_{r,s}(t) = t$ for all other $t \in I_n$. In the next chapter VIII, we shall show a simple fact that every permutation is a product of such transpositions. Therefore, a polynomial f is symmetric if it does not change by any transposition of two indeterminates.

Exercise VII.6.1. Prove the following simple theorem.

Theorem VII.6.1. *The set of all symmetric polynomials in $F[\mathbf{x}]$ is a subring of the integral domain $F[\mathbf{x}]$.*

Natural building blocks of the symmetric polynomials are the **simple symmetric polynomials**.

Definition VII.6.2. *A* **simple symmetric polynomial** *in* $F[\mathbf{x}]$ *is a symmetric polynomial that can be expressed as a sum of the monomials that are obtained by means of all permutations of the indeterminates of one of them (that can be any one of the monomials).*

Example VII.6.2. If $f(x_1, x_2, x_3) = x_1^2 x_2^2 x_3^2 + x_1^3 + x_2^3 + x_1 x_3 + x_2 x_3 \in F[x_1, x_2, x_3]$, then $f^\sigma(x_1, x_2, x_3) = x_1^2 x_2^2 x_3^2 + x_1 x_2 + x_1 x_3 + x_2^3 + x_3^3$ for $\sigma \in \Sigma_3$ satisfying $\sigma(1) = 2, \sigma(2) = 3, \sigma(3) = 1$. Hence f is not symmetric. However $g(x_1, x_2, x_3) = f(x_1, x_2, x_3) + x_1 x_2 + x_3^3$ is a symmetric polynomial and is a sum of three simple symmetric polynomials, namely, $x_1^2 x_2^2 x_3^2$, $x_1^3 + x_2^3 + x_3^3$ and $x_1 x_2 + x_1 x_3 + x_2 x_3$.

We leave it to the reader to justify the following statement.

Theorem VII.6.2. *Every symmetric polynomial is a sum of simple symmetric polynomials.*

Using Definition VII.6.2, a simple symmetric polynomial is thus fully determined by a **leading member**, that is, by the member $a_{k_1,k_2,\ldots,k_n} x_1^{k_1} x_2^{k_2} \cdots x_n^{k_n}$ satisfying the following inequalities $k_1 \geq k_2 \geq \cdots \geq k_n$. Such a simple symmetric polynomial will be denoted by the symbol

$$f = \sum a\, x_1^{k_1} x_2^{k_2} \cdots x_n^{k_n} \text{ and its degree } (k_1, k_2, \ldots, k_n) \text{ by the symbol } \deg f.$$

The set of all simple symmetric polynomials can easily be linearly ordered by means of lexicographical order of their defining leading members:

$$\Sigma a x_1^{k_1} x_2^{k_2} \cdots x_n^{k_n} \prec \Sigma b x_1^{l_1} x_2^{l_2} \cdots x_n^{l_n},$$

if $k_1 = l_1, k_2 = l_2, \ldots, k_s = l_s, k_{s+1} < l_{s+1}$ for some $s \leq n$.

Definition VII.6.3. *We define the following two types of symmetric polynomials in* $F[\mathbf{x}]$:

$$S_t(x_1, x_2, \ldots, x_n) = \Sigma x_i^t = x_1^t + x_2^t + \cdots + x_n^t \quad \text{for all } t \in \mathbb{N}$$

and the **elementary symmetric polynomials**

$$E_t(x_1, x_2, \ldots, x_n) = \sum\nolimits_{1 \leq i_1 < i_2 < \cdots < i_t \leq n} x_{i_1} x_{i_2} \cdots x_{i_t} \quad \text{for all } t = 1, 2, \ldots, n.$$

Remark VII.6.2. Recall that the values of the elementary symmetric polynomials at the roots $\alpha_1, \alpha_2, \ldots, \alpha_n$ of an algebraic equation

$$x^n + a_1 x^{n-1} + a_2 x^{n-2} + \cdots + a_{n-2} x^2 + a_{n-1} x + a_n = 0$$

satisfy the **formulas of Viète (François Viète (1540 - 1603))**

$$a_t = (-1)^t E_t(\alpha_1, \alpha_2, \ldots, \alpha_n),\ 1 \leq t \leq n.$$

We now turn our attention to the fundamental property of elementary symmetric polynomials. This property shows how all symmetric polynomials can be generated in a unique manner from the elementary symmetric polynomials, see Theorem VII.6.5. Before proving Theorem VII.6.5, we first prove two useful auxiliary theorems.

Theorem VII.6.3. *Let $f \in F[\mathbf{x}]$ be a simple symmetric polynomial. Then there exists a symmetric polynomial $g \in F[E_1, E_2, \ldots, E_n]$ such that the polynomial $f_1 \in F[\mathbf{x}]$ defined by the relation*

$$f_1(\mathbf{x}) = f(\mathbf{x}) - g\,(E_1(\mathbf{x}),\, E_2(\mathbf{x}),\, \ldots,\, E_n(\mathbf{x})\,)$$

is a sum of simple symmetric polynomials each of which is of degree smaller than that of the polynomial f.

Proof. Let $f = \Sigma x_1^{k_1} x_2^{k_2} \ldots x_n^{k_n}$, where $k_r \neq 0,\ k_{r+1} = 0$. Put

$$g(E_1, E_2, \ldots, E_n) = E_r^{k_r} E_{r-1}^{k_{r-1}-k_r} \ldots E_1^{k_1-k_2},$$

that is, $g(x_1, x_2, \ldots, x_n) =$

$$(\Sigma x_1 x_2 \ldots x_r)^{k_r}\ (\Sigma x_1 x_2 \ldots x_{r-1})^{k_{r-1}-k_r}\ (\Sigma x_1 x_2 \ldots x_{r-2})^{k_{r-2}-k_{r-1}} \ldots\ldots (\Sigma x_1)^{k_1-k_2}.$$

Subtracting g from f, the highest degree terms vanish and the theorem follows. □

Theorem VII.6.4. *The elementary symmetric polynomials are* **algebraically independent** *over F, that is, if $h \in F[\mathbf{x}]$ satisfies*

$$h(E_1(\mathbf{x}), E_2(\mathbf{x}), \ldots, E_n(\mathbf{x})) = 0, \qquad \text{(VII.6.1)}$$

then h is the zero polynomial.

Proof. We are going to give an indirect proof. Let h be a nonzero polynomial of least degree that does not satisfy (VII.6.1). Write down the polynomial h as a polynomial in one indeterminate x_n with coefficients h_t from $F[x_1, x_2, \ldots, x_{n-1}]$:

$$h(x_1, x_2, \ldots, x_n) = h_0(x_1, x_2, \ldots, x_{n-1}) + h_1(x_1, x_2, \ldots, x_{n-1})x_n + \cdots + h_s(x_1, x_2, \ldots, x_{n-1})x_n^s. \qquad \text{(VII.6.2)}$$

The coefficient $h_0 \neq 0$. For, otherwise

$$h(x_1, x_2, \ldots, x_n) = x_n \tilde{h}(x_1, x_2, \ldots, x_n)$$

and thus $E_n \tilde{h}(E_1, E_2, \ldots, E_n) = 0$. This would mean that $\tilde{h}(E_1, E_2, \ldots, E_n) = 0$, with $\deg(\tilde{h}) < \deg(h)$, in contradiction to the choice of the polynomial h.

Taking into account (VII.6.1), we substitute in (VII.6.2) the elementary symmetric polynomials E_t for the indeterminates x_t to obtain

$$0 = h_0(E_1, E_2, \ldots, E_{n-1}) + h_1(E_1, E_2, \ldots, E_{n-1})E_n + \cdots + h_s(E_1, E_2, \ldots, E_{n-1})E_n^s.$$

This is an equality expressing a relation among the indeterminates $x_1, x_2, \ldots, x_n$. Choosing $x_n = 0$, we deduce

$$h_0(\tilde{E}_1, \tilde{E}_2, \ldots, \tilde{E}_{n-1}) = 0, \ \text{ where } \tilde{E}_t(x_1, x_2, \ldots, x_{n-1}) = E_t(x_1, x_2, \ldots, x_{n-1}, 0),$$

are just the elementary symmetric polynomials in $n-1$ indeterminates $x_1, x_2, \ldots, x_{n-1}$.

The proof is now completed by applying mathematical induction. □

We are now ready to formulate the **principal theorem** on symmetric polynomials.

Theorem VII.6.5. *For every symmetric polynomial $f \in F[\mathbf{x}]$, there is a* **unique** *polynomial $g \in F[E_1, E_2, \dots, E_n]$ such that*

$$f(\mathbf{x}) = g(E_1(\mathbf{x}), E_2(\mathbf{x}), \dots, E_n(\mathbf{x})).$$

Proof. Using mathematical induction, Theorem VII.6.3 yields the existence of the polynomial g. The fact that g is determined uniquely follows from Theorem VII.6.4. □

Theorem VII.6.5 implies the following corollary.

Corollary VII.6.1. *The integral subdomain of all symmetric polynomials in $F[\mathbf{x}]$ is isomorphic to the integral domain $F[E_1, E_2, \dots, E_n]$.*

We add a few remarks concerning the simple symmetric polynomials. We can formulate, in a similar way as for the elementary symmetric polynomials, the following statement for the simple symmetric polynomials S_t.

Theorem VII.6.6. *For every symmetric polynomial $f \in F[\mathbf{x}]$, there is a* **unique** *polynomial $h \in F[S_1, S_2, \dots, S_n]$ such that*

$$f(\mathbf{x}) = h(S_1(\mathbf{x}), S_2(\mathbf{x}), \dots, S_n(\mathbf{x})).$$

Remark VII.6.3. A proof of Theorem VII.6.6 follows from the fact that there is a very close relation between the polynomials S_t and polynomials E_t from $F[\mathbf{x}]$ expressed for every $d = 1, 2, \dots$ in terms of **Newton's formula**

$$S_d = E_1 S_{d-1} - E_2 S_{d-2} + \cdots + (-1)^{k+1} E_k S_{d-k} + \cdots + (-1)^d E_{d-1} S_1 + (-1)^{d+1} S_d E_0,$$

where $S_0 = d$ and $E_t = 0$ for $t > n$.

Hence, we can express, recursively, every polynomial S_d in terms of polynomials E_t and every polynomial E_d in terms of polynomials S_t. For example, for $n = 3$, we have

$$\begin{aligned} S_1 &= E_1, \\ S_2 &= E_1 S_1 - 2E_2 = E_1^2 - 2\,E_2, \\ S_3 &= E_1 S_2 - E_2 S_1 + 3\,E_3 = E_1^3 - 3\,E_1 E_2 + 3\,E_3, \\ S_4 &= E_1 S_3 - E_2 S_2 + E_3 S_1 = E_1^4 - 4\,E_1^2 E_2 + 4\,E_1 E_3 + 2\,E_2^2, \\ S_5 &= E_1 S_4 - E_2 S_3 + E_3 S_2 = E_1^5 - 5\,E_1^3 E_2 + 5\,E_1^2 E_3 + 5\,E_1 E_2^2 - 5\,E_2 E_3, \text{ etc.} \end{aligned}$$

Exercise VII.6.2. For $n = 4$, express E_4 in terms of S_t.

Exercise VII.6.3. (i) Express the following sums in term of the elementary symmetric polynomials

$$(a)\ \frac{1}{x_1} + \frac{1}{x_2} + \frac{1}{x_3}, \quad (b)\ \frac{1}{x_1^2} + \frac{1}{x_2^2} + \frac{1}{x_3^2}, \quad (c)\ x_1^3 + x_2^3 + x_3^3 \ \text{ and } \ (d)\ x_1^4 + x_2^4 + x_3^4.$$

(ii) If α, β and γ are the roots of the equation $x^3 = px + q$, find the sum of the fifth powers of the roots of this equation.

(iii) If α, β and γ are the roots of the equation $x^3 + 8x^2 - 3x - 2 = 0$, find a polynomial whose roots are $\alpha + \beta, \beta + \gamma$ and $\gamma + \alpha$. Find such polynomial in the case that the equation for the roots α, β and γ is $x^3 + ax^2 + bx + c = 0$.

Exercise VII.6.4. Show that the polynomial $f = f(x_1, x_2, \dots, x_n) = \sum_{1 \le r,s \le n} x_r^2 x_s$ can be expressed by means of the elementary symmetric polynomials in the form

$$f \;=\; a\, E_1^3 + b\, E_1\, E_2 + c\, E_3, \quad \text{where } \; a, b, c \; \text{ are integers.}$$

Determine the integers a, b and c by choosing $x_1 = 1, x_2 = x_3 = 0,$ then $x_1 = x_2 = 1, x_3 = 0,$ etc.

VII.7 Additional exercises and remarks

In this short section, we include several exercises and remarks, some of them more demanding.

Exercise VII.7.1. The polynomials

$$\begin{aligned} f_1(x) &= x^n + x^{n-1} + \cdots + x^2 + x + 1, \\ f_2(x) &= (n+1)x^n + nx^{n-1} + \cdots + 3x^2 + 2x + 1 \;\text{ and} \\ f_3(x) &= (n+1)^2x^n + n^2x^{n-1} + \cdots + 3^2x^2 + 2^2x + 1 \end{aligned}$$

are elements of the integral domain $\mathbb{Q}[x]$. Determine the coefficient of the term x^n of the product $f_1\, f_2\, f_3$.

[Hint: The coefficients of the product $f_1(x)f_2(x)f_3(x) = a_0 + a_1x + a_2x_2 + \cdots$ form an arithmetical progression of the fifth order.]

Exercise VII.7.2. Let $n \in \mathbb{N}$. Express the rational function

$$\frac{(x+1)(x+2)\dots(x+n) - n!}{x(x+1)(x+2)(x+3)\dots(x+n)}$$

as a sum of partial fractions. Hence, derive the formula

$$\sum_{t=1}^{n} \frac{(-1)^{t+1}}{t}\binom{n}{t} \;=\; 1 + \frac{1}{2} + \frac{1}{3} + \cdots + \frac{1}{n}. \tag{VII.7.1}$$

Hint:

$$\frac{(x+1)(x+2)\dots(x+n) - n!}{x(x+1)(x+2)(x+3)\dots(x+n)} = \frac{1}{x} - \frac{n!}{x(x+1)(x+2)(x+3)\dots(x+n)}$$

$$= \frac{1}{x} - \binom{n}{0}\frac{1}{x} + \binom{n}{1}\frac{1}{x+1} + \cdots + (-1)^{k+1}\binom{n}{k}\frac{1}{x+k} + \cdots + (-1)^{n+1}\binom{n}{n}\frac{1}{x+n}.$$

If we denote the product $1 \times 2 \times 3 \times \cdots \times (k-1) \times (k+1) \times \cdots \times n$ by $k^o(n)$,

$$\frac{(x+1)(x+2)\dots(x+n) - n!}{(x+1)(x+2)(x+3)\dots(x+n)}$$
$$= \frac{x^{n-1} + A_{n-2}x^{n-2} + \cdots + A_2x^2 + A_1x + (k^o(1) + k^o(2) + \cdots + k^o(n))}{(x+1)(x+2)(x+3)\dots(x+n)}.$$

Taking $x = 0$, we obtain

$$\frac{k^o(1) + k^o(2) + \cdots + k^o(n)}{n!} = 1 + \frac{1}{2} + \frac{1}{3} + \cdots + \frac{1}{n},$$

from which the formula (VII.7.1) follows.

Remark VII.7.1. The formula (VII.7.1) is usually derived by means of integration. For example,

$$\begin{aligned} 1 + \tfrac{1}{2} + \tfrac{1}{3} + \cdots + \tfrac{1}{n} &= \textstyle\int_0^1 (1 + x + x^2 + \cdots + x^{n-1})\, dx \\ &= \textstyle\int_0^1 \frac{1-x^n}{1-x}\, dx = \int_0^1 \frac{1-(1-x)^n}{x}\, dx \\ &= \textstyle\int_0^1 \left(\sum_{t=1}^n (-1)^{t+1}\binom{n}{t} x^{t-1}\right) dx \\ &= \textstyle\sum_{t=1}^n \frac{(-1)^{t+1}}{t}\binom{n}{t}. \end{aligned}$$

Exercise VII.7.3. Let $f \in \mathbb{Q}[x]$ be an arbitrary polynomial. Prove that there exists a unique polynomial $g \in \mathbb{Q}[x]$ such that

$$\frac{1}{2}\left[g(x+1) + g(x-1)\right] = f(x).$$

Hint: For every $n \in \mathbb{N}_0$, choose $f_n(x) = x^n$ and define the polynomials $g_n(x) \in \mathbb{Q}[x]$ by means of

$$\frac{1}{2}\left[g_n(x+1) + g_n(x-1)\right] = x^n.$$

Then

$$g_0(x) = 1 \text{ and the } \mathbf{derivative}\ g_n'(x) = n\, g_{n-1}(x) \text{ for } n \geq 1.$$

Hence

$$g_n(x) = n! \int_{a_n}^{x} (\cdots (\int_{a_2}^{x_2} (\int_{a_1}^{x_1} dx)\, dx_1) \dots) dx_{n-1},$$

where $a_t = 1$ for even t and $a_t = 0$ for odd t.

The polynomials $g_n(x)$ for $n = 0, 1, 2, 3, 4, 5, 6, 7$ and 8 are

$$\begin{aligned} g_0(x) &= 1, \\ g_1(x) &= x, \\ g_2(x) &= x^2 - 1, \\ g_3(x) &= x^3 - 3x, \\ g_4(x) &= x^4 - 6x^2 + 5, \\ g_5(x) &= x^5 - 10x^3 + 25x, \\ g_6(x) &= x^6 - 15x^4 + 75x^2 - 61, \\ g_7(x) &= x^7 - 21x^5 + 175x^3 - 475x, \\ g_8(x) &= x^8 - 28x^6 + 350x^4 - 1708x^2 + 1385. \end{aligned}$$

We remark that the numbers $g_n(0)$ are **Euler's numbers** E_n, which are defined by means of the following Taylor's series:

$$\frac{2}{e^x + e^{-x}} = \sum_{n=0}^{\infty} \frac{E_n}{n!} x^n.$$

Exercise VII.7.4. Let $f \in \mathbb{Q}[x]$ be an arbitrary polynomial. Prove that there is a uniquely determined polynomial $g \in \mathbb{Q}[x]$ such that

$$g(0) = 0 \text{ and } g(x) - g(x-1) = f(x).$$

[Hint: If $f(x) = \sum_{t=0}^{n} a_t x^t$, then the coefficients of the polynomial $g(x) = \sum_{t=1}^{n+1} b_t x^t$ are determined as the unique solution of the system of $n+1$ linear equations $A\mathbf{x} = \mathbf{a}$, where $\mathbf{a} = (a_o, a_1, \ldots, a_n)^T$ and the determinant of the matrix A is $(n+1)! \prod_{t=1}^{n+1} b_t$.

We have $g(1) = f(1)$ and $g(-1) = -f(0)$. In general, we have

$$g(n) = \sum_{t=1}^{n} f(t) \text{ and } g(-n) = -\sum_{t=0}^{n-1} f(-t).$$

In particular, in the case $f(x) = x^k$, the polynomial g is denoted S_k. Hence

$$S_k(n) = \sum_{t=1}^{n} t^k.]$$

Exercise VII.7.5. For every real number x and every natural number k, we define

$$\binom{x}{k} = \frac{1}{k!} x(x-1)\ldots(x-k+1).$$

Prove the following surprising statements:

(i) Let $f \in \mathbb{Z}[x]$ be an (integral) polynomial of degree n whose value at every integral point is an integral multiple of the factorial $n!$. Then

$$f(x) = \sum_{k=0}^{n} n!\, a_{n-k+1} \binom{x}{k}, \tag{VII.7.2}$$

where a_{n-k+1} are integers.

(ii) The value of a polynomial of the form (VII.7.2) with **arbitrary** integers a_{n-k+1} is an integral multiple of the factorial $n!$ at every integral point.

(iii) Show that the values of $a\,x^4 + (6a-4b)x^3 + (11a-12b+12c)x^2 + (6a-8b+12c)x$ with $a, b, c \in \mathbb{Z}$ are at **all integral points** integral multiples of the number 24.

Remark VII.7.2. The following formulation may help in better understanding the previous statements: Let $f(x) \in \mathbb{Z}[x]$ be an (integral) polynomial of degree n whose value for $x = 0, 1, \ldots, n-1$ are integral multiples of $n!$. Then the values $f(z)$ are integral multiples of $n!$ for all integers $z \in \mathbb{Z}$ and f has the form (VII.7.2) with integers a_{n-k+1}.

Remark VII.7.3 (Interpolation of functions by polynomials). Given a sequence $\mathbf{a} = (a_1, a_2, \ldots, a_n, \ldots)$ of real numbers, define the **difference sequence** $\mathcal{D}(\mathbf{a})$ of the sequence $\mathbf{a}$ by means of

$$\mathcal{D}(\mathbf{a}) = (a_2 - a_1, a_3 - a_2, \ldots, a_{n+1} - a_n, \ldots).$$

Define recursively $\mathcal{D}_1(\mathbf{a}) = \mathcal{D}(\mathbf{a})$ and $\mathcal{D}_{k+1}(\mathbf{a}) = \mathcal{D}(\mathcal{D}_k(\mathbf{a}))$ for $k \geq 1$; write $\mathbf{a} = \mathcal{D}_0(\mathbf{a})$. Denote by b_{n+1} the first term of the sequence $\mathcal{D}_n(\mathbf{a})$. Thus, $b_1 = a_1$, $b_2 = a_2 - a_1, \ldots$. Finally write $\mathbf{b} = \mathbf{\Delta}(\mathbf{a}) = (b_1, b_2, \ldots, b_n, \ldots)$.

The real **semi-polynomials** $P_k(x)$, $k = 1, 2, \ldots$ are defined recursively as follows: $P_1(x) = 0$ for $x \leq 0$ and $P_1(x) = 1$ for $x > 0$; and for $k > 0$,

$$P_{k+1}(x) = 0 \text{ for } x \leq k \text{ and } P_{k+1}(x) = \frac{x-k}{k} P_k(x) \text{ for } x > k.$$

Exercise VII.7.6. Prove:

(i) Let f be a real function defined for $x \geq 0$. Consider the sequence

$$\mathbf{a}_f = (a_1 = f(1), a_2 = f(2), \ldots, a_n = f(n), \ldots).$$

Then the **interpolation function** I_f of the function f, defined by

$$I_f(x) = \sum_{k=1}^{\infty} b_k \, P_k(x),$$

where $\mathbf{b}_f = \mathbf{\Delta}(\mathbf{a}_f) = (b_1, b_2, \ldots, b_n, \ldots)$ satisfies the equalities $I_f(n) = f(n)$ for all natural numbers n. Show how this is related to Lagrange's interpolation, see Theorem VII.2.4 and Example VII.2.1.

(ii) Verify: If $f(x) = 2^{x-1}$, then $\mathbf{a} = (1, 2, 4, 8, \ldots, 2^{n-1}, \ldots)$, $\mathbf{b} = \Delta(\mathbf{a}) = (1, 1, 1, 1, \ldots, 1 \ldots)$ and thus

$$I_f(n) = \sum_{k=1}^{\infty} P_k(n) = 2^{n-1}.$$

Exercise VII.7.7. Verify: If $f(x) = \sin \frac{\pi x}{2}$, then $\mathbf{a}_f = (1, 0, -1, 0, 1, 0, -1, 0, -1, 0, \ldots)$, and $\mathbf{b}_f = (1, -1, 0, 2^{1+4k}, -2^{2+4k}, 2^{2+4k}, 0, -2^{3+4k}, 2^{4+4k}, -2^{4+4k}, 0, \ldots)$, where $k = 0, 1, 2, \ldots$. Show that the sequence $\mathbf{b}_f = (b_1, b_2, \ldots, b_n, \ldots)$ can be described recursively by

$$b_1 = 1, \; b_2 = -1, \; b_n = -2(b_{n-2} + b_{n-1}) \text{ for } n \geq 3$$

and that

$$b_n = \frac{1}{2}\left[(-1-i)^{n-1} + (-1+i)^{n-1}\right].$$

(In fact, b_n are the coefficients of the Taylor expansion of the rational function

$$\frac{1+x}{1+2x+2x^2}).$$

Hence, for $n \in \mathbb{N}$, we have

$$\begin{aligned} I_f(n) = 2 - n - \sum_{k=1}^{\infty}[2^{1+4k}\, P_{4+8k}(n) - 2^{2+4k}\, P_{5+8k}(n) + 2^{2+4k}\, P_{6+8k}(n) \\ - 2^{3+4k}\, P_{8+8k}(n) + 2^{4+4k}\, P_{9+8k}(n) - 2^{4+4k}\, P_{10+8k}(n)] \;=\; \sin\frac{n\pi}{2}. \end{aligned}$$

Remark VII.7.4. The following two examples describe the approximation presented in Exercise VII.7.7 explicitly. The polynomial

$$q(x) = 2 - x + 2P_4(x) - 4P_5(x) + 4P_6(x) \;=\; \frac{1}{30}\,(x^5 - 20x^4 + 145x^3 - 460x^2 + 604x - 240)$$

satisfies

$$q(n) \;=\; \sin\frac{n\pi}{2} \quad \text{for } n \;=\; 1,2,3,4,5,6,7,$$

while the polynomial

$$\begin{aligned} q(x) &= 2 - x + 2P_4(x) - 4P_5(x) + 4P_6(x) - 8P_8(x) + 16P_9(x) - 16P_{10}(x) + 32P_{12}(x) \\ &= \frac{1}{1247400}\,(x^{11} - 66x^{10} + 1870x^9 - 29700x^8 + 289773x^7 - 1792098x^6 \\ &\quad + 7018220x^5 - 17021400x^4 + 24776576x^3 - 21073536x^2 + 9077760x) \end{aligned}$$

satisfies

$$q(n) \;=\; \sin\frac{n\pi}{2} \text{ for } n \;=\; 1,2,3,4,5,6,7,8,9,10,11,12\,.$$

You may like to check the way the graphs of these functions converge to the graph of the function $\sin\frac{n\pi}{2}$.

Remark VII.7.5 (Discriminant). Let

$$x_n + a_{n-1}x^{n-1} + \cdots + a_1x + a_0 \;=\; 0 \tag{VII.7.3}$$

be an algebraic equation with coefficients from a field F and let $\alpha_1, \alpha_2, \ldots, \alpha_n$ be the roots of (VII.7.3) in some extension field of F. The discriminant **Disc** of this equation is defined by

$$\mathbf{Disc} \;=\; \prod_{1\le r<s\le n} (\alpha_r - \alpha_s)^2.$$

Thus the equation (VII.7.3) has at least one double root if and only if its discriminant is zero.

Exercise VII.7.8. Show that there exists a polynomial $g \in F[\mathbf{x}]$ in n indeterminates such that

$$\mathbf{Disc} = g(a_0, a_1, \ldots, a_{n-1}).$$

Exercise VII.7.9. (i) Calculate the discriminants of equations of degree 2, 3 and 4.

(ii) Calculate the discriminant of the equation $x^5 + ax + b = 0$ and determine all real pairs (a, b) for which this equation has at least one double root.

CHAPTER VIII. GROUPS

In the previous section, we have started to cover certain simple general concepts (such as relations or binary algebraic operations) in an "abstract" manner - this was the reasoning when we did not specify particular sets and their elements, but rather treated these objects merely as being subject to certain rules.

This approach will now be our "modus operandi" for the rest of the book. We will experiment with certain well-chosen particular examples, and then treat some aspects of these examples in a "general" way (valid for all particular cases). This approach has proved to be very efficient and successful, for the **"abstract"** results can be applied to many different (**"concrete"**) situations avoiding considering and proving them in any particular case. "Mathematicians do not study objects, but relations between objects." Thus, they are free to replace some objects by others so long as the relations remain unchanged. Content to them is irrelevant: they are interested in form only. This quote by **Henri Poincaré** has been summarized as "Mathematics is the art of giving the same name to different things." (In contrast to: Poetry is the art of giving different names to the same thing.) It is a rather surprising fact that dealing with abstract situations is generally simpler than dealing with specific concrete situations. **James Joseph Sylvester** is credited with pointing out as early as 1876 that **general statements are simpler than their particular cases.** Experience obtained by looking at specific cases can result in directing investigations along a fruitful path, however the final success seems to be in the correct choice of concepts and their efficient exploitation.

In the preceding chapters, we have already met a number of different examples of groups. We consider yet the following two important classes of groups (depending on the natural parameter n).

VIII.1 Symmetric groups $\mathcal{S}_n$

The elements of our groups will be bijective transformations of sets.

Definition VIII.1.1. *Let X be a (nonempty) set. A bijective transformation $\pi : X \to X$ will be called a* **permutation** *of the set X. The set of all permutations of X together with binary operation of composition $\circ$ of transformations forms a group that is called the* **symmetric group** $\mathcal{S}(X)$ *of permutations of the set X.*

The neutral element of the group $\mathcal{S}(X)$ is the identical mapping 1_X of X and the inverse element to a permutation π is the permutation μ satisfying $\pi \circ \mu = 1_X$. It will be denoted by $\mu = \pi^{-1}$. Of course, $\pi \circ \pi^{-1} = \pi^{-1} \circ \pi = 1_X$.

Here, we shall be interested in permutations of finite sets X. The structure of the symmetric group of a set X only depends on the number of elements of the set X. We will choose to represent a set X of n elements by the set $I_n = \{1, 2, \dots, n\}$, denote the respective symmetric group by $\mathcal{S}_n$ and describe the permutation π mapping t to $\pi(t) = i_t$ by

$$\pi = \begin{pmatrix} 1 & 2 & \dots & t & \dots & n \\ i_1 & i_2 & \dots & i_t & \dots & i_n \end{pmatrix}. \qquad \text{(VIII.1.1)}$$

The composition of permutations will be called multiplication of permutations. For the sake of simplicity we shall frequently drop the notation $\circ$. We point out the well known fact that the number of elements (that is, the **order**) of $\mathcal{S}_n$ is $n!$ (see Section 2 of Chapter I).

Exercise VIII.1.1. It is evident that for each $m = 1, 2, \dots, n$, there are subgroups of $\mathcal{S}_n$ that are isomorphic to $\mathcal{S}_m$. Show that there are 4 subgroups of $\mathcal{S}_4$ that are isomorphic to $\mathcal{S}_3$ and 10 subgroups of $\mathcal{S}_5$ that are isomorphic to $\mathcal{S}_3$.

Of course, for every permutation $\pi \in \mathcal{S}_n$, there are natural numbers r such that $\pi^r = 1$; the smallest one of such numbers, say d, is the **order** of the permutation π : $\pi^d = 1$ and all permutations $\pi^t \neq 1$ for $1 \leq t \leq d-1$. In other words, π generates a **cyclic group** $C(d)$ of order d. The notation (VIII.1.1) seems to be convenient if n is small. Thus

$$\begin{pmatrix} 1 & 2 & 3 & 4 \\ 2 & 3 & 1 & 4 \end{pmatrix} \circ \begin{pmatrix} 1 & 2 & 3 & 4 \\ 3 & 4 & 1 & 2 \end{pmatrix} = \begin{pmatrix} 1 & 2 & 3 & 4 \\ 1 & 4 & 2 & 3 \end{pmatrix}. \qquad \text{(VIII.1.2)}$$

Observe that the product of two permutations π_1 and π_2 is defined by

$$(\pi_1 \circ \pi_2)(t) = \pi_1(\pi_2(t)).$$

In general, $\pi_1 \circ \pi_2 \neq \pi_2 \circ \pi_1$, that is, $\mathcal{S}_n$ is for $n \geq 2$ noncommutative. A more convenient way to describe π is to express it as a product of **cycles**.

Definition VIII.1.2. *Let* (VIII.1.1) *be a permutation of* $\mathcal{S}_n$. *A subset* $\{k_1, k_2, \dots, k_d\}$ *of distinct numbers from* I_n *is called a* **cycle** *of* π *if*

$$\pi(k_1) = k_2,\ \pi(k_2) = k_3, \dots,\ \pi(k_t) = k_{t+1}, \dots,\ \pi(k_{d-1}) = k_d,\ \pi(k_d) = k_1.$$

Such a cycle will be denoted by $(k_1\, k_2\, \dots\, k_d)$ *and* d *will be called its* **length**. *Cycles of length two will be called* **transpositions.**

Remark VIII.1.1. The notation $(k_1\, k_2\, \dots\, k_d)$ for a given cycle is "circular" in the sense that it may start at any point (see Figure VIII.1.1):

$$(k_1\, k_2\, \dots\, k_d) = (k_2\, k_3\, \dots\, k_d\, k_1) = \dots = (k_d\, k_1\, \dots\, k_{d-1}).$$

The cycle $(k_1\, k_2\, \ldots\, k_d)$ with $\{k_1, k_2, \ldots, k_d\} \subseteq I_n$ can be considered as a permutation σ of $\mathcal{S}_n$ for which $\sigma(k) = k$ for all $k \in I_n, k \neq k_t, 1 \leq t \leq d$. In this sense we may write every permutation in $\mathcal{S}_n$ as a product of cycles that are disjoint, that is, each $k \in I_n$ appears just once. In fact, we should say "at most once" since we usually do not write down cycles of length 1. For instance the product (VIII.1.2) will be written simply as $(1\,2\,3)\,(1\,3)\,(2\,4) = (2\,4\,3)$.

The expression of a permutation in terms of disjoint cycles is, up to an order of the cycles, **unique**. Thus $1, (12), (13), (23), (123)$ and (132) are the elements of $\mathcal{S}_3$. It is very easy to calculate products of permutations using the cycle presentation. For instance, the product of $\pi_1 = (128)(37)(46)$ and $\pi_2 = (23)(4876)$ equals $\pi_2\pi_1 = (1368)(27)$ while $\pi_1\pi_2 = (1274)(38)$. Importantly, the **inverse** of a cycle $(k_1\, k_2\, \ldots\, k_{d-1}\, k_d)$ is simply the cycle $(k_d\, k_{d-1}\, \ldots\, k_2\, k_1)$. Thus

$$(\pi_1\, \pi_2)^{-1} = [(1\,2\,7\,4)\,(3\,8)]^{-1} = (1\,4\,7\,2)\,(3\,8)$$
$$[= \pi_2{}^{-1}\, \pi_1{}^{-1} = (2\,3)\,(4\,6\,7\,8)\,(1\,8\,2)\,(3\,7)\,(4\,6)]\,.$$

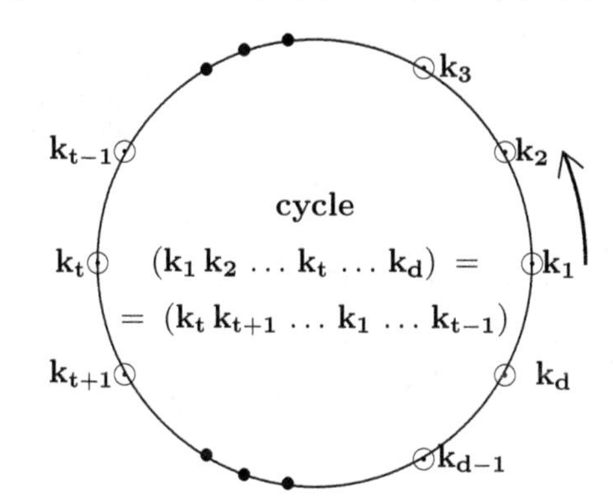

Figure VIII.1.1 Cycle of a permutation

The structure of $\mathcal{S}_n$ can be studied for small n easily by means of the **multiplicative table** of the group. For instance

mult	1	(12)	(13)	(23)	(123)	(132)
1	1	(12)	(13)	(23)	(123)	(132)
(12)	(12)	1	(132)	(123)	(23)	(13)
(13)	(13)	(123)	1	(132)	(12)	(23)
(23)	(23)	(132)	(123)	1	(13)	(12)
(123)	(123)	(13)	(23)	(12)	(132)	1
(132)	(132)	(23)	(12)	(13)	1	(123)

is the multiplicative table for the group $\mathcal{S}_3$. Writing $a = (123)$ and $b = (12)$, we can see that the group $\mathcal{S}_3$ can be generated by a and b satisfying the relations $a^3 = b^2 = (ab)^2 = 1$.

The multiplicative table is sufficient to describe the structure of $\mathcal{S}_3$ in this situation and, in particular, the lattice of its subgroups (see Figure VIII.1.2). Generators and relations are defined formally in Section 3 of this chapter.

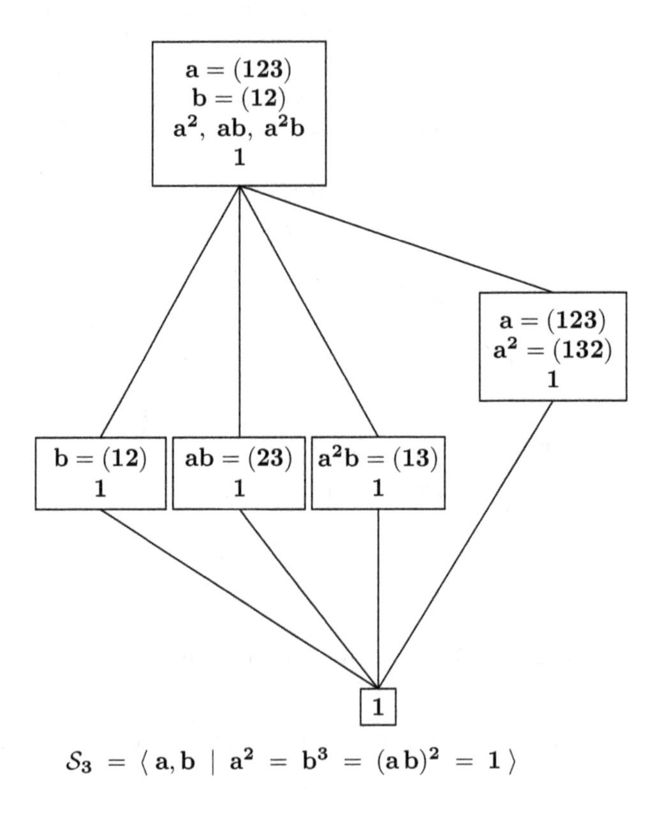

Figure VIII.1.2 Lattice of subgroups of $\mathcal{S}_3$

Exercise VIII.1.2. Write the multiplication table of $\mathcal{S}_4$ in terms of the cycles. Can we write every element of $\mathcal{S}_4$ in terms of $a = (1234)$ and $b = (123)$? Use the relations $a^4 = b^3 = (ab^2)^2 = 1$.

In the last exercise we see that as the size of the order of a group increases, its multiplication table is no longer a suitable tool for describing the structure of the group. Figure VIII.1.3 shows that the subgroup lattices of symmetric groups $\mathcal{S}_n$ become quite complicated with increasing n. In order to describe the structure of groups, an (abstract) theory has to be developed. A few initial steps of such an approach are presented in this chapter.

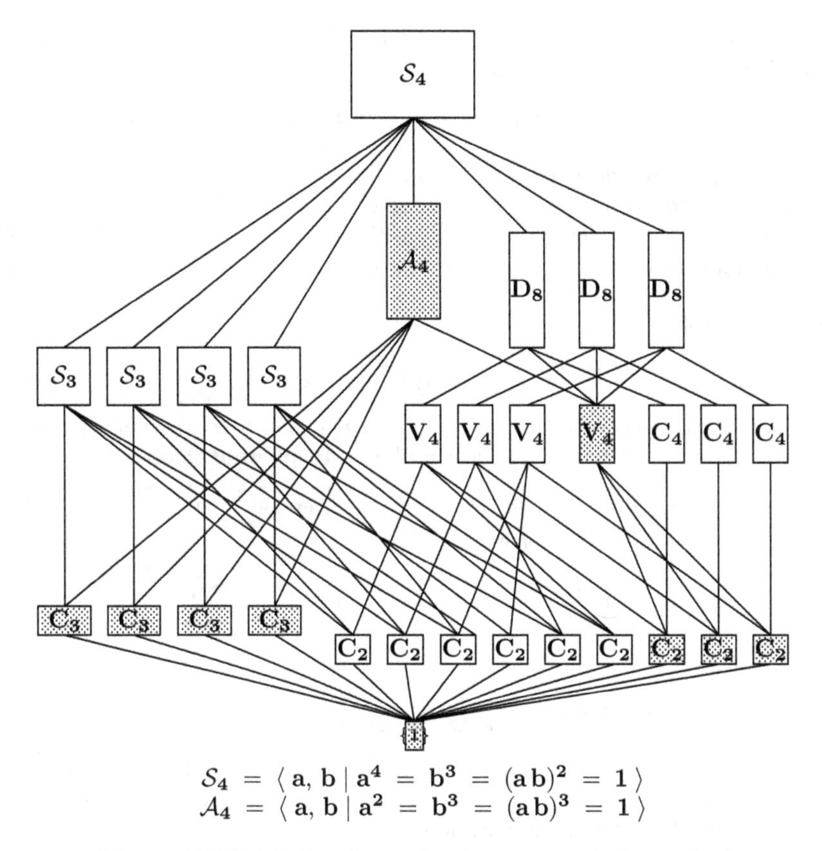

Figure VIII.1.3 Lattice of subgroups of $\mathcal{S}_4$ and $\mathcal{A}_4$

Theorem VIII.1.1. *Every permutation is a product of transpositions. In fact, every permutation is a product of* **adjacent transpositions**, *that is, transpositions of the form* $(k\,k+1)$ *for* $k \in I_n$.

Proof. Every permutation is a product of cycles and every cycle $(k_1\,k_2\,\ldots\,k_d)$ of length d is a product of $d-1$ transpositions:

$$(k_1\,k_2\,\ldots\,k_d) = (k_1\,k_d)\,(k_1\,k_{d-1})\,\ldots\,(k_1\,k_3)\,(k_1\,k_2).$$

Finally, every transposition $(k\,m)$, $k < m$ is a product of $2(m-k)-1$ adjacent transpositions

$$(m\,m-1)\,(m-1\,m-2)\ldots(k+2\,k+1)\,(k+1\,k)\,(k+2\,k+1)\ldots(m-1\,m-2)\,(m\,m-1).$$

This completes the proof. □

Corollary VIII.1.1. *The symmetric group $\mathcal{S}_n$ can be generated by $n-1$ generators*

$$\tau_k = (k\,k+1), 1 \le t \le n-1\,.$$

We can see that the expressions for permutations in terms of cycles or transpositions are not unique. For instance, we already have

$$(1\,2\,3) \;=\; (1\,3)\,(1\,2) \;=\; (2\,3)\,(1\,3) \;=\; (1\,3)\,(2\,3)\,(1\,2)\,(2\,3)\,(1\,3)\,(1\,2) \;=\; \cdots\cdots$$

However, we are going to show that there is something common to **all** decompositions of a given permutation into various products of transpositions, namely the **parity** of the number of factors: The numbers of factors of decompositions of a given permutation are either all even, or all are odd. Now, we have to choose a proper concept to establish this statement.

Definition VIII.1.3. *Let $\pi \in \mathcal{S}_n$. A pair $\{k, m\}$ is said to be an* **inversion** *of the permutation π if*

$$(m-k)\,(\pi(m)-\pi(k)) \;< 0.$$

Denote by $\mathbf{inv}(\pi)$ *the number of all inversions of π and call the permutation π an* **even permutation** *if* $\mathbf{inv}(\pi)$ *is an even number and an* **odd permutation** *if* $\mathbf{inv}(\pi)$ *is odd.*

Remark VIII.1.2. Since for $k < m$, $\{k, m\}$ is an inversion if $\pi(k) > \pi(m)$, it is easy to count the inversions of π in the notation

$$\pi = \begin{pmatrix} 1 & 2 & \dots & k & \dots & m & \dots & n \\ i_1 & i_2 & \dots & i_k & \dots & i_m & \dots & i_n \end{pmatrix}.$$

For example, if

$$\sigma = \begin{pmatrix} 1 & 2 & 3 & 4 & 5 & 6 & 7 & 8 \\ 8 & 4 & 2 & 6 & 1 & 3 & 7 & 5 \end{pmatrix},$$

then the inversions of σ are $\{1, r\}$ $(r = 2, 3, 4, 5, 6, 7, 8)$, $\{2, r\}$ $(r = 3, 5, 6)$, $\{3, 5\}$, $\{4, r\}$ $(r = 5, 6, 8)$ and $\{7, 8\}$. Hence $\mathbf{inv}(\sigma) = 15$ and σ is an odd permutation. Of course, $\mathbf{inv}(\pi) = 0$ if and only if $\pi = 1$ (the identity permutation).

We collect some properties of $\mathbf{inv}(\pi)$.

Theorem VIII.1.2. *Let τ be an adjacent transposition from $\mathcal{S}_n$. Then*

$$|\,\mathbf{inv}(\pi\,\tau) - \mathbf{inv}(\pi)\,| \;=\; 1 \quad \textit{for any } \pi \in \mathcal{S}_n.$$

Thus, a permutation π is even if and only if the permutation $\pi\,\tau$ is odd.

Proof. A proof of this basic fact is surprisingly easy. Indeed, if $\tau = (k\;\; k+1)$, then using the notation (VIII.1.1), the second line of $\pi\,\tau$ differs from the second line of π only in the interchange of the neighboring values of $\pi(k) = i_k$ and $\pi(k+1) = i_{k+1}$. Thus, if $i_k < i_{k+1}$, then $\mathbf{inv}(\pi\,\tau) \;=\; \mathbf{inv}(\pi) + 1$; otherwise $\mathbf{inv}(\pi\,\tau) \;=\; \mathbf{inv}(\pi) - 1$. □

Corollary VIII.1.2. *The number of even permutations (as well as the number of odd permutations) in $\mathcal{S}_n$ is $\dfrac{1}{2}n!$.*

Proof. Let $\{\pi_1, \pi_2, \dots, \pi_s\}$ be the set of **all** even permutations from $\mathcal{S}_n$ and let τ be an adjacent transposition. Then the set $\{\pi_1\,\tau, \pi_2\,\tau, \dots, \pi_s\,\tau\}$ is the set of all (distinct) odd permutations. Hence $2k = n!$ and the statement follows. □

In the proof of Theorem VIII.1.2 we have seen that every transposition is the product of an odd number of adjacent transpositions. Furthermore, in the same proof, we have seen that a cycle of length d is a product of $d-1$ transpositions. We can thus formulate additional corollaries.

Corollary VIII.1.3. *Transpositions are odd permutations. Cycles of odd length are even permutations and cycles of even length are odd permutations. A permutation is even if and only if it is a product of even number of transpositions and thus the set of all even permutations in $\mathcal{S}_n$ $(n > 1)$ forms a subgroup of $\mathcal{S}_n$ of order $\frac{1}{2}n!$, the* **alternating group** *$\mathcal{A}_n$ (see* Figure VIII.1.3).

The last statement can be translated into the language of homomorphisms.

Corollary VIII.1.4. *Let $n \in \mathbb{N}$ satisfy $n \geq 2$. Then there is a group homomorphism*

$$\Pi : \mathcal{S}_n \to (\mathbb{Z}_2, +)$$

of the symmetric group $\mathcal{S}_n$ onto the additive group of the integers modulo 2 defined by

$$\Pi(\pi) = \begin{cases} [0] \text{ for even permutations } \pi \\ [1] \text{ for odd permutations } \pi. \end{cases}$$

The kernel of this epimorphism is the alternating group $\mathcal{A}_n$.

Remark VIII.1.3. The additive group $(\mathbb{Z}_2, +)$ is isomorphic to the multiplicative group of units $\mathbf{U} = \{-1, 1\}$ of $\mathbb{Z}$. We could therefore also define $\Pi : \mathcal{S}_n \to \mathbf{U}$ by putting $\Pi(\pi) = 1$ for even and $\Pi(\pi) = -1$ for odd permutations as it is often done. This homomorphism is often denoted by **sgn**, as the **signum** (sign) of a permutation.

Exercise VIII.1.3. Let a permutation $\pi = \gamma_1 \gamma_2 \ldots \gamma_s$ be (up to order) a unique product of disjoint cycles γ_t, $1 \leq t \leq s$, including all cycles of length one (that is, those cycles $g_t = (k)$, for which $\pi(k) = k$). Show that $\sum_{t=1}^{s} \mathbf{length}(\gamma_t) = n$ and that π is even if and only if the number $n - s$ is even.

Exercise VIII.1.4. Prove: Let $\{\gamma_1, \gamma_2, \ldots, \gamma_r\}$ be a set of **disjoint** cycles. Then

$$\sum_{t=1}^{r} \mathbf{length}(\gamma_t) \equiv \mathbf{inv}(\gamma_1 \gamma_2 \ldots \gamma_r) \pmod 2.$$

Explain why this relation provides an expedient way to decide on the parity of a permutation.

Exercise VIII.1.5. (i) Let γ_1 and γ_2 be two cycles that have just one entry in common. Prove that the product $\gamma_1\gamma_2$ is a cycle. What is its order? Compare the cycles $\gamma_1\gamma_2$ and $\gamma_2\gamma_1$.

(ii) Applying (i) to two transpositions, prove that any even permutation can be written as a product of cycles of length 3.

Exercise VIII.1.6. Determine the subgroups of $\mathcal{S}_5$ consisting of permutations of $I = \{1, 2, 3, 4, 5\}$ that

(i) leave one of the elements of I fixed;

(ii) leave two of the elements of I fixed;

(iii) leave a subset of two elements of I fixed.

Generalize these results.

VIII.2 Dihedral groups $\mathcal{D}_{2n}$

Dihedral groups describe the symmetries of regular n-gons. We start with a square.

Example VIII.2.1 (Dihedral group $\mathcal{D}_8$). Consider a square "A B C D" and its symmetries. Let α and β be two **transformations** mapping the square $ABCD$ to the square $DABC$, and to square $CBAD$, respectively. They describe the following two symmetries of the square $ABCD$: α describes the "**rotation**" of the square about its center by $\frac{\pi}{2}$ in the counterclockwise direction and β is the "**reflection**" of the square about the diagonal from upper left to lower right (see the following illustration).

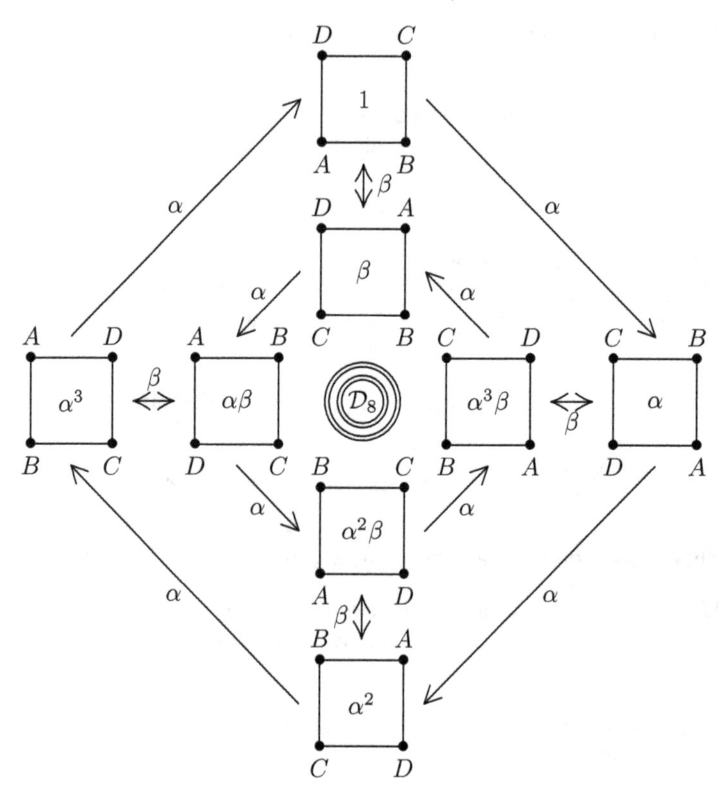

Now, we can compose the transformations to get (in the outside cycle) the transformations (and the respective symmetries) α, α^2 and α^3, and (in the inner cycle) the transformations (and the respective symmetries) $\beta, \alpha\beta, \alpha^2\beta$ and $\alpha^3\beta$. It is easy to see that these 8 symmetries (including the identity transformation) form the **group of (all) symmetries of the square** $ABCD$, called the **dihedral group** $\mathcal{D}_8$.

It is also easy to see that there is an isomorphism of the dihedral group $\mathcal{D}_8$ and the image of the homomorphism $\varphi : \mathcal{D}_8 \to \mathcal{S}_4$ given by

$$\alpha \mapsto \begin{pmatrix} 1 & 2 & 3 & 4 \\ 2 & 3 & 4 & 1 \end{pmatrix} = (1\,2\,3\,4) \quad \text{and} \quad \beta \mapsto \begin{pmatrix} 1 & 2 & 3 & 4 \\ 3 & 2 & 1 & 4 \end{pmatrix} = (1\,3)\,.$$

Exercise VIII.2.1. Describe, in a similar manner, the group of all symmetries of the regular n−gon (that is, the **dihedral group**) $\mathcal{D}_{2n}$ of $2n$ elements and its embedding into the symmetric group $\mathcal{S}_n$. Distinguish the case when n is even and when n is odd.

We can see easily that $\mathcal{S}_3 \simeq \mathcal{D}_6$. In fact the structure of $\mathcal{D}_{2n}$ is quite simple in the case that n is a prime.

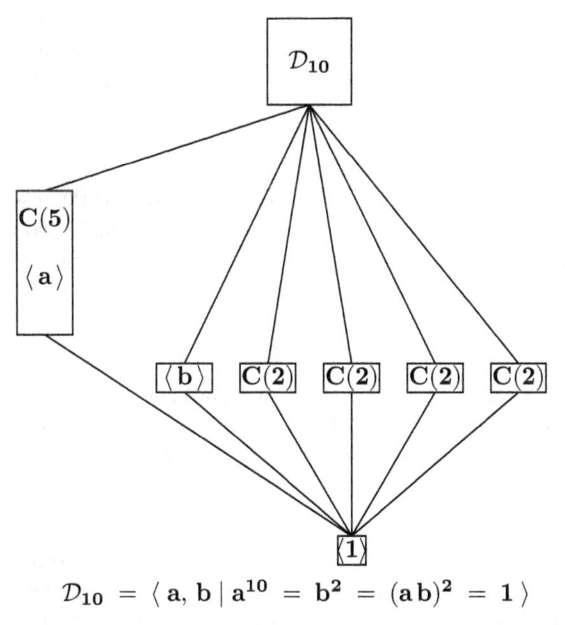

$$\mathcal{D}_{10} = \langle\, \mathbf{a}, \mathbf{b} \mid \mathbf{a}^{10} = \mathbf{b}^2 = (\mathbf{a}\,\mathbf{b})^2 = 1 \,\rangle$$

Figure VIII.2.1 Lattice of subgroups of $\mathcal{D}_{10}$

Figure VIII.2.1 provides a hint for Exercise VIII.2.2 as well as for Exercise VIII.2.3. It will be fully explained later in this chapter.

Exercise VIII.2.2. Prove that $\mathcal{D}_{2p}$, where p is a prime, has $p+1$ nontrivial subgroups and that all are cyclic groups. One of them is a cyclic group $C(p)$ of order p and all others are cyclic groups $C(2)$ of order 2.

Exercise VIII.2.3. Prove that the group $\mathcal{D}_{2n}$ can be generated by two elements a, b satisfying the conditions $a^n = b^2 = (a\,b)^2 = 1$. The information $\mathcal{D}_{2n} = \langle\ a,\ b \mid a^n = b^2 = (a\,b)^2 = 1\ \rangle$ is called a **presentation** of the group $\mathcal{D}_{2n}$. The dihedral groups $\mathcal{D}_2$ and $\mathcal{D}_4$ are abelian: $\mathcal{D}_4$ is isomorphic to the four-group V_4 (see Example III.9.1) and $\mathcal{D}_2$ to the cyclic group $C(2)$.

The concept of presentation of a group will be fully explained in connection with the construction of a quotient group (see Theorem VIII.3.8). We just say here that the notation that has been used in the previous exercise (and earlier in describing some of the Figures) carries the following meaning: All elements of $\mathcal{D}_{2n}$ are products of powers of a and b subject to the listed identities. We note the fact that such expressions are by no means unique. This however does not prevent us from selecting in some groups a unique representation of individual elements. For instance, the elements of the group $\mathcal{D}_{2n}$ may be described as follows

$$\mathcal{D}_{2n} = \{1, a, a^2, \ldots, a^t, \ldots, a^{n-1}, b, a\,b, a^2\,b, \ldots, a^t\,b, \ldots, a^{n-1}\,b\}.$$

The dihedral groups $\mathcal{D}_{2^n} = \langle a, b \mid a^{2^{n-1}} = b^2 = (ab)^2 = 1\rangle$ contain smaller dihedral groups as subgroups, something we have already seen with the symmetric group. This is well illustrated in Figure VIII.2.2 and may also help in solving the next exercises (where we assume that $n \geq 3$).

Exercise VIII.2.4. Prove that $\mathcal{D}_{2^n}$ has three maximal subgroups (that is, proper subgroups such that the only subgroup that contains each of them is the whole group). One of them is isomorphic to the cyclic group $C = C(2^{n-1})$ of order 2^{n-1} and the other two are isomorphic to the dihedral group $\mathcal{D}_{2^{n-1}}$. Show that the intersection of these two dihedral subgroups is a cyclic group of order 2^{n-2} that is a subgroup of the group C.

Exercise VIII.2.5. Prove that $\mathcal{D}_{2^n}$ has $n-1$ (nontrivial) cyclic subgroups that form a chain of subgroups of the group C of Exercise VIII.2.2. Describe their generators and prove that the element of order 2 in C is the only element of the group $\mathcal{D}_{2^n}$ that commutes with all elements.

Exercise VIII.2.6. Prove that beside the $n-1$ cyclic subgroups of Exercise VIII.2.5 all other proper subgroups are dihedral groups of the form $\mathcal{D}_{2^t}, 1 \leq t \leq n-1$ and that the number of subgroups isomorphic to $\mathcal{D}_{2^t}$ is 2^{n-t}. Thus show that the number of all subgroups, including the trivial group $\langle 1\rangle$ and the whole group $\mathcal{D}_{2^n}$ is $2^n + n - 1$ (see Figure VIII.2.2).

Exercise VIII.2.7. Show that for every $t \in \{1, 2, \ldots, n-1\}$, the 2^{n-t} subgroups

$$\langle\, a^{2^{n-t}},\ a^s\,b\rangle,\ \text{ where } s = 0, 1, \ldots, 2^{n-t} - 1,$$

are precisely all the subgroups of $\mathcal{D}_{2^n}$ that are isomorphic to $\mathcal{D}_{2^t}$.

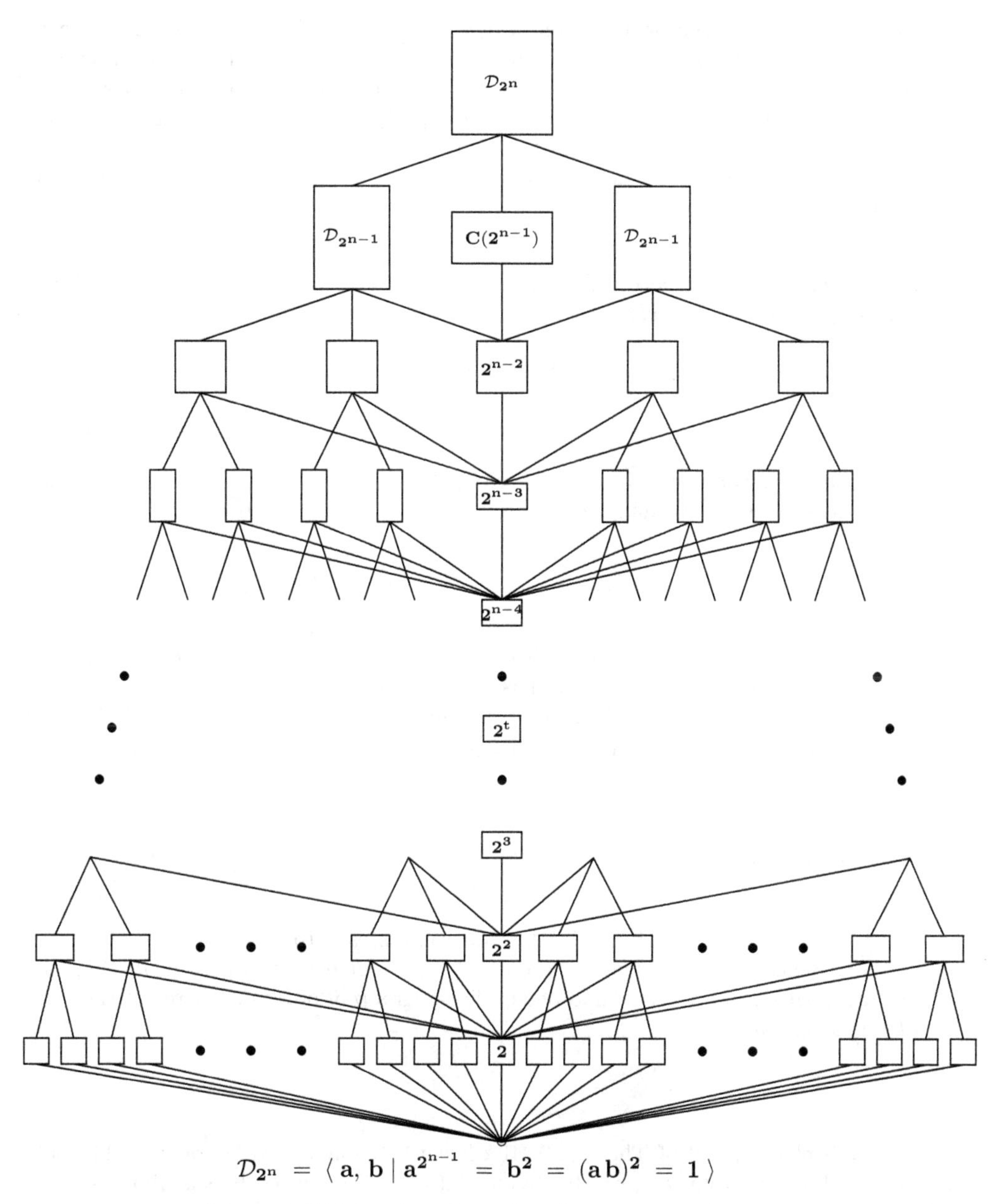

$$\mathcal{D}_{2^n} = \langle\, a, b \mid a^{2^{n-1}} = b^2 = (ab)^2 = 1 \,\rangle$$

Figure VIII.2.2 Lattice of subgroups of $\mathcal{D}_{2^n}$

Exercise VIII.3.12 of the next section suggests that not only subgroups, but also all homomorphic images of $\mathcal{D}_{2^n}$ are dihedral groups of type $\mathcal{D}_{2^t}$. In fact, the previous series of exercises and Figure VIII.2.2 suggest that there may be conclusions concerning the lattice of subgroups that could hold for every dihedral group $\mathcal{D}_{2n}$, see Exercise VIII.3.12. Indeed, one can see in Figure VIII.2.2 that a particular type of subgroup is always attached to a particular cyclic subgroup that is defined by a divisor of the number n. This provides a lead to Theorem VIII.2.1.

Theorem VIII.2.1. *The number of subgroups (including the trivial ones) of the dihedral group $\mathcal{D}_{2n}$ is given by the formula*

$$\sum_{d\,|\,n}(d+1).$$

Proof. The proof is surprisingly simple. The group $\mathcal{D}_{2n}$ is generated by two elements a and b that satisfy the following (defining) relations: $a^n = 1, b^2 = 1$ and $(a\,b)^2 = 1$ and thus

$$\mathcal{D}_{2n} = \{\,1,\, a,\, a^2, \dots,\, a^r, \dots,\, a^{n-1},\, b,\, a\,b,\, a^2\,b, \dots,\, a^r\,b, \dots,\, a^{n-1}\,b\}.$$

Now, $\mathcal{D}_{2n}$ contains the cyclic subgroup $C(n) = \langle a\rangle = \{1,\, a,\, a^2, \dots,\, a^r, \dots,\, a^{n-1}\}$ of order n. For every divisor d of n, $n = d\,e$, C has a unique (cyclic) subgroup $C(d) = \langle a^e\rangle = \{1,\, a^e,\, a^{2e}, \dots,\, a^{re}, \dots,\, a^{(n-1)e}\}$. We describe **all** subgroups G of $\mathcal{D}_{2n}$ satisfying $G \cap C(n) = C(d)$. One can see that these are just e subgroups $G = \mathcal{D}_{2d}^{(t)}, 0 \le t \le e-1$, containing the following elements

$$\mathcal{D}_{2d}^{(t)} = \{1,\, a^e,\, a^{2\,e}, \dots,\, a^{r\,e}, \dots,\, a^{(d-1)\,e},\, a^t\,b,\, a^{t+e}\,b,\, a^{t+2\,e}\,b, \dots,\, a^{t+r\,e}\,b, \dots,\, a^{t+(d-1)\,e}\,b\}.$$

The key product $a^{t+r\,e}\,b$ with $a^{s\,e}$ in this subgroup is easy to evaluate: $a^{t+r\,e}\,b\;a^{s\,e} = a^{t+(\,d+r-s)\,e}\,b$. Hence, we can list all subgroups of $\mathcal{D}_{2n}$: For each divisor d of n, there are n/d dihedral subgroups of order $2d$ and one cyclic subgroup of order d. Therefore there are $\sum_{d\,|\,n}(d+1)$ subgroups of $\mathcal{D}_{2n}$. □

An analysis of the proof of Theorem VIII.2.1 can be summarized as follows. The subgroups of the dihedral group $\mathcal{D}_{2n}$ can be divided into families labeled by the divisors of the number n. For each divisor d of n, the respective family of subgroups contain one cyclic group of order d and n/d subgroups $\mathcal{D}_{2\,d}$. If n is even, the center of $\mathcal{D}_{2n}$ is the cyclic group of order 2; otherwise the center of the group $\mathcal{D}_{2n}$ is trivial (see Exercise VIII.3.2).

This process is illustrated in Figure VIII.2.3 dealing with $\mathcal{D}_{20}$. The subgroups of $C(10)$ are shaded, of distinct shape. The groups attached to $C(d)$ have the same shape as the subgroup $C(d)$.

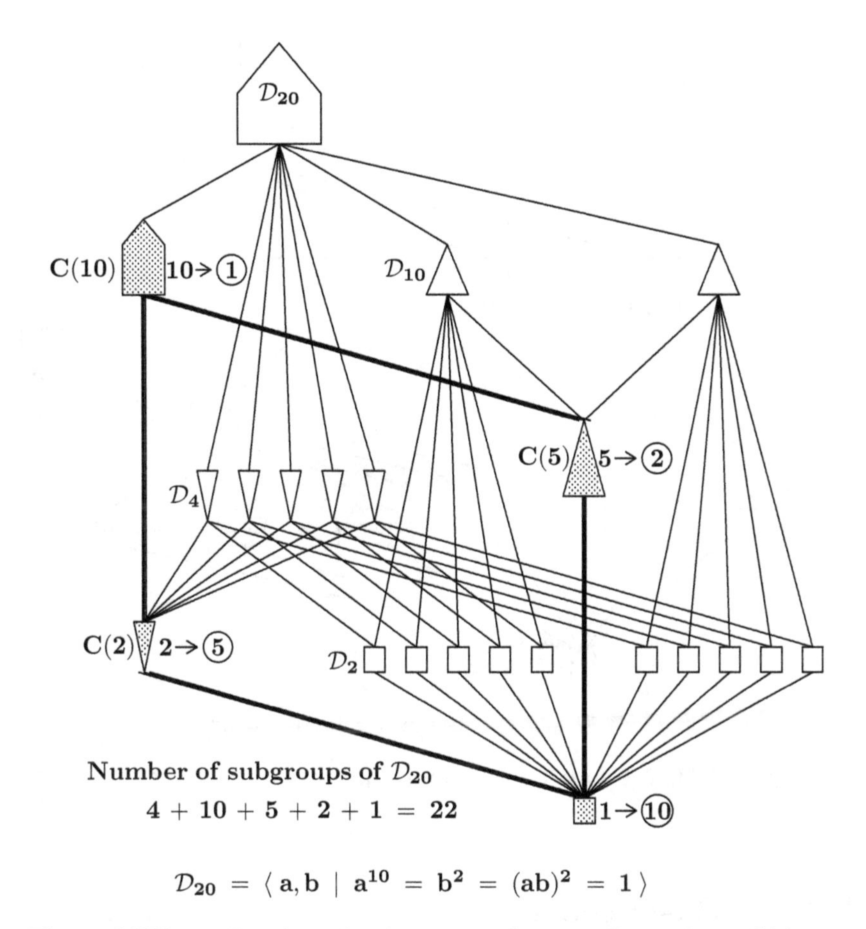

$$\mathcal{D}_{20} = \langle\, \mathbf{a}, \mathbf{b} \mid \mathbf{a}^{10} = \mathbf{b}^2 = (\mathbf{ab})^2 = 1 \,\rangle$$

Figure VIII.2.3 Lattice of subgroups of group $\mathcal{D}_{20} \simeq \mathcal{D}_{10} \times C(2)$

Exercise VIII.2.8. Describe all elements of the individual subgroups of $\mathcal{D}_{12}$ as illustrated in Figure VIII.2.4.

Exercise VIII.2.9. Draw a diagram of the lattice of all subgroups of $\mathcal{D}_{30}$ and describe their elements.

Remark VIII.2.1. We reiterate the geometrical interpretation of the elements of the dihedral group $\mathcal{D}_{2n} = \langle\, a, b \mid a^n = b^2 = (ab)^2 = 1 \,\rangle$ as the symmetries of a regular n-sided polygon P: The elements of the cyclic subgroup $C(n) = \langle a \rangle$ represent the rotations of P and the remaining elements $\{a^t b \mid 0 \le t \le n-1\}$ represent the reflections defined by the lines of reflection passing through vertices or midpoints of edges (depending on the parity of n) and the center of the polygon.

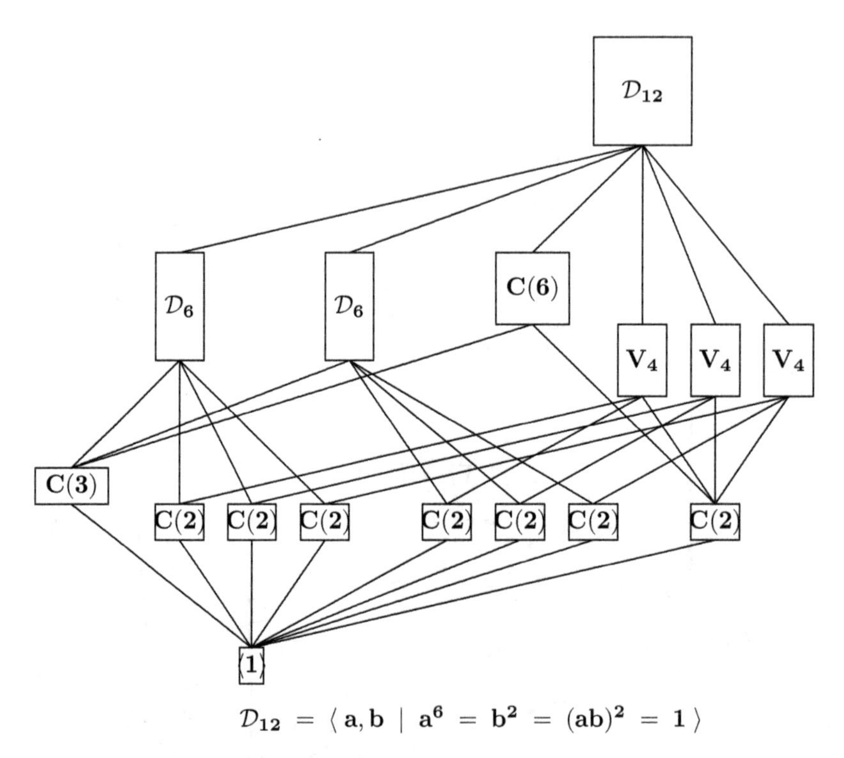

Figure VIII.2.4 Lattice of subgroups of group $\mathcal{D}_{12} \simeq \mathcal{D}_6 \times \mathbf{C}(2)$

VIII.3 Basic theory

In this section, we shall deal with the category **Gr** of groups (see Definition X.6.1). We have already met and described a number of groups, in some cases even the lattices of their subgroups. The category **Gr** consists of objects, the groups and morphisms (the group homomorphisms). Recall that a mapping $\varphi : G \to H$ from a group $G = (G, \cdot)$ to a group $H = (H, \circ)$ is a homomorphism if $\varphi(g_1 \cdot g_2) = \varphi(g_1) \circ \varphi(g_2)$ for all elements $g_1,\ g_2 \in G$. We have carefully denoted the operations in G by $\cdot$ and in H by $\circ$; further in the text, we will usually omit such notations and denote the result of the operation simply by $g_1\, g_2$, or, in the case of additive notation, by $g_1 + g_2$. Additive notation will mainly be used for commutative (abelian) groups. The abelian groups, together with their homomorphisms, form a subcategory **Ab** of **Gr**. Special homomorphisms carry special names: a bijective group homomorphism is called a group isomorphism, a homomorphism $G \to G$ is an **endomorphism** of G and a bijective endomorphism of G is an **automorphism** of G.

Exercise VIII.3.1. Prove: A monoid M is a group if and only if every equation $a\,x = b$ with $a, b \in M$ has a unique solution in M.

Exercise VIII.3.2. Show that the set $Z(G)$ of all commuting elements of a group G, that is, the set

$$Z(G) = \{a \in G \,|\, a\,g = g\,a \text{ for all elements } g \in G\}$$

is a subgroup of G (called the **center** of G). A group G is abelian if and only if $Z(G) = G$. Prove that $Z(\mathcal{S}_n) = \{1\}$.

Recall that the **order** of a finite group G is the number of elements of G and will be denoted by $|G|$, and that the order of an element $g \in G$ is the order of the cyclic subgroup of G generated by g and will be denoted by $O(g)$. Groups of small order can be conveniently described by their **multiplication tables**, which are often called **Cayley tables**. For instance, the group of all symmetries of the square, the dihedral group $\mathcal{D}_8 = \langle a, b \mid a^4 = b^2 = (ab)^2 = e\rangle$ has the following multiplication table of $\mathcal{D}_8$ (from which we can see that the center of $\mathcal{D}_8$ is the cyclic group $Z(\mathcal{D}_{2n}) = \{e, a^2\}$).

mult	e	a	a^2	a^3	b	a^3b	a^2b	ab
e	e	a	a^2	a^3	b	a^3b	a^2b	ab
a	a	a^2	a^3	e	ab	b	a^3b	a^2b
a^2	a^2	a^3	e	a	a^2b	ab	b	a^3b
a^3	a^3	e	a	a^2	a^3b	a^2b	ab	b
b	b	a^3b	a^2b	ab	e	a	a^2	a^3
a^3b	a^3b	a^2b	ab	b	a^3	e	a	a^2
a^2b	a^2b	ab	b	a^3b	a^2	a^3	e	a
ab	ab	b	a^3b	a^2b	a	a^2	a^3	e

Note that in each row and each column of the table each element of G appears exactly once; this is, of course, just a reflection of the statement of Exercise VIII.3.1.

Exercise VIII.3.3. Write down the multiplication table of the group of all symmetries of a hexagon, that is, of the dihedral group $\mathcal{D}_{12}$.

Exercise VIII.3.4. The following table is an incomplete multiplication table of a group of order 6. Complete the table.

mult	a	b	c	d	e	f
a	c	$*$	$*$	e	$*$	$*$
b	$*$	e	$*$	$*$	c	$*$
c	a	$*$	$*$	$*$	$*$	$*$
d	$*$	$*$	$*$	c	$*$	$*$
e	$*$	$*$	$*$	a	$*$	$*$
f	$*$	d	$*$	$*$	$*$	c

Example VIII.3.1. Important examples of groups are the **general linear groups** $GL_n(F)$ of all invertible $n \times n$ matrices (those with a nonzero determinant) with entries from a field F. The group operation is matrix multiplication. For all $n \geq 2$, the groups $GL_n(F)$ are noncommutative. All matrices in $GL_n(F)$ with determinant 1 form a subgroup, the **special linear group** $SL_n(F)$.

Exercise VIII.3.5. Prove that $GL_2(GF_2) \simeq \mathcal{S}_3 \simeq D_6$, where GF_2 is the field of two elements.

Exercise VIII.3.6. Let $\alpha : G \to H$ be a group homomorphism. Show that

(i) the image of the identity element of G is the identity element of H and that $\alpha(g^z) = \alpha(g)^z$ for all integers z;

(ii) for every $g \in G$ the inverse element of the image $\alpha(g)$ satisfies $(\alpha(g))^{-1} = \alpha(g^{-1})$.

(iii) the image of the homomorphism α, $\text{Im}\,\alpha = \alpha(G) = \{\alpha(g) \,|\, g \in G\}$ is a subgroup of the group H.

Exercise VIII.3.7. Show that, for every element a of a group G, the mappings $\rho_a : G \to G$ and $\lambda_a : G \to G$ defined for $g \in G$ by $\rho_a(g) = g\,a$ and $\lambda_a(g) = a\,g$ are bijective mappings of G. Can ρ_a or λ_a be an automorphism of G?

Many original studies in group theory were formulated in terms of **permutation groups** instead of "abstract groups", that is, the elements of groups were certain permutations. A possibility for such a treatment is formulated in the following, purely theoretical theorem. The justification for its statement lies in the preceding Exercise VIII.3.7.

Theorem VIII.3.1 (Cayley's theorem). *Every group G is isomorphic to a subgroup of the group of all permutations of the set G (that is, every group is a permutation group).*

Proof. Define the mapping $f : G \to \mathcal{S}(G)$, where $\mathcal{S}(G)$ denotes the group of all permutations of the set G, as follows: For $a \in G$,

$$f(a) = \lambda_a \in \mathcal{S}(G),$$

where $\lambda_a(g) = a\,g$ for all $g \in G$. It is straightforward to show that f is an embedding of G into $\mathcal{S}(G)$ satisfying $\lambda_{a\,b} = \lambda_a \circ \lambda_b$. □

Example VIII.3.2. As an illustration,

$$\begin{array}{ccc} 1 & \mapsto & 1 \\ a & \mapsto & (1234)(5876) \\ a^2 & \mapsto & (13)(24)(57)(68) \\ a^3 & \mapsto & (1432)(5678) \\ b & \mapsto & (15)(26)(37)(48) \\ a^3b & \mapsto & (16)(27)(38)(45) \\ a^2b & \mapsto & (17)(28)(35)(46) \\ ab & \mapsto & (18)(25)(36)(47) \end{array}$$

is an embedding of the dihedral group $\mathcal{D}_8$ into $\mathcal{S}_8$ described in Cayley's theorem. We hasten to add that, using a description of the group of symmetries of the square as a group of permutations on 4 letters, we can easily see that there are embeddings of $\mathcal{D}_8$ into $\mathcal{S}_4$ given by

$$\begin{array}{cclll} 1 & \mapsto & 1 & 1 & 1 \\ a & \mapsto & (1234) & (1423) & (1342) \\ a^2 & \mapsto & (13)(24) & (12)(34) & (14)(23) \\ a^3 & \mapsto & (1432) & (1324) & (1243) \\ b & \mapsto & (13) & (12) & (14) \\ a^3b & \mapsto & (14)(23) & (13)(24) & (12)(34) \\ a^2b & \mapsto & (24) & (34) & (23) \\ ab & \mapsto & (12)(34) & (14)(23) & (13)(24) \end{array}$$

Exercise VIII.3.8. Check the calculations in Example VIII.3.2.

In the last section of this chapter, we shall see the importance of the concept of an automorphism of a group (in context of a semidirect product).

Recall that a homomorphism $\omega : G \to G$ is an **endomorphism** of a group G if $\omega(a\,b) = \omega(a)\,\omega(b)$ for all elements $a, b \in G$. An endomorphism ω is an **automorphism** of G if ω is a bijective mapping. This implies that there is an inverse mapping $\omega^{-1} : G \to G$ and one can show that it is also an automorphism. Thus, the set $\mathbf{Aut}(G)$ of all automorphisms of a group G forms with respect to composition of mappings a group: the **automorphism group** $\mathbf{Aut}(G)$ of G.

Example VIII.3.3. We are going to show that $\mathbf{Aut}(V_4)$, where $V_4 = \{1, a, b, ab\}$ denotes the (abelian) Klein four-group (with multiplication $a^2 = b^2 = 1$), is isomorphic to $\mathcal{S}_3$. Indeed, every automorphism ω of V_4 is determined by the image $\omega(a)$ of a and the image $\omega(b)$ of b. Consequently, the number of all automorphisms of V_4 is 6. Now, we can easily check that the automorphism α defined by $\alpha(a) = a, \alpha(b) = a\,b$ and the automorphism β defined by $\beta(a) = a\,b, \beta(b) = b$ satisfy $\alpha^2 = \beta^2 = (\alpha\beta)^3 = 1_{V_4}$ the identity automorphism of V_4. Thus $\mathbf{Aut}(V_4) \simeq \mathcal{S}_3$.

We now determine the order of the group $\mathbf{Aut}(\mathcal{D}_{2n})$ for all $n \geq 3$. For, every automorphism ω of $\mathcal{D}_{2n} = \langle a, b \mid a^n = b^2 = (a\,b)^2 = 1\rangle$ is determined by the images $\omega(a)$ and $\omega(b)$. Since $\omega(a)$ can be any generator of the cyclic group $C(n) = \langle a \rangle$ the number of possibilities for $\omega(a)$ is $\varphi(n)$, where φ is Euler's totient function. Furthermore, the number of all possible images $\omega(b)$ is n, since any reflection can be an image of b. Thus

$$|\,\mathbf{Aut}(\mathcal{D}_{2n})\,| \;=\; n\,\varphi(n).$$

Exercise VIII.3.9. Determine the automorphism group $\mathbf{Aut}(C(n))$ of the cyclic group of order n.

Exercise VIII.3.10. Prove that the automorphism group $\mathbf{Aut}(\mathcal{D}_8) \simeq \mathcal{D}_8$.

Exercise VIII.3.11. Let G be a group and, for every $a \in G$, let $\omega_a : G \to G$ be the mapping defined by $\omega_a(g) = a^{-1}\,g\,a$. We say that $a^{-1}\,g\,a$ is the **conjugate** of g by a. Prove that ω_a is an automorphism and that the set $\mathbf{Inn}(G)$ of all these **inner automorphisms** of G is a subgroup of $\mathbf{Aut}(G)$.

The relation "is conjugate to" in a group G

$$g \;\mathrm{conj}_G\; h \quad \text{if and only if} \quad h = a^{-1}\,g\,a \text{ for some } a \in G$$

is an equivalence relation. Thus the group G is partitioned into (disjoint) **conjugacy classes**. The identity element alone forms one of the conjugacy classes. An element alone forms a conjugacy class if and only if it belongs to the center of the group. All conjugacy classes of a group G consist of a single element if and only if G is abelian.

To determine all conjugacy classes in permutation groups is straightforward. This is due to the fact that to obtain the permutation $\sigma = \pi^{-1} \circ \rho \circ \pi$ for given permutations ρ and π we use the following simple relation

$$\sigma(\pi^{-1}(i)) = (\pi^{-1} \circ \rho \circ \pi)(\pi^{-1}(i)) = (\pi^{-1} \circ \rho)(i) = \pi^{-1}(\rho(i)). \tag{VIII.3.1}$$

Thus, writing the permutation ρ as a product of (disjoint) cycles, we get σ by replacing all entries i in the cycles by $\pi^{-1}(i)$. We can see, in particular, that conjugate permutations are of the same "cycle type", that is, the number of the cycles and their lengths are preserved. Hence, for $\rho = (1\,2\,3\,4\,5)$ and $\pi = (1\,5\,3\,2)$, $\pi^{-1} = (1\,2\,3\,5)$ and $\pi^{-1}\,\rho\,\pi = (2\,3\,5\,4\,1)$.

Example VIII.3.4. Use the relation (VIII.3.1) to find all conjugacy classes of the alternating group $\mathcal{A}_4$. Recall that $\mathcal{A}_4$ is the subgroup of the symmetric group $\mathcal{S}_4$ consisting of all even permutations. Thus, $\mathcal{A}_4$ contains the identity element, eight cycles of length 3, namely

$$(1\,2\,3), (1\,2\,4), (1\,3\,4), (2\,3\,4), (1\,3\,2), (1\,4\,2), (1\,4\,3), (2\,4\,3),$$

and three products of two cycles of length 2

$$(1\,2)\,(3\,4), (1\,3)\,(2\,4), (1\,4)\,(2\,3).$$

A short calculation shows that $\mathcal{A}_4$ has **four** conjugacy classes:

$$\{1\},\; \{(1\,2\,3), (1\,2\,4), (1\,3\,4), (2\,3\,4)\}, \{(1\,3\,2), (1\,4\,2), (1\,4\,3), (2\,4\,3)\}$$

and

$$\{(1\,2)\,(3\,4), (1\,3)\,(2\,4), (1\,4)\,(2\,3)\},$$

containing $1, 4, 4, 3$ elements, respectively. In fact, one can also conclude that the alternating group has, in addition to the trivial subgroups $\{1\}$ and $\mathcal{A}_4$, three cyclic subgroups of order 2, four cyclic subgroups of order 3 and the Klein four-group V_4 that we have already met earlier. Draw a diagram of the lattice of these subgroups.

Example VIII.3.5. Here are some further examples and exercises:

(i) The map $x \mapsto e^x$ of the additive group of real numbers onto the multiplicative group of positive real numbers is an isomorphism.

(ii) (a) Denote by $G_\circ$ the set of all rational numbers $\neq 1$ with operation "$\circ$" defined as follows: $a \circ b = a + b - ab$. Then, $G_\circ$ is a group isomorphic to the multiplicative group $\mathbb{Q}^*$ of all non-zero rational numbers. Indeed, define the mapping $\phi : G_\circ \to \mathbb{Q}^*$ by $\phi(a) = 1 - a$ and show that it is an isomorphism. Thus, to find for example $(-1.5) \circ 7$, multiply $2.5 \times (-6) = -15$ and conclude that $(-1.5) \circ 7 = \phi^{-1}(-15) = 16$.

(b) Describe, in a similar way, the structure of the group $G_\star$ of all rational numbers $\neq -1$ with operation $\star$ defined by

$$a \star b = a + b + ab.$$

(c) Show that the previous two examples are special cases of the following isomorphism: Let $r \neq 0$ be a fixed rational number and $G_\sharp$ the group of all rational numbers $\neq r$ with the operation $\sharp$ defined as follows:

$$a \sharp b = a + b - \frac{ab}{r}.$$

Then $G_\sharp$ is isomorphic to $\mathbb{Q}^*$.

(iii) Let $(G, +)$ be an (additive) abelian group. Let $g \in G$ be an arbitrary element. Consider the group G_g that has the same elements as G (that is, as sets $G_g = G$) and whose group operation $*$ is given by $a * b = a + b - g$. Show that the neutral element of G_g is g, the opposite element to a is $-a + g + g$ and the groups G and G_g are isomorphic. Illustrate this phenomenon on $G = \mathbb{Z}$, the additive group of integers.

(iv) The map $z = e^{i\alpha} \mapsto \alpha$ of the multiplicative group of all complex numbers z such that $|z| = 1$ onto the additive group of real numbers modulo 2π is an isomorphism.

Groups are often given, as we have already seen in the case of dihedral groups, by **generators and relations**, that is, by **presentations.** This concept is intimately related to the concept of a **normal subgroup** of a group which, in turn, is related to the concept of a homomorphism of that group. All this is clarified in the following definition and Theorems VIII.3.2 and VIII.3.3.

Definition VIII.3.1. *A subgroup H of a group G is said to be a* **normal** *subgroup of G if every inner automorphism of G maps H onto itself, that is, if, for every $g \in G$ and $h \in H$, $g^{-1} h g \in H$, briefly $g^{-1} H g = H$.*

Exercise VIII.3.12. Determine all normal subgroups of the dihedral group $\mathcal{D}_{2n}$. Show that the intersection and the join of two normal subgroups (that is, the subgroup generated by their set theoretical union) is a normal subgroup. Show that a subgroup H of a group G satisfying $|G| = 2\,|H|$ is normal in G. Draw the lattice of all normal subgroups of $\mathcal{D}_{2n}$. We denote by $d(n)$ the number of all divisors of a natural number n and show that the number of normal subgroups of the group $\mathcal{D}_{2n}$ is $d(n) + 1$ for n is odd and $d(n) + 3$ for n even.

The definition of a normal subgroup H of a group G can be expressed in term of conjugacy classes: A subgroup H of G is normal in G if and only if it is a union of certain conjugacy classes of the group G. For instance, referring to Example VIII.3.3, the subgroup V_4 is the only nontrivial normal subgroup of $\mathcal{A}_4$. This means that $\mathcal{A}_4$ is not simple in the following sense.

Definition VIII.3.2. *A group G is said to be* **simple** *if it does not have any nontrivial normal subgroups.*

Simple groups play a fundamental role in group theory: Every group can be "constructed" from these building blocks. It is easy to describe all abelian simple groups.

Exercise VIII.3.13. Prove that the abelian simple groups are just the cyclic groups of prime order.

In the next chapter, we will indicate the importance of the fact that the alternating group $\mathcal{A}_5$ is simple. We shall see that the existence of algebraic equations of the fifth order that cannot be solved by radicals (by formulas involving coefficients of the equations) is a consequence of this fact. In order to prepare a simple proof of the statement that $\mathcal{A}_5$ is a simple group (Theorem VIII.3.4), return to the considerations we have followed in Example VIII.3.4.

Example VIII.3.6. Consider the group $\mathcal{A}_5$. The order of $\mathcal{A}_5$ is 60. The elements, that is, the even permutation of five symbols are either cycles of length 5 or cycles of length 3 or products of two disjoint transpositions. Thus, there are $\frac{5!}{5} = 24$ cycles of length 5, $\frac{3!}{3}\binom{5}{3} = 20$ cycles of length 3, $\frac{1}{2}\binom{5}{2}\binom{3}{2} = 15$ products of cycles of length 2 and the identity element in $\mathcal{A}_5$.

To determine the conjugacy class of the permutation $(1\ 2\ 3\ 4\ 5)$ in $\mathcal{A}_5$, we perform on it, in accordance with (VIII.3.1), all permutations induced by elements of $\mathcal{A}_5$. We find this way that the conjugacy class has 12 elements:

$$(1\ 2\ 3\ 4\ 5),\ (1\ 2\ 4\ 5\ 3),\ (1\ 2\ 5\ 3\ 4),\ (1\ 3\ 2\ 5\ 4),\ (1\ 3\ 4\ 2\ 5),\ (1\ 3\ 5\ 4\ 2),$$
$$(1\ 4\ 2\ 3\ 5),\ (1\ 4\ 3\ 5\ 2),\ (1\ 4\ 5\ 2\ 3),\ (1\ 5\ 2\ 4\ 3),\ (1\ 5\ 3\ 2\ 4),\ (1\ 5\ 4\ 3\ 2)\,.$$

In a similar way, we can see that there is a conjugacy class formed by the remaining cycles of length 5, that is, the conjugacy class of $(1\ 2\ 3\ 5\ 3)$ having also 12 elements. All 20 cycles of length 3 form a conjugacy class, as well as all 15 products of transpositions do. Thus, in total there are five conjugacy classes of $1, 12, 12, 20$ and 15 elements, respectively.

Trivially, all subgroups of abelian groups are normal. This however does not characterize abelian groups. The **quaternion group** $\mathcal{Q}_8$ usually defined as the multiplicative group of quaternions

$$\{1, -1, i, -i, j, -j, k, -k\}$$

defined in Chapter V, has the property that it is noncommutative and all its subgroups are normal.

Exercise VIII.3.14. Describe all elements of the individual subgroups of $\mathcal{Q}_8$ as illustrated in Figure VIII.3.1 in terms of the generators $\{a, b, c\}$. Complete the following multiplication table of $\mathcal{Q}_8$ and conclude that all subgroups of $\mathcal{Q}_8$ are normal.

1	a	a^2	a^3	b	b^3	c	c^3
a	·	·	·	c	·	b^3	·
a^2	·	·	a	·	b	·	c
a^3	·	a	·	c^3	·	b	·
b	c^3	·	c	·	1	·	a^3
b^3	·	b	·	1	·	a^3	·
c	b	·	b^3	·	a	·	·
c^3	·	c	·	a	·	·	·

Show also that the group presented by

$$\langle\, a, b \mid a^4 = a^2 b^2 = (ab)^2 b^2 = 1 \rangle$$

is isomorphic to $\mathcal{Q}_8$.

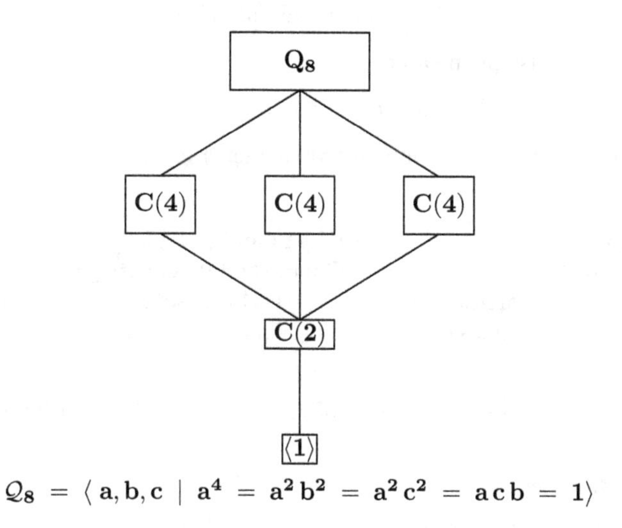

$$\mathcal{Q}_8 = \langle\, \mathbf{a}, \mathbf{b}, \mathbf{c} \mid \mathbf{a}^4 = \mathbf{a}^2 \mathbf{b}^2 = \mathbf{a}^2 \mathbf{c}^2 = \mathbf{a}\,\mathbf{c}\,\mathbf{b} = 1 \rangle$$

Figure VIII.3.1 Lattice of subgroups of group $\mathcal{Q}_8$

Normal subgroups and homomorphisms are closely related. The next two theorems will describe the connection.

Theorem VIII.3.2. *Let $\varphi : G \to H$ be a group homomorphism. Then the* **kernel** *of φ*

$$\text{Ker}\, \varphi = \{g \in G \,|\, \varphi(g) = 1\}$$

is a normal subgroup of G. The **image** *of φ*

$$\textbf{Im}\, \varphi = \{\varphi(g) \,|\, g \in G\}$$

is a subgroup of H. The homomorphism φ can be decomposed into an epimorphism (projection) π followed by a monomorphism (injection) μ:

$$\begin{array}{ccc} G & \xrightarrow{\varphi} & H \\ \| & & \uparrow{\scriptstyle \mu} \\ G & \xrightarrow{\pi} & \operatorname{Im} \varphi \end{array}$$

Proof. Ker φ is a subgroup of G. Since $\varphi(a^{-1}\, g\, a) = \varphi(a^{-1})\, 1\, \varphi(a) = \varphi(a^{-1}\, a) = 1$, **Ker** φ is a normal subgroup in G. The rest is evident.

□

An important component of the relationship between normal subgroups and group homomorphisms is the concept of a **partition** of a group into **cosets**. It is also the essence of Lagrange's theorem. We now present this relationship.

Let H be a subgroup of a group G. We will imitate the process of collecting integers modulo n : The role of the additive group $(\mathbb{Z}, +)$ is taken by G and the role of the subgroup $(n\mathbb{Z}, +)$ by H. Thus, define the following equivalence relation between elements $a, b \in G$:

$$a \equiv b \pmod{H} \text{ if and only if } a^{-1}\, b \in H. \tag{VIII.3.2}$$

Of course, we may also define an equivalence by

$$a \equiv b \pmod{H} \text{ if and only if } a\, b^{-1} \in H. \tag{VIII.3.3}$$

As group multiplication is noncommutative in general, these two equivalences may be different.

Exercise VIII.3.15. Show that both (VIII.3.2) and (VIII.3.3) are equivalence relations on G and that the first defines a partition of G into the **left cosets** $gH = \{g\, h \mid h \in H\}$ of H in G and the second one a partition of G into the **right cosets** $Hg = \{h\, g \mid h \in H\}$ of H in G. Prove that all cosets have the same cardinality equal to the cardinality (that is, equal to the **order**) of H.

The cardinality of the set of all cosets of H in G is called the **index** of the subgroup H in G.

Example VIII.3.7. Let

$$H = \{1, (1234), (13)(24), (1432), (13), (14)(23), (24), (12)(34)\} \subseteq \mathcal{S}_4;$$

then there are three right cosets of H in $\mathcal{S}_4$:

$$\begin{aligned} H,\ H(12) &= \{(12), (134), (1423), (243), (123), (1324), (142), (34)\} \text{ and} \\ H(14) &= \{(14), (234), (1243), (132), (143), (23), (124), (1342)\} \end{aligned}$$

and three left cosets of H in $\mathcal{S}_4$:

$$\begin{aligned} H,\ (12)H &= \{(12), (234), (1324), (143), (132), (1423), (124), (34)\} \text{ and} \\ (14)H &= \{(14), (123), (1342), (243), (134), (23), (142), (1243)\}. \end{aligned}$$

Theorem VIII.3.3. *Let H be a subgroup of a group G.*

(*i*) (**Lagrange's theorem**) *If H is a finite subgroup of order $|H| = k$, then the number of elements in each coset of H in G is k. Thus if G is a finite group of order $|G| = n$, k is necessarily a divisor of n. In particular, order of every subgroup of G and of every element g of G (defined as order of the cyclic group $\langle g \rangle$ generated by g) is a divisor of n.*

(*ii*) (**Quotient group** G/H) *A subgroup H is a normal subgroup of G if and only if the left and the right cosets of H in G coincide: $gH = Hg = \overline{g}$ for all $g \in G$. In this case, the set of all cosets with multiplication $\overline{g}_1\,\overline{g}_2 = \overline{g_1\,g_2}$ (induced by multiplication in the group G) forms a group called the* **quotient group** *(or the factor group) of G by H and denoted by $\overline{G} = G/H$. The identity element of $\overline{G}$ is the coset $1\,H = H$. There is a* **canonical epimorphism** $\Psi : G \to \overline{G}$ *defined by $\Psi(g) = gH$ for all $g \in G$. The epimorphism Ψ induces a bijection between the set of all subgroups K satisfying $H \subseteq K \subseteq G$ and the set of all quotient groups $H/H = \{1\} \subseteq K/H \subseteq G/H$. If G is finite, then*

$$|G/H| \cdot |H| = |G|,$$

that is, the product of the order and index of H in G equals the order of G.

Proof. A proof of (i) is already contained in Exercise VIII.3.15.

In order to prove (ii), note that for any element $g \in G$ and any $h \in H$, always $h\,g = g\,h'$ for $h' = g^{-1}\,h\,g$. Consequently, H is normal in G if and only if

$$H\,g = \{h\,g\,|h \in H\} = \{g\,h'\,|h' \in H\} = g\,H.$$

Observe that $g\,H = g'H$ if and only if $g^{-1}\,g' \in H$. Hence, if H is normal in G, one can define the following multiplication on the set G/H consisting of the cosets of H in G

$$g_1\,H \circ g_2\,H = \{g_1\,h_1\,g_2\,h_2\,|h_1, h_2 \in H\} = \{g_1\,g_2\,h_1'\,h_2\,|h = h_1'\,h_2 \in H\} = g_1\,g_2\,H.$$

The definition does not depend on the choice of the representatives g_1 and g_2 of the cosets. Indeed, if $g_1'\,H = g_1\,H$ and $g_2'\,H = g_2\,H$, that is, if $g_1^{-1}\,g_1' = h_1 \in H$ and $g_2^{-1}\,g_2' = h_2 \in H$, then

$$(g_1\,g_2)^{-1}\,(g_1'\,g_2') = g_2^{-1}\,(g_1^{-1}\,g_1')\,g_2' = g_2^{-1}\,h_1\,g_2' = g_2^{-1}\,g_2'\,h_1' = h_2\,h_1' \in H.$$

Since the multiplication $\circ$ is associative and

$$H \circ g\,H = g\,H \circ H = g\,H \text{ and } (g\,H)^{-1} = g^{-1}\,H,$$

the theorem follows. □

Now we are ready to prove the following statement promised earlier.

Theorem VIII.3.4. *The alternating group $\mathcal{A}_5$ is simple.*

Proof. We have found in Example VIII.3.6 that there are five conjugacy classes of $1, 12, 12, 20$ and 15 elements each. The group $\mathcal{A}_5$ has 60 elements and thus, according to Lagrange's theorem, the order of any subgroup of $\mathcal{A}_5$ has to be a divisor of 60. Any normal subgroup H of G has to be, in addition, a union of some conjugacy classes, including the trivial class $\{1\}$. No sum

$$1 + 12k + 12l + 20m + 15n \text{ with integers } 0 \leq k, l, m, n \leq 1,$$

is a divisor of 60 unless all $k = l = m = n = 0$ or $k = l = m = n = 1$. Thus the only normal subgroups are the trivial subgroup $\{1\}$ and G itself, that is, $\mathcal{A}_5$ is simple.

□

The next theorem asserts that **all** alternating groups $\mathcal{A}_n$ for $n \geq 5$ are simple. In contrast, the alternating group $\mathcal{A}_4$ is not simple. In reference to Example VIII.3.4, one can see that the Klein four-group V_4 is a normal subgroup of $\mathcal{A}_4$.

Theorem VIII.3.5. *All alternating groups $\mathcal{A}_n$ for $n \geq 5$ are simple.*

Proof. We will proceed by induction: We have already shown that $\mathcal{A}_5$ is simple and now, assuming that $\mathcal{A}_{n-1}$ is simple, we will prove that $\mathcal{A}_n$ is simple, as well. Our strategy is very simple: First, we will consider $\mathcal{A}_{n-1}$ canonically embedded in $\mathcal{A}_n$ and show that any proper normal subgroup H of $\mathcal{A}_n$ must intersect $\mathcal{A}_{n-1}$ nontrivially. Consequently, it must contain $\mathcal{A}_{n-1}$; but, then it contains the entire conjugacy class of (all) cycles of length 3, and these generate the whole group $\mathcal{A}_n$. A leading idea of the proof is to explore the simple fact expressed in (VIII.3.1): For any permutations ρ and π

$$(\pi^{-1} \circ \rho \circ \pi)(\pi^{-1}(i)) \;=\; \pi^{-1}(\rho(i)).$$

Thus, let $H \neq \{1\}$ be a normal subgroup of $\mathcal{A}_n, n \geq 6$ and assume that $H \cap \mathcal{A}_n \;=\; \{1\}$. The expression of every element $\rho \neq 1$ of H will thus involve n. If there is $1 \leq d < n$ for which $\rho(d) = d$, then for any i, j distinct from n, d,

$$\sigma \;=\; (i\, j)\,(n\, d)\;\; \rho\;\; (i\, j)\,(n\, d) \in H$$

satisfies $\sigma(n) \;=\; n$ and therefore $\sigma \neq 1$ belongs to $\mathcal{A}_n \cap H$.

If there is no $d \in \{1, 2, \ldots, n-1\}$ for which $\rho(d) = d$, that is, if all numbers $\{1, 2, \ldots, n\}$ appear in the (nontrivial, disjoint) cycles of ρ, then either all such cycles are of length 2 or at least one has length at least $3 :\; (d\; i\; \cdots\; j)$. In the latter case, let m and n be distinct positive integers, which are distinct from d, and consider the element

$$\rho' \;=\; \rho\;\; (i\, j)\,(m\, n)\;\; \rho\;(i\, j)\,(m\, n) \in H$$

to get $\rho' \neq 1$ and $\rho'(d) = d$. If, in the other case, only disjoint transpositions are involved in the expression for ρ and thus $n \geq 8$, let $\rho = (d\, i)\,(j\, k)\,(m\, n)\; \ldots$. Then

$$\rho' \;=\; \rho\;\; (d\, i)\,(j\, m)\;\; \rho\;\; (d\, i)\,(j\, m) \in H$$

and, again $\rho' \neq 1$ and $\rho'(d) = d$.

In any case, $\mathcal{A}_n \cap H \neq \{1\}$ is a normal subgroup of $\mathcal{A}_{n-1}$, and thus $\mathcal{A}_{n-1} \subseteq H \subseteq \mathcal{A}_n$. It is now rather easy to conclude that $H = \mathcal{A}_n$. To this end, we just use a rather trivial fact that a product of two transpositions can be written as a product of one or two cycles of length 3. Indeed, $(i\ j)\,(j\ k) = (i\ j\ k)$ for $i \neq k$ and $(i\ j)\,(m\ n) = (i\ j)\,(j\ m)\,(j\ m)\,(m\ n) = (i\ j\ m)\,(j\ m\ n)$ for i, j distinct from m, n. Therefore, the group $\mathcal{A}_n$ is generated by cycles of length 3. However, **all** cycles of length 3 belong to H! Indeed every cycle that involves n, say $(n\ i\ j)$ can be written as

$$(n\ i\ j) = (k\ n)\,(i\ j)\,(j\ i\ k)\,(k\ n)\,(i\ j) \in H,$$

since $(j\ i\ k) \in \mathcal{A}_{n-1} \subseteq H$. Thus $H = \mathcal{A}_n$, as required. □

There are three results closely related to the concept of a quotient group. They clarify this concept and are usually called the isomorphism theorems. The first one is already known to us from the examples and exercises.

Theorem VIII.3.6 (First isomorphism theorem). *Let $\varphi : G \to H$ be a group homomorphism. Then*

$$G/\mathbf{Ker}\varphi \simeq \mathbf{Im}\varphi.$$

Exercise VIII.3.16. Let $\varphi : G \to H$. Let $W \in G/\mathbf{Ker}\varphi$ be the fiber (that is, the inverse image) of $h \in H$. Let $\varphi(g) = h$. Show that $W = g\,(\mathbf{Ker}\varphi) = (\mathbf{Ker}\varphi)\,g$.

Exercise VIII.3.17. Prove that the group $\mathbf{Inn}(G)$ of all inner automorphisms of a group G is normal in the group $\mathbf{Aut}(G)$. The quotient $\mathbf{Aut}(G)/\mathbf{Inn}(G) = \mathbf{Out}(G)$ is sometimes called the group of outer automorphisms.

Exercise VIII.3.18. Show that $\mathbf{Inn}(G) \simeq G/Z(G)$, where $Z(G)$ is the center of the group G.

Exercise VIII.3.19. Verify that $\mathbf{Aut}(\mathcal{Q}_8) \simeq \mathcal{S}_4$, $\mathbf{Inn}(\mathcal{Q}_8) \simeq V_4$ and $\mathbf{Out}(\mathcal{Q}_8) \simeq \mathcal{D}_6$.

Exercise VIII.3.20. Show that $SL_n(F)$ is a normal subgroup of $GL_n(F)$ and determine the quotient group $GL_n(F)/SL_n(F)$. Consider the determinant map from $GL_n(F)$ to the multiplicative group of nonzero elements of the field F.

Example VIII.3.8. Let $\varphi : \mathbb{C}^\times \to \mathbb{R}^+$ be defined by $\varphi(z) = |z|$. Then φ is a group epimorphism of the multiplicative group of all non-zero complex numbers onto the multiplicative group of all positive real numbers. φ can also be interpreted as an endomorphism of the multiplicative group of all non-zero complex numbers. The kernel of φ is the complex unit circle $\mathbf{K} = \{z \in \mathbb{C}^\times \| z| = 1\}$. Thus the fibers of this homomorphism (that is, the inverse images of individual real positive numbers) are the circles in the Argand-Gauss plane centered at the origin. These circles are the elements of $\mathbb{C}^\times/\mathbb{R}^+$; the product of two circles is the circle whose radius is the product of their radii.

The proofs of the second and third isomorphism theorems are just simple applications of the first isomorphism theorem.

Theorem VIII.3.7 (Second isomorphism theorem). *Let H be a normal subgroup of a group G. Then, for any subgroup A of G, the subset $AH = \{a\,h\,|a \in A, h \in H\}$ is a subgroup of G and there is a canonical isomorphism*

$$AH/H \simeq A/(A \cap H) \text{ given by } a\,(A \cap H) \mapsto a\,H.$$

Proof. It is easy to show that AH is a subgroup of G and that H is normal in AH. Indeed, $a_1\,h_1\;a_2\,h_2 = a_1\,a_2\;(a_2^{-1}\,h_1\,a_2)\,h_2 \in AH$ and, for every $a\,h \in AH$, there is $a^{-1}\,h_1$ with $h_1 = a\,h^{-1}\,a^{-1}$ satisfying $a\,h\;a^{-1}\,h_1 = 1$. Moreover, $(a^{-1}\,h_1)\;h\;(a\,h_3) = a^{-1}\,a\,(a^{-1}\,h_1\,h_2\,a)\,h_3 \in H$.

The mapping $\varphi : A \to AH/H$ defined by $\varphi(a) = a\,H$ is an epimorphism. The elements of its kernel are those elements $a \in A$ that belong to h and thus $\mathbf{Ker}\varphi = A \cap H$. Applying the first isomorphism theorem we get the required isomorphism. □

Theorem VIII.3.8 (Third isomorphism theorem). *Let $A \subseteq H \subseteq G$ be normal subgroups of G. Then there is a canonical isomorphism*

$$(G/A)/(H/A) \simeq G/H \text{ given by } g\,A\,(H/A) \mapsto g\,H.$$

Proof. To apply Theorem VIII.3.6, we define the mapping $\varphi : G/A \to G/H$ by $\varphi(gA) = gH$ for every $g \in G$. This is an epimorphism whose kernel is formed by all gA with $g \in H$. Thus $\mathbf{Ker}\varphi = H/A$ and the required isomorphism follows. □

Example VIII.3.9. We illustrate the last two theorems. In Figure VIII.1.3 choose $H = \mathcal{A}_4$ and $A = \mathcal{S}_3$. Applying Theorem VIII.3.7, we deduce

$$\mathcal{S}_4/\mathcal{A}_4 \simeq \mathcal{S}_3/C(3).$$

Similarly, in Figure VIII.2.3 choose $A = C(2) \subset H = C(10) \subset G = \mathcal{D}_{20}$. Then $\mathcal{D}_{20}/C(2) \simeq \mathcal{D}_{10}, C(10)/C(2) \simeq C(5), \mathcal{D}_{20}/C(10) \simeq C(2)$ and thus $\mathcal{D}_{10}/C(5) \simeq C(2)$.

Theorem VIII.3.3 allows us to properly define the concept of the **presentation** of a group that we have already used in definition of some groups. For example, we have defined the dihedral group $\mathcal{D}_{2n}$ as follows:

$$\mathcal{D}_{2n} = \langle a, b \mid a^n = b^2 = (ab)^2 = 1\rangle.$$

Here, a and b were two elements - **the generators** - that we used to form elements of the group, namely finite words of the form $a^{z_1}\,b^{z_2}\,a^{z_3}\dots a^{z_d}$, where the exponents are integers $z_t, 1 \le t \le d$, subject to certain rules - **the relations** - expressed by $a^n = b^2 = (ab)^2 = 1$. More generally, we can have r generators $a_1, a_2, \dots, a_r$ and s relations $r_1, r_2, \dots, r_s$. For instance, we have described the quaternion group $\mathcal{Q}_8$ by $\langle a, b, c \mid a^2 = b^2 = c^2 = abc = 1\rangle$. We hasten to point out that such presentations are far from unique. After all, the listing (as generators) of all elements of a finite group together with the multiplication table (as relations of the form $xyz^{-1} = 1$) is also a representation of the group (albeit a very poor one). Other efficient presentations of the group $\mathcal{Q}_8$ are $\langle a, b \mid a\,b\,a = b, b\,a\,b = a\rangle$ or $\langle a, b \mid a^4 = 1, a^2 = b^2, aba = b\rangle$. Note that there is no need for an extra symbol 0 here (compare Section 7 of Chapter I).

We formalize this concept. Start with a graph Γ that has one vertex V and r loops $a_1, a_2, \dots a_r$. The path semigroup, that is in this case a monoid is called a **free monoid** $\mathcal{S}(\Gamma)$

on r generators; denote the identity element by 1. Now, to each loop a_t attach an additional loop denoted by a_t^{-1}. This way, we obtain an extension of the graph Γ, namely a graph with one vertex and $2r$ loops. Consider the respective "extended" free monoid $\mathcal{S}(r) = \overline{\mathcal{S}(\Gamma)}$ on $2r$ generators.

On this monoid $\mathcal{S}(r)$, define the following equivalence $\equiv$ induced by $2r$ relations

$$a_t\, a_t^{-1} = 1 \text{ and } a_t^{-1}\, a_t = 1 \text{ for } 1 \leq t \leq r.$$

This means that two paths (words) of $\mathcal{S}(r)$ are $\equiv$-equivalent if one can be obtained from the other by a finite number of insertions or deletions of subpaths of the form $a_t\, a_t^{-1}$ or $a_t^{-1}\, a_t$ for suitable indices t. It follows that in every $\equiv$-equivalence class there is a canonical representative, namely the **reduced path** (reduced word). It is either the trivial path 1, or a path of nonzero length $x_1\, x_2 \ldots x_d$, where each $x_k, 1 \leq k \leq d$ is either certain a_t or a_t^{-1} and no two neighbors $x_k\, x_{k+1}$ are of the form $a_t\, a_t^{-1}$ or $a_t^{-1}\, a_t$ for some $1 \leq t \leq r$. The multiplication of reduced paths is induced by the multiplication of the free monoid $\mathcal{S}(r)$. This way we obtain a group, the **free group** $\mathcal{F}(r)$ on the set of r generators

$$\mathcal{F}(r) = \langle a_1, a_1^{-1}, a_2, a_2^{-1}, \ldots, a_r, a_r^{-1} \mid a_t\, a_t^{-1} = a_t^{-1}\, a_t = 1 \text{ for all } 1 \leq t \leq r \rangle.$$

The inverse of the path $x_1\, x_2 \ldots x_{d-1}\, x_d$ is the path $y_d\, y_{d-1} \cdots y_2\, y_1$, where for all $1 \leq k \leq d$,

$$y_k = \begin{cases} a_t & \text{if } x_k = a_t^{-1}; \\ a_t^{-1} & \text{if } x_k = a_t. \end{cases}$$

We simply write

$$(x_1\, x_2 \ldots x_{d-1}\, x_d)^{-1} = y_d\, y_{d-1} \cdots y_2\, y_1 \text{ or } x_1\, x_2 \ldots x_{d-1}\, x_d = (y_d\, y_{d-1} \cdots y_2\, y_1)^{-1}.$$

Be aware that this notation includes the relations $(a_t^{-1})^{-1} = a_t$.

For every $r \geq 1, \mathcal{F}(r)$ is infinite and for $r \geq 2$, it is noncommutative. Trivially, $\mathcal{F}(1)$ is isomorphic to the infinite cyclic group. We also mention that, conveniently, the elements of a free group are often presented in the form $a_{t_1}^{z_1}\, a_{t_2}^{z_2} \ldots a_{t_n}^{z_n}$ with $a_{t_k} \neq a_{t_{k+1}}$ and nonzero integers z_k.

The importance of the concept of a free group stems from the following fact.

Theorem VIII.3.9. *Every group G generated by r elements $\{g_1, g_2, \ldots, g_r\}$ is a homomorphic image of the free group $\mathcal{F}(r)$. Thus $G \simeq \mathcal{F}(r)/K$ for a suitable (normal) subgroup K of $\mathcal{F}(r)$. The elements of the group K are called the* **relations** *of the group G.*

Proof. Let $\{a_1, a_1^{-1}, \ldots a_r, a_r^{-1}\}$ be the generators of $\mathcal{F}(r)$. Define the mapping $\varphi : \mathcal{F}(r) \to G$ by sending every path (word) $x_1\, x_2 \ldots x_d \in \mathcal{F}(r)$ to the element $u_1\, u_2 \ldots u_d \in G$, where for $1 \leq k \leq d$,

$$u_k = \begin{cases} g_t & \text{if } x_k = a_t; \\ g_t^{-1} & \text{if } x_k = a_t^{-1}. \end{cases}$$

It is a routine calculation to show that φ is a group epimorphism onto G, and thus, in view of the first isomorphism theorem, we get $G \simeq \mathcal{F}(r)/K$ with $K = \mathbf{Ker}\varphi$. □

Remark VIII.3.1. The restriction to **finitely generated** groups (that is, groups with a finite number of generators) is not essential. The construction of the free groups on any set of generators and a statement of Theorem VIII.3.9 for any group G follow the same lines as in the cases that we consider in this book. We mention a rather advanced statement (that we will not need and will not prove in this book) that every subgroup of a free group is again a free group.

Returning now to the presentation

$$G = \langle g_1, g_2, \ldots, g_r \mid w_1 = w_2 = \cdots = w_s = 1 \rangle, \tag{VIII.3.4}$$

it represents the quotient group representation $G = \mathcal{F}(r)/K$ with the set of generators $\{w_q \mid 1 \le g \le s\}$ of the subgroup K, whereby every w_q is a product of some generators g_t and g_t^{-1}. Very often some of the relations $w_q = 1$ are rewritten in the form $w'_q = w''_q$. The relations $w_1 = w_2 = \cdots = w_s = 1$ in (VIII.3.4) allow us to determine all the elements of the group G, that is, the order of G, and the multiplication table of G.

Exercise VIII.3.21. Show that $\langle a, b \mid a\,b\,a = b, b\,a\,b = a \rangle$, as well as $\langle a, b \mid a^4 = 1, a^2 = b^2, aba = b \rangle$ are presentations of the same group, the quaternion group $\mathcal{Q}_8$.

Exercise VIII.3.22. Do the presentations $\langle g, h \mid g^3 = h^2 = (gh)^4 = 1 \rangle$ and $\langle a, b, c \mid a^2 = b^2 = c^2 = (ab)^3 = (bc)^3 = 1,\ ac = ca \rangle$ define the same group?

Exercise VIII.3.23. Explain the following presentations of the cyclic group C_n of order $n = r\,s$:

$$\langle a \mid a^n = 1 \rangle = \langle b, c \mid b^r = c^s = 1,\ b\,c = c\,b \rangle.$$

Exercise VIII.3.24. Describe the subgroup structure of each of the groups

$$G_1 = < a, b \mid a^3 = b^3 = (ab)^2 = 1 >$$

and

$$G_2 = < a, b \mid a^3 = b^5 = (ab)^2 = 1 > .$$

Exercise VIII.3.25. Prove that the symmetric group $\mathcal{S}_n$ can be generated by the transposition $(1\ 2)$ and the cycle $(1\ 2\ 3\ \ldots n)$.

Besides forming quotient groups, there is yet another very important construction of forming new groups from given groups, namely the construction of a **direct product** of groups.

Definition VIII.3.3. *Let $G_1, G_2, \ldots, G_r$ be a set of arbitrary groups with multiplications $\circ_1, \circ_2, \ldots, \circ_r$, respectively. Then the cartesian product $G = G_1 \times G_2 \times \cdots \times G_r$ with the operation*

$$\circ((a_1, a_2, \ldots, a_r), (b_1, b_2, \ldots, b_r)) = (\circ_1(a_1, b_1), \circ_2(a_2, b_2), \ldots, \circ_r(a_r, b_r))$$

is a group called the **direct product** *or* **direct sum** *of the groups $G_1, G_2, \ldots, G_r$. We shall often denote this group briefly by $G = \prod_{s=1}^r G_s$ in the case that $\circ_s$'s are multiplications or $G = \bigoplus_{s=1}^r G_s$ in the case that $\circ_s$'s are additions.*

We have defined the direct product rather formally trying to emphasize the fact that we are dealing with unrelated groups G_s with different operations $\circ_s$. The rule of the new operation is however simple: The product (or sum) is performed coordinatewise. In what follows, our formulations will be mostly in terms of multiplications, that is, products; a translation into the terminology of addition is straightforward.

The definition of the direct product does not depend on the order of the factors in the product G_s, that is,

$$G = \prod_{s=1}^{r} G_s \simeq \prod_{s=1}^{r} G_{\pi(s)} \text{ for any permutation } \pi \in \mathcal{S}_r.$$

We now give a few properties of the direct product G.

The subset $H_s = \{(1, 1, \ldots, a_s, 1, \ldots, 1) \mid a_s \in G_s\}$ is a **normal** subgroup of G isomorphic to G_s. Every element g of the direct product $G = G_1 \times G_2 \times \cdots \times G_r$ has a **unique** representation as a product $g = h_1 h_2 \ldots h_r$, where $h_s \in H_s$ for all $s \in \{1, 2, \ldots, r\}$. The normal subgroups H_s also satisfy the following two properties:

(i) $H_1 H_2 \ldots H_r = \{h_1 h_2 \ldots h_r \mid h_t \in H_t, t = 1, 2, \ldots, r\} = G$;

(ii) $H_1 H_2 \ldots H_{s-1} \cap H_s = \{1\}$ for each $s = 2, 3, \ldots, r$.

Exercise VIII.3.26. Assume that the H_t are normal in G. Prove that under this condition (ii) is equivalent to the formally stronger property

(iii) the normal subgroup $H_s^* = H_1 H_2 \ldots H_{s-1} H_{s+1} \ldots H_r$ satisfies, for each $s \in \{1, 2, \ldots, r\}$, $H_s \cap H_s^* = \{1\}$.

Having defined the direct product of groups, a natural question arises: When is a given group G isomorphic to a direct product of its subgroups? That is, as many books would put it, after defining an "external" direct product, how do we define and relate it to an "internal" direct product? The answer is formulated in the next theorem.

Theorem VIII.3.10. *Let $H_1, H_2, \ldots, H_r$ be normal subgroups of a group G satisfying the conditions* (i) *and* (ii) *above. Then G is isomorphic to the direct product $H_1 \times H_2 \times \cdots \times H_r$.*

Proof. In view of the condition (ii), $h_s h_t = h_t h_s$ for all $h_s \in H_s, h_t \in H_t$ with $s \neq t$. Indeed $c = (h_s h_t h_s^{-1})\, h_t^{-1} = h_s\, (h_t h_s^{-1} h_t^{-1})$ belongs to both H_t and H_s and since $H_t \cap H_s = \{1\}, c = 1$, that is, $h_s h_t = h_t h_s$. Consequently, every element $g \in G$ has the form $g = h_1 h_2 \ldots h_r$ for certain h_s $s = 1, 2, \ldots, r$, and this expression is unique. Hence the mapping $\varphi : G \to H_1 \times H_2 \times \cdots \times H_r$ defined by $\varphi(g) = (h_1, h_2, \ldots, h_r)$ is an isomorphism. □

Example VIII.3.10. Consider the dihedral group $G = \mathcal{D}_{12} = \langle a, b \mid a^6 = b^2 = (a\,b)^2 = 1\rangle$ and its subgroups

$$H_1 = \langle a^2, b \mid (a^2)^3 = b^2 = (a^2\, b)^2 = 1\rangle \simeq \mathcal{D}_6 \text{ and } H_2 = \langle a^3 \mid (a^3)^2 = 1\rangle \simeq C(2).$$

As a subgroup of index 2, H_1 is normal in G. The subgroup H_2 is the center of G and thus a normal subgroup of G. We have $H_1 H_2 = G$ and $H_1 \cap H_2 = \{1\}$. Hence, $\mathcal{D}_{12} \simeq \mathcal{D}_5 \times C(2)$. See Figure VIII.2.4.

Example VIII.3.11. Consider the dihedral group $G = \langle a, b \mid a^3 = b^2 = (a\,b)^2 = 1\rangle \simeq \mathcal{D}_6 \simeq \mathcal{S}_3$ in Figure VIII.1.2. Here, the subgroups

$$H_1 = \langle a \mid a^3 = 1\rangle \simeq C(3) \text{ and } H_2 = \langle b \mid b^2 = 1\rangle \simeq C(2)$$

satisfy $H_1 \cap H_2 = \{1\}$. Nevertheless G is **not** a direct product of H_1 and H_2. This is because H_2 is not a normal subgroup of G. In the last section of this chapter, we will define a **semidirect product** of two groups and show that g is a semidirect product of H_1 and H_2.

Exercise VIII.3.27. Show that $\mathcal{D}_{10}$ (see Figure VIII.2.1) is **indecomposable**, that is, not a direct product of two proper subgroups, while $\mathcal{D}_{20} \simeq \mathcal{D}_{10} \times C(2)$ (see Figure VIII.2.3). Generalize these facts for arbitrary dihedral groups $\mathcal{D}_{2n}$.

Exercise VIII.3.28. Prove that the centers of groups satisfy

$$Z(H_1 \times H_2 \times \cdots \times H_r) = Z(H_1) \times Z(H_2) \times \cdots \times Z(H_r).$$

Example VIII.3.12. Consider a group G of order p^2, where p is an odd prime (that is, > 2). We are going to show that then G must be an abelian group either isomorphic to the cyclic group $C(p^2)$ or to the direct product of two groups isomorphic to the cyclic group $C(p)$. Indeed, either there is and element $g \in G$ of order p^2 in which case $G \simeq C(p^2)$ or, by Lagrange's theorem, all elements $a \neq 1$ of G have order p and thus $H = \langle a\rangle \simeq C(p)$. We are going to prove that H is normal in G. Otherwise, there will be $b \in G$ such that $b^{-1}\,a\,b \notin H$ and thus $\langle a\rangle \cap \langle b^{-1}\,a\,b\rangle = \{1\}$. Now, the left cosets $a^t\langle b^{-1}\,a\,b\rangle, 1 \leq t \leq p$ of the cyclic group $\langle b^{-1}\,a\,b\rangle \simeq C(p)$ form a partition of G and thus there must be r_0 and s_0 such that $b = a^{r_0}\,(b^{-1}\,a\,b)^{s_0} = a^{r_0}\,b^{-1}\,a^{s_0}\,b$. Therefore $a^{r_0}\,b^{-1}\,a^{s_0} = 1$ and thus $b = a^{r_0+s_0}$, contradicting our assumption. Consequently, every proper subgroup of G (isomorphic to $C(p)$) is normal in G and $G = H_1 \times H_2$ for any two such subgroups for which $H_1 \cap H_2 = \{1\}$. Note that in this case, G is isomorphic to the additive group of a 2-dimensional vector space over the finite field $GF(p)$ of p elements.

We mention in passing that there are noncommutative groups of order p^3. Can you give an example?

VIII.4 ⋆ Abelian groups

Just a word of introduction: Results on abelian groups are, generally speaking, presented in additive notation, that is, the operation is denoted by $+$, g^n becomes a "natural multiple" $n\,g = g + g + \cdots + g$ (n summands), a direct product $G_1 \times G_2 \times \cdots \times G_r$ becomes a direct sum $G_1 \oplus G_2 \oplus \cdots \oplus G_r$, etc. Here, we will continue in multiplicative notation assuming that a reader may benefit by becoming fluent in both notations.

There are several basic results on abelian groups that we are going to concentrate on. We start with a reminder that we have touched on the subject by considering the structure of cyclic groups in Section 11 of Chapter III. A summary of that section reads as follows.

Theorem VIII.4.1. *An infinite cyclic group $C(\infty)$ is isomorphic to the additive group $\mathbb{Z}$ of integers. All its subgroups are infinite cyclic groups and their lattice is isomorphic to the divisibility lattice of* $(\mathbb{N}, \preceq)$ *as illustrated in Figure* III.4.1.

A finite cyclic group $C(n)$ is uniquely determined by its order n: It is isomorphic to the additive group $\mathbb{Z}_n$. All subgroups of $C(n)$ are cyclic groups $C(d)$, where d is a divisor of n, and their lattice is isomorphic to the lattice of all divisors of n. In fact,

$$C(n) \simeq \prod_{p\,|\,n,\ p \text{ prime}} C(p^{k_p}), \quad \textit{whereby } n = \prod_{p\,|\,n,\ p \text{ prime}} p^{k_p}.$$

Given two elements of an abelian group G of finite order, it is evident that their product as well as their inverses are of finite order. Thus the set of all elements of finite order in an abelian group G is a subgroup $T(G)$ of G, called the **torsion** (or **periodic**) **subgroup** of G.

Exercise VIII.4.1. Show that the only element of finite order of the abelian group $\overline{G} = G/T(G)$ is the identity element, that is, that $\overline{G}$ is an abelian **torsion-free** group.

Exercise VIII.4.2. Prove that the order of the product of two elements of an abelian group of orders m and n, respectively, is a divisor of $m\,n$. What can we conclude if the order is equal to $m\,n$?

Exercise VIII.4.3. Let $\mathcal{F}_{ab}(r) = C(\infty) \times \cdots\cdots \times C(\infty)$ (r factors) be a **free abelian group**. Applying Theorem VIII.3.9, prove that every abelian group with r generators is a homomorphic image of $\mathcal{F}_{ab}(r)$.

An abelian group is called **mixed** if it is neither torsion nor torsion-free. In general, the torsion subgroup $T(G)$ is not a direct summand of G, that is, G is not a direct product of its torsion subgroup and a torsion-free subgroup (isomorphic to $G/T(G)$). Thus the theory of mixed groups involves more than simply combining the results about torsion and torsion-free groups and is rather involved. Of course, a free abelian group is torsion-free by definition. On the other hand, a torsion-free abelian group with a finite number of generators is free, a result that we will not prove here.

We shall concentrate on torsion groups and on finite abelian groups, in particular. The first basic property of such groups is the following statement which reflects the basic rules of divisibility of integers and its proof is therefore left to the reader.

Theorem VIII.4.2 (Primary decomposition). *Let G be an abelian torsion group. Then for any prime p, the subset*

$$G_p = \{g \in G \mid g \textit{ has order a power of } p\}$$

*is a subgroup G_p of G called the p-**primary component** of G. In fact*

$$G = \prod_p G_p,$$

where the product is taken over all primes p.

We observe that the latter decomposition of G into its p-primary components is **unique**: All subgroups G_p are defined in a unique fashion. Note that our earlier assertion on the structure of a finite cyclic group is also a simple consequence of Theorem VIII.4.2.

Exercise VIII.4.4. Prove that two torsion groups G and H are isomorphic if and only if G_p and H_p are isomorphic for every prime p.

Because of Theorem VIII.4.2 and Exercise VIII.4.4 the study of abelian torsion groups is reduced to the study of p-primary groups. We are going to use the decomposition into p-primary parts in a proof of the following important theorem of **Prüfer** and **Reinhold Baer (1902 - 1979)**.

Theorem VIII.4.3. *An abelian group G of bounded order is a direct product of cyclic groups.*

Proof. In view of Theorem VIII.4.2, we may assume that G is p-primary. We are going to prove the essential part of the proof, namely the following statement:

A primary group G of bounded order has a cyclic direct factor, that is,
$G = C(k) \times H$, *where k is the maximal order of any element of G.*

Thus, let G be an abelian p-primary group of bounded order, $g_1 \in G$ be an element of maximal order p^m and H a maximal subgroup of G such that $\langle g_1\rangle \cap H = \{1\}$. Denote by G_1 the direct product $\langle g_1\rangle \times H \subseteq G$. We are going to prove that $G_1 = G$.

Assume that $G_1 \neq G$ and choose $g \in G$ such that $g \notin G_1$. Due to the maximality of H, $\langle g\rangle \cap G_1 \neq \{1\}$. Thus there is ℓ such that $g_0 = g^{p^{\ell-1}} \notin G_1$, but $g_0^p = g^{p^\ell} \in G_1$. Hence $g_0^p = g_1^r h$ for suitable r and $h \in H$. Then $1 = g_0^{p^n} = (g_1^r h)^{p^{n-1}} = g_1^{r\,p^{n-1}} h^{p^{n-1}}$ and therefore $g_1^{r\,p^{n-1}} \in \langle g_1\rangle \cap H$. From here, $g_1^{r\,p^{n-1}} = 1$ and therefore $r = \ell p$. Consequently, $g_0^p = g_1^{\ell p} h$, that is, $(g_0\, g_1^{-\ell})^p = h$.

We have $g_0\, g_1^{-\ell} \notin G_1 \supset H$ and $(g_0\, g_1^{-\ell})^p \in H$. This means that $H_1 = \langle H, g_0\, g_1^{-\ell}\rangle$ satisfies $H \subsetneq H_1$ and therefore $H_1 \cap \langle g_1\rangle \neq \{1\}$. This implies that there are exponents m, n such that $1 \neq g_1^m = (g_0\, g_1^{-\ell})^n h_1$ for some $h_1 \in H$. Here, the numbers n and p are relatively prime. For, $n = t\,p$ would yield $1 \neq g_1^m \in \langle g_1\rangle \cap H = \{1\}$, a contradiction. Now, we have $g_0^n = g_1^m\, g_1^{\ell n}\, h_1^{-1} \in G_1$, $g_0^p \in G_1$ and $(p, n) = 1$; it follows that $g_0 \in G_1$, in contradiction with our assumption $g_0 \notin G_1$. Thus $G = G_1 = \langle g_1\rangle \times H \subseteq G$.

Now, we can apply, inductively, the same procedure for the subgroup H, and refer to the monograph "Infinite Abelian Groups" (University of Michigan Press, 1954) by **Irving Kaplansky (1917 - 2006)** to use the concept of a "maximal pure independent subset" of G to conclude that

$$G = \prod_{\iota \in I} C(k_\iota),$$

as required. □

We emphasize the fundamental significance of Theorem VIII.4.3: *The direct products of finite cyclic groups having an upper bound on their orders are just the abelian groups of bounded order.*

We obtain an important consequence.

Corollary VIII.4.1. *Every finite abelian group is a direct product of cyclic groups.*

Remark VIII.4.1. We remark that Theorem VIII.4.3 is usually formulated using additive notation. Then the assumption would read $p^n\, g = 0$ for all $g \in G$ and the statement would read G is a direct sum of cyclic groups.

A natural question arises as to what extent is a decomposition of a torsion group into its cyclic factors unique. Such a question is well-founded since the direct summands in the proof of Theorem VIII.4.3 involved a choice. The answer is given here for a special case that will be used in the general situation.

First of all, due to the uniqueness of p-primary decomposition, we may consider separately each p-primary component of a torsion group. Secondly, we get in each p-primary group G a **tower** (a chain) of subgroups $\{1\} \subset G^{(1)} \subset G^{(2)} \subset G^{(3)} \subset \cdots$, where $G^{(t)} = \{g \in G \mid g^{p^t} = 1\}, 1 \leq t \ldots$ and this tower is finite for p-groups of bounded exponent

$$\{1\} \subsetneq G^{(1)} \subsetneq G^{(2)} \subsetneq \cdots \subsetneq G^{(h)} = G. \tag{VIII.4.1}$$

The (highest) exponent h is sometimes referred to as the **height** of the group G. Now, for G of height 1 a direct decomposition into cyclic groups is a decomposition into cyclic groups of order p. Such a group has a structure of a vector space over the field $GF(p)$ of p elements and the number of factors $C(p)$ in the decomposition is just the dimension of that vector space. Thus it is unique up to an isomorphism. This is what we shall exploit in the proof of the general statement.

Theorem VIII.4.4. *Any two decompositions of a group G into a direct product of cyclic groups $C(p^k)$ for suitable primes p and exponents k are isomorphic, that is, the numbers of the direct factors $C(p^k)$ are for all primes p and all exponents $k \geq 1$ unique.*

Proof. As we have already mentioned, we can restrict our considerations to the case of a p-group G. In this case, all cyclic subgroups are of the form $C(p^k)$. Consider the tower (VIII.4.1) (which we may allow even to be infinite). For each $t \geq 0$, consider the quotient group $G^{(t+1)}/G^{(t)}$. By Theorem VIII.4.3, it is isomorphic to a direct product of cyclic groups $C(p)$ whose number is an invariant of the group and is equal to the number of cyclic factors $C(p^{t+1})$ in the direct decomposition of G. Thus the number of such direct factors is, in any decomposition into direct product of cyclic groups, the same. Theorem VIII.4.4 follows. □

Corollary VIII.4.2. *Every abelian group of bounded order (and thus any finite abelian group) can be decomposed into a direct product of indecomposable (that is, p-primary) subgroups and the numbers of factors of a given order in any two such decompositions are equal.*

Remark VIII.4.2. It is worth mentioning that there is neither restriction on the height of G nor on the number of direct factors in Theorem VIII.4.4.

Example VIII.4.1. We are now able to list all abelian groups of a given finite order. Thus the number of nonisomorphic groups of order p^n is equal to the number of partitions $\rho(n)$ of natural number n; here, $\rho(n)$ is simply equal to the number of nonnegative integral solutions of $x_1 + 2x_2 + 3x_3 + \cdots + nx_n = n$. For example, $\rho(4) = 5$ since $4 = 3+1 = 2+2 = 2+1+1 = 1+1+1+1$ and therefore

$$C(16), \;\; C(8) \times C(2), \;\; C(4) \times C(4),$$

$$C(4) \times C(2) \times C(2) \text{ and } C(1) \times C(1) \times C(1) \times C(1)$$

are the only abelian groups of order $16 = 2^4$. There are, of course, just 5 abelian groups of order $3^4 = 81$ or $23^4 = 279841$. The number of abelian groups of order $1024 = 2^{10}$ is $\rho(10) = 42$, while the number of abelian groups of order $65536 = 2^{16}$ is $\rho(16) = 231$.

The number $A(n)$ of abelian groups for an arbitrary n can be found from the primary decomposition of n. If

$$n = p_1^{k_1}\, p_2^{k_2}\, \ldots\, p_r^{k_r}, \text{ then } A(n) = \rho(k_1)\,\rho(k_2)\,\ldots\rho(k_r).$$

Thus, for $n = 288 = 2^5\, 3^2,\ A(288) = \rho(5)\,\rho(2) = 7 \cdot 2 = 14:$

$$C(2) \times C(2) \times C(2) \times C(2) \times C(2) \times C(3) \times C(3)$$

$$C(2) \times C(2) \times C(2) \times C(2) \times C(2) \times C(9)$$

$$C(2) \times C(2) \times C(2) \times C(4) \times C(3) \times C(3)$$

$$C(2) \times C(2) \times C(2) \times C(4) \times C(9)$$

$$C(2) \times C(4) \times C(4) \times C(3) \times C(3) \qquad C(2) \times C(4) \times C(4) \times C(9)$$

$$C(2) \times C(2) \times C(8) \times C(3) \times C(3) \qquad C(2) \times C(2) \times C(8) \times C(9)$$

$$C(4) \times C(8) \times C(3) \times C(3) \qquad C(4) \times C(8) \times C(9)$$

$$C(2) \times C(16) \times C(3) \times C(3) \qquad C(2) \times C(16) \times C(9)$$

$$C(32) \times C(3) \times C(3) \qquad C(32) \times C(9)$$

If n is a product of distinct primes, then $A(n) = 1$; hence, there is only one abelian group of order $30030 = 2 \cdot 3 \cdot 5 \cdot 7 \cdot 11 \cdot 13$: namely $C(30030)$.

Exercise VIII.4.5. Prove that every finite abelian group G has a unique decomposition

$$G = C(n_1) \times C(n_2) \times \cdots \times C(n_r), \text{ where } n_t \geq 2 \text{ and } n_{t+1} | n_t \text{ for all } 1 \leq t \leq r.$$

The sequence $(n_1, n_2, \ldots, n_r)$ is said to be the **type** of the group G and the number r its **rank.**

Exercise VIII.4.6. Let G be a finite abelian group of type $(n_1, n_2, \ldots, n_r)$. Show that G contains an element of order k if and only if k is a divisor of n_1. Conclude that n_1 is the smallest natural number n such that $g^n = 1$ for all $g \in G$.

Although the description of finite abelian groups is rather simple, their lattices of subgroups are, in general, very complex. Figure VIII.4.1 illustrates the **Hasse diagram (Helmut Hasse (1898 - 1979))** of the lattice of all subgroups of the direct product of two cyclic groups of order 9.

Exercise VIII.4.7. Describe the elements of the subgroups of $G = C(9) \times C(9)$ (see Figure VIII.4.1).

Exercise VIII.4.8. Show that the infinite group D of all complex numbers z such that $z^n = 1$ for some natural n is an infinite torsion abelian group. Show that, for every prime p, the p-primary subgroup D_p of D consists of all complex z satisfying $z^{p^k} = 1$ for some k. The subgroup is isomorphic to the **Prüfer's group** $C(p^\infty)$ defined by $C(p^\infty) = \langle g_s \mid g_1^p = 1,\ g_s^p = g_{s-1}, s \geq 2\rangle$. Prove that D is a subgroup of the infinite cartesian product of all D_p consisting of those sequences that have only a finite number of components different from 1 (that is, that D is an infinite direct product of the components D_p).

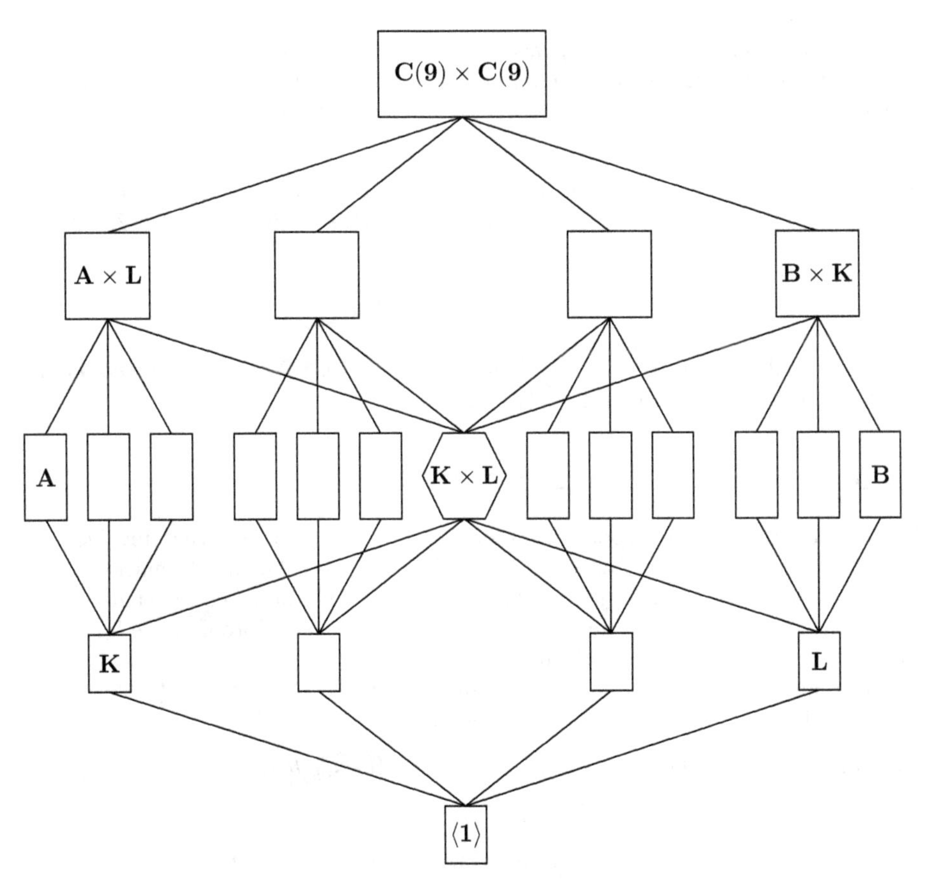

$$\mathbf{G} = \langle\, \mathbf{a}, \mathbf{b} \mid \mathbf{a}^9 = \mathbf{b}^9 = 1,\ \mathbf{a}\mathbf{b} = \mathbf{b}\mathbf{a}\rangle = \mathbf{A} \times \mathbf{B} = \langle \mathbf{a}\rangle \times \langle \mathbf{b}\rangle$$

Figure VIII.4.1 Lattice of subgroups of C(9) × C(9)

We conclude this section by the following important statement (see Section 4 of Chapter VII).

Theorem VIII.4.5. *The multiplicative group of a finite field is cyclic.*

Proof. We make use of the following simple result from group theory: If x and y are two elements of a finite abelian group of orders m and n respectively then there exists an element of the group whose order is the least common multiple of m and n.

Let $F = GF(p^n)$ be the finite field of $q = p^n$ elements. Since the nonzero elements of F form a multiplicative group $F^\times$ of $q-1$ elements, we have

$$x^{q-1} = 1 \text{ for all } x \in F^\times. \tag{VIII.4.2}$$

By applying the above mentioned group-theoretic result, we can find an element $g \in F^\times$ whose order r is the least common multiple of the orders of all the elements of $F^\times$. Thus the order of each element of $F^\times$ divides r, and r divides $q-1$. For all $\alpha \in F^\times$ we have $\alpha^r = 1$. Because the polynomial $x^r - 1$ has at most r roots in F, it follows that the number $q-1$ of elements in $F^\times$ is less than or equal to r. Thus $r = q-1$. But $1, \alpha, \ldots, \alpha^{r-1}$ are all distinct and belong to $F^\times$. Thus $F^\times$ is generated by α, and so cyclic of order $q-1$.

□

Remark VIII.4.3. We mention an important corollary of the last theorem. For $n = 1$, (VIII.4.2) yields Fermat's little Theorem (see Theorem III.9.5) since $GF(p) \simeq \mathbb{Z}_p$. Moreover, since

$$x^{q-1} - 1 = \prod_{r=1}^{q-1}(x - a_r),$$

the product of all nonzero elements a_r from F $(1 \leq r \leq q-1)$ equals -1 and we obtain, again for $n = 1$, Wilson's theorem (see Theorem III.9.9).

VIII.5 Action of a group on a set

In this section, we touch upon an important mathematical idea that involves making some mathematical objects act on other ones. In doing so a lot of information is derived about both objects. Here our objective will be groups acting on a set. This action enables us to perform some nontrivial counting that will culminate in the proof of the famous Sylow theorems together with some of their applications.

We have already met an action of a group on itself in the proof of Cayley's theorem (Theorem VIII.3.1). We now give a formal definition.

Definition VIII.5.1. *Let G be a group and X a set. We say that G* **acts on** *X if, for every $g \in G$, there is a map $\varphi_g : X \to X$ such that*
$\varphi_g \circ \varphi_h = \varphi_{gh}$ for every $g, h \in G$ and
$\varphi_e = 1_X$ (that is, the image of the neutral element of G is the identity mapping 1_X of X).

Remark VIII.5.1. We point out that every φ_g must be a bijection (that is, a **permutation** of X); for,

$$\varphi_g \circ \varphi_{g^{-1}} = \varphi_{gg^{-1}} = \varphi_{g^{-1}g} = \varphi_{g^{-1}} \circ \varphi_g = \varphi_e = 1_X.$$

All bijections of X, that is all permutations of X, form a group $\mathcal{S}(X)$ with respect to composition of maps. We can therefore reformulate Definition VIII.5.1 as follows:

Theorem VIII.5.1. *There is a one-to-one correspondence between the actions of a group G on a set X and the homomorphisms $\varphi : G \to \mathcal{S}(X)$ of the group G into the group $\mathcal{S}(X)$ of all permutations of the set X. If a group G acts on X, then $\varphi : G \to \mathcal{S}(X)$ defined by $\varphi(g) = \varphi_g$ for every $g \in G$ is a homomorphism. Conversely, if φ is a homomorphism of G into $\mathcal{S}(X)$, then there is an action of G on X defined for every $g \in G$ by the map $\varphi_g = \varphi(g)$.*

We simplify the notation by writing $g * x$ instead of $\varphi_g(x)$. Thus G acts on X if and only if the operation $G \times X \to X$ satisfies $(gh) * x = g * (h * x)$ and $e * x = x$ for all $g, h \in G$ and $x \in X$.

Exercise VIII.5.1. Suppose that the group G acts on the set X. Let Y^X be the set (function space) of all mappings (functions) $f : X \to Y$. For $g \in G$ and $f \in Y^X$ define

$$(g * f)(x) = f(g^{-1} * x) \text{ for all } x \in X.$$

Show that this definition yields an action of G on Y^X.

The following definition provides tools to express the basic results.

Definition VIII.5.2. *Suppose that the group G acts on the set X. Let $x \in X$. The subset*

$$\mathcal{O}(x) = \{g * x \in X \mid g \in G\}$$

of the set X is said to be an **orbit** *of the element x. The subgroup*

$$St(x) = \{g \in G \mid g * x = x\}$$

is called the **stabilizer** *of the element x. The normal subgroup*

$$K = \{g \in G \mid g * x = x \text{ for all } x \in X\} = \bigcap_{x \in X} St(x)$$

of G is the **kernel** *of the action of G on X. The action is said to be* **faithful** *if $K = \{1\}$ is trivial. If there is an element x_0 such that $\mathcal{O}(x_0) = X$, the action is said to be* **transitive**.

A transitive action means that $\mathcal{O}(x) = X$ for all elements $x \in X$.

Theorem VIII.5.2. *Suppose that the group G acts on the set X. Two orbits $\mathcal{O}(x)$ and $\mathcal{O}(y)$ are either equal or disjoint. Since the union of all orbits is X, the family of all orbits defines a* **partition of X**.

Proof. If $z \in \mathcal{O}(x) \cap \mathcal{O}(y)$, then $z = g * x = h * y$, and thus $y = (h^{-1}g) * x$ and $x = (g^{-1}h) * y$. Hence, $\mathcal{O}(x) = \mathcal{O}(y)$. The rest follows. □

The following theorem is of particular importance for finite groups.

Theorem VIII.5.3. *Let $St(x)$ be the stabilizer of $x \in X$ and $y \in \mathcal{O}(x)$. Let $g_0 \in G$ be such that $g_0 * x = y$. Then*

$$\{g \in G \mid g * x = y\} = g_0 \, St(x)$$

is a left coset of $St(x)$. Hence, there is a bijection between the family of the left cosets of the stabilizer $St(x)$ and the orbit $\mathcal{O}(x)$, that is, the cardinality $|\mathcal{O}(x)|$ is equal to the index $|G : St(x)|$ of $St(x)$ in G, that is,

$$|\mathcal{O}(x)|\ |St(x)| = |G|.$$

Moreover, $St(y) = g_0 St(x) g_0^{-1}$.

Proof. For $h \in St(x), (g_0 h) * x = g_0 * (h * x) = g_0 * x = y$. On the other hand, if $g * x = y$, then $g * x = g_0 * x$ implies $g_0^{-1} g * x = x$, i. e. $g_0^{-1} g \in St(x)$. Thus $g \in g_0\ St(x)$. Hence, there is a bijection

$$\mathcal{O}(x) \longleftrightarrow \text{left cosets of } St(x) \text{ in } G \quad \text{given by} \quad g * x \longleftrightarrow gSt(x).$$

The remaining statements now follow. □

There are two important consequences of Theorem VIII.5.3.

Corollary VIII.5.1. *Suppose that a finite group G acts on a set. Then the number of elements in every orbit is a divisor of the order $|G|$ of the group. Moreover, for any two elements x and y belonging to the same orbit, $|St(x)| = |St(y)|$.*

The following theorem, which is very important in dealing with combinatorial problems, is often attributed to **William Burnside (1852 - 1927)**; it is also known as a theorem of **Augustin Louis Cauchy** and **Ferdinand Georg Frobenius**.

Theorem VIII.5.4 (Counting orbits). *Let G be a finite group acting on a set X. Then the number N of orbits of G is equal to*

$$N = \frac{1}{|G|} \sum_{g \in G} |X_g|, \quad \textit{where } X_g = \{x \in X \mid g \star x = x\}.$$

In other words, N is equal to the average number of fixed points of an element of G.

Proof. Let X_0 be a set of representatives of the orbits of G, so $|X_0| = N$. The number of elements of the subset $M = \{(g, x) | g * x = x\}$ of the cartesian product $G \times X$ can be expressed in two distinct ways:

$$|M| = \sum_{x \in X} |St(x)| \qquad \text{and} \qquad |M| = \sum_{g \in G} |X_g|.$$

Then, using Theorem VIII.5.3 and Corollary VIII.5.1, we obtain

$$\sum_{x \in X} |St(x)| = \sum_{r \in X_0} \sum_{x \in \mathcal{O}(r)} |St(x)| = \sum_{r \in X_0} |\mathcal{O}(r)||St(r)| = |X_0||G|.$$

□

We are going to illustrate an application of the Counting orbits theorem on two examples.

Example VIII.5.1. In how many ways can the sides of a square be painted by four colors in such a way that each color can be used several times and that two colorings are considered equivalent if one can be obtained from the other by action of the dihedral group

$$\mathcal{D}_8 = \langle \alpha, \beta \mid \alpha^4 = \beta^2 = (\alpha\beta)^2 = 1 \rangle.$$

To solve this problem, denote the colors by a, b, c, d. The overall number of possibilities of painting the four sides (without any equivalence) is $4^4 = 256$. Hence the set X of all colorings satisfies $|X| = 256$.

We describe now the sets $X_1, X_\alpha, X_{\alpha^2}, X_{\alpha^3}, X_\beta, X_{\alpha\beta}, X_{\alpha^2\beta}$ and $X_{\alpha^3\beta}$. Trivially, $|X_1| = |X| = 256$. The sets X_α and X_{α^3} contain only the squares whose four sides have the same color; thus, $|X_\alpha| = |X_{\alpha^3}| = 4$. The opposite sides of the squares in the set X_{α^2} have the same color and thus $|X_{\alpha^2}| = 4^2 = 16$. If we interpret β as the transformation that exchanges a pair of opposite sides of the square, then the sets X_β and $X_{\alpha^2\beta}$ contain the squares colored in such a way that one of the pairs of the opposite sides have the same color; therefore $|X_\beta| = |X_{\alpha^2\beta}| = 4^3 = 64$. Finally, the sets $X_{\alpha\beta}$ and $X_{\alpha^3\beta}$ contain those squares whose pairs of sides with a common vertex have the same color and thus $|X_{\alpha\beta}| = |X_{\alpha^3\beta}| = 4^2 = 16$. Now, applying Theorem VIII.5.4, we obtain

$$|\,X_0\,| = \frac{1}{8}(256 + 8 + 16 + 128 + 32) = 55.$$

We list here all these colorings of the square:

$4 = 4\times$ using a single color ($aaaa$),
$24 = 6\times$ using two colors ($aaab$, $aabb$, $abab$, $abbb$),
$24 = 4\times$ using three colors ($aabc$, $abbc$, $abcc$, $abac$, $babc$, $cacb$),
3 using all four colors ($abcd$, $abdc$, $acbd$).

The sizes of the individual orbits are as follows:

$|\mathcal{O}(aaaa)| = 1,$
$|\mathcal{O}(aaab)| = |\mathcal{O}(aabb)| = |\mathcal{O}(abbb)| = 4,$
$|\mathcal{O}(abab)| = 2,$
$|\mathcal{O}(aabc)| = |\mathcal{O}(abbc)| = |\mathcal{O}(abcc)| = 8,$
$|\mathcal{O}(abac)| = |\mathcal{O}(babc)| = |\mathcal{O}(cacb)| = 4,$
$|\mathcal{O}(abcd)| = |\mathcal{O}(abdc)| = |\mathcal{O}(acbd)| = 8.$

Hence, $4 \times 1 + 6 \times 12 + 6 \times 2 + 4 \times 24 + 4 \times 12 + 3 \times 8 = 256$.

Example VIII.5.2. A necklace contains 3 red beads and 20 blue ones. How many such different necklaces exist?

A mathematical model of such a necklace is a regular 23-gon. The set of all possible triples of red vertices has $\binom{23}{3} = 1771$ elements. Two such triples represent the same necklace if there is possible to transform one of the triple into the other by action of the dihedral group

$$\mathcal{D}_{46} = \langle \alpha, \beta \mid \alpha^{23} = \beta^2 = (\alpha\beta)^2 = 1 \rangle.$$

This group contains, aside the identity element, 22 elements of order 23 and 23 elements of order 2. Trivially, $X_1 = 1$ and $X_a = 0$ for every element $a \in \mathcal{D}_{46}$. For every element $b \in \mathcal{D}_{46}$ of order 2 (which represents a reflection symmetry) $|X_b| = 11$. Hence

$$\sum_{g \in \mathcal{D}_{46}} |X_g| = 1771 + 23 \times 11 = 2024, \qquad \text{and thus} \qquad |X_0| = \frac{2\,024}{46} = 44.$$

Note that all 44 necklaces can be described by the integral partitions of the number 20, that is, by specifying the numbers of blue beads in between the individual red beads:

$$\begin{array}{llllll} 20 = 0+0+20 & = 0+1+19 & = 0+2+18 & = \ldots\ldots & & = 0+10+10 \\ \quad = 1+1+18 & = 1+2+17 & = 1+3+16 & = \ldots\ldots & & = 1+9+10 \\ \quad = 2+2+16 & = 2+3+15 & = 2+4+14 & = \ldots\ldots & & = 2+9+9 \\ \quad = 3+3+14 & = 3+4+13 & = 3+5+12 & = \ldots\ldots & & = 3+8+9 \\ \quad = 4+4+12 & = 4+5+11 & = 4+6+10 & = \ldots\ldots & & = 4+8+8 \\ \quad = 5+5+10 & = 5+6+9 & = 5+7+8 & = 6+6+8 & & = 6+7+7. \end{array}$$

There are 11 orbits that have 23 elements (corresponding to those necklaces that have equal numbers of beads in two out of the three chains of blue beads) and 33 orbits that have 46 elements.

Exercise VIII.5.2. Justify the following table related to Example VIII.5.1.

number of red beads	number of different necklaces
1 or 22	1
2 or 21	11
3 or 20	44
4 or 19	220
5 or 18	759
6 or 17	2277
7 or 16	5412
8 or 15	10824
9 or 14	17930
10 or 13	25102
11 or 12	29624

Exercise VIII.5.3. Let $p = 2n+1$ be a prime and let $t \in \{1, 2, \ldots, n\}$. Prove that the number of different necklaces with $2t$ red beads and $2(n-t)+1$ blue beads equals

$$\frac{1}{2}\left[\frac{1}{2t}\binom{2n}{2t-1} + \binom{n}{t}\right].$$

Exercise VIII.5.4 (Coloring of a set). Suppose that finite group G acts on a finite set X. Let Y be a finite set of (different) colors. The elements of the function space $Y^X = \{f : X \to Y\}$ will be called colorings of X. Two colorings f_1 and f_2 are indistiguishable if there is $g \in G$ such that $f_1(x) = f_2(g * x)$ for all $x \in X$. Show that the number N of recognizable colorings of X is the number of orbits of the induced action of G on Y^X (see Exercise VIII.5.1). Hence, $N\,|G| = \sum_{g \in G} |(Y^X)_g|$. Show that the number of distinct colorings of the vertices of a square by n colors is $\frac{1}{8}\,(n^4 + 2n^3 + 3n^2 + 2n)$.

The following simple theorem describes the relationship between an action of a group and the induced action of its subgroups. Let H be a subgroup of a group G that acts on a set X by $\varphi : G \to \mathcal{S}(X)$. Then φ defines the **induced action** $\varphi^{(H)} : H \to \mathcal{S}(X)$ of the subgroup H on X. We denote the respective orbits of an element $x \in X$ by these actions $\mathcal{O}_G(x)$ and $\mathcal{O}_H(x)$.

Theorem VIII.5.5. *We consider an action of a group G on a set X and the induced action of a subgroup H of G on X. Then the partition of X by the orbits of the H-action is a* *refinement of the partition of X by the orbits of the G-action, that is, every orbit $\mathcal{O}_G(x)$ is partitioned by the orbits $\mathcal{O}_H(y)$, where $y \in \mathcal{O}_G(x)$.*

Proof. Since $\mathcal{O}_H(x) \subseteq \mathcal{O}_G(x)$ for all $x \in X$, everything now follows. □

Remark VIII.5.2. In the case that the number of H-orbits is finite (for instance in the important case when a finite group acts on a finite set), we can simply display the orbits as the union of families of orbits

$$\{\mathcal{O}_H(x_{11}), \mathcal{O}_H(x_{12}), \ldots, \mathcal{O}_H(x_{1l_1})\} \cup \cdots \cup \{\mathcal{O}_H(x_{t1}), \mathcal{O}_H(x_{t2}), \ldots, \mathcal{O}_H(x_{tl_t})\} \cup \cdots$$

$$\cup \{\mathcal{O}_H(x_{k1}), \mathcal{O}_H(x_{k2}), \ldots, \mathcal{O}_H(x_{kl_k})\},$$

where $\{\mathcal{O}_H(x_{t1}), \mathcal{O}_H(x_{t2}), \ldots, \mathcal{O}_H(x_{tl_t})\}$ is a partition of $\mathcal{O}_G(x_{t1})$ for all $t \in \{1, 2, \ldots, k\}$.

We conclude this section with a few applications:

(1) **Cayley's theorem** revisited (see Theorem VIII.3.1). Let G be a group acting on $X = G$ by **left multiplication**, that is, $\varphi_g : G \to G$ is defined by $g * x = \varphi_g(x) = gx \in G$ for all $g, x \in G$.

Here, for any $x, y \in G$, there is $g = yx^{-1}$ such that $\varphi_g(x) = yx^{-1}x = y$, and thus there is a **unique** orbit $\mathcal{O}(x) = X(= G)$ for every $x \in G$. For every $x \in G$, the stabilizer $St(x)$ is the trivial subgroup $\{1\}$ of G. Hence the corresponding homomorphism $\varphi : G \to \text{Im}\varphi \subseteq \mathcal{S}(G)$ is an isomorphism. This is the essence of Cayley's theorem: Every group G is isomorphic to a particular subgroup of the group of all permutations of G.

(2) Let G act on the set X of **all** subgroups K of G by conjugation $K \to gKg^{-1}$. Then, for each subgroup $K \subseteq G, \mathcal{O}(K)$ is the set of all subgroups of G conjugate to K, and $St(K) = \{g \in G \mid gKg^{-1} = h\}$ is the **normalizer** $N_G(K)$ of K in G. Recall again that $|\mathcal{O}(K)| = |G : N_G(K)|$ (Theorem VIII.5.3). Of course, $\mathcal{O}(K) = \{K\}$ if and only if K is a **normal subgroup** of G. In particular, $\mathcal{O}(Z(G)) = \{Z(G)\}$.

(3) Let H be a subgroup of a group G. Let $X = \{g_1H = H, g_2H, \ldots, g_dH\}$ be the set of all left cosets of H in G. Define the action of G on X by **left multiplication**: For $g \in G$,

$$g * g_tH = \varphi_g(g_tH) = gg_tH.$$

Then, there is a **unique** orbit $\mathcal{O}(g_tH) = X$ for every g_tH. The stabilizer

$$St(g_tH) = \{g \in G \mid gg_tH = g_tH\} = \{g \in G \mid g = g_thg_t^{-1} \text{ for some } h \in H\} = g_tHg_t^{-1}$$

is a subgroup of G conjugate to H. Of course, $H = St(H)$ is the stabilizer of $H \in X$. Consequently, if H is a normal subgroup of G, then H is the stabilizer of **all** $g_tH \in X$.

The Class Equation (Action by conjugation). Let G act on $X = G$ by conjugation: $\varphi_g(x) = g * x = gxg^{-1}$. It is easy to check that this is a group action of G on G. Here, for each $x \in G, \mathcal{O}(x) = \{gxg^{-1} \mid g \in G\}$ is the **conjugacy class** of x in G. Furthermore, $St(x) = \{g \in G \mid gxg^{-1} = x\} = \{g \in G \mid gx = xg\}$ is the **centralizer** $C_G(x)$ of x in G. Now, $\mathcal{O}(x) = \{x\}$ if and only if x belongs to the **center** $Z(G)$ of G, and we get, using Theorem VIII.5.3, ($|\mathcal{O}(x)| = |G : C_G(x)|$), the following assertion.

Theorem VIII.5.6. *Let $\{g_1, g_2, \ldots, g_d\}$ be a complete set of representatives of the conjugacy classes of a* **finite** *group G that do not belong to the center $Z(G)$ of G. Then*

$$|G| = |Z(G)| + \sum_{t=1}^{d} |G : C_G(g_t)|.$$

A consequence is the following statement.

Corollary VIII.5.2. *Let p be a prime. If G is a nontrivial finite p-group, that is, if $|G| = p^n$ for $n > 0$, then the center $Z(G)$ is non-trivial.*

Here, it is worth mentioning that there exist infinite p-groups (that is groups such that the orders of their elements are powers of the prime p) with a trivial center. Such groups were first constructed by **Aleksandr Gennadievich Kurosh (1908 - 1971)** in 1939.

Exercise VIII.5.5. Prove that if the factor group $G/Z(G)$ is a cyclic group, then the group G is abelian. Note that a special case of this statement asserts that every group of order p^2 is abelian (see Example VIII.3.12).

Remark VIII.5.3. We point out that there are noncommutative groups G such that $G/Z(G)$ is abelian. The quaternion group $\mathcal{Q}_8$ is an example: The center $Z(G)$ is a cyclic group $C(2)$ and the factor group $\mathcal{Q}_8/Z(\mathcal{Q}_8)$ is isomorphic to the four-group $C(2) \times C(2)$. The quaternion group also shows that a group of order p^3 does not have to be abelian. The dihedral group $\mathcal{D}_8$ of order 2^3 is also nonabelian.

Exercise VIII.5.6. Let p be a prime. Show that

$$G_1(p^3) = \langle a, b \mid a^{p^2} = b^p = 1, ba = a^{p+1}b \rangle$$

is a noncommutative group of order p^3. Verify the validity of the lattice of its subgroups as illustrated in Figure VIII.5.1.

Remark VIII.5.4. The group

$$G_2(p^3) = \langle a, b, c \mid a^p = b^p = c^p = 1, ba = ab,\ ca = ac,\ cb = abc \rangle,$$

where p is a prime, is also a noncommutative group of order p^3. For $p \geq 3$, $G_1(p^3)$ and $G_2(p^3)$ are non-isomorphic. Figure VIII.5.2 illustrates the lattice of all subgroups of $G_2(27)$. It can be shown that for $p \geq 3$,

$$C(p^3),\ C(p^2) \times C(p),\ C(2) \times C(2) \times C(2),\ G_1(p^3) \text{ and } G_2(p^3)$$

are the only (non-isomorphic) groups of order p^3.

Exercise VIII.5.7. Show that $G_1(2^3) \simeq G_2(2^3) \simeq \mathcal{D}_8$.

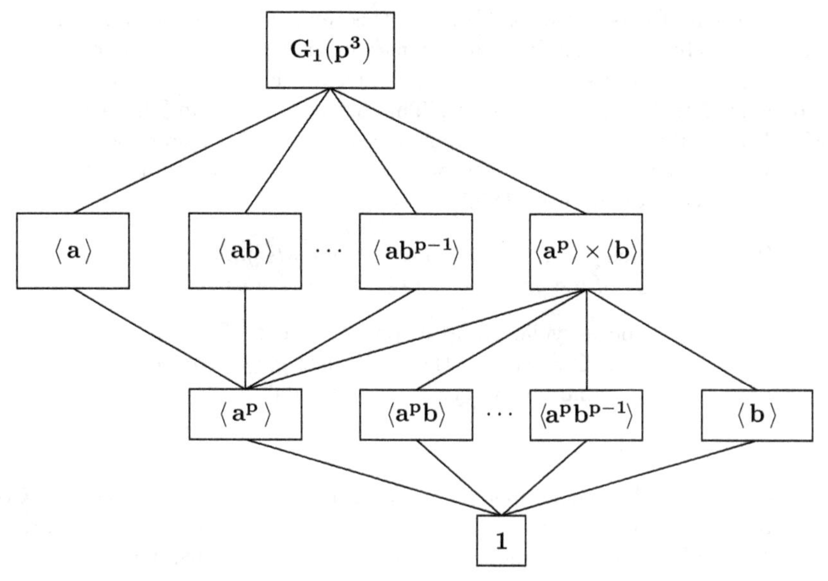

$$G_1(p^3) = \langle\, a, b \mid a^{p^2} = b^p = 1,\ ba = a^{p+1}b \,\rangle$$

Figure VIII.5.1 Lattice of $2\,(p+2)$ subgroups of $G_1(p^3)$

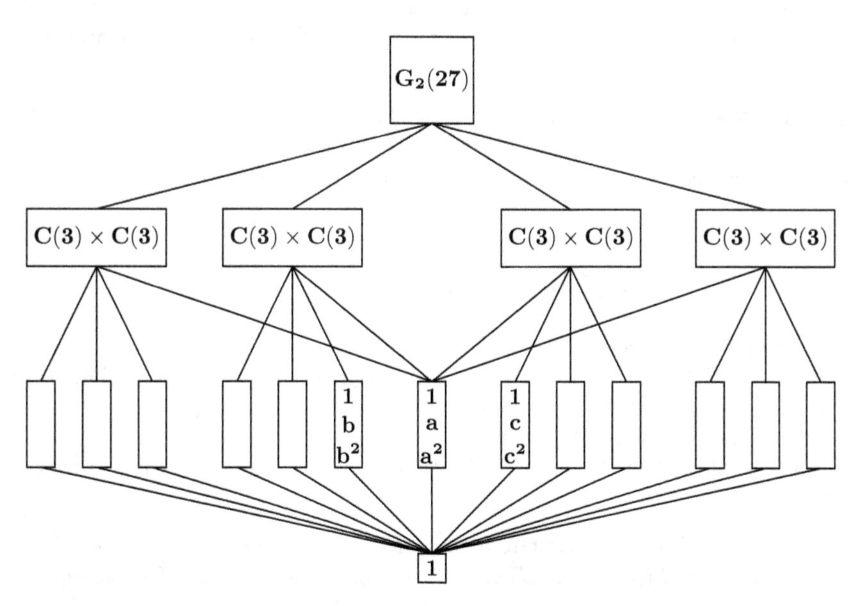

$$G_2(27) = \langle a,\ b,\ c \mid a^3 = b^3 = c^3 = aba^2b^2 = aca^2c^2 = abcb^2c^2 = 1\rangle$$

Figure VIII.5.2 Lattice of subgroups of $G_2(27)$

The last theorem of this section is Cauchy's theorem. The original proof given by **Augustin Louis Cauchy** was a hundred pages long and unprecedented in its complex computations involving permutational group theory. We give an elegant proof of Cauchy's theorem due to **James H. McKay (1928 - 2012)**. This proof appeared in The American Mathematical Monthly 66 (1959), p. 119. The importance of Cauchy's theorem lies in the fact that it paved the way to subsequent generalizations by other algebraists culminating in the work of **Peter Ludwig Mejdell Sylow (1832 - 1918)**.

Theorem VIII.5.7 (Cauchy's theorem). *Let G be a finite group. Let p be a prime number which divides the order $|G|$ of the group. Then G has an element of order p.*

Proof. Let $H = < h >$ be a cyclic group of order p. Let X be the set of all p-tuples $(g_1, g_2, \ldots, g_p)$ of elements of G such that their products $g_1 g_2 \ldots g_p$ equal to identity. The group action of H on X is defined by a "cyclic permutation"

$$h * (g_1, g_2, \ldots, g_p) = (g_2, g_3, \ldots, g_p, g_1).$$

It is easy to check that this cyclic operation is a group action. Now, in view of Corollary VIII.5.1, the orbits have length either 1 or p. Since the orbit of the sequence $(1, 1, \ldots, 1)$ is of length 1, the number k of all sequences whose orbits have length 1 is non-zero. At the same time, the number of all sequences is $|G|^{p-1}$ and thus

$$|G|^{p-1} = r + sp,$$

where s is the number of orbits of length p. Consequently, r must be a multiple of p, and thus $r \geq 2$. Therefore, there is a sequence $(g, g, \ldots, g)$ with $g \in G, g \neq 1$ and $g^p = 1$, as required. □

Exercise VIII.5.8. Let p be a prime. Prove that a finite group has order p^n for some $n \in \mathbb{N}_0$ if and only if all its elements have orders that are powers of p. Give an example of a group that is generated by elements of order p despite its order not beeng a power of p.

VIII.6 Sylow's theorems

At the end of the last section, using Cauchy's theorem, we were able to answer a rather innocent question: If the orders of all elements of a group G are powers of a prime p, must the order of G be a power of p? To learn more about the subgroups of prime power orders of a finite group we need a finer insight. A major development in this direction are the following two theorems established by Sylow in 1872; they have become indispensable tools for the study of finite groups.

Theorem VIII.6.1 (First Sylow theorem). *Let G be a group of order $|G| = p^k r$, where p is a prime number and $(p, r) = 1$. Then G has a subgroup of order p^k (a so-called* **Sylow p-subgroup***).*

Proof. Let G acts on the set $X = \{S_1, S_2, \ldots, S_s\}$ of **all subsets** (that is, **not** only on subgroups) S_t of G that have p^k elements, by left multiplication: $g * S_t = gS_t$ for all $1 \leq t \leq s$. Observe that for every element $x \in G, |Gx| = p^k r$; this means that $|\mathcal{O}(S_t)| \geq r$ and thus, by Theorem VIII.5.3, $|St(S_t)| \leq p^k$ for all $1 \leq t \leq s$. On the other hand, the number s of elements in X is equal to $\binom{|G|}{p^k} = \binom{p^k r}{p^k}$ and thus s is **not divisible** by p (see Exercise VIII.5.7). As a consequence, there must be an orbit $\mathcal{O}(S_{t_0})$ whose length $|\mathcal{O}(S_{t_0})|$ is not divisible by p, and thus $|St(S_{t_0})| \geq p^k$. Consequently, the order of $St(S_{t_0})$ is p^k, that is, G contains a Sylow p-subgroup. □

Remark VIII.6.1. The first Sylow theorem should be viewed as a partial converse of Lagrange's theorem that asserts that orders of subgroups of a finite group divide the group order. There exist however groups G that have, for a particular divisor d of their order, no subgroup of order d. An example of such a group is the alternating group $\mathcal{A}_4 = \langle a, b \,|\, a^3 = b^2 = (a\,b)^3 = 1\rangle$ of order 12 that possess no subgroup of order 6. The first Sylow theorem only guarantees the existence of subgroups whose order is the highest prime power dividing the order of the group. Building on Exercise VIII.5.8, the next two exercises prove a generalization that guarantees the existence of subgroups whose order is any prime power dividing the order of the group.

Exercise VIII.6.1. Let p be a prime and k a natural number. Prove:

(i) If $0 < m \leq p^k - 1$ and p and r are relatively prime, then
$$p^s \text{ is a divisor of } m \text{ if and only if } p^s \text{ is a divisor of } p^k r - m.$$

(ii) If p^t is the highest power of p that divides r, then p^t is also the highest power of p that divides the binomial coefficient $\binom{p^k r}{p^k}$.

Exercise VIII.6.2. Use Exercise VIII.6.1 to prove the following generalization of the first Sylow theorem:

Let G be a finite group and p a prime number. If p^k divides the order of G, then G has a subgroup of order p^k. The case $k = 1$ is Cauchy's theorem (Theorem VIII.5.7).

The second Sylow theorem brings further information concerning the Sylow p-subgroups, in particular concerning their number. For the proof, we need some refined statements concerning conjugations. In particular, we will consider a subgroup H of a group G acting on a set X of subgroups of G by conjugation.

Let G be a finite group of order $|G| = p^k r$, where p is prime and $d(p, r) = 1$. Let $X = \{H_1, H_2, \ldots, H_s\}$ be the set of **all** Sylow p-subgroups of the group G.

(i) First, let $H = G$ act on the set X by conjugation. Since for every subgroup H_t, we have $St(H_t) \supseteq H_t$, we obtain by Lagranges's theorem $|St(H_t)| = p^k s$, where s is a divisor of r and thus by Theorem VIII.5.3

the number $|\mathcal{O}(H_t)|$ of subgroups in the G-orbit of each H_t divides the number r.

(ii) Second, let the subgroup $H = H_1$ act on X by conjugation. Trivially, $\mathcal{O}(H_1) = \{H_1\}$. We claim that **the number s of all subgroups of order p^k** satisfies

$$s \equiv 1 \pmod{p}. \qquad \text{(VIII.6.1)}$$

This is a consequence of the following two results (Theorem VIII.6.2 and Remark VIII.6.2).

Theorem VIII.6.2. *Let H and K be two subgroups of a group G. Suppose $hKh^{-1} = K$ for all $h \in H$. Then the subgroup $\langle H, K\rangle$ generated by H and K is $HK(= KH)$. In particular, if both H and K are p-groups, then HK is also a p-group. Moreover, if K is a maximal p-subgroup of G, then $H \subseteq K$.*

Proof. By our assumption, for every $h \in H$ and $k \in K$, there is $k' \in K$ such that $kh = hk'$. Hence, $h_1k_1 \cdot h_2k_2 = h_1h_2k_1'k_2$ and $(hk)^{-1} = h^{-1}k''$. It follows that HK is a subgroup. The order of every element hk is a power of p. Indeed, if $h^{p^n} = 1$, then $(hk)^{p^n} = h^{p^n}(k')^{p^n-1}k = k''$ is of prime power order. Thus, HK is a p-group. Finally, if K is a maximal p-subgroup of G, then $HK = K$, and thus $H \subseteq K$, as required. □

Remark VIII.6.2. Note that, given a p-subgroup H and a maximal p-subgroup K of a group G, then either $H \subseteq K$ or there is $h_0 \in H$ such that $h_0Kh_0^{-1} \neq K$. This fact completes the proof of the condition (VIII.6.1). For, as already mentioned, $\mathcal{O}(H_1) = \{H_1\}$ and, for each $2 \leq t \leq s, |\mathcal{O}(H_t)| \geq 2$ since $H_1 \neq H_t$. Furthermore, the length $|\mathcal{O}(H_t)|$ of the orbit $\mathcal{O}(H_t)$ equals the index of the stabilizer of H_t in H_1, and is therefore equal to a power of p.

We are now ready to formulate and prove the second Sylow theorem using Theorem VIII.6.2 and the statements (i) and (ii) following Exercise VIII.6.2.

Theorem VIII.6.3 (Second Sylow theorem). *Let the order of G satisfy $|G| = p^k r$, where p is a prime number and $d(p, r) = 1$. Let $\{H_1, H_2, \ldots, H_s\}$ be the set of all Sylow p-subgroups (that is, all subgroups of order p^k). Then*

$$s \equiv 1 \pmod{p},$$

all subgroups $\{H_1, H_2, \ldots, H_s\}$ are conjugate, and

s is a divisor of r.

Proof. The first assertion is proved in (ii) above.

The content of (i) can be reformulated, in the spirit of Remark VIII.5.4, as follows: All G-orbits (that is, all conjugacy classes of the Sylow p-subgroups) have their lengths relatively prime to p (in fact, they are divisors of r). Each of these G-orbits is partitioned by the respective H-orbits. With the exception of the singleton orbit $\{H_1\}$, every other H-orbit has length p^k with a suitable $k \geq 1$. Consequently, there can be no G-orbit that would not contain the H-orbit $\{H_1\}$, and thus there is a unique G-orbit containing all Sylow p-subgroups.

Finally, since there is a unique G-orbit of all s Sylow p-subgroups, s divides r by (i). □

Remark VIII.6.3. If G is a finite abelian group, then the Sylow p-subgroup of G is its p-primary component G_p. There are only finitely many such nontrivial components $G_{p_s}, 1 \leq s \leq r$, and $G = \prod_{s=1}^r G_{p_s}$ (see Theorem VIII.4.2). A group (not necessarily abelian) that is a direct product of its Sylow p-subgroups is called **nilpotent**. Nilpotent groups are groups that are "nearly abelian" in the sense that they are characterized by the existence of a finite **central series**

$$\{1\} = H_0 \subsetneq H_1 \subsetneq \cdot \subsetneq H_t \subsetneq H_{t+1} \subsetneq \cdots \subsetneq H_l = G,$$

where $H_{t+1}/H_t \subseteq Z(G/H_t)$ for all $0 \leq t \leq l-1$. The shortest possible length of a central series of G is called the **nilpotency class** of G. Thus the nontrivial abelian groups are just the nilpotent groups of nilpotency class 1.

Here are a few applications of Sylow's theorems.

Example VIII.6.1. Determine all groups of order pq, where $p < q$ are two distinct primes. Denote by Q the Sylow q-subgroup; thus Q is a cyclic group of order q. The number n_q of such groups is $1 + kq$ for some $k \geq 0$. This number must also be a divisor of the prime p, and therefore necessarily $k = 0$, that is, $n_q = 1$. This means that Q is a normal subgroup of G and hence a group G of order pq is never simple.

Let P be a Sylow p-subgroup of G. The number $n_p = 1 + kp$ of such groups is a divisor of q and thus equals either to 1 or q. If p is not a divisor of the number $q - 1$, then P is the only Sylow p-subgroup and thus normal in G. Since, in this case, both P and Q are normal subgroups of G and $P \cap Q = \{1\}, G = P \times Q \simeq C(pq)$ is a cyclic group of order pq.

The case when p is a divisor of the number $q - 1$ is more intricate and will be fully explained only in the next section dealing with semidirect products. In this case there exists, besides the cyclic group, one additional (noncommutative) group of order pq with q Sylow p-subgroups. It can be defined by

$$G = \langle a, b \mid a^q = b^p = 1,\ ba = a^r b,\ \text{where } r^p \equiv 1 \pmod{q} \rangle.$$

Such a number $r \neq 1$ always exists: Take $r = t^s \pmod{q}$ for a suitable t, where $ps = q - 1$ (using Euler's theorem).

Consequently, every group of order equal to one of the following numbers

$$15, 33, 35, 51, 65, 69, 77, 85, 87, 91, 95, 115, 119, 133, 143, 145, 161, 187, 209, 221,$$
$$247, 299, 319, 323, 377, 391, 437, 493, 551, 667, \ldots \text{ is necessarily cyclic.}$$

Example VIII.6.2. There are only two (non-isomorphic) groups of order 99 and both are abelian. In fact, a group G of order 99 satisfies either $G \simeq C(99)$ or $G \simeq C(3) \times C(33) = C(3) \times C(3) \times C(11)$. This can be derived very easily. For, the number n_{11} of Sylow 11-subgroups is a number $1 + 11k$ that is a divisor of 9; thus $k = 0$ and G contains a normal subgroup isomorphic to $C(11)$. Similarly, there is only one Sylow 3-subgroup of G of order 9 that is necessarily abelian and thus is either isomorphic to $C(9)$ or to $C(3) \times C(3)$.

Example VIII.6.3. There are just 3 (nonisomorphic) groups G of order 75. The group G is ether isomorphic to $C(75) = C(3) \times C(25)$ or to $C(15) \times C(5) = C(3) \times C(5) \times C(5)$, or to the noncommutative group

$$G = \langle a, b, c \mid a^5 = b^5 = c^3 = 1,\ ba = ab,\ ca = ab^2c,\ cb = ab^3c \rangle \tag{VIII.6.2}$$

It is easy to see that there is only one Sylow 5-subgroup. This is a normal subgroup of G of order 25 and thus necessarily abelian. In the case that the Sylow 3-subgroup is also normal in G, we conclude that G is abelian isomorphic either to $C(75)$ or to $C(15) \times C(5)$. In the case that there are more Sylow 3-subgroups, their number n_3 must be a divisor of 25 of the form $1 + 3k$. Thus $n_3 = 1 + 3 \cdot 8 = 25$. Such a group G whose Sylow 5-subgroup is isomorphic

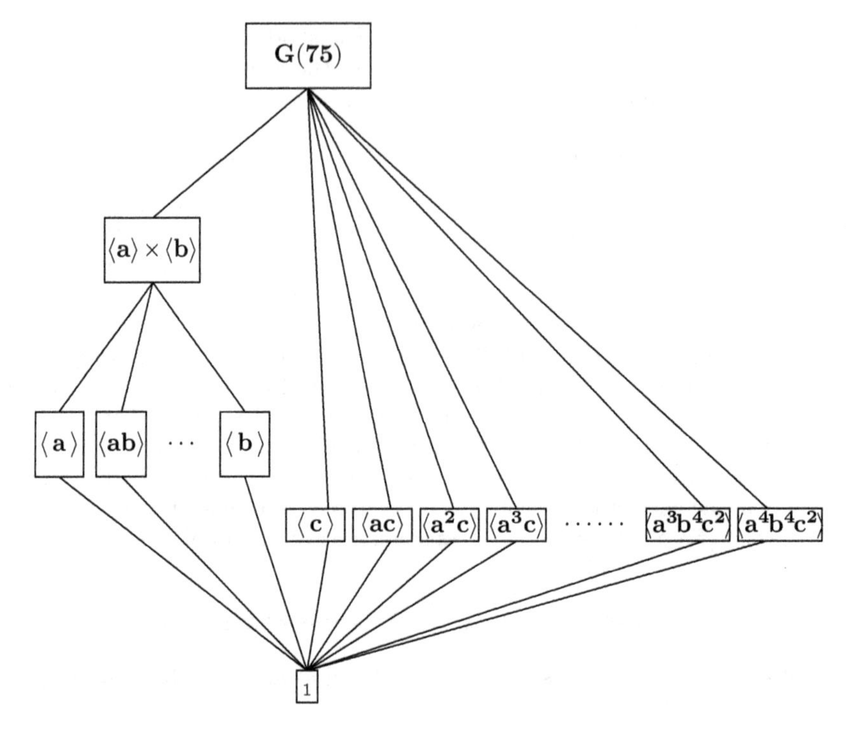

$$\mathbf{G(75)} = \langle\, \mathbf{a, b, c} \mid \mathbf{a^5 = b^5 = c^3 = 1,\ ba = ab,\ ca = ab^2c,\ cb = ab^3c} \,\rangle$$

Figure VIII.6.1 Lattice of 34 subgroups of G(75)

to $C(5) \times C(5)$ exists and is unique. It is defined by (VIII.6.2) and the structure of its subgroups is illustrated in Figure VIII.6.1. The reason for this "mysterious" definition will be the subject matter of the following section on the **semidirect product** of two groups.

Exercise VIII.6.3. Prove that groups of order 56 or 132 have a normal Sylow p-subgroup for some prime divisor p of their orders, that is, that there are no simple groups of order 56 or 132.

Exercise VIII.6.4. Let G be a group of order $p^2 q$, where p and q are distinct primes. Assume that p is not a divisor of $q-1$ and q is not a divisor of p^2-1. Prove that G is isomorphic either to $C(p^2 q)$ or to $C(p\,q) \times C(p)$ (see Example VIII.6.1).

Exercise VIII.6.5. Let G be a group of order $p\,q\,r$, where $p < q < r$ are primes. Prove that G has a normal subgroup isomorphic to either $C(p)$ or $C(q)$ or $C(r)$. In fact, deeper reasoning shows that G contains a normal group of order r (see Example VIII.6.1).

VIII.7 Semidirect products

The dihedral group $G = \mathcal{D}_{2n} = \langle a, b \,|\, a^n = b^2 = (a\,b)^2 = 1\rangle$ is not a direct product of its normal (cyclic) subgroup $A = \langle a \,|\, a^n = 1\rangle$ and a cyclic subgroup $H = \langle h\rangle$ of order 2, where $h = a^t b$ for $0 \le t \le n-1$. However, $G = AH = \{ah \,|\, a \in A, h \in H\}$ and the expressions $g = a\,h$ are for the elements $g \in G$ unique. Many other groups G allow such an "almost direct" decompositions of G and this provides a motivation for the new concept of a **semidirect product** that we want to introduce here. As we shall see it will enable us to describe (and classify) groups of small orders.

A motivation for the construction is as follows. Assume that a group G contains two subgroups A and H such that

(i) $A \cap H = \{1\}$,

(ii) A is a normal subgroup of G and

(iii) the group G is generated by the subgroups A and H : $G = \langle A, H\rangle$.

As a consequence, every element $g \in G$ is possible to express in a unique way in the form $g = a\,h$ with $a \in A, h \in H$. This form follows from the fact that for two arbitrary elements $a \in A$ and $h \in H, h\,a = a'\,h$ for a suitable $a' \in A$. Moreover,

$$(a_1\,h_1)\,(a_2\,h_2) \;=\; a_1\,h_1\,a_2\,(h_1^{-1}\,h_1)\,h_2 \;=\; a_1\,(h_1\,a_2\,h_1^{-1})\,h_1\,h_2 \;=\; a_3\,h_3,$$

where $a_3 \;=\; a_1\,(h_1\,a_2\,h_1^{-1})$ and $h_3 \;=\; h_1\,h_2$. Here, $a_3 \in A$ because A is normal in G. Hence, the multiplication is determined by an **inner automorphism** of the normal subgroup A, that is, by action of H on the subgroup A defined by conjugation

$$h * a = h\,a\,h^{-1}, \text{ and thus } (a_1\,h_1)\,(a_2\,h_2) \;=\; a_1\,(h_1 *\, a_2)\,h_1\,h_2.$$

The action $*$ of the subgroup H on the normal subgroup A establishes a **homomorphism** $\varphi\ : H \to \mathbf{Aut}(A)$ of the subgroup H into the group $\mathbf{Aut}(A)$ of automorphisms of the subgroup A by

$$h \longmapsto \varphi_h\ : a \mapsto h\,a\,h^{-1}.$$

Now, acquainted with this motivation, we can formulate and easily prove the following theorem that defines the **semidirect product** $A \rtimes H$ **of two given groups** A and H.

Theorem VIII.7.1. *Let A and H be two groups and φ be a homomorphism of the group H into the automorphism group $\mathbf{Aut}(A)$ of A : $\varphi(h) = \varphi_h$ for $h \in H$. On the cartesian product $G = \{(a,h) \mid a \in A, h \in H\}$ define the following binary operation $\circ$ by*

$$(a_1, h_1) \circ (a_2, h_2) = (a_1\,\varphi_{h_1}(a_2), h_1\,h_2).$$

This operation is associative with a neutral element $1_G = (1_A, 1_H)$ and for every element (a,h), there is an inverse element $(\varphi_{h^{-1}}(a^{-1}), h^{-1})$. Hence, $(G, \circ)$ is a group that possesses the following properties:

(i) *$A_0 = \{(a, 1_H) \mid a \in A\}$ is a normal subgroup of G isomorphic to the group A;*

(ii) *$H_0 = \{(1_A, h) \mid h \in H\}$ is a subgroup of G isomorphic to the group H;*

(iii) *$A_0 \cap H_0 = \{(1_A, 1_H)\}$;*

(iv) $(a,h) = (a, 1_H) \circ (1_A, h)$ *and*

(v) $(1_A, h) \circ (a, 1_H) \circ (1_A, h^{-1}) = (\varphi_h(a), 1_H)$.

The group G is said to be a semidirect product of the groups A and H and will be denoted $G = A \rtimes H$.

Proof. All statements are just results of routine calculations. For example,

$$\begin{aligned}
&[(a_1,h_1)\circ(a_2,h_2)]\circ(a_3,h_3) = (a_1\varphi_{h_1}(a_2), h_1h_2)\circ(a_3,h_3)\\
&= (a_1\varphi_{h_1}(a_2)\varphi_{h_1h_2}(a_3), h_1h_2h_3) = (a_1\varphi_{h_1}(a_2)\varphi_{h_1}(\varphi_{h_2}(a_3)), h_1h_2h_3)\\
&= (a_1\varphi_{h_1}((a_2)\varphi_{h_2}(a_3)), h_1h_2h_3) = (a_1,h_1)\circ(a_2\varphi_{h_2}(a_3), h_2h_3)\\
&= (a_1,h_1)\circ[(a_2,h_2)\circ(a_3,h_3)]
\end{aligned}$$

shows that the operation $\circ$ is associative. Similarly

$$(a,h)\circ(\varphi_{h^{-1}}(a^{-1}), h^{-1}) = (a\,\varphi_h(\varphi_{h^{-1}}(a^{-1})), 1_H) = (1_A, 1_H)$$

and

$$(1_A,h)\circ(a,1_H)\circ(1_A,h^{-1}) = (\varphi_h(a),h)\circ(1_A,h^{-1}) = (\varphi_h(a)\,\varphi_h(1_A),1_H) = (\varphi_h(a),1_H).$$

□

Using the definition of a semidirect product as presented in the previous theorem, we can formulate our introductory comments as a theorem that provides a characterization of a semidirect product.

Theorem VIII.7.2. *Let A be a (nontrivial) normal subgroup of a group G and H a subgroup of G such that $A \cap H = \{1\}$ and $\langle A, H\rangle = G$. Then $G \simeq A \rtimes H$. In this case, H is said to be a* **complement** *for the normal subgroup A of G.*

Exercise VIII.7.1. Give a formal proof of Theorem VIII.7.2. In particular, describe the homomorphism $\varphi : H \to \mathbf{Aut}(A)$.

We observe that a direct product $G = A\times H$ is a particular case of the semidirect product $A \rtimes H$. We can express this relationship in the following simple way.

Theorem VIII.7.3. *Let A and H be two groups and $\varphi : H \to \mathbf{Aut}(A)$ a homomorphism of H into $\mathbf{Aut}(A)$. Then $A\times H \simeq A \rtimes H$ if and only if $\mathbf{Im}\varphi = \{1_A\}$.*

Exercise VIII.7.2. Give a formal proof of Theorem VIII.7.3 and show that $A\times H = A\rtimes H$ is also equivalent to the fact that H is a normal subgroup of $A \rtimes H$. The equality $=$ should be understood as an isomorphism that is identity on the subgroups A and H.

Remark VIII.7.1. We now return to Example VIII.6.3 to illustrate the construction in the previous theorem on the example of a group of order 75 considered there. We deal with the groups

$$A = \langle a\rangle \times \langle b\rangle \cong C(5)\times C(5) \quad \text{and} \quad H = \langle c\rangle \cong C(3)$$

and the homomorphism φ mapping the generator c of the group H into the automorphism φ_c of the group A defined by $\varphi_c(a) = ab^2$ and $\varphi_c(b) = ab^3$.

Exercise VIII.7.3. Check that φ_c of Remark VIII.7.1 is an automorphism of A and that $\varphi_c^3 = 1$. Hence, derive the relations in (VIII.6.2) and the isomorphism $G(75) \simeq (C(5) \times C(5)) \rtimes C(3)$.

Exercise VIII.7.4. Using Theorem VIII.7.3, show that a group G of order p^2, where p is a prime, is isomorphic either to $C(p^2)$ or to $C(p) \times C(p)$ (see Exercise VIII.5.5).

Remark VIII.7.2. It is appropriate now to emphasize the fact that a semidirect product depends substantially on the homomorphism $\varphi : H \to \mathbf{Aut}(A)$! There exist **non-isomorphic** semidirect products of the same two groups (in dependence of the choice of the mentioned homomorphism φ) as illustrated by the following example of the group $G(20)$ and the Frobenius group $F(20)$:

$$G(20) = \langle\, a, b \mid a^5 = b^4 = abab^3 = 1 \,\rangle;$$

$$F(20) = \langle\, a, b \mid a^5 = b^4 = a^2bab^3 = 1 \,\rangle.$$

Both groups are of the form $C(5) \rtimes C(4)$ (see Figures VIII.7.1 and VIII.7.2). These are two of the 5 non-isomorphic groups of order 20. The other three groups are

$$\mathcal{D}_{20} \simeq C(10) \rtimes C(2) \simeq \mathcal{D}_{10} \times C(2) \simeq C(5) \rtimes V_4$$

(see Figure VIII.2.3), $C(20)$ and $C(10) \times C(2)$. The lattice of the subgroups of the last one is given in Figure VIII.7.3.

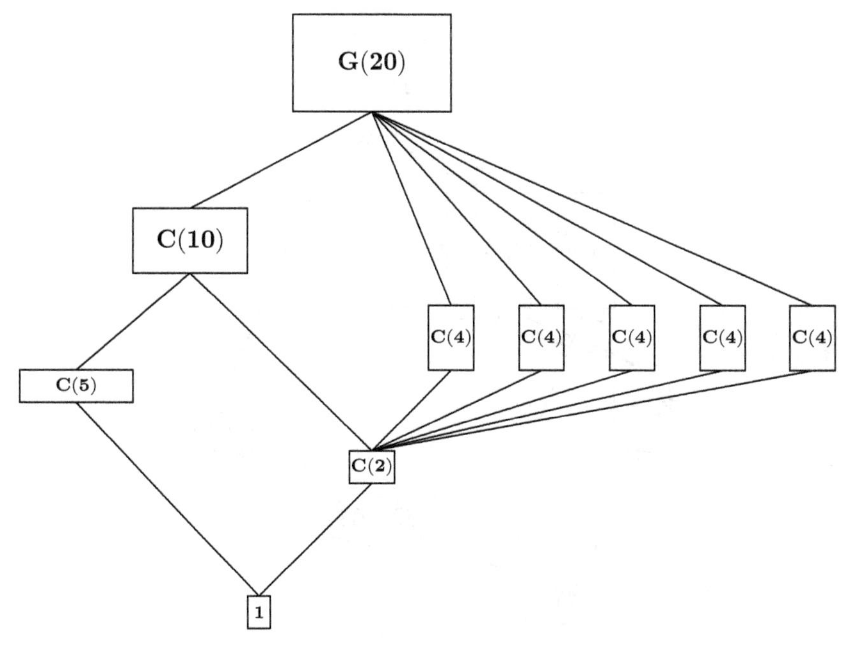

$$\mathbf{G(20) = \langle\, a, b \mid a^5 = b^4 = abab^3 = 1 \,\rangle \simeq C(5) \rtimes C(4)}$$

Figure VIII.7.1 Lattice of subgroups of G(20)

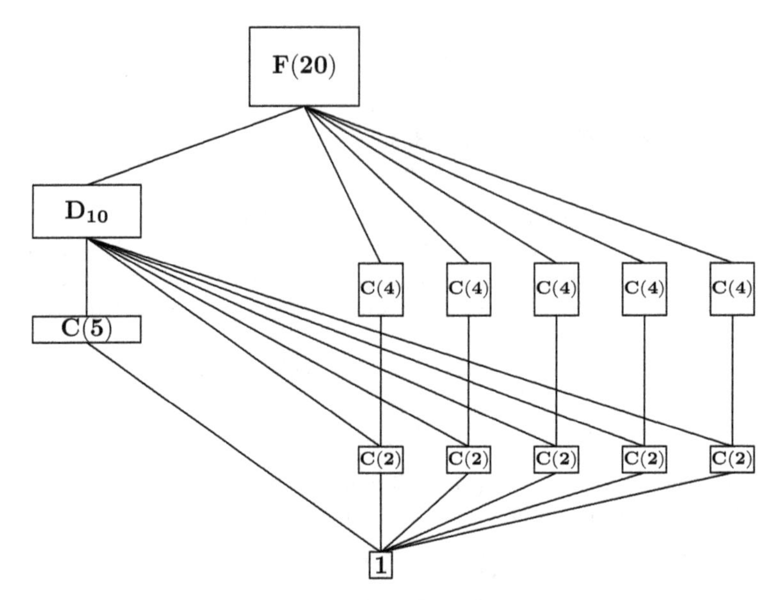

$$\mathbf{F(20)} = \langle\, \mathbf{a,b} \mid \mathbf{a^5 = b^4 = a^2bab^3 = 1} \,\rangle \simeq \mathbf{C(5) \rtimes C(4)}$$

Figure VIII.7.2 Lattice of subgroups of F(20)

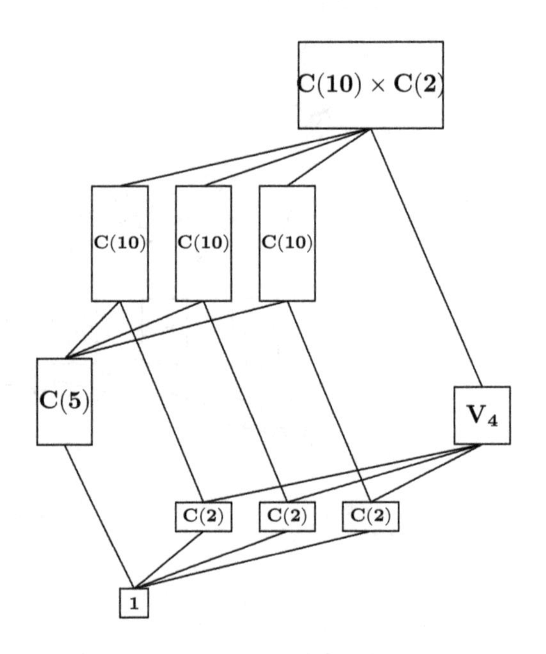

$$\mathbf{G} = \langle\, \mathbf{a,b} \mid \mathbf{a^{10} = b^2 = baba^9 = 1} \,\rangle \simeq \mathbf{C(5) \times V_4}$$

Figure VIII.7.3 Lattice of subgroups of G = C(10) × C(2)

To show that there are only five non-isomorphic groups of order 20 is rather simple. Sylow's theorems imply that the Sylow 5-subgroup of a group G of order 20 must be normal in G (and is isomorphic to $C(5)$). The Sylow 2-subgroup is isomorphic either to $C(4)$ or to V_4. If it is normal in G, then we deal with an abelian group isomorphic either to $C(20)$ or to the direct product $C(10) \times C(2)$. Otherwise, we check all possible homomorphisms of $C(4)$ and V_4 into $\mathbf{Aut}(C(5)) \simeq V_4$ and conclude that they lead to the three noncommutative groups of order 20 listed above.

Exercise VIII.7.5. Show that $\mathbf{Aut}(C(5)) \simeq C(4)$ and describe the homomorphisms $\varphi : C(4) \to \mathbf{Aut}(C(5))$ for the groups $G(20)$ and $F(20)$.

Remark VIII.7.3. Every cyclic group $C(n) = \langle a \rangle, n \geq 3$, has an automorphism ω of order 2 defined by $\omega(a) = a^{n-1}$. For every group H that contain a (normal) subgroup N of index 2 exists therefore a nontrivial (that is, noncommutative) semidirect product $G = C(n) \rtimes H$. Hereby the subgroup N is the center of the group and the quotient group G/N is isomorphic to the dihedral group $\mathcal{D}_{2n}$.

Exercise VIII.7.6. Prove all the assertions of Remark VIII.7.3 and conclude that, for every $n \geq 3$ and $m \geq 1$, there exists a noncommutative group $G = C(n) \rtimes C(2m)$ and write down its presentation.

Example VIII.7.1. There are just 5 non-isomorphic groups G of order 12. The situation is rather similar to the case of groups of order 20. The Sylow 2-subgroup P_2 is isomorphic either to $C(4)$ or V_4 and the Sylow 3-group P_3 is isomorphic to $C(3)$. If both P_2 and P_3 are normal in G, then G is abelian and isomorphic either to $C(12)$ or to $C(6) \times C(2) = C(3) \times V_4$. We already know that the alternating group $\mathcal{A}_4 \simeq V_4 \rtimes C(3)$ (see Figure VIII.1.3) is a nonabelian group of order 12; this is the case when $P_2 \simeq V_4$ is a normal subgroup of G. We also know that the dihedral group $\mathcal{D}_{12} \simeq \mathcal{D}_6 \rtimes C(2) \simeq C(3) \rtimes V_4$ is of order 12; this is the case when $P_3 \simeq C(3)$ is a normal subgroup of G and $P_2 \simeq V_4$ (see Figure VIII.2.4; note that $\mathbf{Aut}(P_3) \simeq C(2)$). The only possible remaining case is when P_3 is a normal subgroup of G and $P_2 \simeq C(4)$. In this case the only nontrivial semidirect product is the **dicyclic group** Dic_3 as defined in Figure VIII.7.4.

Exercise VIII.7.7. Show that the dicyclic group Dic_3 is isomorphic to the subgroup of $GL_2(\mathbb{C})$ generated by the matrices

$$\begin{pmatrix} 0 & i \\ i & 0 \end{pmatrix} \quad \text{and} \quad \begin{pmatrix} \omega & 0 \\ 0 & \omega^2 \end{pmatrix},$$

where $\omega \neq 1$ and $\omega^3 = 1$. Write down a presentation of Dic_3 in terms of two generators a and b, a of order 6 and b satisfying $b^2 = a^3$.

Exercise VIII.7.8. Prove that there are only four non-isomorphic groups G of order 28, that is, show that aside from the abelian groups $G \simeq C(28)$ or $G \simeq C(14) \times C(2) = C(7) \times V_4$, there are only two non-isomorphic nonabelian groups of order 28: the dihedral group

$$\mathcal{D}_{28} \simeq C(14) \rtimes C(2) \simeq \mathcal{D}_{14} \times C(2) \simeq C(7) \rtimes V_4$$

and the group $G(28)$ defined in Figure VIII.7.5.

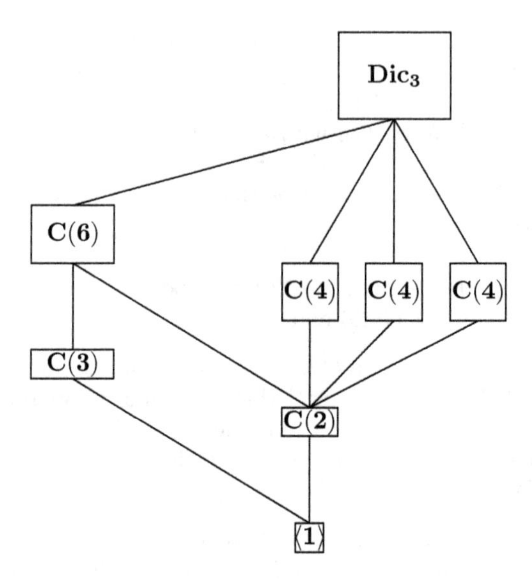

$$\mathbf{Dic_3} = \langle\, \mathbf{a}, \mathbf{b} \mid \mathbf{a}^3 = \mathbf{b}^4 = \mathbf{abab}^3 = \mathbf{1} \,\rangle \simeq \mathbf{C(3)} \rtimes \mathbf{C(4)}$$

Figure VIII.7.4 Lattice of subgroups of $\mathbf{Dic_3}$

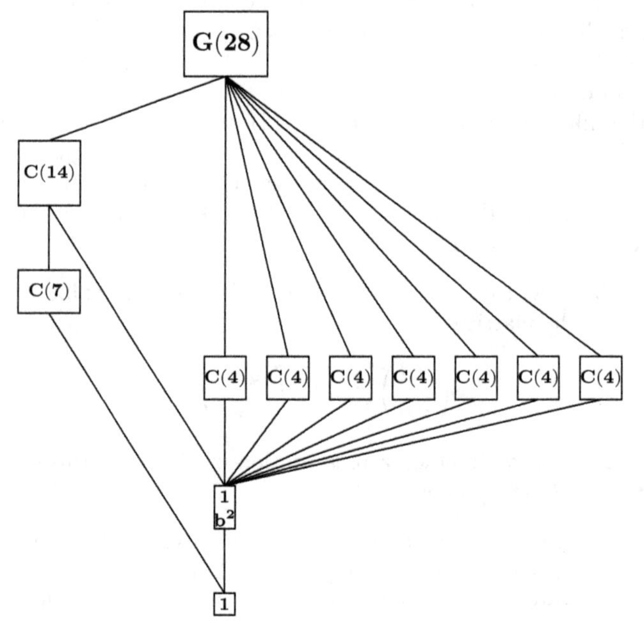

$$\mathbf{G(28)} = \langle\, \mathbf{a}, \mathbf{b} \mid \mathbf{a}^7 = \mathbf{b}^4 = \mathbf{abab}^3 = \mathbf{1} \,\rangle \simeq \mathbf{C(7)} \rtimes \mathbf{C(4)}$$

Figure VIII.7.5 Lattice of subgroups of G(28)

After gaining experience with the previous exercises, it should be quite simple to determine all non-isomorphic groups of order 18. Here is a hint.

Exercise VIII.7.9. Justify that the abelian groups $C(18)$ and $C(6) \times C(3)$ together with three nonabelian groups $\mathcal{D}_{18} \simeq C(9) \rtimes C(2), G_1(18) \simeq ((C(3) \times C(3)) \rtimes C(2)$ and $G_2(18) \simeq ((C(3) \times C(3)) \rtimes C(2) \simeq C(3) \rtimes \mathcal{D}_6$ defined in Figures VIII.7.6, VIII.7.7 and VIII.7.8 are the only non-isomorphic groups of order 18.

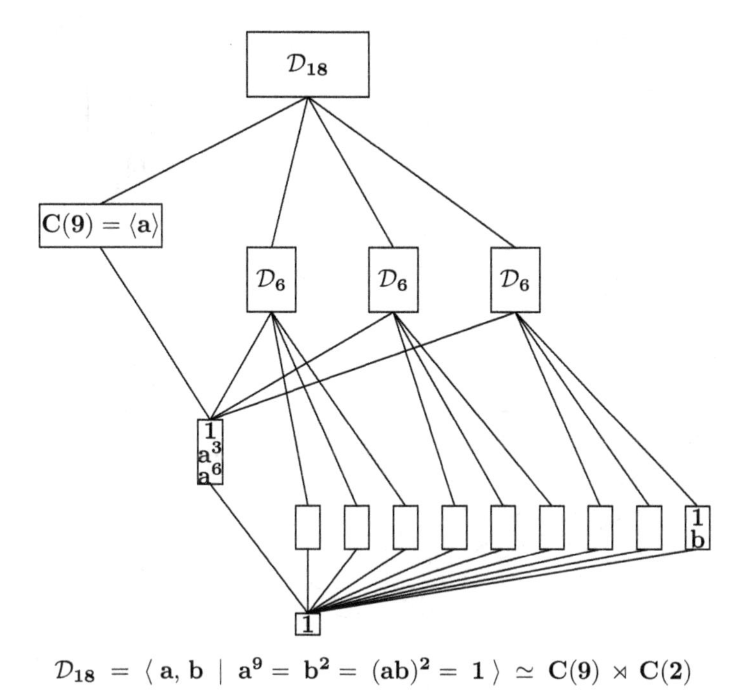

$$\mathcal{D}_{18} = \langle\, \mathbf{a}, \mathbf{b} \mid \mathbf{a}^9 = \mathbf{b}^2 = (\mathbf{ab})^2 = \mathbf{1} \,\rangle \simeq \mathbf{C(9)} \rtimes \mathbf{C(2)}$$

Figure VIII.7.6 Lattice of subgroups of $\mathcal{D}_{18}$

We point out that, by now, we can easily determine non-isomorphic groups of all orders up to 20 with the **exception of order** 16. For the prime orders $n = 2, 3, 5, 7, 11, 13, 17, 19$ and $n = 15$, there is only one group of that order, namely $C(n)$. For even products of two (possibly equal) primes $2n = 4, 6, 10$ and 14, there are two groups of that order, namely $\mathcal{D}_{2n}$ and $C(2n)$, for $n = 9, 15$ there are also two groups of that order, namely $C(9)$ and $C(3) \times C(3)$. We can easily establish that there are only five non-isomorphic groups of order 8: the abelian groups $C(8), C(4) \times C(2)$ and $C(2) \times C(2) \times C(2)$ together with two nonabelian groups $\mathcal{D}_8$ and $\mathcal{Q}_8$. Furthermore, we have exhibited in the previous examples and exercises all non-isomorphic groups of orders 12, 18 and 20. So our task is to exhibit 14 non-isomorphic groups of order 16. There are 5 such abelian groups:

$$C(16), C(8) \times C(2), C(4) \times C(4), C(4) \times C(2) \times C(2) \text{ and } C(2) \times C(2) \times C(2) \times C(2).$$

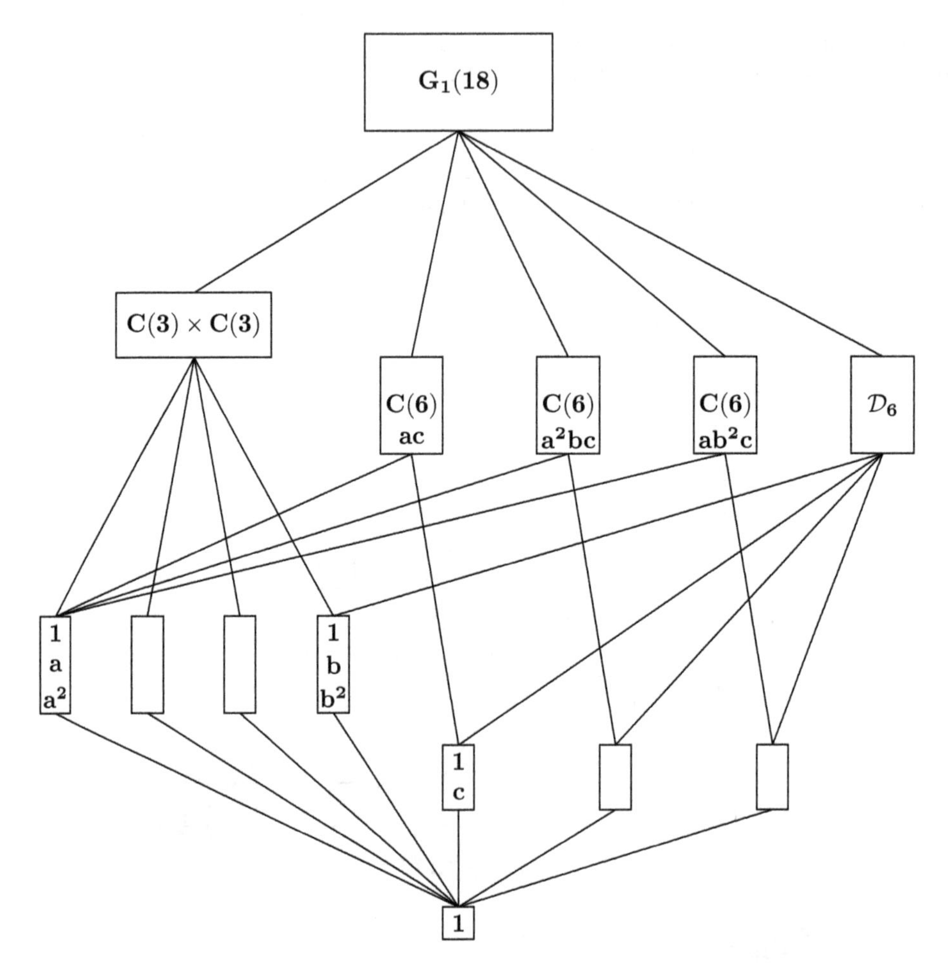

$$\begin{aligned}\mathbf{G_1(18)} &= \langle \mathbf{a, b, c} \mid \mathbf{a^3 = b^3 = c^2 = aba^2b^2 = aca^2c = (bc)^2 = 1} \rangle \\ &\simeq \mathbf{C(3)} \times \mathcal{D}_6 \simeq (\mathbf{C(3)} \times \mathbf{C(3)}) \rtimes \mathbf{C(2)}\end{aligned}$$

Figure VIII.7.7 Lattice of subgroups of $\mathbf{G_1(18)}$

Exercise VIII.7.10. Determine the homomorphisms φ that define the semidirect products of the groups $G_1(16), G_2(16)$ and $G_3(16)$ that are described in Figures VIII.7.9, VIII.7.10 and VIII.7.11.

Exercise VIII.7.11. Describe, in detail, the structure of the groups $Q\mathcal{D}_{16}$ and $G\mathcal{Q}_{16}$ defined in Figures VIII.7.12 and VIII.7.13.

Exercise VIII.7.12. Prove that the group $G = \langle\, a, b \mid a^4 = b^2 = abab \,\rangle$ is isomorphic to the **generalized quaternion group** $G\mathcal{Q}_{16} = \langle\, a, b \mid a^8 = a^4b^2 = abab^3 = 1 \,\rangle$.

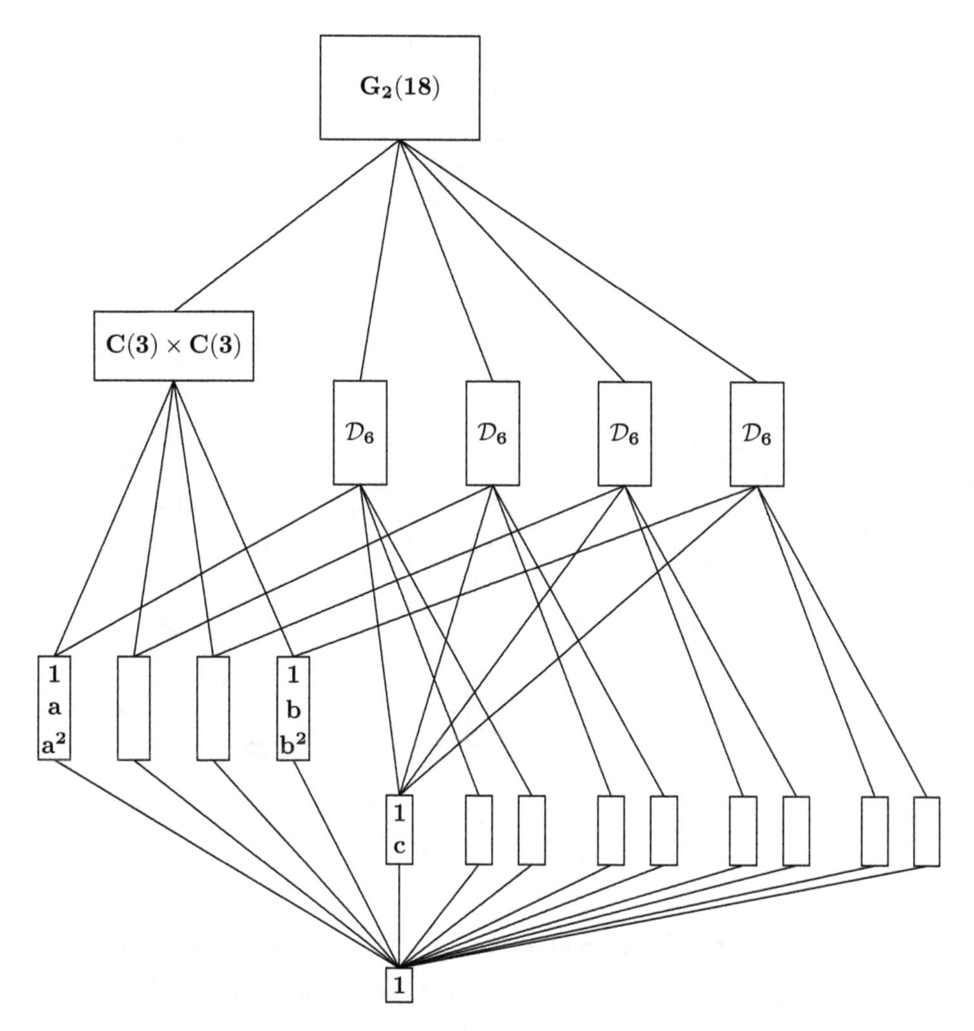

$$\begin{aligned} \mathbf{G_2(18)} &= \langle \mathbf{a, b, c} \mid \mathbf{a^3 = b^3 = c^2 = aba^2b^2 = (ac)^2 = (bc)^2 = 1} \rangle \\ &\simeq \mathbf{C(3)} \rtimes \mathcal{D}_6 \simeq (\mathbf{C(3) \times C(3)}) \rtimes \mathbf{C(2)} \end{aligned}$$

Figure VIII.7.8 Lattice of subgroups of $\mathbf{G_2(18)}$

Exercise VIII.7.13. Prove that the group G presented by

$$G = \langle a, b \,|\, a^3 = b^6 = (a\,b^3)^2 = a\,b^2\,a^2\,b^4 = 1 \rangle$$

is isomorphic to $G_1(18) \simeq (C(3) \times C(3)) \rtimes C(2)$. Consider $(C(3) \times C(3)) = \langle a \rangle \times \langle b^4 \rangle$ and $C(2) = \langle b^3 \rangle$.

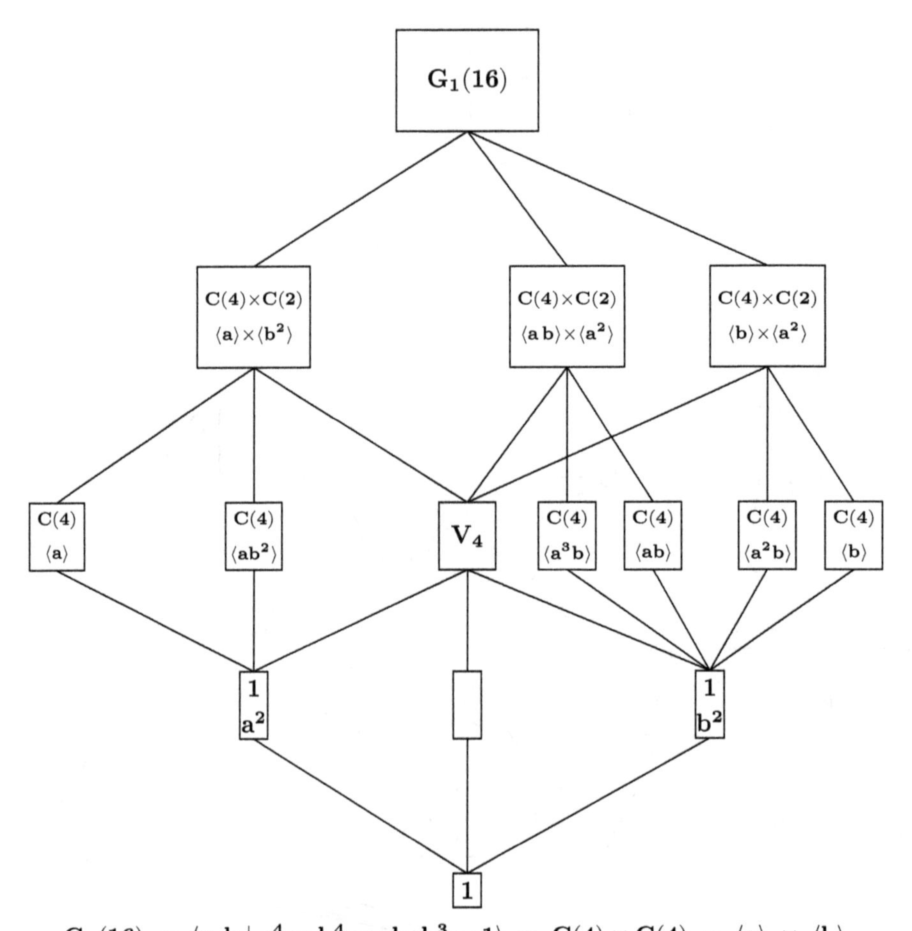

$$\mathbf{G_1(16)} = \langle \mathbf{a}, \mathbf{b} \mid \mathbf{a^4} = \mathbf{b^4} = \mathbf{abab^3} = \mathbf{1} \rangle \simeq \mathbf{C(4)} \rtimes \mathbf{C(4)} = \langle \mathbf{a} \rangle \rtimes \langle \mathbf{b} \rangle$$

Figure VIII.7.9 Lattice of subgroups of $\mathbf{G_1(16)}$

Exercise VIII.7.14. Decide whether the group $G_1(18)$ of Figure VIII.7.7 can be expressed as a semidirect product of the cyclic groups $C(3)$ and $C(6)$.

Figures VIII.7.9, VIII.7.10, VIII.7.11, VIII.7.12 and VIII.7.13 define five non-isomorphic nonabelian groups of order 16.

Exercise VIII.7.15. In each case, describe the homomorphisms φ that define the groups $G_1(16), G_2(16), G_3(16), \mathcal{QD}_{16}$ and $G\mathcal{Q}_{16}$, and describe the elements in the subgroups of these groups as they are illustrated in the above mentioned figures.

Exercise VIII.7.16. Describe the groups in Figure VIII.7.14, write down their presentations and prove that these non-isomorphic groups have isomorphic lattices of subgroups: The modular group $M(16)$ is noncommutative, while $G(16)$ is abelian.

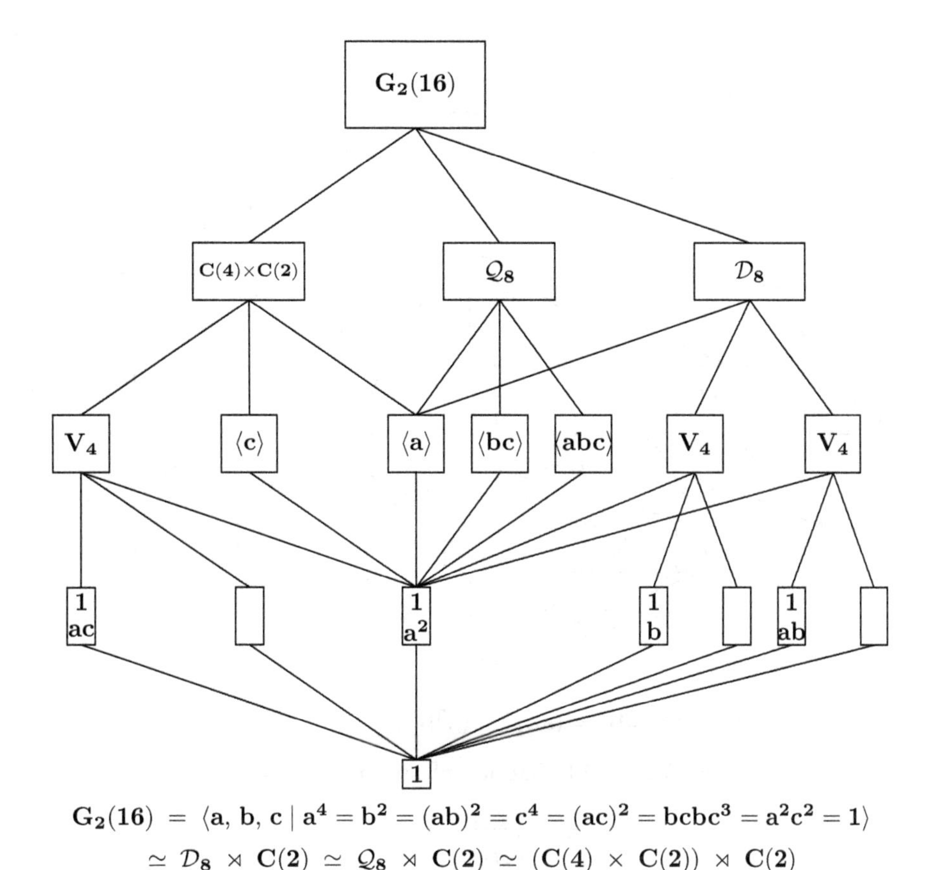

$$\begin{aligned}\mathbf{G_2(16)} &= \langle \mathbf{a, b, c} \mid \mathbf{a^4 = b^2 = (ab)^2 = c^4 = (ac)^2 = bcbc^3 = a^2c^2 = 1} \rangle \\ &\simeq \mathcal{D}_8 \rtimes \mathbf{C(2)} \simeq \mathcal{Q}_8 \rtimes \mathbf{C(2)} \simeq (\mathbf{C(4)} \times \mathbf{C(2)}) \rtimes \mathbf{C(2)}\end{aligned}$$

Figure VIII.7.10 Lattice of subgroups of $\mathbf{G_2(16)}$

Remark VIII.7.4. Observe that both the groups of Figure VIII.7.14 have in their subgroups exactly the same elements! However, while all subgroups of the abelian group $G(16)$ are (trivially) all normal in $G(16)$, show that two of the subgroups of the modular group $M(16)$ are not normal in $M(16)$. Write down formal presentations of these two groups.

To prove that the above 14 non-isomorphic groups of order 16 are the only ones is at this stage a rather tedious and lengthy exercise. The following remark shortcuts a number of arguments.

Remark VIII.7.5. The semidirect products $A \rtimes_{\varphi_1} H$ and $A \rtimes_{\varphi_2} H$ such that

$$H \xrightarrow{\varphi_1} \mathbf{Aut}(A) = H \xrightarrow{\alpha} H \xrightarrow{\varphi_2} \mathbf{Aut}(A),$$

where α is an automorphism of the group H, are isomorphic.

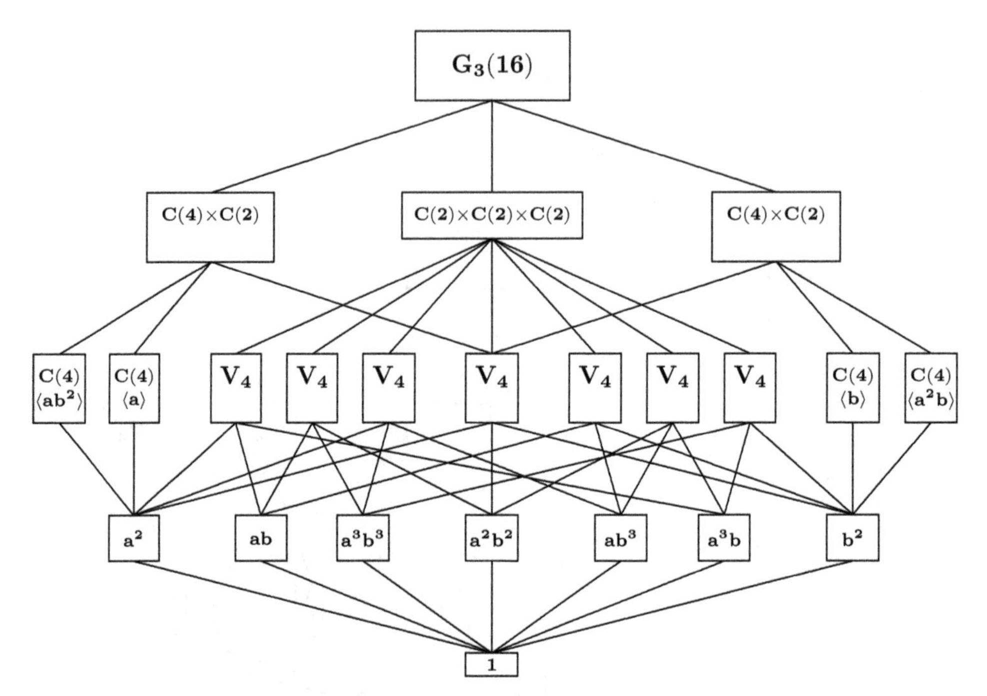

$\mathbf{G_3(16)} = \langle \mathrm{a, b} \mid \mathrm{a}^4 = \mathrm{b}^4 = (\mathrm{ab})^4 = (\mathrm{a}^3\mathrm{b})^2 = 1\rangle \simeq (\mathbf{C(4)} \times \mathbf{C(2)}) \rtimes \mathbf{C(2)}$

Figure VIII.7.11 Lattice of subgroups of $\mathbf{G_3(16)}$

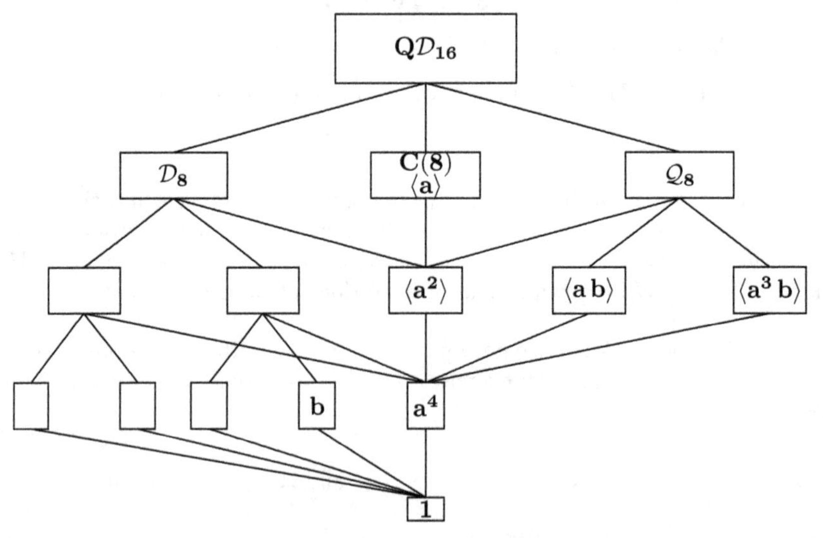

$\mathbf{Q}\mathcal{D}_{16} = \langle \mathrm{a, b} \mid \mathrm{a}^8 = \mathrm{b}^2 = \mathrm{baba}^5 = 1\rangle \simeq \mathbf{C(8)} \rtimes \mathbf{C(2)} \simeq \mathcal{Q}_8 \rtimes \mathbf{C(2)}$

Figure VIII.7.12 Lattice of subgroups of the quasidihedral group $\mathbf{Q}\mathcal{D}_{16}$

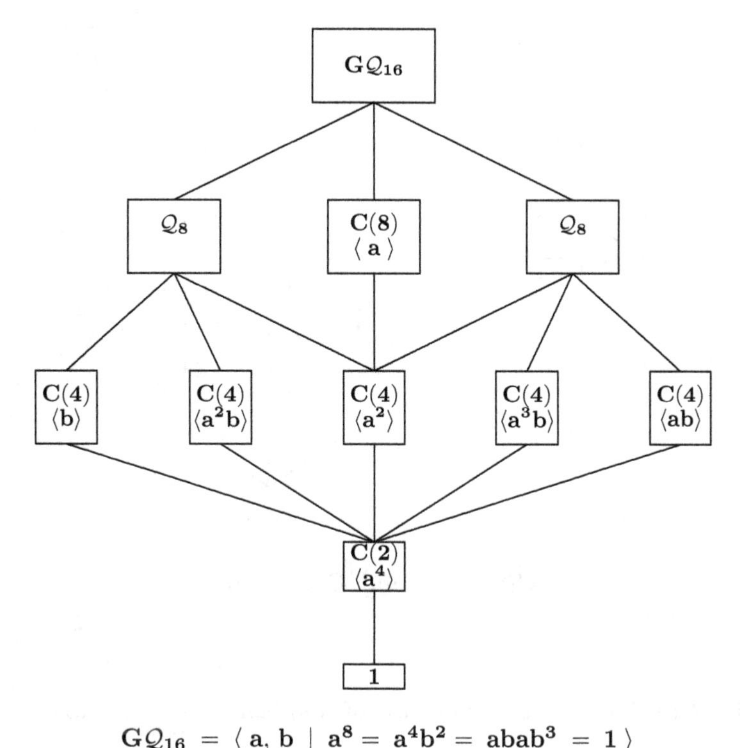

$$\mathbf{G}\mathcal{Q}_{16} = \langle\, \mathbf{a}, \mathbf{b} \mid \mathbf{a}^8 = \mathbf{a}^4\mathbf{b}^2 = \mathbf{abab}^3 = 1\, \rangle$$

Figure VIII.7.13 Lattice of subgroups of the group $\mathbf{G}\mathcal{Q}_{16}$

Exercise VIII.7.17. Prove the statement of Remark VIII.7.5.

Exercise VIII.7.18. Prove that

$$\mathcal{D}_{4n} \simeq \mathcal{D}_{2n} \times C(2)$$

for every odd number n.

Exercise VIII.7.19. Prove that there are just four non-isomorphic groups of order 66: the cyclic group $C(66)$, the dihedral group $\mathcal{D}_{66}$, the direct product of the dihedral group $\mathcal{D}_{22}$ and the cyclic group $C(3)$ and the direct product of the dihedral group $\mathcal{D}_6$ and the cyclic group $C(11)$. Could you present some of them as nontrivial semidirect products?

Exercise VIII.7.20. Prove the following criterion of Sylow for the nonexistence of simple groups of certain orders: *Let n be a natural number that is not a prime. If there is a prime divisor p of n such that n is not a multiple of any of the number $1 + kp, k = 1, 2, \ldots,$ then there is no simple group of order n.* Test this criterion for $n = 15, 18, 20, 42, 54, \ldots$.

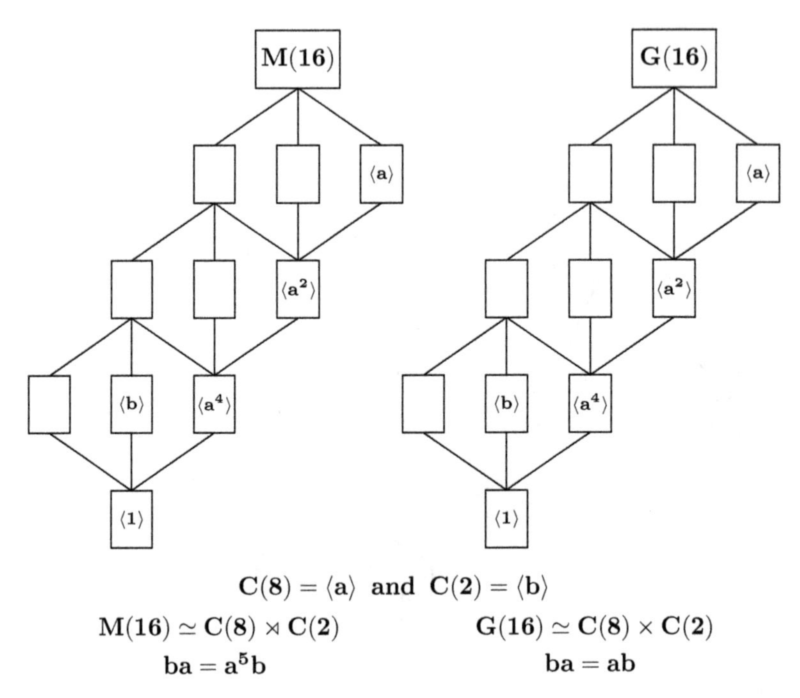

Figure VIII.7.14 Lattices of subgroups of non-isomorphic groups of order 16

To appreciate diversity of group structures, we display a catalog of all groups of order ≤ 100: The groups of the following orders are the uniquely determined (cyclic) groups:

1, 2, 3, 5, 7, 11, 13, 15, 17, 19, 23, 29, 31, 33, 35, 37, 41, 43, 47, 51, 53, 59, 61, 65, 67, 69, 71, 73, 77, 79, 83, 85, 87, 89, 91, 95 and 97.

There are just two non-isomorphic groups for the following orders:

4, 6, 9, 10, 14, 21, 22, 25, 26, 34, 38, 39, 45, 46, 49, 55, 57, 58, 62, 74, 82, 86, 93, 94 and 99.

There are just

three non-isomorphic groups of order 75;

four non-isomorphic groups of orders 28, 30, 44, 63, 66, 70, 76 and 92;

five non-isomorphic groups of orders 8, 12, 18, 20, 27, 50, 52, 68 and 98;

six non-isomorphic groups of orders 42 and 78;

ten non-isomorphic groups of orders 90;

twelve non-isomorphic groups of orders 88;

thirteen non-isomorphic groups of orders 56 and 60;

fourteen non-isomorphic groups of orders 16, 36 and 40;

fifteen non-isomorphic groups of orders 24, 54, 81 and 84;

sixteen non-isomorphic groups of orders 100;
50 non-isomorphic groups of orders 72;
51 non-isomorphic groups of orders 32;
52 non-isomorphic groups of orders 48 and 80;
260 non-isomorphic groups of orders 96;
267 non-isomorphic groups of orders 64.

We add that there are 2,328 non-isomorphic groups of order 128, 56,092 non-isomorphic groups of order 256 and more than 8 millions non-isomorphic groups of order 512 etc.

Of course, creating a catalog of groups is certainly not the way to study groups. It is important to describe the simple groups and the process of their extensions.

History of the classification of simple groups (**Hölder's program** of the late nineteenth century) is rather dramatic. It mainly concerns the **sporadic simple groups**, that is, those simple groups that do not belong to any infinite sequence of simple groups (such as for example the sequence of alternating groups $\mathcal{A}_n$ for $n \geq 5$). There are 26 such groups. The order of the largest one, the **monster**, equals

$$\begin{aligned}&2^{46} \cdot 3^{20} \cdot 5^{9} \cdot 7^{6} \cdot 11^{2} \cdot 13^{3} \cdot 17 \cdot 19 \cdot 23 \cdot 29 \cdot 31 \cdot 41 \cdot 47 \cdot 59 \cdot 71\\&= 808\,017\,424\,794\,512\,875\,886\,459\,904\,961\,710\,757\,005\,754\,368\,000\,000\,000,\end{aligned}$$

which is $\approx 8 \times 10^{53}$, and thus has more elements than the number of atoms of our Earth. This group was constructed in 1980 by **Robert Louis Griess (1945 -)** as a group of rotations of a 196,883-dimensional space. Hence, every element of the group is described by a $196,883 \times 196,883$ matrix. This was a culmination of Hölder's program which attracted more than 100 mathematicians and produced several hundred publications of almost 10,000 pages. We summarize the outcome.

Theorem VIII.7.4. *There exists a catalog of* 18 *infinite sequences of simple groups and* 26 *sporadic simple groups (that is, of simple groups that do not belong to any of the infinite sequences) such that every simple group is isomorphic to a group in this catalog.*

CHAPTER IX. RINGS AND FIELDS

In previous chapters, we frequently met examples of rings and fields, for example, integral domains of integers and polynomials, fields of rational numbers, real numbers and complex numbers, finite fields, rings of matrices, path algebras, etc. In this chapter we further our previous experience with rings and fields by adding some motivation, returning to the theory of field extensions and touching upon Galois theory. We also add some comments on the subject of algebraically closed fields, and in particular the proof of the fundamental theorem of algebra concerning the field of complex numbers.

IX.1 Several examples

As motivation for further study, we start with three important ring constructions.

First ring construction: Endomorphism rings of abelian groups

We recall that an **endomorphism** of a group G is a mapping $\varphi : G \to G$ satisfying the condition $\varphi(ab) = \varphi(a)\varphi(b)$ for any two elements a and b of the group G, that is, it is a homomorphism of a group into itself. We have noted that the set of all endomorphisms of a group G forms, with respect to the composition of mappings, a monoid: the **endomorphism monoid End** G **of the group** G. Indeed, a (composition) product of two endomorphisms is again an endomorphism.

We now consider **abelian** groups $G = (G, +)$ using **additive** notation; accordingly, the "endomorphism" condition reads $\varphi(a + b) = \varphi(a) + \varphi(b)$.

In addition to the **multiplication** $\varphi\,\psi$ (that is, composition $\varphi \circ \psi$) of endomorphisms φ and ψ, we can also define **addition** $\oplus$ by

$$(\varphi \oplus \psi)(a) = \varphi(a) + \psi(a)$$

for every φ and ψ of **End** G, and every $a \in G$. Clearly, **End** G is, with the operation $\oplus$, an **abelian group**.

Since

$$[(\varphi \oplus \psi)\eta](a) = \varphi(\eta(a)) \oplus \psi(\eta(a)) = (\varphi\eta \oplus \varphi\eta)(a)$$

for every $a \in G$, and since, similarly, $[\eta(\varphi \oplus \psi)](a) = (\eta\varphi \oplus \eta\varphi)(a)$, the (right) and (left) **distribution laws** hold:

$$(\varphi \oplus \psi)\eta = \varphi\eta \oplus \psi\eta \text{ and } \eta(\varphi \oplus \psi) = \eta\varphi \oplus \eta\varphi.$$

Thus, **End** G is a **ring** (with identity). Of course, in general, **End** G is a non-commutative ring (see (iii) in the next exercise).

Exercise IX.1.1. Prove:

(i) **End** $(\mathbb{Z}, +) \simeq \mathbb{Z}$;

(ii) **End** $(\mathbb{Z}_m, +) \simeq \mathbb{Z}_m$;

(iii) **End** $(G \oplus G) \simeq \mathrm{Mat}_{2\times 2}(\textbf{End } G) = \{\begin{pmatrix} \varphi_{11} & \phi_{21} \\ \varphi_{12} & \phi_{22} \end{pmatrix} \mid \varphi_{ij} \in \textbf{End } G\}$.

(iv) Let p be a prime. Recall that the Prüfer p-group is an abelian group generated by a sequence of elements $\{g_1, g_2, \dots, g_n, \dots\}$ that satisfy the relations $g_1^p = 1, g_n^p = g_{n-1}$ for $n = 2, 3, \dots$. This group is named after **Heinz Prüfer**. We rewrite this definition in additive notation and describe its endomorphism ring: the **domain of integral** p**-adic numbers.**

Exercise IX.1.2. Show that the endomorphism ring of the additive group of rational numbers is isomorphic to the field $\mathbb{Q}$. Can you describe the endomorphism ring of the multiplicative group of all positive rational numbers?

Second ring construction: Group rings

Let G be a finite group and F a field. Consider the vector space FG over the field F whose basis is the set of all elements of G. Thus

$$FG = \{\sum_{g\in G} a_g\, g \mid a_g \in F\}.$$

There is an injective mapping $\varphi : F \to FG$ given by $\varphi(a) = a\,1_G$ for all $a \in F$, where 1_G is the identity element of the group G (that is, all $a_g = 0$, but $a_{1_G} = a$). Thus, we can identify F with the image **Im**φ, that is, consider $F \subseteq FG$.

Now, define a multiplication in FG as follows: If $f_1 = \sum_{g\in G} a_g\, g$ and $f_2 = \sum_{g\in G} b_g\, g$, then

$$f_1\, f_2 = \sum_{g\in G} c_g\, g, \text{ where } c_g = \sum_{g=g_1\, g_2} a_{g_1}\, b_{g_2}. \qquad \text{(IX.1.1)}$$

Intuitively, the product (IX.1.1) is simply obtained by performing all possible simplifications (additions) in the set of all $|G|^2$ products obtained by (distributive) multiplication of $|G|$ summands of f_1 with $|G|$ summands of f_2. We may interpret the elements $f \in FG$ as functions $f : G \to F$ and explain the meaning of the operations of addition and multiplication in FG. The multiplication of f_1 and f_2 is a **discrete convolution** : the value of $f_1\, f_2$ at

$g \in G$ (that is, the coefficient c_g) is obtained as the sum of products $a_x b_{x^{-1}g}$, $x \in G$. Such operations have important applications in computer science, in digital signal analysis.

It is a routine matter to check the validity of all rules including the equality $a\,f = f\,a$ for all $a \in F$ and $f \in FG$, to see that FG is an F-**algebra**, that is, a ring with a central field F : a field contained in its center $Z(FG)$. The center of a ring R is, in analogy to groups, a subring

$$Z(R) = \{c \in R \mid cx = xc \text{ for all } x \in R\}.$$

Exercise IX.1.3. Extend the concept of a group algebra

(i) to a **group ring** RG over a commutative ring R;

(ii) to a group algebra FG for an infinite group G.

The subset A of FG of all elements $f_0 = \sum_{g\in G} a_g\, g$ satisfying $\sum_{g\in G} a_g = 0$ is closed both with respect to addition and multiplication. In fact, both $f\, f_0$ and $f_0\, f$ belong to A for arbitrary $f_0 \in A$ and $f \in FG$; thus A is a (two-sided) ideal of FG called the **augmentation ideal** of FG. The quotient ring FG/A is isomorphic to F.

Exercise IX.1.4. Describe the (canonical) isomorphism of FG/A and F.

Exercise IX.1.5. Show that the algebra of all matrices

$$\begin{pmatrix} a_1 & a_2 & 0 & 0 \\ a_2 & a_1 & 0 & 0 \\ 0 & 0 & a_3 & a_4 \\ 0 & 0 & a_4 & a_3 \end{pmatrix}$$

where $a_1, a_2, a_3, a_4 \in F$ subject to $a_1 + a_2 = a_3 + a_4$, is isomorphic to the group algebra FV_4 of the four-group $V_4 = \langle a, b \mid a^2 = b^2 = (ab)^2 = 1\rangle$. Determine the augmentation ideal and define the canonical map $\varphi : FV_4 \to F$.

We add that the concept of the group algebra is pivotal for the theory of representations of groups, that is, for the theory that studies presentations of groups as groups of matrices over a field.

$\star$ Third ring construction: Path algebras

The concept of a path semigroup presented in Chapter I was later enhanced to that of a path algebra in Chapter VI. An illustration of this basic concept was provided by a few examples there. Importantly, the concept was used in the (formal) definition of a polynomial in Chapter VII.

Here we want to emphasize the fundamental importance of the role of path algebras in the theory of representation. We are going to do so on a geometrical problem that has far reaching consequences for a number of developments in mathematics. It concerns the quadruples of subspaces of a vector space.

Recall that a nonzero subspace U of a direct sum of two nonzero vector spaces $V = V_1 \oplus V_2$ may have zero intersections $V_1 \cap U = V_2 \cap U = \{0\}$. Indeed, just take a nonzero vector $v_1 \in V_1$ and a nonzero vector $v_2 \in V_2$, then the nonzero subspace generated by the vector $v_1 + v_2$ has zero intersection with both V_1 and V_2. However, it may happen that the subspace U satisfies $U = (V_1 \cap U) \oplus (V_2 \cap U)$. We say in such a case that the structure $(V; U)$ of a vector space V together with a subspace U decomposes. In general, we say that a structure $(V; U_1, U_2, \ldots, U_k)$ is (directly) **decomposable** if there is a nonzero decomposition of the vector space $V = V_1 \oplus V_2$ such that $U_t = (V_1 \cap U_t) \oplus (V_2 \cap U_t)$ for all $t = 1, 2, \ldots, k$. If it is not possible, we say that $(V; U_1, U_2, \ldots, U_k)$ is **indecomposable**. The problem of determination of all indecomposable structures of this kind is closely related to the path algebras of the graphs in Figure IX.1.1.

Now, if the set of subspaces of a vector space V_F over a field F is void, then elementary linear algebra asserts that, up to an isomorphism, there is only one indecomposable vector space, namely the one-dimensional space F_F.

Furthermore, for $k = 1$, there are two indecomposable structures $(F_F, \{0\})$ and $(F_F; F_F)$. For $k = 2$, there are four indecomposable structures: $(F_F; \{0\}, \{0\})$, $(F_F; F_F, \{0\})$, $(F_F; \{0\}, F_F)$ and $(F_F; F_F, F_F)$. There are 9 indecomposable structures for $k = 3$:

$$(F_F; \{0\}, \{0\}, \{0\}),\ (F_F; F_F, \{0\}, \{0\}),\ (F_F; \{0\}, F_F, \{0\}),\ (F_F; \{0\}, \{0\}, F_F),$$
$$(F_F; F_F, F_F, \{0\}),\ (F_F; F_F, \{0\}, F_F),\ (F_F; \{0\}, F_F, F_F),\ (F_F; F_F, F_F, F_F)\ \text{and}$$
$$(V = v_1 F \oplus v_2 F; v_1 F, v_2 F, (v_1 + v_2) F).$$

Thus there is a two-dimensional indecomposable space endowed with 3 subspaces: the "diagonal" subspace $U_3 = (v_1 + v_2) F$ does the trick. Just to be sure, we are talking about different structures up to isomorphisms. So the structure $(V = v_1 F \oplus v_2 F; v_1 F, v_2 F, (v_1 + v_2) F)$ is "the same" as (that is, isomorphic to) $(V = v_1 F \oplus v_2 F; v_2 F, (v_1 + v_2) F, v_1 F)$ or, if the characteristic of the field is $\neq 2$, to $(V = v_1 F \oplus v_2 F; v_1 F, (v_1 - v_2) F, (v_1 + v_2) F)$ etc.

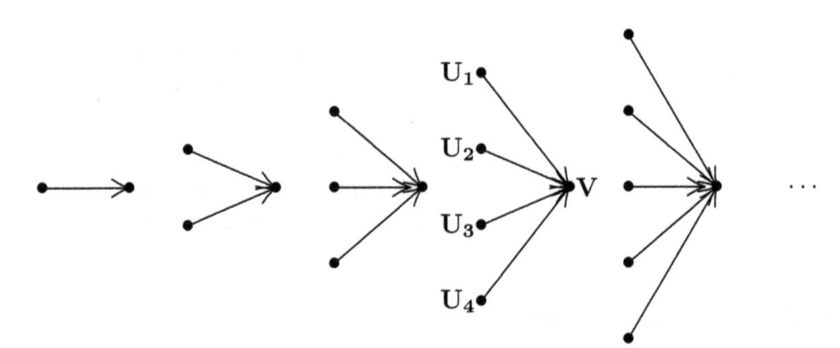

Figure IX.1.1 Graphs defining subspace structures

Perhaps unexpectedly in the next step, when $k = 4$, the number of indecomposable structures becomes infinite. Moreover, they are indecomposable structures of arbitrarily large dimension. To describe all indecomposable in this "tame" case is possible due to the fact that these structures are indecomposable representations of the graph marked in Figure IX.1.1 by $(V; U_1, U_2, U_3, U_4)$. The respective path algebra is isomorphic to the algebra consisting of all matrices of the form

$$A = \begin{pmatrix} a_1 & 0 & 0 & 0 & b_1 \\ 0 & a_2 & 0 & 0 & b_2 \\ 0 & 0 & a_3 & 0 & b_3 \\ 0 & 0 & 0 & a_4 & b_4 \\ 0 & 0 & 0 & 0 & a_5 \end{pmatrix}.$$

One of the series of indecomposable structures can be described as follows:

$$V = \bigoplus_{t=1}^{n} u_t F \oplus w_0 F \oplus \bigoplus_{t=1}^{n} v_t F; \; U_1 = \bigoplus_{t=1}^{n} u_t F, \; U_4 = \bigoplus_{t=1}^{n} v_t F,$$

$$U_2 = (w_0 + v_1) F \oplus \bigoplus_{t=1}^{n-1} (u_t + v_{t+1}) F, \; U_3 = (w_0 + u_1) F \oplus \bigoplus_{t=1}^{n-1} (v_t + u_{t+1}) F.$$

Figure IX.1.2 provides a visual presentation of this structure for $n = 5$, that is, 11-dimensional space.

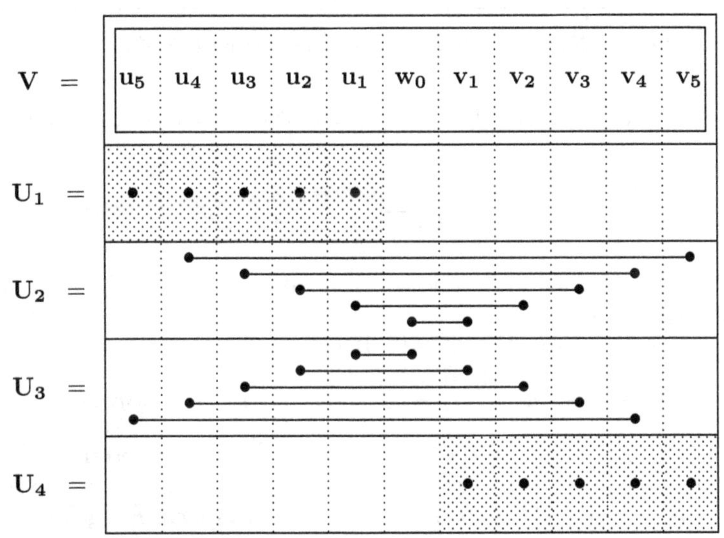

Figure IX.1.2 Four subspace structure (11; 5, 5, 5, 5)

Exercise IX.1.6. Write down matrix algebras isomorphic to the path algebras of the graphs in Figure IX.1.1.

Exercise IX.1.7. Show, using mathematical induction, that the structure depicted in Figure IX.1.2 is indecomposable.

Exercise IX.1.8. Compare the structures in Figure IX.1.2 and Figure IX.1.3 to conclude that the structure $(11; 6, 6, 6, 6)$ is also indecomposable. Use the orthogonal complements of the respective subspaces.

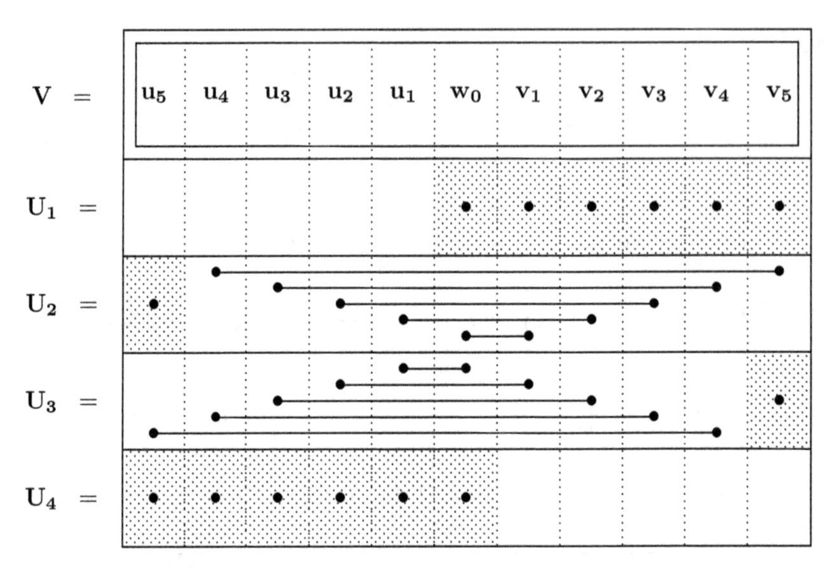

Figure IX.1.3 Four subspace structure (11; 6, 6, 6, 6)

Return to the F-algebra A_3 of Example VI.3.3. The elements of this infinite dimensional path algebra of the graph

$$1\bullet \underset{\mathbf{y}}{\overset{\mathbf{x}}{\rightleftarrows}} \bullet 2$$

can be described as the algebra of all polynomial expressions $f(x, y)$ in noncommuting indeterminates e_1, e_2, x, y subject to $e_1\,x = x\,e_2 = x,\ x\,e_1 = e_2\,x = 0,\ e_2\,y = y\,e_1 = y,\ e_1\,y = y\,e_2 = 0$ and $e_1^2 = e_1,\ e_2^2 = e_2,\ x^2 = y^2 = 0$. Hence,

$$f(x, y) = a_0e_1 + b_0e_2 + a_1x + b_1y + a_2xy + b_2yx + a_3xyx + b_3yxy + \cdots, \qquad \text{(IX.1.2)}$$

with a_t, b_t $(t = 0, 1, 2, 3, \dots)$ from the field F that are all, from a certain $t = n$, zero.

The element $yx \in A_3$ generates the ideal $I = \langle yx \rangle$ of all expressions (IX.1.2) satisfying $a_0 = b_0 = a_1 = b_1 = a_2 = 0$. Thus $A_3/I = \overline{A}_3$ is a 5-dimensional F-algebra.

Exercise IX.1.9. Show that the mapping $\varphi : \overline{A}_3 \to T$, where T is the F-algebra of all matrices

$$\begin{pmatrix} a & b & c \\ 0 & d & e \\ 0 & 0 & a \end{pmatrix}$$

with coefficients from the field F, given by

$$\varphi(a_0e_1 + b_0e_2 + a_1x + b_1y + a_2xy) = \begin{pmatrix} a_0 & a_1 & a_2 \\ 0 & b_0 & b_1 \\ 0 & 0 & a_0 \end{pmatrix}$$

is an isomorphism $\overline{A}_3 \simeq T$.

Remark IX.1.1. We conclude this section with a short remark clarifying the relationship of the path algebras to the previously displayed vector spaces endowed with a structure of subspaces.

Given a ring R, define a (left) **module** M over the ring R, briefly an $R-$module M, as an abelian group $(M,+)$ together with a mapping $R \times M \to M$, i. e. "multiplication" of elements of $m \in M$ by elements of $r \in R$ (similar to the operation of a field in the case of a vector space) to get $r\,m \in M$ satisfying the following requirements (axioms):

$$\begin{aligned} r\,(m_1 + m_2) &= r\,m_1 + r\,m_2, \\ (r_1 + r_2)\,m &= r_1\,m + r_2\,m, \\ (r_1 r_2)\,m &= r_1\,(r_2\,m), \\ 1\,m &= m. \end{aligned}$$

Similarly as in the case of other algebraic objects, $R-$modules can be compared by means of homomorphisms: A homomorphism from an $R-$module M to an $R-$module N is a (group) homomorphism $\varphi : M \to N$ of the respective abelian groups $(M,+)$ and $(N,+)$ satisfying the following additional condition:

$$(r\,m)\,\varphi = r\,(m\,\varphi) \text{ for all elements } m \in M \text{ and } r \in R.$$

We have just defined the category of (left) $R-$modules. Recall the related concepts of an injective homomorphism (a monomorphism), a surjective homomorphism (an epimorphism) and an isomorphism of two $R-$modules.

If a ring A is an algebra over a field F (briefly, an $F-$algebra), that is, if A is a vector space over the field F satisfying the equalities

$$(a_1\,f)\,a_2 = a_1\,(a_2\,f) = (a_1\,a_2)\,f \text{ for all } a_1, a_2 \in A \text{ and } f \in F,$$

we can identify the field F with the subring of all multiples $\{1\,f \mid f \in F\}$ of the identity element of the ring A and thus consider the field F to be a subring of A. If the dimension of the vector space A_F is finite, we say that A is a **finite dimensional algebra**. Such objects play an important role in algebra.

Observe that every module M over an $F-$algebra A is a vector space over F and that the homomorphisms of modules are linear transformations of the respective vector spaces. In this sense, one speaks about the **representations** of the algebra A. Thus, a representation of an $F-$algebra A is a homomorphism

$$T : A \to \mathrm{Lin}(M_F)$$

of the algebra A into the algebra of all linear transformations (endomorphisms) of the vector space M_F: For all $a_1, a_2 \in A$ and $f \in F$,

$$\begin{aligned} T(a_1 + a_2) &= T(a_1) + T(a_2), \\ T(a\,f) &= T(a)\,f, \\ T(a_1\,a_2) &= T(a_1)\,T(a_2), \\ T(1) &= I_M. \end{aligned}$$

The case when the dimension of the vector space M_F is finite is of particular importance. If T is a monomorphism, A is isomorphic to a subalgebra of the algebra $\mathrm{Lin}(M_F)$ and we say that the representation is **faithful**. Every algebra possesses a faithful representation. This fact is expressed in Cayley's theorem for algebras that can be obtained, in analogy to a proof of the related theorem for groups (see Theorem VIII.3.1), by a construction of the **regular** representation.

Given a finite dimensional representation $T : A \to \mathrm{Lin}(M_F)$, we can choose a basis of the vector space M_F, and by assigning to every linear transformation $T(a), a \in A$, the corresponding $n \times n$ matrix over the field F, a **matrix representation** of the algebra A. Of course, such a representation depends on the choice of the basis of the space M_F. An elementary concept of **matrix similarity** assures that a representation is described in a unique way.

We briefly return to the case of path algebras $A = F(\Gamma)$, when the finite oriented graph Γ has no oriented cycles and is simple in the sense that there is at most one arrow between any two vertices. Recall that there is an idempotent $e_i \in A$ (that is, $e_i^2 = e_i$) attached to each vertex $V_i \in \Gamma$. Then every A−module, that is, every representation $T : A \to \mathrm{Lin}(M_F)$, is described by a system $M_i = e_i M e_i$ of vector spaces over the fields $F_i = e_i A e_i$ that are isomorphic to F and by linear transformations ${}_i\varphi_j : M_j \to M_i$.

In the above illustration concerning the graph with 4 arrows, the linear transformations ${}_i\varphi_j$ were monomorphisms and were therefore interpreted as embeddings of vector spaces. Such a geometric interpretation (that represents a linearization of the problem) allows a deeper understanding of the structure of representations and clarifies functorial methods (using Coxeter's functors) used in construction of indecomposable representations of the **hereditary (tensor) algebras** described in a general situation of **valued graphs** in *Indecomposable representations of graphs and algebras*, Memoirs AMS, No. 173 (1976). There, Dynkin and extended Dynkin (Euclidean) diagrams play a fundamental role. The first three graphs in Figure IX.1.1 are simple Dynkin diagrams (and correspond to the case where the corresponding path algebras have a finite number of indecomposable representations), the fourth graph is a Euclidean diagram (corresponding to a "tame" algebra) where it is possible to describe all (infinite number of) indecomposable representations.

IX.2 Basic facts revisited

We start with a reference to the Definition VI.2.1 of a ring $(R, +, \cdot)$. We will find some parts of this chapter parallel to that on groups.

Recall that we have defined a ring R to possess the identity element 1 and that a subring of R is required to contain 1.

Remark IX.2.1. In some applications, it is convenient also to consider **rings without an identity element**. An example of such a ring is the ring of all even integers. There exists however a simple embedding of such a ring S into a ring with the identity element (i. e. into a ring as we have defined it). Indeed, put $R = \{(x, n) \mid x \in S, n \in \mathbb{Z}\}$ and define

$$(x, n) + (y, m) = (x + y, n + m) \text{ and } (x, n)(y, m) = (xy + ny + mx, nm).$$

Exercise IX.2.1. Verify that R is a ring with the identity element $(0, 1)$ and that the subset R' of all elements of R of the form $(x, 0)$ is isomorphic to S.

Remark IX.2.2. In a similar way, as in the group theory, rings (in particular, finite rings) can be defined by means of their additive and multiplicative tables. For instance, $R = \{0, 1, a, b, c, d, e, f\}$ may be presented by the following tables:

add	0	1	*a*	*b*	*c*	*d*	*e*	*f*
0	0	1	*a*	*b*	*c*	*d*	*e*	*f*
1	1	0	*b*	*a*	*f*	*e*	*d*	*c*
a	*a*	*b*	0	1	*d*	*c*	*f*	*e*
b	*b*	*a*	1	0	*e*	*f*	*c*	*d*
c	*c*	*f*	*d*	*e*	0	*a*	*b*	1
d	*d*	*e*	*c*	*f*	*a*	0	1	*b*
e	*e*	*d*	*f*	*c*	*b*	1	0	*a*
f	*f*	*c*	*e*	*d*	1	*b*	*a*	0

and

mult	0	1	*a*	*b*	*c*	*d*	*e*	*f*
0	0	0	0	0	0	0	0	0
1	0	1	*a*	*b*	*c*	*d*	*e*	*f*
a	0	*a*	*a*	0	*c*	*d*	*c*	*d*
b	0	*b*	0	*b*	0	0	*b*	*b*
c	0	*c*	0	*c*	0	0	*c*	*c*
d	0	*d*	*a*	0	*c*	*d*	0	*d*
e	0	*e*	0	*b*	0	0	*e*	*b*
f	0	*f*	*a*	*b*	*c*	*d*	*b*	*f*

Of course, to check the validity of the ring axioms is very lengthy and tedious; therefore such a presentation is impractical.

Exercise IX.2.2. Prove that the ring R of Remark IX.2.2 is isomorphic to the ring of certain 2×2 matrices over the field $\mathbb{Z}_2$, namely that

$$R \simeq \{\begin{pmatrix} x\,y \\ 0\,z \end{pmatrix} \mid x, y, z \in \mathbb{Z}_2\}.$$

We should also be aware of the fact that even familiar rings or fields like the integers or rational numbers can be presented in an "unusual" form. The following formulations appear very often in textbooks in some special form as exercises and are usually not fully explained. Here is such a typical exercise:

Exercise IX.2.3. Let $\mathbb{Q}$ be the set of all rational numbers with the following operations of "addition" $\oplus$ and "multiplication" $\odot$:

$$x \oplus y = x + y + 1 \text{ and } x \odot y = xy + x + y.$$

Prove that $(\mathbb{Q}, \oplus, \odot)$ is a field with "zero" -1 and "identity" 0.

Two points should be raised: First, the resulting field is, in fact, isomorphic to the field $\mathbb{Q}$. Secondly, the particular choice of $\mathbb{Q}$ does not play any role. Hence formulate the following exercise.

Exercise IX.2.4. Let F be a field and $a, b \in F$ arbitrary elements such that $a \neq b$. Define the following two operations (of addition and multiplication) on the set $R = F$:

$$x \oplus y = x + y - a \text{ and } x \odot y = \frac{1}{b-a}(xy - a(x+y) + ab).$$

(i) Prove that $(R, \oplus, \odot)$ is a field isomorphic to the field F (with "zero" a and "identity" b).

(ii) Formulate the special cases : (**1**) zero $= 0$, identity $= b \neq 0$; (**2**) zero $= a \neq 1$, identity $= 1$ and (**3**) zero $= 1$, identity $= 0$.

(iii) Also show that the previous Exercise IX.2.3 is a special case.

Remark IX.2.3. We sketch a solution of Exercise IX.2.4. One can proceed the way most of the textbooks would suggest (even in their formulations of the exercise):

Both operations $\oplus$ and $\odot$ are commutative. Furthermore, since

$$(x \oplus y) \oplus z = x + y + z - 2a = x \oplus (y \oplus z) \quad \text{and}$$

$$(x \odot y) \odot z = \frac{1}{(b-a)^2}[xyz - a(xy + xz + yz) + a^2(x + y + z) + ab(b - 2a)] = x \odot (y \odot z),$$

they are also associative. The distributive law can also be easily checked. The next step is to calculate the additive neutral element $a : a \oplus x = x \oplus a = x$ and the opposite element $2a - x$ of $x : x \oplus (2a - x) = a$. Similarly, one can calculate the multiplicative neutral element b (observe $b \odot x = x \odot b = x$) and the inverse $\frac{b(b-a)+a(x-b)}{x-a}$ of the element x since

$$x \odot \frac{b(b-a) + a(x-b)}{x-a} = b.$$

Thus, we have shown that $(R, \oplus, \odot)$ is a field.

However, all these calculations can be replaced by the following approach that shows the essence and deeper understanding of the problem:

Define the mapping $\Phi : F \to R$ by $\Phi(x) = (b-a)x + a$. Then,

$$\Phi(x + y) = (b - a)(x + y) + a = (b - a)x + a + (b - a)y + a - a = \Phi(x) + \Phi(y) - a.$$

Thus Φ is an isomorphism of the additive groups $(F, +)$ and $(R, \oplus)$:

$$\Phi(x) + \Phi(y) - a = \Phi(x) \oplus \Phi(y).$$

Similarly,

$$\Phi(xy) = (b-a)xy + a = \frac{1}{b-a}[\Phi(x)\Phi(y) - a(\Phi(x)+\Phi(y)) + ab] = \Phi(x) \odot \Phi(y).$$

Thus, $(R, \oplus, \odot)$ is a field isomorphic to F. Note also that $\Phi^{-1} : R \to F$ is given by

$$\Phi^{-1}(x) = \frac{x-a}{b-a}.$$

We can therefore easily calculate the "zero" $\Phi(0) = a$ and the "identity" $\Phi(1) = b$ in R. Moreover, the opposite to x is

$$\Phi[-\Phi^{-1}(x)] = \Phi\left[\frac{a-x}{b-a}\right] = a - x + a = 2a - x$$

and the inverse to x is

$$\Phi\left[\frac{1}{\Phi^{-1}(x)}\right] = \Phi\left[\frac{b-a}{x-a}\right] = \frac{b(b-a)+a(x-b)}{x-a}.$$

Finally, the three special cases are

(i) $R = F : x \oplus y = x + y, x \odot y = x\,y\,b^{-1}$;

(ii) $R = F : x \oplus y = x + y - a, x \odot y = (x-a)(y-a)(1-a)^{-1} + a$ and

(iii) $R = F : x \oplus y = x + y - 1, x \odot y = x + y - x\,y$.

Exercise IX.2.5. Define on the set $\mathbb{Z}$ of integers operations of addition $\oplus$ and multiplication $\odot$ in such a way that additive neutral element "zero" will be 5, the multiplicative neutral element "identity" will be -5 and $(\mathbb{Z}, +, \cdot) \simeq (\mathbb{Z}, \oplus, \odot)$.

Exercise IX.2.6. In order to refresh earlier knowledge, here are several elementary statements for the reader to prove:

(i) $a \cdot 0 = 0 \cdot a = 0$ for all elements a of a ring R.

(ii) $(-1) \cdot a = a \cdot (-1) = -a$ for all elements a of a ring R.

(iii) $(-a) \cdot b = -(a \cdot b)$ and $(-a) \cdot (-b) = a \cdot b$ for all elements a and b of a ring R.

(iv) $\sum_{r=1}^{m} a_r \cdot \sum_{s=1}^{n} b_s = \sum_{r=1}^{m} \sum_{s=1}^{n} a_r \cdot b_s$.

(v) If $a^2 = a$ for all elements of a ring R, then R is a commutative ring.

(vi) Describe the ring of all subsets $A, B, \ldots$ of a given (fixed) set X with the addition defined by

$$A \oplus B = (A \setminus B) \cup (B \setminus A) \text{ and } A \odot B = A \cap B.$$

(vii) The intersection of any set of subrings of a ring R is a subring of R. The subrings of a ring R form a lattice.

(viii) The subset $\mathbf{U}(R)$ of all units of a ring R, that is, of all elements $a \in R$ that have a multiplicative inverse $a^{-1} \in R$, is a multiplicative group. Show that

$$\mathbf{U}(\mathbb{Z}_n) = \{[k] \mid (k,n) = 1\}.$$

Thus the order of $\mathbf{U}(\mathbb{Z}_n)$ equals to the value of Euler's function $\varphi(n)$.

(ix) The following two subsets

$$S_1 = \{a + b\sqrt{2} \mid a, b \in \mathbb{Z}\} \subset S_2 = \{a + b\sqrt{2} \mid a, b \in \mathbb{Q}\}$$

of the field $\mathbb{R}$ of real numbers are subrings of $\mathbb{R}$ and, in fact, S_2 is a field, that is, $\mathbf{U}(S_2) = S_2 \setminus \{0\}$.

The group of units of $\mathbb{Z}_n$ is possible to determine quite simply for small values of n (just by mere listing of the units and considering the cyclic groups that they generate). One can see that $\mathbf{U}(\mathbb{Z}_6) = \{[1],[5]\}$ and thus $\mathbf{U}(\mathbb{Z}_6) \simeq \mathbb{C}(2)$. It is easy to check that

$$\mathbf{U}(\mathbb{Z}_{14}) = \{[1],[9],[11]\} \times \{[1],[13]\} \simeq C(3) \times C(2),$$

$$\mathbf{U}(\mathbb{Z}_{15}) = \{[1],[2],[4],[8]\} \times \{[1],[11]\} \simeq C(4) \times C(2) \text{ and}$$
$$\mathbf{U}(\mathbb{Z}_{18}) = \{[1],[7],[13]\} \times \{[1],[17]\} \simeq C(3) \times C(2).$$

Exercise IX.2.7. Verify that
$\mathbf{U}(\mathbb{Z}_{24}) = \{[1],[5]\} \times \{[1],[7]\} \times \{[1],[13]\} \simeq C(2) \times C(2) \times C(2)$;
$\mathbf{U}(\mathbb{Z}_{60}) = \{[1],[7],[49],[43]\} \times \{[1],[11]\} \times \{[1],[19]\} \simeq C(4) \times C(2) \times C(2)$;
$\mathbf{U}(\mathbb{Z}_{100}) = \{[1],[21],[41],[61],[81]\} \times \{[1],[7],[43],[49]\} \times \{[1],[51]\} \simeq C(5) \times C(4) \times C(2)$;
$\mathbf{U}(\mathbb{Z}_{360}) = \{[1], 241],[121]\} \times \{[1],[343],[289],[127]\} \times \{[1],[19]\} \times \{[1],[71]\} \times \{[1],[251]\} \simeq C(3) \times C(4) \times C(2) \times C(2) \times C(2)$ etc.

The orders of these groups are given, as we have already mentioned, by the values of Euler's function φ. Recall that for $n = p_1^{k_1} p_2^{k_2} \dots p_d^{k_d}$, where p_t are distinct primes and $k_t \in \mathbb{N}$ for all $t \in \{1, 2, \dots, d\}$,

$$\varphi(n) = n\left(1 - \frac{1}{p_1}\right)\left(1 - \frac{1}{p_2}\right) \cdots \left(1 - \frac{1}{p_d}\right).$$

This indicates a way in which we can determine $\mathbf{U}(\mathbb{Z}_n)$ for large n.

The following construction of a finite direct product of rings is similar to that of a finite direct product of groups.

Definition IX.2.1. *A cartesian product* $R = R_1 \times R_2 \times \cdots \times R_d = \prod_{t=1}^{d} R_t$ *of* d *rings with coordinate-wise operations of addition and multiplication is a ring, called the* **direct product** *of the rings* $R_1, R_2, \dots, R_d$.

Exercise IX.2.8. Prove that

$$\mathbf{U}(\prod_{t=1}^{d} R_t) = \prod_{t=1}^{d} \mathbf{U}(R_t).$$

As a consequence, we can formulate the following theorem:

Theorem IX.2.1. *Let $n = p_1^{r_1} p_2^{r_2} \dots p_d^{r_d}$ be the (unique) prime decomposition of a given natural number n. Then*

$$\mathbf{U}(\mathbb{Z}_n) = \mathbf{U}(\mathbb{Z}_{p_1^{r_1}}) \times \mathbf{U}(\mathbb{Z}_{p_2^{r_2}}) \times \cdots \times \mathbf{U}(\mathbb{Z}_{p_d^{r_d}}).$$

Here, we are using the fact that

$$\mathbb{Z}_n = \mathbb{Z}_{p_1^{r_1}} \times \mathbb{Z}_{p_2^{r_2}} \times \cdots \times \mathbb{Z}_{p_d^{r_d}}.$$

Exercise IX.2.9. Show that $\mathbf{U}(\mathbb{Z}_2) \simeq C(1) = \{1\}$,

for $r \geq 2$,

$$\mathbf{U}(\mathbb{Z}_{2^r}) \simeq C(2) \times C(2^{r-2})$$

and for $p \geq 3, r \geq 1$,

$$\mathbf{U}(\mathbb{Z}_{p^r}) \simeq C(p-1) \times C(p^{r-1}).$$

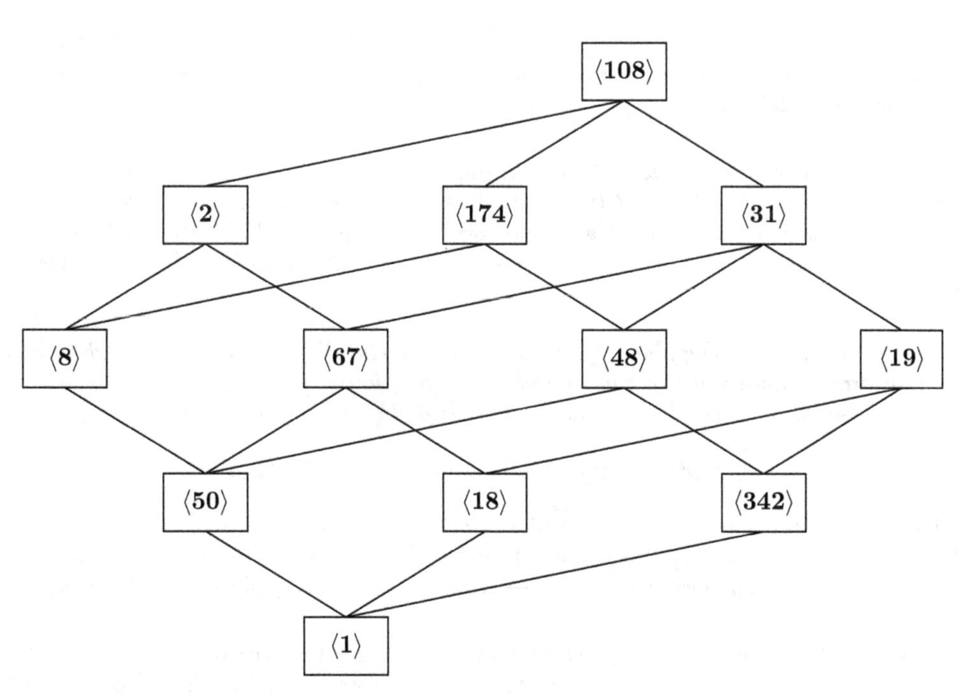

Figure IX.2.1 Lattice of $\mathbf{U}(\mathbb{Z}_{343}) \simeq \mathbf{C}(2) \times \mathbf{C}(3) \times \mathbf{C}(49)$

Figure IX.2.1 gives an explicit description of the lattice of all subgroups of $\mathbf{U}(\mathbb{Z}_{7^3}) = \{[1],[2],[3],[4],[5],[6],[8],\ldots,[342]\}$ of order 294:

Here,

$$\langle[108]\rangle \simeq C(294), \quad \langle[2]\rangle \simeq C(147), \quad \langle[174]\rangle \simeq C(98),$$
$$\langle[31]\rangle \simeq C(42), \quad \langle[8]\rangle \simeq C(49), \quad \langle[67]\rangle \simeq C(21),$$
$$\langle[48]\rangle \simeq C(14), \quad \langle[19]\rangle \simeq C(6), \quad \langle[50]\rangle \simeq C(7),$$
$$\langle[18]\rangle \simeq C(3) \quad \text{and} \quad \langle[342]\rangle \simeq C(2).$$

In Chapter VI, in connection with homomorphisms of rings, we have also defined a **two-sided ideal** of a ring (see Definition VI.2.2). Consider, for a given non-zero element $a \in R$ the subset $I = \{a\,x \mid x \in R\}$ of R. Observe that either $I = R$ which is in the case that a is a unit of R (that is, there is a multiplicative inverse a^{-1} in R), or I is not a subring (since it does not contain 1). However, it is a subgroup of the additive group $(R,+)$ in this case and with every $b \in I$ and every $x \in R$ the product $b \cdot x$ belongs to I. We say that I is a **right ideal** of the ring R. Similarly, a subset J of R that is a subgroup of $(R,+)$ and has the property that $x \cdot b \in J$ for every $b \in J$ and $x \in R$ is said to be a **left ideal** of the ring R.

Exercise IX.2.10. Show that the **left annihilator** of

(i) an element a of a ring R, that is, the subset $\{x \in R \,|\, x\,a = 0\} \subseteq R$, is a left ideal of R;

(ii) a left ideal L of a ring R, that is, the set $\{x \in R \,|\, x\,a = 0 \text{ for all } a \in L\}$, is a two-sided ideal of R.

(iii) Define the right annihilator of a set and of a right ideal of R, and state the respective conclusions as in (i) and (ii).

Every homomorphism $\varphi : R \to S$ from a ring R to a ring S defines a (two-sided) ideal $\mathbf{Ker}\varphi = \{a \in R \,|\, \varphi(a) = 0\}$ of R (the **kernel** of the homomorphism φ). Here we like to underline a particular feature of a ring homomorphism (as promised already in Chapter VI) that is related to our assumption that rings have an identity element. Here is a formal definition.

Definition IX.2.2. *A mapping $\varphi : R \to S$ from a ring R to a ring S is said to be a* **ring homomorphism** *if it is a homomorphism of the additive groups $(R,+)$ and $(S,+)$ and homomorphism of the multiplicative monoids $(R,\cdot)$ and $(S,\cdot)$, that is, if for all $a, b \in R$,*

$$\varphi(a+b) = \varphi(a) + \varphi(b), \ \varphi(ab) = \varphi(a)\cdot\varphi(b) \ \text{ and } \ \varphi(1_R) = 1_S.$$

The last requirement means, in particular, that every homomorphism $\varphi : R \to S$ defines a subring $\mathbf{Im}\varphi = \{\varphi(a) \,|\, a \in R\}$ (the **image** of the homomorphism φ). Recall also the fact that $\varphi = \varphi_2 \circ \varphi_1$ with an epimorphism $\varphi_1 : R \to \mathbf{Im}\varphi$ and a monomorphism $\varphi_2 : \mathbf{Im}\varphi \to S$.

Exercise IX.2.11. Let $R = \mathrm{Mat}_{2\times 2}(F)$ be a ring of all 2×2 matrices over a field F. Let $\varphi : R \to S$ be a ring homomorphism **onto** the ring S. Prove that either $R \simeq S$ or $S = \{0\}$. Can you prove the same conclusion when $R = \mathrm{Mat}_{n\times n}(F)$ for any natural number n?

The property of the matrix rings over fields formulated in Exercise IX.2.11 means that these rings are **simple** in the sense that they do not contain any nontrivial two-sided ideal. In fact, the rings with this property that satisfy certain finiteness conditions (such as the existence of minimal one-sided ideals) are just the matrix ring over fields; this is the essence of the Molien - Wedderburn - Artin theorem **(Theodor Molien (1861 - 1941), Joseph Wedderburn (1882 - 1948) and Emil Artin (1898 - 1962))**.

Having experience acquired in Chapters II, VII and VIII studying congruences of integers, polynomials and groups, it is easy to formulate a general definition and the related constructions for arbitrary rings.

Definition IX.2.3. *Let I be a (two-sided) ideal of a ring R. Define the I-equivalence $\equiv$ for elements of the ring R as follows:*

$$a \equiv b \pmod{I} \text{ if and only if } a - b \in I.$$

Exercise IX.2.12. Describe the I-equivalence class $[a]$ of an element $a \in R$ and show that the I-equivalence is a **congruence** in the following sense: If $a \equiv b$ and $c \equiv d$, then $a + c \equiv b + d$ and $ac \equiv bd$.

At this stage, we should recall the construction of the quotient group modulo a normal subgroup and the following theorems from Chapter VIII. We are going to proceed similarly forming a quotient ring modulo an ideal.

Since the I-equivalence is a congruence, we may define on the set of all I-equivalence classes the operations of addition and multiplication derived from the operations in the ring R. Thus

$$[a] \oplus [b] = [a+b] \text{ and } [a] \odot [b] = [ab]. \qquad \text{(IX.2.1)}$$

Exercise IX.2.13. Justify that the operations in (IX.2.1) are well-defined, that is, that they did not depend on the choice of representatives in the I-equivalence classes.

Exercise IX.2.14. Prove the following theorem:

Theorem IX.2.2. *Let I be an ideal of a ring R. Then the set of all I-equivalence classes with the operations* (IX.2.1) *forms a ring, the* **quotient ring** $R/I = (R/I, \oplus, \odot)$ *of the ring R by the ideal I. There is a canonical* **surjective homomorphism** $\varphi_I : R \to R/I$ *defined by $\varphi_I(a) = [a]$ for all $a \in R$.*

In analogy to group theory mentioned above, we can formulate the following three isomorphism theorems.

Theorem IX.2.3 (First isomorphism theorem). *Every ring homomorphism $\varphi : R \to S$ yields an isomorphism*

$$R/\mathbf{Ker}\varphi \simeq \mathbf{Im}\varphi.$$

Thus, for a surjective homomorphism φ, $S \simeq R/I$, where $I = \mathbf{Ker}\varphi$.

Proof. Referring to Section 4 of Chapter I, consider the partition of R formed by the inverse images (fibres) $A_c = \varphi^{-1}(c)$ of the elements $c \in \mathbf{Im}\varphi \subseteq S$. In particular, $A_0 = \mathbf{Ker}\varphi$ is an ideal of the ring R, and two elements $a, b \in R$ belong to the same A_c (that is, satisfy $\varphi(a) = \varphi(b)$) if and only if $a - b \in A_0$. This follows from the fact that $\varphi(a-b) = \varphi(a) - \varphi(b) = 0$. Hence, the elements of the partition $\{\varphi^{-1}(c) \mid c \in \mathbf{Im}\varphi\}$ of R are just the A_0-equivalence classes, and the homomorphism φ induces the required isomorphism. □

Theorem IX.2.4 (Second isomorphism theorem - parallelogram rule). *Let I be an ideal and T a subring of a ring R. Then the set*

$$I + T = \{a + b \mid a \in I,\ b \in T\}$$

is a subring of the ring R, I is an ideal of the ring $I+T$ and $I \cap T$ is an ideal of the ring T. Furthermore, there is an isomorphism

$$(I+T)/I \simeq T/(I \cap T).$$

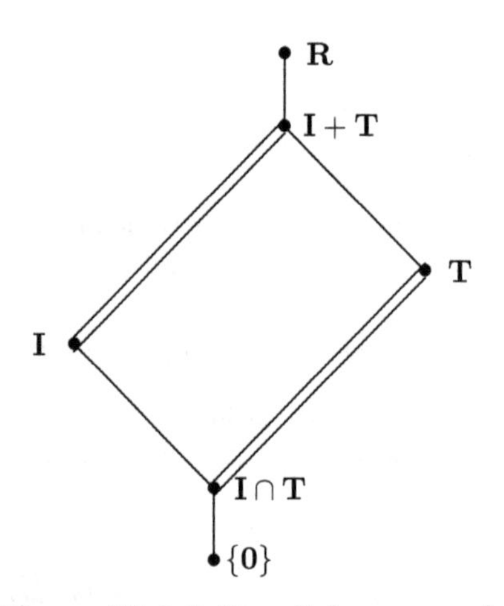

Figure IX.2.2 Parallelogram rule

Proof. For any $a_1, a_2 \in I,\ b_1, b_2 \in T$,

$$(a_1 + b_1) + (a_2 + b_2) = (a_1 + a_2) + (b_1 + b_2),$$

$$(a_1 + b_1)(a_2 + b_2) = (a_1 a_2 + a_1 b_2 + b_1 a_2) + b_1 b_2, \quad \text{and } 1 = 0 + 1;$$

thus $I + T$ is a subring of R. Equally evident are the facts that I is an ideal of the ring $I + T$ and that $I \cap T$ is an ideal of the ring T.

Now, every element of the quotient ring $(I+T)/I$ has the form $a + b + I, a \in I, b \in T$. Define a ring homomorphism

$$\varphi : T \to (I+T)/I \text{ by } \varphi(b) = b + I \in (I+T)/I.$$

Since $a+b+I = b+I \in (I+T)/I$, φ is a surjective homomorphism. Consider the kernel of φ. If $c \in \mathbf{Ker}\varphi$, then $\varphi(c) = c+I = I$ and thus $c \in I$. On the other hand, every $c \in I \cap T$ satisfies $\varphi(c) = c+I = I$, and therefore $\mathbf{Ker}\varphi = I \cap T$. Applying Theorem IX.2.2, we obtain

$$T/\mathbf{Ker}\varphi \;=\; T/(I \cap T) \;\simeq\; (I+T)/I \;=\; \mathbf{Im}\varphi.$$

□

For the formulation of the third isomorphism theorem, we have to extend the construction of a quotient ring to the ring without the identity element. This will allow us to talk about a quotient ideal by a subideal.

Exercise IX.2.15. Let R be a ring **without** the identity element. Let I be an ideal of R. Show that we can define the I-equivalence and form the quotient ring R/I. Thus, in particular, if $J \subseteq I$ are two ideals of a ring R, we can form the ring I/J; show that it is an ideal of the quotient ring R/J.

Theorem IX.2.5 (Third isomorphism theorem). *Let I and J be ideals of a ring R such that $J \subseteq I$. Then the quotient ideal $I/J = \{a+J \mid a \in I\}$ is an ideal of the quotient ring R/J and*

$$R/J \Big/ I/J \;\simeq\; R/I.$$

Proof. Evidently, I/J is a subring of R/J. In fact we can verify that I/J is an ideal of R/J. If $a \in I, c \in R$, we can see that

$$(a+J)\,(c+J) = ac+J \in I/J \;\text{ and }\; (c+J)\,(a+J) = ca+J \in I/J,$$

since both ac and ca belong to I.

Now, define the mapping

$$\varphi : R/J \to R/I \;\text{ by }\; \varphi(c+J) = c+I \in R/I.$$

We have $\varphi(c_1+J)+\varphi(c_2+J) = (c_1+I)+(c_2+I) = (c_1+c_2)+I = \varphi((c_1+c_2)+J) = \varphi((c_1+J)+(c_2+J))$, and similarly $\varphi(c_1+J)\,\varphi(c_2+J) = \varphi((c_1+J)\,(c_2+J))$. Thus, φ is a ring homomorphism that is surjective (the partition of R defined by J is a refinement of the partition of R defined by I).

A class $c+J$ belongs to $\mathbf{Ker}\varphi$ if and only if $\varphi(c+J) = c+I = I$, that is, if and only if $c \in I$, and hence $\mathbf{Ker}\varphi = I/J$. The required isomorphism then follows using Theorem IX.2.3. □

Exercise IX.2.16. Let I and J be two ideals of a ring R. Show that the mapping

$$\varphi : R \to S = R/I \times R/J \;\text{ defined by }\; \varphi(a) = (a+I, a+J)$$

is a ring homomorphism. Using Theorem IX.2.5, prove that S contains a subring isomorphic to the quotient ring $R/(I \cap J)$.

Exercise IX.2.17. Use the previous exercise to show that, for relatively prime m and n,

$$\mathbb{Z}/mn\mathbb{Z} \;\simeq\; \mathbb{Z}/m\mathbb{Z} \times \mathbb{Z}/n\mathbb{Z}.$$

Exercise IX.2.18. Let I be a proper ideal of an integral domain D. Prove that the quotient ring D/I is an integral domain if and only if the following condition is satisfied by the ideal I: Whenever, for $a, b \in D$, the product $a\,b \in I$, then either $a \in I$ or $b \in I$. An ideal satisfying this condition is called a **prime ideal**. In this sense, a ring is an integral domain if and only if the zero ideal $\{0\}$ is a prime ideal. Determine all prime ideals of the integral domain $\mathbb{Z}$ of integers.

Exercise IX.2.19. Let I be an ideal of an integrity domain D. Prove that the quotient ring D/I is a field if and only if I is a **maximal ideal** of the integral domain D, that is, I satisfy the following condition: $I \neq D$ and if J is an ideal of D such that $I \neq J$ and $I \subset J$, then $J = D$. Thus, every maximal ideal of an integral domain D is a prime ideal of D.

Exercise IX.2.20. Prove: An integral domain is a field if and only if it does not contain an infinite number of ideals. We point out that, in fact, a field F has no proper ideals, that is, $\{0\}$ and F are the only ideals of a field F. Conclude: Every finite integral domain is a field.

We add that every integral domain has maximal ideals. More generally: If $J \neq D$ is an ideal of an integral domain D, then there exists a maximal ideal I of D such that $J \subseteq I$. A proof of this statement requires an application of **Zorn's lemma**: Suppose a partially ordered set P has the property that every chain has an upper bound in P. Then the set P contains at least one maximal element. This result is named after **Max August Zorn (1906 - 1993)**. Zorn's lemma is equivalent to the **axiom of choice**: The Cartesian product of a non-empty family of non-empty sets is non-empty. An exercise indicating the proof of this equivalence is given on page 65 of the book *Naive Set Theory* by Paul R. Halmos. Another formulation of the axiom of choice is given in Section 2 of Chapter X.

IX.3 Divisibility in integral domains

The study of divisibility of integers and polynomials occupied a great part of Chapters III and VII. Moreover, the basic concepts and results were summarized in Chapter VI. Here we will concentrate on a rather large class of integral domains with unique factorization, that is, integral domains in which every non-zero element that is not a unit has a unique presentation as a product of irreducible elements (atoms). We have already seen in Chapter VI that the question of divisibility in integral domains is rather subtle.

Recall that an **irreducible element** a of an integral domain D is a non-zero element that is not a unit (that is, does not posses a multiplicative inverse) and allows only trivial factorizations, that is, in every factorization $a = b\,c$ either a or b is a unit.

Example IX.3.1 (Quadratic integral domains $\mathbb{Z}[\sqrt{d}]$). Let d be a non-zero integer, $d \neq 1$, that is not divisible by a square of any prime. Trivially, the subring $\mathbb{Z}[\sqrt{d}] = \{a + b\sqrt{d} \mid a, b \in \mathbb{Z}\}$ of the field $\mathbb{C}$ of complex numbers is an integral domain. We point out that for a negative $d = -e$, we have denoted in Chapter VI these domains by $\mathbb{Z}[\sqrt{e}\,i]$. In the case $d = -1$, that is, in the case $\mathbb{Z}[i]$, we deal with the **domain of Gaussian numbers**. Define a **norm** of an element $z = a + b\sqrt{d} \in \mathbb{Z}[\sqrt{d}]$ by $N(z) = |a^2 - d\,b^2|$. The norm $N(z)$ is a nonnegative integer that equals 0 if and only if $z = 0$. The norm of a Gaussian number $z = a + b\,i$ is thus $N(z) = a^2 + b^2$.

Exercise IX.3.1. Show that for every $z_1, z_2 \in \mathbb{Z}[\sqrt{d}]$, $N(z_1\, z_2) = N(z_1)\, N(z_2)$, and use this equality to prove the following theorem.

Theorem IX.3.1. *Every nonzero element of the integral domain $\mathbb{Z}[\sqrt{d}]$ that is not a unit, is a product of irreducible elements. The units of $\mathbb{Z}[\sqrt{d}]$ are the following elements:*

(i) For $d \leq -2$: 1 and -1;

(ii) For $d = -1$: $1, -1, i$ and $-i$;

(iii) For $d \geq 2$: $1, -1$ and an infinite number of others.

In (iii), the condition for $u = r + s\sqrt{d}$ to be a unit is $r^2 - d\, s^2 = \pm 1$. The equation $d\, s^2 + 1 = r^2$ is Pell's equation, which was mentioned in Section 1 of Chapter II and discussed in Section 6 of Chapter IV.

The decomposition of the elements into a product of irreducible elements mentioned in Theorem IX.3.1 does not have to be unique. There were several examples mentioned in Chapter VI showing such different factorizations. One of them presented two different factorizations of $4 \in \mathbb{Z}[\sqrt{-3}] : 4 = 2 \times 2 = (1+\sqrt{3}\, i) \times (1-\sqrt{3}\, i)$. This justifies the following clarification of the concept of unique factorization.

Definition IX.3.1. *An integral domain D is said to be a UFD or* **unique factorization domain** *(sometimes called a Gaussian domain) if every non-zero element $a \in D$ that is not a unit of D, has a unique decomposition as a product of irreducible elements of D, that is, if*

$$a = p_1\, p_2 \cdots p_r = q_1\, q_2 \cdots q_s,$$

where p_m and q_n are irreducible elements, then $r = s$ and, after a suitable permutation of the elements q_n, we have $p_m = u_m q_m$, where the u_m are units for all $m \in \{1, 2, \ldots, r\}$, so that p_m and q_m are associates.

In such a domain, we can single out a subset $P = \{p_\lambda \mid \lambda \in \Lambda\}$ of irreducible elements such that every irreducible element of D is associated precisely to one of the elements of P. Consequently, we can express every element $a \in D$ uniquely in the form

$$a = u \prod_{\lambda \in \Lambda} p_\lambda^{k_\lambda}, \tag{IX.3.1}$$

where u is a unit, $k_\lambda \geq 0$ and all k_λ but for a finite number are equal to 0. Such expressions are often used to determine the greatest common divisor $d = d(a, b)$ and the least common multiple $m = m(a, b)$ of two elements a and b of D.

Exercise IX.3.2. Prove the following theorem.

Theorem IX.3.2. *Let D be a unique factorization domain. Let*

$$a = u \prod_{\lambda \in \Lambda} p_\lambda^{k_\lambda} \text{ and } b = v \prod_{\lambda \in \Lambda} p_\lambda^{l_\lambda}, \tag{IX.3.2}$$

be the (unique) factorizations of the elements a and b of D into irreducibles. Then

$$d = d(a,b) = \prod_{\lambda\in\Lambda} p_\lambda^{r_\lambda}, \quad \textit{where } r_\lambda = \mathrm{Min}(k_\lambda, l_\lambda)$$

has the following property: $d|a, d|b$ *and if* $d'|a$ *and* $d'|b$*, then* $d'|d$.

Moreover,

$$m = m(a,b) = \prod_{\lambda\in\Lambda} p_\lambda^{s_\lambda}, \quad \textit{where } s_\lambda = \mathrm{Max}(k_\lambda, l_\lambda)$$

has the following property: $a|m, b|m$ *and if* $a|m'$ *and* $b|m'$*, then* $m|m'$.

The product $d(a,b)\, m(a,b)$ *is associated to the product* ab.

As we can see, to find $d(a,b)$ and $m(a,b)$ by means of Theorem IX.3.1 is rather trivial. However, it assumes that the factorizations (IX.3.2) are known, and we should be aware of the fact that to establish such factorizations is, in general, a very difficult task (recall R.S.A. protocol in Chapter III). Therefore it is very important to have an algorithm that we can perform in some integral domains, namely in the **Euclidean domains** (see Definition VI.2.5) to establish the greatest common divisor of two elements **without factorization** of the elements. We have performed this procedure for the integers and polynomials in the earlier chapters and so now, in general, we are going to present it as an exercise.

Exercise IX.3.3. Prove the following theorem.

Theorem IX.3.3 (Euclidean algorithm and Bézout's equality). *Let a and b be two nonzero elements of a Euclidean domain D with a norm* ν. *Then there exists* $q_s, r_s, 0 \le s$ *defining the following finite sequence*

$a = bq_0 + r_0,\ \nu(r_0) < \nu(b);$

$b = r_0 q_1 + r_1,\ \nu(r_1) < \nu(r_0);$

$r_0 = r_1 q_2 + r_2,\ \nu(r_2) < \nu(r_1);$

....

$r_{t-1} = r_t q_{t+1} + r_{t+1},\ \nu(r_{t+1}) < \nu(r_t)$

....

that terminates at the moment when $r_t = 0$. *The last nonzero* r_{t-1} *is the greatest common divisor* $d = d(a,b)$ *of the elements a and b if* $t > 0$ *and* $d = d(a,b) = b$ *if* $r_0 = 0$.

Moreover, there are elements u and v in D such that

$$d = au + bv.$$

A consequence is the following important Gauss' lemma.

Corollary IX.3.1. *Let* a, b *and* c *be elements of a Euclidean domain such that a is a divisor of* bc *and a and b are relatively prime. Then a is a divisor of c. In particular, an irreducible element* $p \in D$ *has the following property: If* $p \mid bc$ *and* $p \nmid b$*, then* $p \mid c$.

Recall that a nonzero element p of a general integral domain D that is not a unit and satisfies Gauss' lemma is said to be **prime** (see Exercise VI.2.6).

Exercise IX.3.4. Prove: Every prime element of an integral domain is irreducible. In a Euclidean domain, these two concepts coincide.

In general, the concepts of a prime element and an irreducible element differ. Referring to an earlier example of $\mathbb{Z}[\sqrt{-3}]$, we see that $2 \in \mathbb{Z}[\sqrt{-3}]$ is irreducible, but **not prime**. Nevertheless, the conclusion of the previous exercise can be extended.

Exercise IX.3.5. In a unique factorization domain, the concepts of prime and irreducible elements coincide.

Example IX.3.2. As we have already mentioned, an important example of a Euclidean domain is the domain $\mathbb{Z}[i]$ of Gaussian numbers. We show that the norm $N(a+b\,i) = a^2+b^2$ may serve as the "Euclidean norm" ν. Let $z = a+b\,i$ and $w = c+d\,i$. We are going to show that there are g and r in $\mathbb{Z}[i]$ such that

$$z = w\,q + r \text{ with } \nu(r) \leq \frac{1}{2}\,\nu(w) \text{ or } r = 0.$$

The proof follows readily from the following simple inequalities:

$$\begin{aligned}\frac{a+bi}{c+di} &= \frac{(a+bi)(c-di)}{c^2+d^2} = \frac{(ac+bd)+(bc-ad)i}{c^2+d^2}\\ &= \frac{A(c^2+d^2)+R+[(B(c^2+d^2)+S]i}{c^2+d^2},\end{aligned}$$

where $|R| \leq \frac{1}{2}(c^2+d^2)$ and $|S| \leq \frac{1}{2}(c^2+d^2)$. From here, $z = wq+r$ with

$$q = A+Bi \in \mathbb{Z}[i] \ \text{ and } \ r = \frac{(c+di)(R+Si)}{c^2+d^2} \in \mathbb{Z}[i].$$

Now,

$$\begin{aligned}\nu(r) &= \nu\Big(\frac{(c+di)(R+Si)}{c^2+d^2}\Big) = \frac{(c^2+d^2)(R^2+S^2)}{(c^2+d^2)^2} = \frac{R^2+S^2}{c^2+d^2}\\ &\leq \frac{1}{c^2+d^2}\Big[\frac{1}{4}(c^2+d^2)^2+\frac{1}{4}(c^2+d^2)^2\Big].\end{aligned}$$

And the last expression is just

$$\frac{1}{2}\,(c^2+d^2) = \frac{1}{2}\,\nu(w).$$

Example IX.3.3. Let $-8+10i$ and $7+9i$ belong to $\mathbb{Z}[i]$. To find the greatest common divisor, calculate successively

$$\begin{aligned}-8+10i &= (7+9i)i + (1+3i),\\ 7+9i &= (1+3i)(3-i) + (1+i) \text{ and}\\ 1+3i &= (1+i)(2+i).\end{aligned}$$

Thus $d(-8+10i, 7+9i) = 1+i$ and, moreover

$$\begin{aligned}1+i &= (7+9i)-(1+3i)(3-i) = (7+9i) - [(-8+10i)-(7+9i)\,i]\,(3-i)\\ &= (-8+10i)(-3+i)+(7+9i)(2+3i).\end{aligned}$$

Exercise IX.3.6. (i) Find $d = d(-8+5i, 7+9i)$ and express d as a linear combination of $-8+5i$ and $7+9i$.

(ii) The integral domain $\mathbb{Z}[\sqrt{2}\,]$ can be shown to be a Euclidean domain in a similar way as $\mathbb{Z}[i]$. In this domain

$$(5+\sqrt{2})(2-\sqrt{2}) = (11-7\sqrt{2})(2+\sqrt{2}).$$

Since their norms are prime numbers, all factors are irreducible elements. Is the equality in contradiction with the statement that the factorizations in Euclidean domains are unique?

In Example VI.2.2 we have indicated a proof of the fact that every Euclidean domain is a **principal ideal domain**, that is, that every ideal can be generated by a single element. We formulate this statement as a theorem.

Theorem IX.3.4. *Every Euclidean domain is a principal ideal domain.*

Proof. Let I be an ideal of a Euclidean domain D. If $I \neq \{0\}$, denote by a an element of I whose norm $\nu(a)$ is minimal. Thus $\langle a\rangle \subseteq I$. If $\langle a\rangle \neq I$, then there is an element $b \in I \setminus \langle a\rangle$ and elements q and $r \neq 0$ such that $b = aq + r$ and $\nu(r) < \nu(a)$. Since $r = b - aq$ is an element of the ideal I, we get a contradiction to the minimal value of $\nu(a)$. Hence $I = \langle a\rangle$, as required. □

The following property of principal ideal domains is important.

Theorem IX.3.5. *Every principal ideal domain is a unique factorization domain.*

Proof. We are going to prove the theorem in four steps.

(1) *Every irreducible element of a principal ideal domain is prime.* This has already been proved in Exercise IX.3.1. However, it follows from the fact that any two elements of D generate a principal ideal $\langle a, b\rangle = \langle d\rangle$, and thus the greatest common divisor $d = au + bv$. If a is irreducible and $a \mid b\,c$, $a \nmid b$, then, for suitable u, v, w, we have $auc + bvc = c$ and $aw = bc$, that is, $a\,(uc + vw) = c$.

(2) *Every increasing chain of ideals in a principal ideal domain is finite.* If

$$\langle a_1\rangle \subseteq \langle a_2\rangle \subseteq \langle a_3\rangle \subseteq \cdots \subseteq \langle a_n\rangle \subseteq \cdots \tag{IX.3.3}$$

is such a chain, then the union $I = \bigcup_n \langle a_n\rangle$ is a principal ideal $I = \langle b\rangle$ and $b \in \langle a_{n_0}\rangle$ for some n_0. Hence, $\langle a_n\rangle = \langle a_{n_0}\rangle$ for all $n \geq n_0$.

(3) *Every nonzero element of a principal ideal domain that is not a unit, is a product of irreducible elements.* Let a be such an element. First, we show that there is an irreducible element p such that $a = pb$. This is a consequence of the statement 2. Indeed, the statement is trivially true if $a = a_1$ is an irreducible element. Otherwise, $a_1 = a_2b_2$ with neither a_2 nor b_2 an unit. If a_2 is not irreducible, we can write similarly as before $a_2 = a_3b_3$, and if a_3 is not irreducible, continue $a_3 = a_4b_4$ etc. Now, we get a chain of ideals (IX.3.3) and according to 2, there is n such that $\langle a_n\rangle = \langle a_{n+1}\rangle$ with $a_n = a_{n+1}b_{n+1}$, where neither a_{n+1} nor b_{n+1} being units. However, since $a_{n+1} = a_nc$ for some $c \in D$, we get $a_n = a_ncb_{n+1}$ showing that b_{n+1} is a unit. This contradiction implies that a_n is an irreducible element. We have established the product $a = pb$ with $p = a_n$ and $b = b_nb_{n-1}\cdots b_2$.

Thus, we can gradually write $a = p_1 a_1$ with p_1 irreducible and unless a_1 is irreducible $a_1 = p_2 a_2$ with p_2 irreducible and continue to get, by statement 2 again, that the principal ideals $\langle a_n \rangle$ and $\langle a_{n+1} \rangle$ coincide. As before, we have the following equalities

$$a = p_1 p_2 \cdots p_n a_n, \; a_n = p_{n+1} a_{n+1}, \; a_{n+1} = a_n c \text{ and thus } a_n = a_n c p_{n+1}.$$

This means that p_{n+1} is a unit and thus a_n is irreducible. In conclusion, a is a **finite** product of irreducible elements.

(4) *If* $a = p_1 p_2 \cdots p_r = q_1 q_2 \cdots q_s$, *where all* p_m *and* q_n *are irreducible elements, then* $r = s$ *and each element* p_m *is associated with* $q_{\pi(m)}$ *for a suitable permutation* π *of* $\{1, 2, \ldots, r\}$. Assume that $r \leq s$. We are going to proceed by induction. Assume that the statement is not, in general, true. Of course, it is true trivially for $r = 1$. Let k be the least number of irreducible factors for which the statement does not hold:

$$a = p_1 p_2 \ldots p_k = q_1 q_2 \ldots q_t.$$

By statement 1, p_k is a prime element and since $p_k \mid q_1 q_2 \ldots q_t$, necessarily there is $1 \leq j \leq t$ such that p_k and q_j are associates (that is, $p_k \mid q_j$). Therefore, for a certain unit u, $p_k u = q_j$ and thus

$$a^* = p_1 p_2 \ldots p_{k-1} = q_1 q_2 \ldots q_{j-1} u q_{j+1} \cdots q_t.$$

Now, since a^* is a product of $k - 1$ irreducible elements, the factorization of a^* is unique, and therefore also factorization of a is unique. The statement 4 and thus Theorem IX.3.5 is established. □

Figure VI.2.1 illustrates the inclusions that we have just proved. These are proper inclusions, as indicated in the figure.

Exercise IX.3.7. Similarly as in the case of the domain $\mathbb{Z}[i]$, prove that the domains $\mathbb{Z}[\sqrt{3}\,]$ and $\mathbb{Z}[\sqrt{2}\, i]$ are Euclidean domains.

Exercise IX.3.8. Prove: *If* D *is a unique factorization domain, then* $D[x]$ *is also a unique factorization domain.* First, prove the statement for $D = \mathbb{Z}$, and then rewrite it for a general case.

[Hint: Call a polynomial **primitive** if the greatest common divisor of its coefficients is 1. Show that the product of two primitive polynomials is a primitive polynomial. Make use of the fact that $F[x]$, where F is the quotient field of the domain D, is a Euclidean domain.]

IX.4 Field extensions

Recall and extend our previous terminology of Chapter VII. Given two fields $F \subseteq E$, we say that E is an **extension field**, or simply an **extension**, of the field F and the field F is sometimes called a **base field**. In this case, E can be considered as a vector space E_F over the field F; the dimension of this space is called the **degree** of the extension of E over F and denoted by $[F : E]$. Two extensions $F \subseteq E_1$ and $F \subseteq E_2$ are said to be **equivalent** if there exists an isomorphism $\varphi : E_1 \to E_2$ such that $\varphi(a) = a$ for all elements of the field F.

Exercise IX.4.1. Show that any nonzero homomorphism $\varphi : F_1 \to F_2$ of the fields F_1 and F_2 is a monomorphism that is injective. Thus any nonzero homomorphism of two fields that is surjective is necessarily an isomorphism.

Let E be an extension of F and $A = \{\mathbf{a}_1, \mathbf{a}_2, \dots, \mathbf{a}_d\}$ a subset of elements of E. The field that is the intersection of all subfields containing both F and A will be denoted by $F(A)$ or $F(\mathbf{a}_1, \mathbf{a}_2, \dots, \mathbf{a}_d)$ and is obtained by the (field) **adjunction** of the elements $\mathbf{a}_1, \mathbf{a}_2, \dots, \mathbf{a}_d$ to F. Of course, in general, A may be an infinite set.

Exercise IX.4.2. Show that $F(A)$ consists of all rational expressions (that is, those obtained by addition, subtraction, multiplication and division) from elements of F and A. For $A = A_1 \cup A_2$, prove that $F(A) = F(A_1)(A_2)$.

The exercise shows that the finite adjunction can be reduced to the adjunctions of single elements. The extensions obtained by the adjunction of a single element are called **simple field extensions.** We recall Theorem VII.4.3 (Kronecker's theorem) that described a simple field extension $F \subseteq \overline{F[x]} = F[x]/\langle p \rangle = F(\bar{x})$. We have seen that the degree of this extension is equals to the degree d of the irreducible polynomial p. Indeed, the set $\{\bar{1}, \bar{x}, \bar{x}^2, \dots, \bar{x}^{d-1}\}$ is a basis of the vector space $F[\bar{x}]$ over F.

We describe the simple fields extensions $K = F(\mathbf{a})$ of a field F. First, K contains the ring R of all finite polynomial expressions $\sum a_t \mathbf{a}^t$ with $a_t \in F$ and there is a homomorphism $\varphi : F[x] \to R$ defined by $\varphi(\sum a_t x^t) = \sum a_t \mathbf{a}^t$ (compare Exercise VII.1.5 and Remark VII.1.4). Thus

$$R \simeq F[x]/I,$$

where I is the ideal of all polynomials $f \in F[x]$ satisfying $f(\mathbf{a}) = 0$.

Now, we have to distinguish two cases: If $I \neq \{0\}$, then by Theorem IX.3.4, I is a principal ideal generated by a polynomial $p \in F[x]$, and since there are no divisors of zero in R, p must be irreducible. But then $F[x]/\langle p \rangle$ is a field and, as a result, $R = F(\mathbf{a})$. If $I = \{0\}$, then $R \simeq F[x]$ and $F(\mathbf{a})$ (which is the field of fractions of R) is isomorphic to the field of rational functions of an indeterminate x.

In the first case, $\mathbf{a}$ is a root of a polynomial from $F[x]$ and is called **algebraic over** F; moreover, $F(\mathbf{a})$ is said to be a **simple algebraic extension** of F. In the second case, $\mathbf{a}$ is called **transcendental** over the field F and $f(\mathbf{a})$ a **simple transcendental extension** of F.

Exercise IX.4.3. Show that there is a unique monic polynomial that generates the ideal I, that is, such that it has $\mathbf{a}$ as a root.

Exercise IX.4.4. Show that any two simple transcendental extensions of a field F are equivalent. Thus, up to equivalence, there is a unique simple transcendental extension of a given field F.

Remark IX.4.1. An important example of a transcendental extension of the field $\mathbb{Q}$ of rational numbers is the field $\mathbb{Q}(\pi)$. This is a consequence of the fact that the number π is not a root of any polynomial from $\mathbb{Q}[x]$, proved by **Carl Louis Ferdinand von Lindemann** (1852 - 1939) in 1882.

We already know from Theorem VII.4.3 (Kronecker's theorem) that there exist simple algebraic extensions $F(\mathbf{a})$ of a field F, when $\mathbf{a}$ is a root of an irreducible polynomial $p \in F[x]$. We may drop the assumption of irreducibility.

Exercise IX.4.5. Prove: *Let $f \in F[x]$ be a polynomial of positive degree that has no roots in the field F. Then there exists an extension E of F that contains an element $\mathbf{a}$ such that $f(\mathbf{a}) = 0$.*

In the case of a simple algebraic extension, we are going to prove the following fact.

Theorem IX.4.1. *Given a field F. There is, up to equivalence, a unique simple algebraic extension $F(\mathbf{a})$ such that $p(\mathbf{a}) = 0$, where p is an irreducible polynomial of $F[x]$.*

Theorem IX.4.1 is a consequence of the following statement that is a rephrase of earlier considerations.

Theorem IX.4.2. *Let p be an irreducible polynomial from $F[x]$ and E an extension of F containing a root of the polynomial p, that is, an element $\mathbf{a} \in E$ such that $p(\mathbf{a}) = 0$. Then*

$$F(\mathbf{a}) \simeq F[x]/\langle p\rangle.$$

Proof. As before, we consider the polynomials $f \in F[x]$ as polynomials of the integral domain $E[x]$; thus, $f(\mathbf{a}) \in E$ for every polynomial $f \in F[x]$. By setting $\Phi(f) = f(\mathbf{a})$, we define a homomorphism $\Phi : F[x] \to E$ (see Exercise VII.1.4 and Remark VII.1.4). Then Ker $\Phi = \{f \in F[x] \mid f(a) = 0\} \subseteq F[x]$ is a principal ideal Ker $\Phi = \langle p\rangle$ and thus

$$\bar{F} = F[x]/\langle p\rangle \simeq \text{Im } \Phi = F(\mathbf{a}) \subseteq E.$$

The proof is illustrated in Figure IX.4.1, where the homomorphism $\Phi = \pi\varphi$ is expressed as a product of the canonical epimorphism $\pi : F[x] \to \bar{F}$ and the monomorphism $\varphi : \bar{F} \to E$. Moreover, μ denotes the standard inclusion of F into $F[x]$.

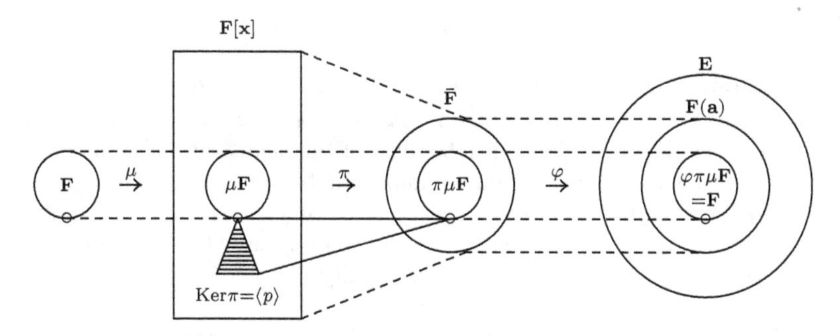

Figure IX.4.1 Simple algebraic extension F(a)

□

Remark IX.4.2. Observe that, in contrast to Kronecker's theorem (Theorem VII.4.3) that constructs a simple algebraic extension $F[x]/\langle p\rangle$ of a field F, the last theorem assumes the existence of an extension and shows that any field containing a root of the irreducible polynomial $p \in F[x]$ contains a subfield isomorphic (in fact, equivalent) to $F[x]/\langle p\rangle$.

Exercise IX.4.6. Prove:

(i) *If α belongs to an extension of F whose degree is n, then it is a root of a polynomial of degree at most n over F.*

(ii) *If α is a root of a polynomial of degree n over F, then $[F(\alpha):F] \leq n$.*

(iii) *The element α is algebraic over F if and only if $[F(\alpha):F] < \infty$.*

(iv) *If the extension E over F satisfies $[E:F] < \infty$, then it is algebraic in the sense that every element of E is algebraic over F.*

Exercise IX.4.7. Show that the polynomial $p(x) = x^3 - 2 \in \mathbb{Q}[x]$ is irreducible. Let α be a root of p and $E = F(\alpha) = \{a_0 + a_1\alpha + a_2\alpha^2 \mid a_t \in \mathbb{Q}\}$. Determine $\alpha^{-1}, (b_0 + b_1\alpha)^{-1}$ and $(b_0 + b_1\alpha + b_2\alpha^2)^{-1}$ for $b_0, b_1, b_2 \in \mathbb{Q}$.

Example IX.4.1. (An algorithm to determine the inverse of $\gamma \in F(\alpha)$). Let $p \in F[x]$ be an irreducible polynomial of degree n and $E = F(\alpha) = F[x]/\langle p\rangle$. Let $\gamma = f(\alpha) = c_0 + c_1\alpha + \cdots + c_k\alpha^k$, where $f \in F[x]$ is a polynomial of degree $k, 1 \leq k \leq n-1$. We are going to determine γ^{-1} inductively as follows: If $k = 1$, then

$$p(x) = (c_0 + c_1x)\,g(x) + r \text{ with } r \in F \text{ and thus } (c_0 + c_1\alpha)^{-1} = -\frac{g(\alpha)}{r}.$$

Assume that we already know the inverses of all $r(\alpha)$, where $r \in F[x]$ is a polynomial of degree $l < k$. Again apply the Euclidean division to get

$$p(x) = f(x)\,h(x) + r(x), \text{ where } \mathbf{deg}\, r < k \text{ and thus } \gamma^{-1} = -\frac{h(\alpha)}{r(\alpha)}.$$

By assumption, we have $(r(\alpha))^{-1} = q(\alpha)$ for some $q(\alpha) \in F(\alpha)$, and thus $\gamma^{-1} = -\,h(\alpha)\,q(\alpha)$.

Exercise IX.4.8. Determine the inverse of $\alpha^4 + 1$ in the field $\mathbb{Q}(\alpha)$, where α is a root of the cyclotomic polynomial $1 + x + x^2 + x^3 + x^4 + x^5 + x^6 \in \mathbb{Q}[x]$.

Remark IX.4.3. Consider the equation $x^3 - 2 = 0$ of Exercise IX.4.7. It has a real root $\alpha_1 = \sqrt[3]{2}$ and two complex roots $\alpha_2 = \omega\,\sqrt[3]{2}$ and $\alpha_3 = \omega^2\,\sqrt[3]{2}$, where $\omega = \frac{-1+i\sqrt{3}}{2}$ is a cubic root of unity. The field $\mathbb{Q}(\alpha_1)$ is a subfield of the field $\mathbb{R}$ of real numbers, while $\mathbb{Q}(\alpha_2)$ and $\mathbb{Q}(\alpha_3)$ are subfields of the complex numbers $\mathbb{C}$ but not of $\mathbb{R}$. These three extensions $\mathbb{Q}(\alpha_1)$, $\mathbb{Q}(\alpha_2)$ and $\mathbb{Q}(\alpha_3)$ of $\mathbb{Q}$ are equivalent, that is, they are algebraically indistinguishable. Of course, the field $\mathbb{Q}(\alpha_1)$ is distinguished by the non-algebraic fact that its elements are real (which involves a non-algebraic concept of continuity).

Exercise IX.4.9. Prove the existence of the following isomorphism: *Let $\varphi : F_1 \to F_2$ be an isomorphism of the fields F_1 and F_2. Extend φ to the isomorphism $\varphi : F_1[x] \to F_2[x]$ by setting $\varphi(\sum a_t\,x^t) = \sum \varphi(a_t)\,x^t \in F_2[x]$ for all $\sum a_t\,x^t \in F_1[x]$. Let $p_1 \in F_1[x]$ be irreducible. Then $p_2 = \varphi(p_1) \in F_2[x]$ is irreducible. Let $p_1(\alpha) = 0$ with $\alpha \in E_1 \supseteq F_1$ and $p_2(\beta) = 0$ with $\beta \in E_2 \supseteq F_2$. Then there is an isomorphism $\psi : F_1(\alpha) \to F_2(\beta)$ such that*

$$\psi(\alpha) = \beta \text{ and } \psi(a) = \varphi(a) \text{ for all } a \in F_1.$$

By Kronecker's theorem, we are able to construct for any polynomial $f \in F[x]$ an algebraic extension E of F that contains at least one root of f. Naturally, we would like to construct an extension of F that will contain **all roots** of f. Just keep in mind that a polynomial of degree n has at most n distinct roots; in fact, it has precisely n roots once we count them with their multiplicity. This fact enables us, by repeating Kronecker's construction of simple extensions to reach in a finite number of steps an extension that will contain all roots of a given polynomial f, that is, to construct the **splitting field** of the polynomial f. We give a formal definition.

Definition IX.4.1. *An extension E of a field F is said to be a* **splitting field** *for a polynomial $f \in F[x]$ if, as a polynomial of $E[x]$, f factors completely into a product of linear factors*

$$f(x) = a_n(x - \alpha_1)(x - \alpha_2)\cdots(x - \alpha_n) \text{ with } \alpha_t \in E,\ t = 1, 2, \ldots, n,$$

and no proper subfield of E containing F has this property.

Theorem IX.4.3. *Let F be a field. Then for any polynomial $f \in F$, there exists an extension E that is a splitting field for f. Any two splitting fields for a polynomial f are equivalent. If the degree of $f \in F[x]$ is n then the degree $[E : F]$ of its splitting field E is at most $n!$.*

Proof. If all roots of f belong to F, then f splits into a product of linear factors in $F[x]$ and there is nothing to prove: $F = E_0$ is the splitting field of f. Otherwise, denote one of the nonlinear irreducible factors of f by $p_1 \in F[x]$ and construct the extension $E_1 = E_0[x]/\langle p_1\rangle$ that contain a root of p_1 and thus an additional root of f in E_1. Thus the number of linear polynomial factors of f over E increased (at least by one). If f as a polynomial over E_1 factors into linear polynomials, E_1 is a splitting field of f. Otherwise, we proceed to construct, in a similar way, an extension E_2 of E_1 and thus of F that contains at least one additional root of f: the number of linear factors of f over E_2 increased again (at least by one). Continuing, inductively, in this process, we will built a tower of extensions

$$F = E_0 \subset E_1 \subset E_2 \subset \cdots \subset E_k = E$$

to reach in a finite number ($k \leq \mathbf{deg}\ f$) of steps the field E such that f splits into linear factors of $E[x]$, that is, the splitting field of f.

To show that any two splitting fields of f are equivalent, we may proceed again inductively; in fact, we may apply step by step, from E_t to E_{t+1} a general statement of Exercise IX.4.10 to get, as a consequence, the desired isomorphism.

In the first step from F to the extension E_1, the degree $[E_1 : F] \leq n = \mathbf{deg}\ f$; this may happen if f is irreducible. In the second step from E_1 to the extension E_2, the extreme situation occurs if f if a product of one linear factor and an irreducible factor (of degree $n-1$). Here, $[E_2 : E_1] \leq n - 1$. In general, for the extension E_{t+1} of E_t, we have $[E_{t+1} : E_t] \leq n - t$ for all $t \leq k - 1$. Since

$$[E : F] = \prod_{t=0}^{k-1} [E_{t+1} : E_t] \leq n(n-1)(n-2)\ldots(n-k+1) \leq n!\,.$$

The last equality follows inductively from the fact that, given fields $F \subset K \subset E$, we have $[E : F] = [E : K][K : F]$ (see Exercise IX.4.10). □

Exercise IX.4.10. Let $F \subset K \subset E$ be field extensions. Prove that if $\{a_1, a_2, \ldots, a_r\}$ is a basis of the vector space E_K and $\{b_1, b_2, \ldots, b_s\}$ a basis of the vector space K_F, it is evident that $\{c_{i,j} = a_i b_j \mid 1 \le i \le r, 1 \le j \le s\}$ is a basis of the vector space E_F. Conclude that both degrees $[K : F]$ and $[E : F]$ are divisors of the degree $[E : F]$.

Exercise IX.4.11. Let $c < d$ be two primes. Justify that the field $\mathbb{Q}(\sqrt{c}, \sqrt{d})$ is the splitting field for the polynomial $(x^2 - c)(x^2 - d) \in \mathbb{Q}[x]$. Prove that $\mathbb{Q}(\sqrt{c}, \sqrt{d}) = \mathbb{Q}(\sqrt{c} + \sqrt{d})$. Find an irreducible polynomial $p \in \mathbb{Q}[x]$ such that $p(\sqrt{c} + \sqrt{d}) = 0$ so that $\mathbb{Q}(\sqrt{c}, \sqrt{d})$ is also the splitting field for p. Determine the degree $[\mathbb{Q}(\sqrt{c}, \sqrt{d}) : \mathbb{Q}]$.

Exercise IX.4.12. Generalize the statements of Exercise IX.4.11 for natural numbers c and d such that none of the numbers c, d and cd is a square of an integer. Apply the results for $\mathbb{Q}(4\sqrt{3}, 5\sqrt{2})$.

Exercise IX.4.13. Show that the splitting field for the polynomial $x^3 - 2 \in \mathbb{Q}[x]$ is the extension $E = \mathbb{Q}(\sqrt[3]{2}, i\sqrt{3})$ and prove that the only fields K satisfying $\mathbb{Q} \subsetneqq K \subsetneqq E$ are the fields $\mathbb{Q}(\alpha_1), \mathbb{Q}(\alpha_2), \mathbb{Q}(\alpha_3)$ and $\mathbb{Q}(i\sqrt{3})$ (see Remark IX.4.3). Determine $[\mathbb{Q}(\alpha_t) : \mathbb{Q}]$, $[\mathbb{Q}(i\sqrt{3}) : \mathbb{Q}]$ and $[E : \mathbb{Q}]$.

Exercise IX.4.14. Let $a > c$ be natural numbers such that neither $a + c$ nor $b = a^2 - c^2$ is a square of an integer. By expressing the extension field $\mathbb{Q}(\sqrt{a + \sqrt{b}})$ in the form $\mathbb{Q}(\sqrt{r}, \sqrt{s})$ for suitable numbers r and s, or otherwise, determine the degree $[\mathbb{Q}(\sqrt{a + \sqrt{b}}) : \mathbb{Q}]$. Find an irreducible polynomial $p \in \mathbb{Q}[x]$ such that $p(\sqrt{a + \sqrt{b}}) = 0$. Apply the results to the particular case of the extension $\mathbb{Q}(\sqrt{7 + 2\sqrt{10}})$.

Exercise IX.4.15. Prove that the splitting field of the polynomial $x^4 + 4 \in \mathbb{Q}[x]$ is the field $\mathbb{Q}(i)$ of degree 2 over $\mathbb{Q}$.

Exercise IX.4.16. Determine the (cyclotomic) splitting field E and the degree $[E : \mathbb{Q}]$ of the polynomial $x^n - 1 \in \mathbb{Q}[x]$, when n is a prime number (see Corollary VII.5.3).

Exercise IX.4.17. Construct a polynomial in $\mathbb{Q}[x]$ having $\sqrt{2}$ and $2 + 3i$ as roots. Show that $E = \mathbb{Q}(\sqrt{2}, 2 + 3i) = \mathbb{Q}(\sqrt{2} + i)$ and find an irreducible polynomial $p \in \mathbb{Q}[x]$ such that E is its splitting field. Decompose p into irreducible factors over $\mathbb{Q}(\sqrt{2})$ and over $\mathbb{Q}(i)$.

A deeper study of the algebraic extensions of fields requires us to introduce the concepts of **separable** and **inseparable** extensions (defined in terms of multiplicities of roots of irreducible polynomials) whose deeper discussion is beyond the scope of this book. Let us just mention that all extensions of any field of characteristic 0 (and thus, in particular, of $\mathbb{Q}$) are separable. A crucial property of such algebraic extensions $E = F(\alpha_1, \alpha_2, \ldots, \alpha_d)$ of F is that they are all simple: $F(\alpha_1, \alpha_2, \ldots, \alpha_d) = F(\sigma)$ for a suitable **primitive** element $\sigma \in E$. This result provides a simple representation $\gamma = \sum a_t \sigma^t, a_t \in F$, for all elements $\gamma \in E$ and thus greatly simplifies investigations of finite extensions.

IX.5 Radical extensions (Solvability of algebraic equations by radicals)

The role of the last three sections of this compendium is to apply the results of the last section to briefly inform the reader about some of the classical questions of algebra related to the solution of algebraic equations. These questions historically belong to general education in mathematics and every mathematician, teachers of mathematics in particular, should be aware of them. Most of the statements will be presented without proofs, some because of their complexity, some because of space limitations.

In order to avoid additional conditions that are necessary in a general situation, we shall assume in this section that the characteristic of the fields is zero. This carries no restriction for the equations with rational (real or complex) coefficients.

First, we introduce the concept of **solvability** of algebraic equations

$$f(x) = 0, \quad \text{where} \quad f \in F[x] \tag{IX.5.1}$$

by **radicals**. By such solvability, we understand a presentation of all solutions of (IX.5.1) by a formula that contains expressions of the form

$$\sqrt[r]{\cdots + \sqrt[s]{\cdots + \sqrt[t]{\cdots + \ldots}}}$$

connected by rational operations for any choice of the radicals $\sqrt[s]{a}$ involved. Specifically, if a radical $\sqrt[s]{a}$ appears in such a formula several times, each time the (chosen) value has to be the same. We now present the relevant formal definitions.

Definition IX.5.1. *Let F be a field. The splitting field of the polynomial $f(x) = x^n - a \in F[x]$ is called a radical extension of the field F; it will be denoted by $F(\sqrt[n]{a})$. If the number n is a prime, we speak about a* **prime radical extension**.

If $f(x) = a_0 + a_1 x + a_2 x^2 + \cdots + a_n x^n \in F[x]$ and $P \simeq \mathbb{Q}$ is the prime subfield of F, denote by $P_f = P(a_0, a_1, a_2, \ldots, a_n)$ the finite extension of the field F by the coefficients of the polynomial f.

Definition IX.5.2. *An algebraic equation* (IX.5.1) *is said to be* **solvable by radicals** *if there exists a (prime radical) chain*

$$F_0 \subseteq F_1 \subseteq F_2 \subseteq \cdots \subseteq F_d \tag{IX.5.2}$$

such that $F_0 = P_f$, the field F_t is a prime radical extension of the field F_{t-1} for every $t = 1, 2, \ldots, d$ and $F_d \supseteq E$, where E is the splitting field of the polynomial f (over the field F_0).

We point out that every radical chain can be refined to a prime radical chain, since a radical extension $F \subseteq F(\sqrt[rs]{a})$ satisfies trivially $F \subseteq F(\sqrt[s]{a}) \subseteq F(\sqrt[s]{a}, \sqrt[r]{\sqrt[s]{a}})$ for arbitrary positive integers r and s.

Example IX.5.1. Every equation of the form

$$x^{2n} + ax^n + b = 0, \text{ where } a, b \in F,$$

is solvable by radicals: $x^{2n} + ax^n + b = (x^n - c_1)(x^n - c_2)$, where

$$c_1 = \frac{-a + \sqrt{a^2 - 4b}}{2} \text{ and } c_2 = \frac{-a - \sqrt{a^2 - 4b}}{2}.$$

Here, $c_1, c_2 \in F_1 = F(\sqrt{d})$, with the discriminant $d = a^2 - 4b \in F$. Thus, all roots belong to the splitting fields of the polynomials $x^n - c_1 \in F_1[x]$ and $x^n - c_2 \in F_1[x]$.

For the next exercise, recall Tartaglia's formula for the solution of cubic equations at the beginning of Chapter IV in order to construct a radical chain.

Exercise IX.5.1. Consider the equation $f(x) = 0$, where $f(x) = a_0 + a_1x + a_2x^2 + x^3$ belongs to $\mathbb{Q}[x]$. Check that by the substitution $x \longmapsto x - \frac{1}{3}a_2$ the equation becomes $g(x) = 0$ with $g(x) = x^3 + px + q \in \mathbb{Q}[x]$ and that f and g have the same splitting field E. Put

$$F_0 = \mathbb{Q},\ F_1 = \mathbb{Q}(i\sqrt{3}),\ F_2 = F_1(\sqrt{d}), \text{ where } d = \frac{q^2}{4} + \frac{p^3}{27} \text{ and}$$

$$F_3 = F_2(\sqrt[3]{c}) = \mathbb{Q}(i\sqrt{3}, \sqrt{d}, \sqrt[3]{c}), \text{ where } c = \frac{q}{2} - \sqrt{d}.$$

Show that $F_3 \supseteq E$ and that, in some cases the radical chain can be shorter. Consider, in particular, the equation $h(x) = x^3 + 6x + 2 = 0$, determine the splitting field of h over $\mathbb{Q}$ and verify that the numbers

$$a(1-a), \frac{1}{2}\left(a(a-1) + a(a+1)\sqrt{3}\, i\right) \text{ and } \frac{1}{2}\left(a(a-1) - a(a+1)\sqrt{3}\, i\right),$$

where $a = \sqrt[3]{2}$, are the roots of this equation.

Example IX.5.2 (Ferrari - Cardano formula). Consider the equation $f(x) = 0$, where $f(x) = a_0 + a_1x + a_2x^2 + a_3x^3 + x^4 \in \mathbb{Q}[x]$. Check that by the substitution $x \longmapsto x - \frac{1}{4}a_3$ the equation becomes $g(x) = 0$ with $g(x) = x^4 + px^2 + qx + r \in \mathbb{Q}[x]$ and that f and g have the same splitting field E. Now $g(x) = (x^2 + cx + a)\,(x^2 - cx + b)$, where

$$ab = r, \quad a - b = -\frac{q}{c} \text{ and } a + b = c^2 + p. \tag{IX.5.3}$$

Now, one can check that c^2 is a root of the cubic equation $h(x) = 0$, where

$$h(x) = x^3 + 2px^2 + (p^2 - 4r)x - q^2 \in \mathbb{Q}[x].$$

Thus, as in Exercise IX.5.1, there is a radical chain

$$F_0 = \mathbb{Q} \subseteq F_1 \subseteq F_2 \subseteq F_3$$

with F_3 containing the splitting field of the polynomial h over $\mathbb{Q}$. From (IX.5.3), we have

$$a = \frac{1}{2}(c^2 + p - \frac{q}{c}) \text{ and } b = \frac{1}{2}(c^2 + p + \frac{q}{c}).$$

Thus, $F_4 = F_3(c)$ contains both a and b. Hence, $F_5 = F_4(\sqrt{c^2 - 4a})$ is a splitting field of the quadratic polynomial $x^2 + cx + a$ over F_4 and consequently, the field $F_6 = F_5(\sqrt{c^2 - 4b})$ contains the splitting field E of the polynomial f over $\mathbb{Q}$.

The developments concerning solvability of algebraic equations of the form (IX.5.1) have a long history. Solutions of quadratic equations (together with the related principle of "completing a square") are in the realm of the system of natural numbers already recorded in the pivotal work **Al-kitab al-mukhtasar fi hisab al-jabr w'al-muqabalab** of **Abu Jafar Muhammed ibn Musa al-Khwarizmi (about 790 - about 850)** in the 9-th century (translated into Latin as **Liber Algebrae et Almucabola**). Algebraic formulas for solutions of cubic and quadratic equations, derived by **Nicolo Fontana Tartaglia** and **Ludovico Ferrari (1522 - 1565)** and documented in **Ars Magna** by **Girolamo Cardano** in 1545, initiated a search for similar formulas for equations of higher degrees, for quintic equations in particular. Following a number of unsuccessful attempts, **Paolo Ruffini (1765 - 1822)** questioned the existence of such formulas and offered the first, albeit incomplete, proof of impossibility to solve a general quintic equation by radicals. The first mathematician who proved that it is impossible to solve quintic equations by radicals, that is, who rigorously proved the following theorem, was **Niels Henrik Abel (1802 - 1829)**.

Theorem IX.5.1 (Abel-Ruffini theorem). *For any $n \geq 5$, there are algebraic equations of the form* (IX.5.1) *that are not solvable by radicals.*

Of course the crucial case is that of a quintic equation. The proof that is beyond the scope of this book, is based on a brilliant idea of **Éveriste Galois (1811 - 1832)**, who discovered a close relation between the solvability of algebraic equations and the group theory that he has originated. The proof of Theorem IX.5.1 is reduced to the question of **"solvability" of symmetric groups $\mathcal{S}_n$**:

The symmetric groups $\mathcal{S}_3$ and $\mathcal{S}_4$ are solvable while the groups $\mathcal{S}_n$ for $n \geq 5$ are not solvable.

Here, a group G is said to be solvable if there is a chain of subgroups

$$G = G_0 \supseteq G_1 \supseteq G_2 \supseteq \cdots \supseteq G_n = \langle 1 \rangle$$

of the group G such that , for every $1 \leq t \leq n$, G_t is a normal subgroup of the group G_{t-1} and the factor group G_{t-1}/G_t is a cyclic group. It is evident that we may, equivalently, require that the orders of all quotient groups G_{t-1}/G_t be prime numbers. Thus

$$\mathcal{S}(3) \supseteq A(3) \supseteq \langle 1 \rangle \text{ and } \mathcal{S}(4) \supseteq A(4) \supseteq V_4 \supseteq C(2) \supseteq \langle 1 \rangle$$

are such chains for the groups $\mathcal{S}_3$ and $\mathcal{S}_4$, and thus these groups are solvable. For $n \geq 5$, as we have proved, the alternating group $\mathcal{A}_n$ is simple and thus have no nontrivial normal subgroup. Consequently, $\mathcal{S}_n$ is not solvable.

The mentioned relation between the solvability of algebraic equations (IX.5.1) and the theory of finite groups is the subject of **Galois theory**. Here, we can only indicate the main ideas of this relationship, that is, the relation between the chains of groups of certain automorphisms and chains of subfields of the splitting field of the polynomial defining the equation.

We already know that an **automorphism of a field** E is an injective (and thus bijective) mapping $\pi : E \to E$ that respects the field operations:

$$\pi(a + b) = \pi(a) + \pi(b) \text{ and } \pi(ab) = \pi(a)\pi(b) \text{ for all } a, b \in E.$$

The set of all automorphisms of the field E forms with respect to composition of mappings a group $\mathrm{Aut}(E)$. Now, for every extension $E \supseteq F$ of the field F consider the subgroup of $\mathrm{Aut}(E)$ of all automorphisms $\pi \in \mathrm{Aut}(E)$ such that $\pi(a) = a$ for all elements a of F. Denote this subgroup by $\mathrm{Aut}(E,F)$. Note that the order of the group $\mathrm{Aut}(E,F)$ satisfies for every finite extension $E \supseteq F$ the inequality

$$|\mathrm{Aut}(E,F)| \leq [E:F].$$

This is a consequence of a simple fact that for any polynomial $f \in F[x]$ and its root a, and any automorphism $\pi \in \mathrm{Aut}(E,F)$, $\pi(a)$ is also a root of the polynomial $f \in F[x]$.

Here, we shall restrict our considerations to the case when E is a splitting field of a polynomial $f \in F[x]$. Then $|\mathrm{Aut}(E,F)| \leq [E:F]$ and we will call $\mathrm{Aut}(E,F)$ the **Galois group** (of the polynomial f or of the equation $f(x) = 0$) and denote it by $\mathrm{Gal}(E/F)$.

To provide simple examples, $\mathrm{Gal}(\mathbb{C}/\mathbb{R}) = \mathrm{Gal}(\mathbb{R}(i)/\mathbb{R})$ is a cyclic group of order 2 generated by the conjugation $\kappa(z) = \bar{z}$. If a field F contains the p-th roots of unity and $f(x) = x^p - a \in F[x]$ is irreducible, then $\mathrm{Gal}(F(\sqrt[p]{a})/F)$ is a cyclic group $C(p)$ of order p.

Considering the lattice $\mathcal{L}(E/F)$ of all intermediate finite extension fields $F_t, F \subseteq E$, the **Fundamental theorem of Galois theory** describes the bijection between this lattice and the lattice $\mathcal{L}(\mathrm{Gal}(E/F))$ of all subgroups of the Galois group $\mathrm{Gal}(E/F)$. Every $F_t \in \mathcal{L}(E/F)$ is mapped onto the Galois group $\mathrm{Gal}(E/F_t)$, and inversely, every $H_t \in \mathcal{L}(\mathrm{Gal}(E/F))$ is mapped onto the extension $\mathrm{Fix}(H_t)$ of F defined as the set of all elements $a \in E$ such that $\alpha(a) = a$ for every automorphism $\alpha \in H_t$. The maps

$$F_t \longmapsto \mathrm{Gal}(E/F_t) \quad \text{and} \quad H_t \longmapsto \mathrm{Fix}(H_t)$$

form an order-reversing **Galois correspondence** whereby

$$|\mathrm{Gal}(E/F_t)| = [E:F_t] \quad \text{and} \quad [\mathrm{Gal}(E/F):\mathrm{Gal}(E/F_t)] = [F_t:F],$$

as illustrated in Figure IX.5.1.

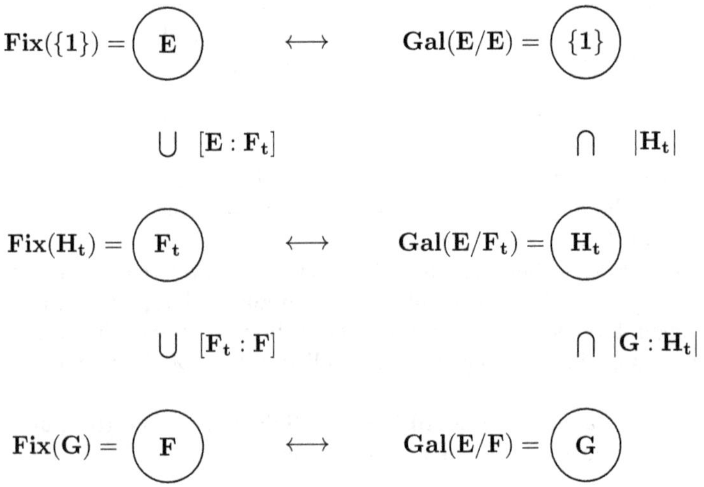

$$\begin{array}{ccc}
\mathbf{Fix(\{1\})} = \mathbf{E} & \longleftrightarrow & \mathbf{Gal(E/E)} = \{\mathbf{1}\} \\
\cup\ [\mathbf{E}:\mathbf{F_t}] & & \cap\ |\mathbf{H_t}| \\
\mathbf{Fix(H_t)} = \mathbf{F_t} & \longleftrightarrow & \mathbf{Gal(E/F_t)} = \mathbf{H_t} \\
\cup\ [\mathbf{F_t}:\mathbf{F}] & & \cap\ |\mathbf{G}:\mathbf{H_t}| \\
\mathbf{Fix(G)} = \mathbf{F} & \longleftrightarrow & \mathbf{Gal(E/F)} = \mathbf{G}
\end{array}$$

Figure IX.5.1 Galois correspondence

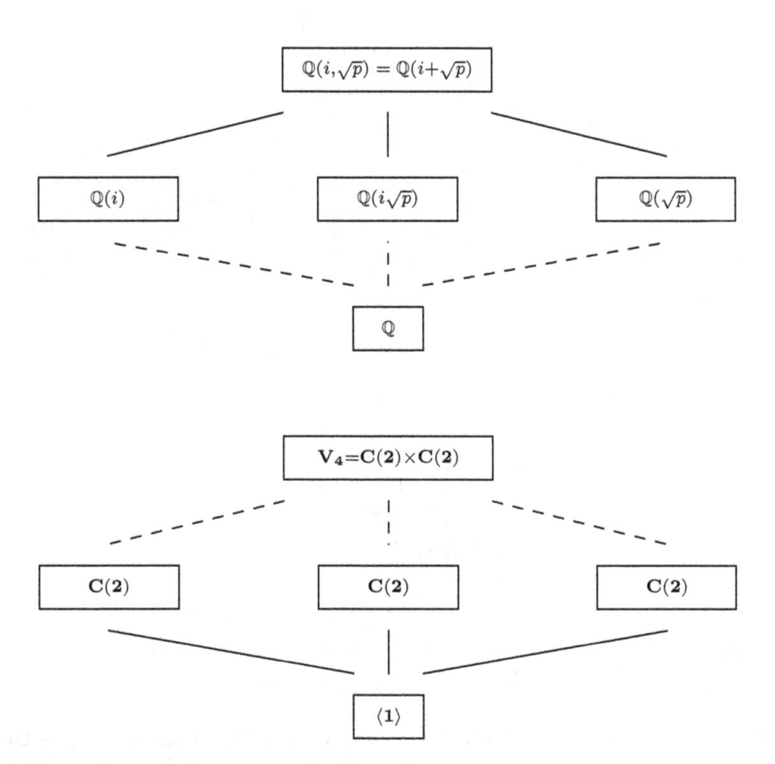

Figure IX.5.2 Galois correspondence of $\mathbf{x^4 - (p-1)x^2 - p = 0}$, p prime

ACT	$r = \sqrt[4]{2}$	$-r$	$r\,i$	$-r\,i$	$i = \frac{r\,i}{r}$	$\sqrt{2} = r^2$	$\sqrt{2}\,i = r\,i \cdot r$
1	r	$-r$	$r\,i$	$-r\,i$	i	r^2	$r^2\,i$
a	$r\,i$	$-r\,i$	$-r$	r	i	$-r^2$	$-r^2\,i$
$\mathbf{a^2}$	$-r$	r	$-r\,i$	$r\,i$	i	r^2	$r^2\,i$
$\mathbf{a^3}$	$-r\,i$	$r\,i$	r	$-r$	i	$-r^2$	$-r^2\,i$
b	r	$-r$	$-r\,i$	$r\,i$	$-i$	r^2	$-r^2\,i$
a b	$r\,i$	$-r\,i$	r	$-r$	$-i$	$-r^2$	$r^2\,i$
$\mathbf{a^2\,b}$	$-r$	r	$r\,i$	$-r\,i$	$-i$	r^2	$-r^2\,i$
$\mathbf{a^3\,b}$	$-r\,i$	$r\,i$	$-r$	r	$-i$	$-r^2$	$r^2\,i$

Figure IX.5.3 Table of the automorphism actions $(\mathrm{Gal}(\mathbb{Q}(\sqrt[4]{2}, i)/\mathbb{Q}))$

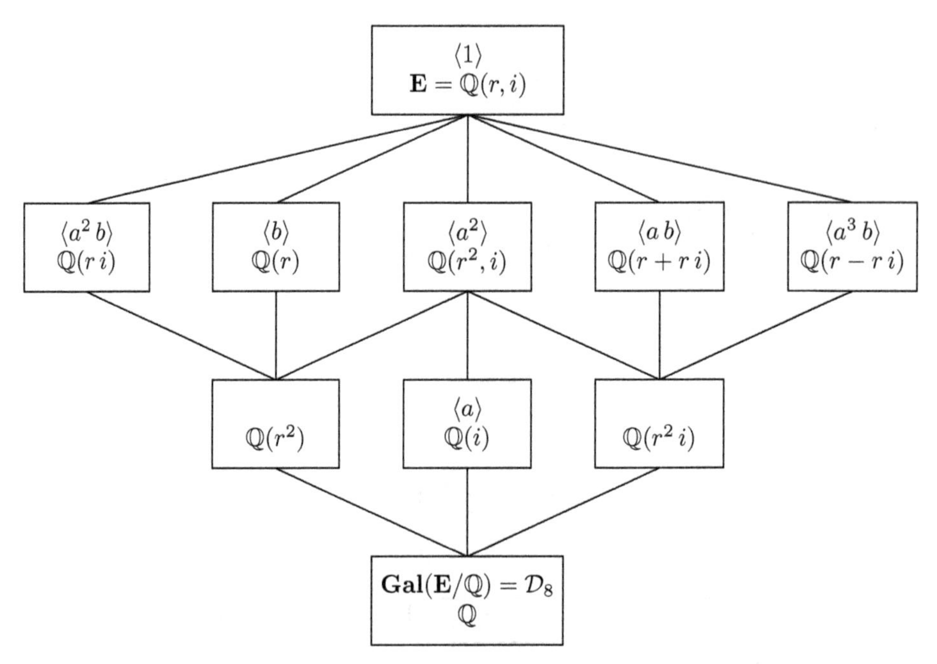

Figure IX.5.4 Galois correspondence ($\mathbf{x^4 - 2 = 0}$, $\mathbf{r = \sqrt[4]{2}}$)

Example IX.5.3. Figure IX.5.2 illustrates the structure of the Galois group of the equation $x^4 - (p-1)x^2 - p = 0$, where p; is a prime, and the respective Galois correspondence.

Since $x^4 - (p-1)x^2 - p = (x^2+1)(x^2-p)$, the roots of the equation are numbers i, $-i$, $\sqrt{p}$ and $-\sqrt{p}$. Thus we obtain a simple lattice of the fields. It is easy to see that there are only four automorphisms that induce the identity automorphism of the base field $\mathbb{Q}$ of the rational numbers: $1, \alpha, \beta$ a $\gamma = \alpha\beta = \beta\alpha$, $\alpha(i + \sqrt{p}) = -i + \sqrt{p}$, $\beta(i + \sqrt{p}) = i - \sqrt{p}$, $\gamma(i + \sqrt{p}) = -i - \sqrt{p}$. Thus $\mathrm{Gal}(\mathbb{Q}(i + \sqrt{p})/\mathbb{Q})$ is isomorphic to the Klein four-group V_4.

Exercise IX.5.2. Describe the generators of the cyclic subgroups of $\mathrm{Gal}(\mathbb{Q}(i+\sqrt{p})/\mathbb{Q})$ and their images in the Galois correspondence of the previous Example.

Exercise IX.5.3. Describe the splitting field E of the equation $x^4 - 2 = 0$. Show that the order of the Galois group $\mathrm{Gal}(E/\mathbb{Q})$ is 8 and check the action of the automorphisms of $\mathrm{Gal}(E/\mathbb{Q})$ as indicated in Figure IX.5.3. Describe the Galois correspondence as illustrated in Figure IX.5.4.

Exercise IX.5.4. Explain the Galois correspondence as indicated in Figure IX.5.5 (follow the questions asked in Exercise IX.5.3).

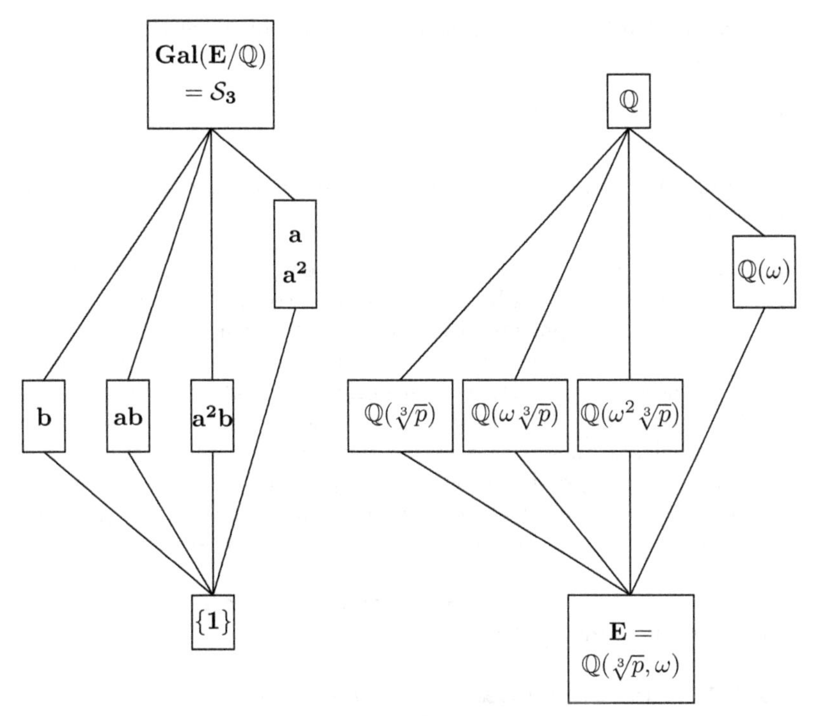

Figure IX.5.5 Galois correspondence ($\mathbf{x^3 - p = 0}$, p prime)

A final remark: To a chain

$$\mathbb{Q} = F_0 \subseteq F_1 \subseteq F_2 \subseteq \cdots \subseteq F_t \subseteq \cdots F_d = E,$$

there corresponds a chain

$$\langle 1 \rangle = \text{Gal}(E/E) \subseteq \text{Gal}(E/F_{d-1}) \subseteq \text{Gal}(E/F_{d-2}) \subseteq \cdots \subseteq \text{Gal}(E/F_1) \subseteq \text{Gal}(E/F)$$

of subgroups of $\text{Gal}(E/F)$. This group is, for every n and a suitable equation of degree n isomorphic to the symmetric group $\mathcal{S}_n$. We have seen this in Exercise IX.5.4 for $n = 3$. We have also seen that for $n \geq 5$, the symmetric group is not solvable. Hence, there are equations of fifth degree (and thus also of higher degrees) that are not solvable by radicals.

IX.6 Constructions with straightedge and compass (Euclidean constructions)

We apply some of our previous results to the famous classical Greek problems, namely to the question of the possibility of performing the following geometrical constructions using only straightedge and compass:

(1) **The trisection of an angle**, that is, a construction of the angle that is one third of the given angle;

(2) **The duplication of a cube**, that is, a construction of a cube with volume twice the volume of a given cube;

(3) **The squaring of a circle**, that is, a construction of a square with area equal to the area of a given circle.

It may be appropriate here to make the following remark. Greek mathematics did not answer the question whether it is possible to achieve these constructions only by straightedge and compass. The rapid developments of the renaissance mathematics did not contribute towards the solution either. In spite of the fact that these problems were very popular and in spite of the fact that there was a longstanding suspicion that such constructions do not exist, it was only in the 19-th century that there appears a definite proof that such construction are indeed by means of a straightedge and compass alone impossible. And we will now indicate such a (quite simple) proof. The idea of the proof is to translate geometric objects into algebraic ones. We shall see that the constructions using straightedge correspond to solutions of linear equations, while the constructions using compass correspond to solutions of quadratic equations.

First, give a rigorous definition of a **Euclidean construction** or, equivalently, of a **constructible point** and **constructible number**.

Definition IX.6.1. *A (plane) Euclidean construction, that is, a construction using only straightedge and compass, is a construction that can be obtained by a successive application of the following* **basic Euclidean constructions***:*

(1) *Drawing a straight line through two given points;*

(2) *Drawing a circle given by center and radius;*

(3) *Finding a point of intersection of two straight lines, points of intersection of a straight line and a circle and points of intersection of two circles.*

The following constructions are Euclidean constructions: A construction of a bisector of a given angle, of the axis of symmetry of two points, of a parallel straight line to a given straight line passing through a point, of a perpendicular straight line to a given straight line passing through a given point etc.

Definition IX.6.2. *Let A and B be two distinct points of a plane. Write $S_0 = \{A, B\}$. Denote by S_1 the set of all points obtained by the basic Euclidean constructions applied to the points of S_0. Inductively, define and denote by $S_n, n \geq 2$ the set of all points obtained by the basic Euclidean constructions applied to the points of S_{n-1}. The points of the union S of the tower of subsets*

$$S_0 \subset S_1 \subset S_2 \subset \cdots \subset S_n \subset \cdots$$

are said to be (Euclidean) **constructible points** *(determined by the points A and B).*

In a parallel fashion, we define constructible numbers.

Definition IX.6.3. *Choose a cartesian coordinate system in such a way that the coordinates of the points A and B of Definition* IX.6.1 *are* $(0,0)$ *and* $(1,0)$. *Then the coordinates of the constructible points, that is, the points of the set S of Definition* IX.6.2 *are said to be* (*Euclidean*) **constructible numbers**.

Thus, for example, S_1 has 6 points whose coordinates are

$$(-1,0),(0,0),(1,0),(2,0),(\frac{1}{2},\frac{\sqrt{3}}{2}) \text{ and } (\frac{1}{2},-\frac{\sqrt{3}}{2})$$

and the numbers $-1,0,1,2,\frac{1}{2},\frac{\sqrt{3}}{2}$ and $-\frac{\sqrt{3}}{2}$ are constructible. We can see that all integers are constructible numbers, and thus so are all lattice points, that is, also all points with both coordinates that are integers.

Exercise IX.6.1. Prove that a point $P = (r,s)$ is constructible if and only if both numbers r and s are constructible. Hence show that all points with rational coordinates are constructible.

Now, all that is necessary to establish (and, in fact, easy to see) is that to determine a point of intersection of two straight lines requires a solution of two linear equations (and thus the coordinates of the point of intersection belong to the field of the coefficients of the given straight lines), while the coordinates of points of intersection of a circle with a straight line or with another circle belong to a quadratic extension of the field that contains the coefficients describing the circles and (or) the straight line. This provides a proof of the second part of the following theorem.

Exercise IX.6.2. Prove the first part of the following theorem, that is, if a and b are constructible numbers, so are the numbers $a+b, a-b, ab, \frac{a}{b}$, if $b \neq 0$, and $\sqrt{a}$.

Theorem IX.6.1. *The set $\mathcal{S}$ of all constructible numbers form a subfield of the field $\mathbb{R}$ of real numbers $\mathbb{Q} \subseteq \mathcal{S} \subset \mathbb{R}$ having the property that $a \in \mathcal{S}$ if and only if $\sqrt{a} \in \mathcal{S}$. If $a \in \mathcal{S}$, then $a \in E \subset \mathcal{S}$, where E is an extension of the field $\mathbb{Q}$ of rational numbers such that, for a certain n, $[E:\mathbb{Q}] = 2^n$.*

Now, the fact the three Greek classical constructions cannot be achieved by using straightedge and compass alone, is a consequence of Theorem IX.6.1.

Indeed:

(1) Doubling of a cube requires a construction of the number $\sqrt[3]{2}$. But $[\mathbb{Q}(\sqrt[3]{2}):\mathbb{Q}] = 3$ and thus this number cannot belong to any finite extension $E \subset \mathcal{S}$ of the field $\mathbb{Q}$, since degree of any such extension $[E:\mathbb{Q}]$ is a power of 2.

(2) Trisection of an angle also cannot be achieved for a general angle ω by a Euclidean construction. For example, the angle $\omega = 60^o$ is defined by the points $(0,0),(1,0)$ and $(\frac{1}{2},\frac{\sqrt{3}}{2})$. Here, a possibility of Euclidean trisecting the angle ω requires that the point $(\cos\frac{\omega}{3},\sin\frac{\omega}{3})$ be constructible, and thus the number $a = \cos 20^o$ be constructible. Since $\cos 60^o = 4\cos^3 20^o - 3\cos 20^o$ and $\cos 60^o = \frac{1}{2}$, a is a root of the equation $8x^3 - 6x - 1 = 0$. However, the polynomial $8x^3 - 6x - 1$ is irreducible over $\mathbb{Q}$, and thus $[\mathbb{Q}(a):\mathbb{Q}] = 3$. Hence, $a \notin \mathcal{S}$.

(3) To show that squaring the circle is impossible by a Euclidean construction requires again to show that the number $\sqrt{\pi}$ and thus the number π as well, is not constructible. Since every constructable number is a root of an algebraic equation over $\mathbb{Q}$ and we have already mentioned earlier (see Remark IX.4.1) that Lindemann proved in 1882 that π is not a root of any algebraic equation with rational coefficients, π cannot belong to $\mathcal{S}$. Of course, Lindemann's proof is well beyond the material of this compendium.

IX.7 ⋆ Fundamental theorem of algebra (Theorem of Argand and D'Alembert)

The goal of this section is to sketch a background of a statement that played a significant role in the early period of Algebra, but lost its central position with further developments. The statement in question is the fundamental theorem of algebra. Here are some equivalent formulations of this theorem describing a profound feature of the field $\mathbb{C}$ of complex numbers.

Theorem IX.7.1. (1) *Every algebraic equation*

$$a_0 + a_1x + a_2x^2 + \cdots + a_nx^n = 0 \text{ with } n \in \mathbb{N} \text{ and all } a_t \in \mathbb{C}, a_n \neq 0,$$

has in $\mathbb{C}$ a solution (a root).

(2) *Every algebraic equation*

$$a_0 + a_1x + a_2x^2 + \cdots + a_nx^n = 0 \text{ with } n \in \mathbb{N} \text{ and all } a_t \in \mathbb{R}, a_n \neq 0,$$

has in $\mathbb{C}$ a solution (a root).

(3) *The range of a complex polynomial function $f(z) = a_0 + a_1z + a_2z^2 + \cdots + a_nz^n$, where $n \in \mathbb{N}$, $a_t \in \mathbb{C}$ for $t = 0, 1, \ldots, n$ and $a_n \neq 0$, is the entire field $\mathbb{C}$ of complex numbers, that is, the range contains every complex number.*

These formulations are often expressed in a variety of derived forms:

(4) *The field $\mathbb{C}$ of complex numbers is algebraically closed.*

(5) *Every polynomial $f \in \mathbb{C}[z]$ of degree $n \in \mathbb{N}$ can be expressed in the form*

$$f(z) = a(z - c_1)(z - c_2)\ldots(z - c_n), \text{ where } a \in \mathbb{C}, \text{ and } c_t \in \mathbb{C} \text{ for } t = 1, \ldots, n.$$

(6) *Every irreducible polynomial $f \in \mathbb{C}[x]$ is linear.*

(7) *Every monic polynomial $f \in \mathbb{R}[z]$ of degree ≥ 1 can be expressed as the product of real linear and real quadratic polynomials.*

Exercise IX.7.1. Write down a proof of equivalence of the previous seven statements.

We start with a rather blind statement: The fundamental theorem of algebra is really not a theorem of algebra, but a theorem on functions of complex variable, that is a theorem of analysis. It has its grounds in the **analytical** properties of complex numbers. Almost every book dealing with complex functions, elementary or advanced, contains a proof of this theorem. There are literally tens of proofs of this theorem. It may be worth mentioning that almost every simple proof of the fundamental theorem of algebra follows from the fact that for a polynomial f that satisfies that $f(z) \neq 0$ for all complex numbers z, the function $f^{-1}(z) = \frac{1}{f(z)}$ is regular and bounded in the entire Argand-Gauss plane and thus, as a consequence of a theorem whose author is **Joseph Liouville (1809 - 1882)**, it must be constant.

As we have already mentioned, the fundamental theorem of algebra lost its significance that it possessed during the time when the central feature of algebra were investigations of algebraic equations with number coefficients. Today, it is simply a fundamental theorem on complex numbers, and as such it has its roots in analytic properties of complex numbers. Since understanding of properties of continuous functions on real numbers such as the existence of minima, maxima or their mean values were not properly defined until the time of **Karl Weierstrass** and **Richard Dedekind**, the first proofs of the fundamental theorem of algebra showed related shortcomings.

Today the widely accepted proof of **Carl Friedrich Gauss** contained in his dissertation in 1799 is regarded as the first proof of this theorem. We mention that a great part of this dissertation deals with a critique of some earlier proofs, that is, of the proofs of **Jean Le Rond d'Alembert** in 1746, **Leonhard Euler** in 1749, **Joseph-Louis Lagrange** in 1772 and of **Pierre-Simon Laplace (1749 - 1827)** in 1795, just to mention some. It is though fair to remark that Gauss' proof in 1799 has some serious gaps that were filled only as late as in 1920 by **Alexander Markowich Ostrowski.**

Browsing through the history, one finds that **René Descartes (1596 - 1650)** may have been one of the first who contributed to the understanding of the fundamental theorem of algebra by expressing (without a proof) in his "La géométrie" in 1637 a real polynomial as a product of linear polynomials (in the case that the roots of the polynomials are "proper"). Of course, it is possible to mention also some other mathematicians who have touched on this subject. However the proof of **Jean Robert Argand** given in 1806 and improved in 1814 (although not entirely rigorous by present standards - just because there were some formal definitions unavailable) should be considered as the first and simplest proof of this theorem. It is based on the idea of Jean Le Rond d'Alembert and, in particular, on the following statement.

Theorem IX.7.2 (D'Alembert's theorem). *If $f \in \mathbb{C}[z]$ is a nonconstant polynomial and $f(z_1) \neq 0$, then for any $\epsilon > 0$ there is $z_2 \in \mathbb{C}$ such that $|z_2 - z_1| < \epsilon$ and $|f(z_2)| < |f(z_1)|$.*

Proof. This theorem is a motivation for the following proof of the fundamental theorem of algebra. Since $f(z_0) = 0$ if and only if $|f(z_0)| = 0$, we shall deal mainly with the real function $|f(z)|$. The idea of our proof is to show that there is a minimum value of $|f(z)|$ for some z_0 and then to show that $f(z_0) = 0$. In the course of the proof there will be only two instances that we shall use analytical properties of complex numbers.

The first step of the proof.

Let $f(z) = a_0 + a_1 z + a_2 z^2 + \cdots + a_n z^n \in \mathbb{C}[z]$, where $a_n \neq 0$ and $n \geq 1$. We are going to exhibit real numbers A and B such that for all $z \in \mathbb{C}$ satisfying $|z| \geq A$, $|f(z)| \geq B > |a_0|$. To this end put $B = |a_0| + |a_1| + \cdots + |a_t| + \cdots + |a_n|$ and $A = 2B\,|a_n|^{-1}$. Note that $A > 1$ and thus $|z|^t > |z|$ for $|z| \geq A$. Now, for $|z| \geq A$, we can easily see that, for each $0 \leq t \leq n-1$,

$$\frac{|a_t|}{|z|^{n-t-1}} \leq \frac{|a_t|}{|z|} \leq \frac{|a_t|\,|a_n|}{2B - |a_n|}$$

and thus, gradually

$$|f(z)| \geq |a_n|\,|z|^n \;-\; |z|^{n-1}\left(\frac{|a_0|}{|z|^{n-1}} + \cdots + \frac{|a_t|}{|z|^{n-t-1}} + \cdots + \frac{|a_{n-2}|}{|z|} + |a_{n-1}|\right)$$

$$\geq |a_n|\,|z|^n \;-\; |z|^{n-1}\,\frac{|a_n|\,(B - |a_n|)}{2\,(B - |a_n|)} \;\geq\; |z|^n\,\frac{|a_n|}{2} \;\geq\; |z|\,\frac{|a_n|}{2} \;\geq\; B \;>\; |a_0|.$$

The second step of the proof (using an analytical argument).

The (real) continuous function $|f(z)|$ attains a minimum $|f(z_0)|$ at a certain point z_0 of the compact disc $\{z \mid |z| \leq A\}$. Since $|f(z_0)| \leq |f(0)| = |a_0| < |f(z)|$ for all z such that $|z| \geq A$, $|f(z_0)|$ is the minimum of the function $|f(z)|$ for the entire Argand-Gauss plane.

The final third step of the proof (again using an analytical argument).

Referring to Exercise VII.2.4, we can write the polynomial f in the form of a new polynomial g defined by $f(z) = g(z - z_0)$. Thus, the (real) function $|g(z)|$ attains its minimum at the origin and the (complex) polynomial g can be written in the form

$$g(z) = b_0 + b_k z^k + \cdots b_n z^n \;=\; b_0 + b_k z^k + z^{k+1}\,h(z), \quad \text{where } \; b_k \neq 0,\; h(z) \in \mathbb{C}[z]$$

and $|b_0|$ is the minimum of $|g(z)|$. Denote by z_* one of the k complex numbers satisfying $z_*^k = -\frac{b_0}{b_k}$. Now, let $\epsilon > 0$ be arbitrary real number and consider the polynomial

$$g(\epsilon z_*) \;=\; b_0 - b_0 \epsilon^k + \epsilon^{k+1} z_*^{k+1}\,h(\epsilon z_*).$$

Since polynomials are bounded on a bounded set, there exists a number K satisfying $|h(\epsilon z_*)| \leq K$ for every ϵ satisfying $0 \leq \epsilon \leq 1$. We want to show that $b_0 = |b_0| = 0$. Assume the opposite, that is, that $b_0 \neq 0$ and choose ϵ_* satisfying

$$0 \;<\; \epsilon_* \;<\; \frac{|b_0|}{|z_*|^{k+1}\,K}.$$

Then

$$|g(\epsilon_*\, z_*)| \;\leq |b_0|\,(1 - \epsilon_*^k) + \epsilon_*^{k+1}\,|z_*|^{k+1}\,K \;<\; |b_0|\,(1 - \epsilon_*^k) + \epsilon_*^k |b_0| \;=\; |b_0|,$$

contradicting the fact that $|b_0|$ is a minimum. Therefore $|b_0| = |g(0)| = 0$ and hence $g(0) = f(z_0) = 0$, as required.

□

CHAPTER X. APPENDIX

X.1 Venn diagrams

The aim of this section is to provide a definition of the **Venn diagram**, which does not seem to be available in the literature. In Chapter I, some simple examples of Venn diagrams

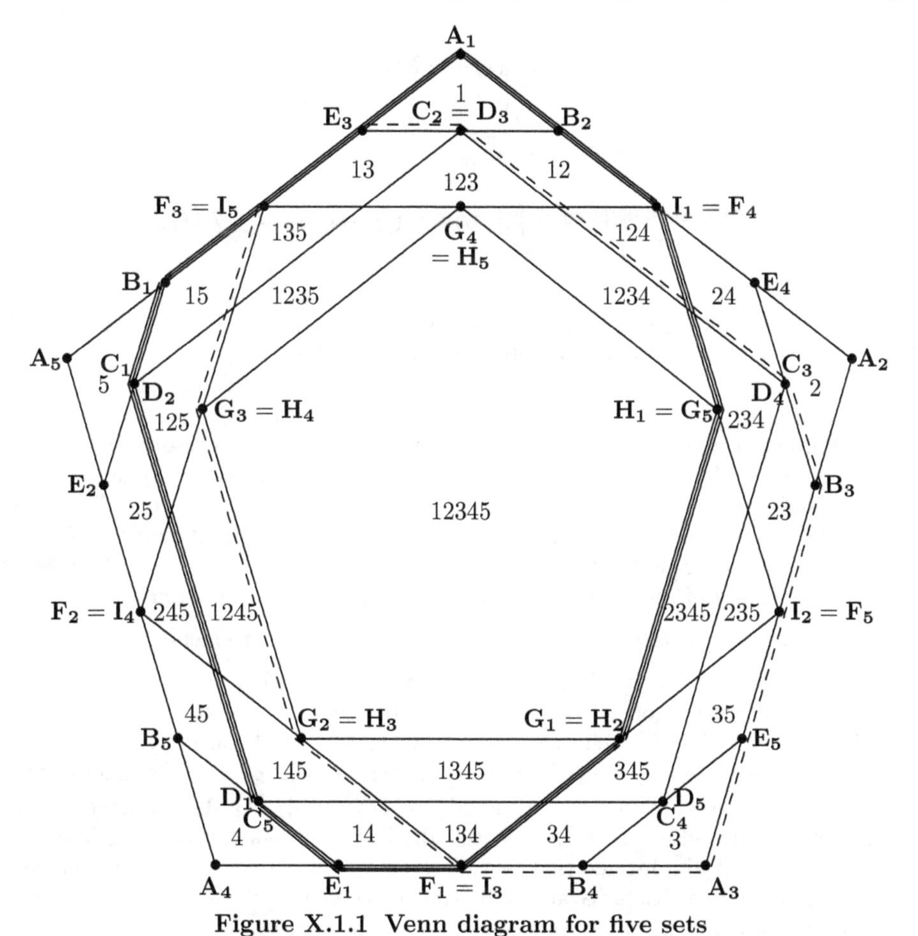

Figure X.1.1 Venn diagram for five sets

were exhibited. Figure X.1.1 illustrates that to exhibit a Venn diagram for five subsets in general position of a universe U is more demanding. In this case we have five nonagons $\mathcal{N}_t = A_t\, B_t\, C_t\, D_t\, E_t\, F_t\, G_t\, H_t\, I_t$, $(t = 1, 2, 3, 4, 5)$ and $32 = 2^5$ subsets. For example in Figure X.1.1 the subset of the plane U denoted by 5 consists of all elements of $\mathcal{N}_5$ that do not belong to any other $\mathcal{N}_t$ $(t \neq 5)$ and the subset denoted by 145 consists of all elements of $\mathcal{N}_t, t = 1, 4, 5$ that do not belong to $\mathcal{N}_t, t = 2, 3$ and the area outside the pentagon $A_1A_2A_3A_4A_5$ consists of all elements of U that do not belong to any polygon $\mathcal{N}_t$.

All these examples provide a lead to a general definition of a Venn diagram. Let $X_1, \ldots, X_n$ be any n subsets of the universal set U, and let $\mathcal{S} = \{X_t |\ t = 1, \ldots, n\} \subseteq \mathcal{P}(U)$. Note that an **index set** is a set used to label objects in another set.

Definition X.1.1. *Let $I = \{t |\ 1 \leq t \leq n\}$ be an index set, and as before $\mathcal{P}(I)$ denotes the set of all subsets of I. Let $\mathcal{P}(U)$ be the set of all subsets of U and $\mathcal{S} = \{X_t |\ t = 1, 2, \ldots, n\} \subseteq \mathcal{P}(U)$. Define the mapping (embedding)*

$$\varphi : \mathcal{P}(I) \longrightarrow \mathcal{P}(U)$$

by

$$\varphi(T) = S_T = \{x |\ x \in \bigcap_{t \in T} X_t, x \notin \bigcup_{t \notin T} X_t\} \ \ \text{for } T \in \mathcal{P}(I).$$

The map φ defines a so-called **indexed decomposition** *(or an* **ordered partition***) $\{S_T |\ T \in \mathcal{P}(I)\}$ of the universe U called a* **Venn diagram** *of the set $\mathcal{S}$.*

The definition of a general Venn diagram allows for the case when some of the sets S_T are empty; however, the important situations are those where all subsets S_T are non-empty. Note the following two features of our concept of a general Venn diagram. First, a Venn diagram is not a mere partition of a universe U; it is an "indexed" partition $\{S_T | T \in \mathcal{P}(I)\}$ determined by the map φ. Second, an arbitrary partition of an universe U can be presented, using a suitable choice of the indices, as a (possibly infinite) Venn diagram.

We leave it to the reader to explain and prove the previous statements.

A Venn diagram of n subsets $X_1, X_2, \ldots, X_n$ of a universe U that are in "general position" creates a **particular** partition of U; this partition consists of 2^n (mutually disjoint) subsets indexed by elements of $\mathcal{P}(I)$. Thus, for $n = 2$ we have $2^2 = 4$ mutually disjoint subsets $A_0 = U \setminus (X_1 \cup X_2), A_1 = X_1 \setminus X_2, A_2 = X_2 \setminus X_1$ and $A_{12} = X_1 \cap X_2$, the unions of which give $2^{(2^2)} = 16$ different combinations of the original subsets A_1, A_2, A_3: 16 different subsets of the universe U (including the empty subset $\emptyset$). These subsets can be organized into an algebraic structure called a **lattice** that reflects the property of one subset being included in another subset in the partition. Graphically, the lattice can be illustrated as illustrated in Figure X.1.2 (a formal definition of this concept is given in Definition I.6.3).

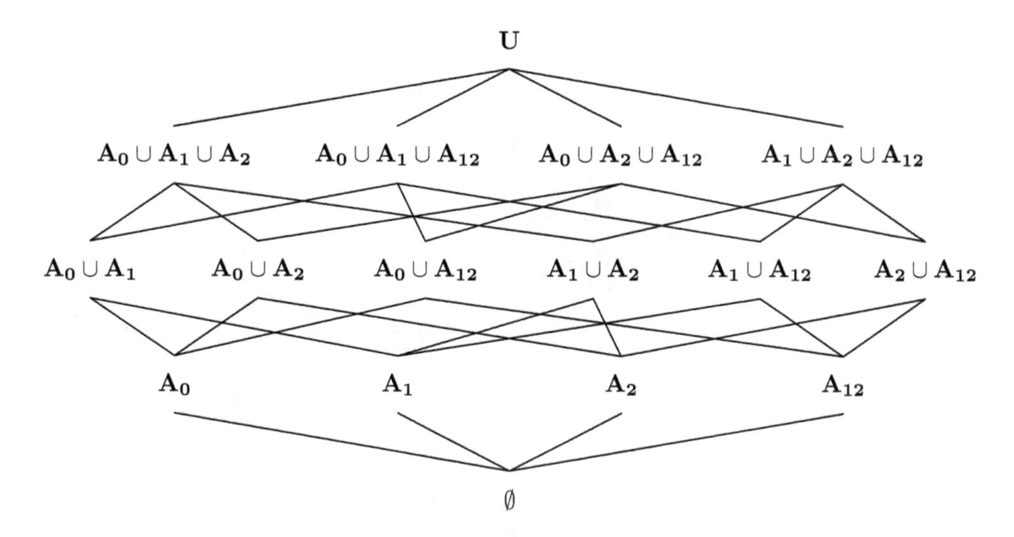

Figure X.1.2 Lattice of all subsets generated by two subsets of a universe

We observe that $X_1 = A_1 \cup A_{12}, X_2 = A_2 \cup A_{12}$ and $X_1 \cup X_2 = A_1 \cup A_2 \cup A_{12}$. Moreover, letting $B_0 = X_1 \cup X_2, B_1 = U \setminus (X_1 \setminus X_2), B_2 = U \setminus (X_2 \setminus X_1)$ and $B_{12} = U \setminus (X_1 \cap X_2)$, we can describe the lattice structure in terms of intersections of the sets B_0, B_1, B_2 and B_{12}. This reflects a **duality** possessed by this lattice; after all, this lattice is in a one-to-one correspondence with the lattice (often called a **Boolean algebra**) of all subsets of a set with four elements. The term Boolean algebra reflects the seminal work of **George Boole (1815 - 1864)** on such algebras.

For three subsets, the same process provides a partition of U into eight subsets whose unions define $2^8 = 256$ different subsets of U; for four subsets, a partition of U into 16 subsets whose unions define $2^{16} = 65,536$ different subsets of U; and, in general, for n subsets, a partition of U into 2^n subsets whose unions define 2^{2^n} different subsets of U. The corresponding lattice has 2^n "generators"

$$A_J = (\bigcap_{t \in J} X_t) \setminus (\bigcup_{t \in (I \setminus J)} X_t),$$

where J is a subset of $I = \{1, 2, \ldots, n\}$.

According to our definition, the diagram in Figure X.1.3 is a Venn diagram of three sets S_1, S_2 and S_3. The triangle ABC is not a component of the diagram, but the union of ABC and DEF is a component.

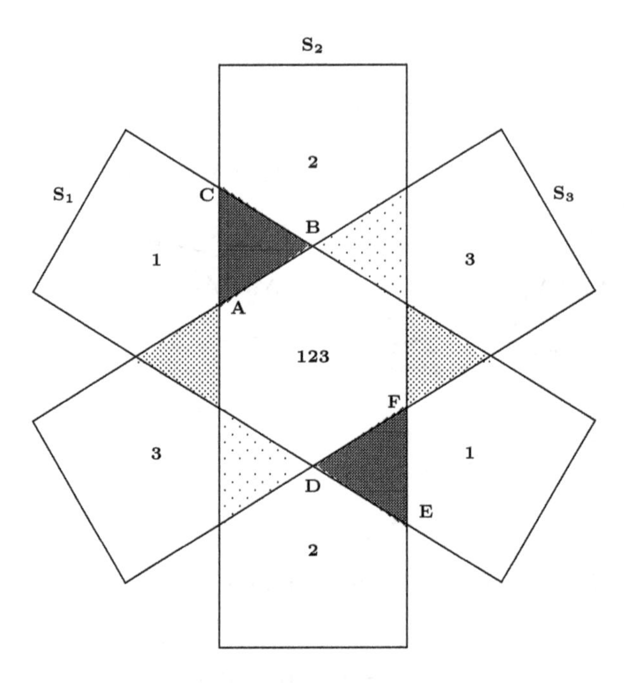

Figure X.1.3 Venn diagram of three sets

X.2 Cardinality, Cantor-Bernstein-Schröder theorem

In Section 2 of Chapter I, we defined the **cardinality** of a set. We recall that two sets have the same cardinality if and only if there exists a bijection between them. The relation "to have the same cardinality" is an equivalence relation. We call the equivalence class determined by a set S the cardinality of S and denote it by $|S|$. The cardinality of a finite set denotes the number of its elements, that is, it is a natural number. The linear order of the natural numbers can be extended to all cardinalities.

Definition X.2.1. *The cardinality of a set A is smaller or equal to the cardinality of a set B, that is $|A| \preceq |B|$ if there is an injective mapping $f : A \to B$.*

The fact that this relation $\preceq$ is a linear order (see Definition I.5.1) is equivalent to the **axiom of choice** that can be expressed in a number of equivalent forms. Here we just mention the following one:

Given two sets A and B, there is either an injective mapping $f : A \to B$ or an injective mapping $g : B \to A$ or both, that is, there is either $|A| \preceq |B|$ or $|B| \preceq |A|$ or both inequalities.

The fact that $|A| \preceq |B|$ and $|B| \preceq |A|$ implies $|A| = |B|$ is a consequence of the following well-known basic statement of the theory of sets. It is also known under various combinations of the three names (**Georg Cantor, Felix Bernstein (1878 - 1956),** and **Ernst Schröder (1841 - 1902)**).

Theorem X.2.1. *Let $f : A \to B$ and $g : B \to A$ be injective mappings. Then there exists a bijective mapping $h : A \to B$.*

This theorem can be proved in a number of ways. We choose one that is quite simple, summarizing in a certain sense original proofs of the above mentioned mathematicians. The main point lies in the following statement usually attributed to **Bronisław Knaster (1893 - 1980)**.

Theorem X.2.2. *Let S be a nonempty set and let $\mathcal{P}(S)$ be the power set of S consisting of all subsets of S, ordered by inclusion. Let $\varphi : \mathcal{P}(S) \longrightarrow \mathcal{P}(S)$ be an order-preserving mapping, that is, a mapping satisfying*

$$X \subseteq Y \text{ implies } \varphi(X) \subseteq \varphi(Y) \text{ for all } X, Y \in \mathcal{P}(S).$$

Then there is a set $T \in \mathcal{P}(S)$ such that $\varphi(T) = T$.

Proof. Consider the subset $\mathcal{M} \subseteq \mathcal{P}(S)$ of all $X \subseteq S$ such that $\varphi(X) \subseteq X$. Clearly $\mathcal{M} \neq \emptyset$ since $S \in \mathcal{M}$. Let

$$T = \bigcap_{X \in \mathcal{M}} X.$$

Now, for every $X \in \mathcal{M}$, $\varphi(T) \subseteq \varphi(X) \subseteq X$, and thus $\varphi(T) \subseteq T$. Furthermore, since $\varphi(\varphi(T)) \subseteq \varphi(T)$, we see that $\varphi(T) \subseteq \mathcal{M}$. This means that $T \subseteq \varphi(T)$ and therefore $\varphi(T) = T$. □

Now we can complete the proof of Theorem X.2.1. Consider the mapping $\varphi : \mathcal{P}(A) \longrightarrow \mathcal{P}(A)$ defined by $\varphi(X) = g(B \setminus f(A \setminus X))$ for every $X \subseteq A$. The mapping φ is order-preserving. Therefore, by Theorem X.2.1, there is a subset $T \subseteq A$ such that $T = g(B \setminus f(A \setminus T))$, and the mapping $h : A \longrightarrow B$ described in Figure X.2.1 can be verified to be one-to-one.

$\mathbf{A \setminus T}$	$\xrightarrow{\mathbf{h = f}}$	$\mathbf{f(A \setminus T)}$
$\mathbf{T}$	$\xrightarrow{\mathbf{h = g^{-1}}}$	$\mathbf{B \setminus f(A \setminus T)}$

Figure X.2.1 Proof of the Cantor-Bernstein-Schröder theorem

Remark X.2.1 (on infinite sets). Richard Dedekind introduced in his monograph *"Was sind und was sollen die Zahlen"* (1888) infinite sets in the following way:

A set S is said to be **infinite** *if there is an injective mapping $f : S \to S$ such that $f(S) \neq S$.*

As a consequence, an infinite set exists if and only if there is a set $\mathbb{N}$ with the following properties: *There is a special element $1 \in \mathbb{N}$ and an injective* **successor function** $\sigma : \mathbb{N} \to \mathbb{N}$ *such that*

(1) $1 \neq \sigma(\mathbb{N})$

and

(2) *whenever a subset $X \subseteq \mathbb{N}$ contains 1 and $\sigma(X) \subseteq X$, then $X = \mathbb{N}$.*

Write $\sigma(1) = 2$, $\sigma^2(1) = \sigma(\sigma(1)) = \sigma(2) = 3, \ldots, \sigma^n(1) = \sigma(\sigma(n-1)) = \sigma(n) = n+1, \ldots$ and formulate the following **iteration theorem**.

Let X be a set containing an element $a \in X$ and $f : X \to X$ an arbitrary mapping. Then there exists a unique mapping $\varphi : \mathbb{N} \to X$ (that is, a **sequence***) such that $\varphi(1) = a$ and $\varphi \circ \sigma = f \circ \varphi$, that is,*

$$\varphi(n+1) = f^n(a),$$

as indicated by the following commuting diagram

$$\begin{array}{ccccccccccccc} \mathbb{N} & \xrightarrow{\sigma} & \mathbb{N} & \xrightarrow{\sigma} & \mathbb{N} & \xrightarrow{\sigma} & \cdots & \xrightarrow{\sigma} & \mathbb{N} & \xrightarrow{\sigma} & \mathbb{N} & \xrightarrow{\sigma} & \cdots \\ \downarrow{\scriptstyle\varphi} & & \downarrow{\scriptstyle\varphi} & & \downarrow{\scriptstyle\varphi} & & & & \downarrow{\scriptstyle\varphi} & & \downarrow{\scriptstyle\varphi} & & \\ X & \xrightarrow{f} & X & \xrightarrow{f} & X & \xrightarrow{f} & \cdots & \xrightarrow{f} & X & \xrightarrow{f} & X & \xrightarrow{f} & \cdots \end{array}$$

and the images of the elements 1 and a

$$\begin{array}{ccccccccccccc} 1 & \xmapsto{\sigma} & 2 & \xmapsto{\sigma} & 3 & \xmapsto{\sigma} & \cdots & \xmapsto{\sigma} & n & \xmapsto{\sigma} & n+1 & \xmapsto{\sigma} & \cdots \\ \downarrow{\scriptstyle\varphi} & & \downarrow{\scriptstyle\varphi} & & \downarrow{\scriptstyle\varphi} & & & & \downarrow{\scriptstyle\varphi} & & \downarrow{\scriptstyle\varphi} & & \\ a & \xmapsto{f} & f(a) & \xmapsto{f} & f^2(a) & \xmapsto{f} & \cdots & \xmapsto{f} & f^{n-1}(a) & \xmapsto{f} & f^n(a) & \xmapsto{\sigma} & \cdots \end{array}$$

The example of defining the n-th power r^n of a real number r illustrates this theorem well: $X = \mathbb{R}, a = r$ and $f : \mathbb{R} \to \mathbb{R}$ is defined by $f(x) = rx$. We conclude this remark with a statement asserting the uniqueness of Dedekind's set of natural numbers $\mathbb{N}$ constructed above.

Let $X = N^*$ be a set and $f = \sigma^* : N^* \to N^*$ an injective mapping satisfying the requirements (1) and (2). Then the sets $\mathbb{N}$ and N^* are canonically isomorphic, that is, there is a unique bijection $\alpha : \mathbb{N} \to N^*$ such that $\alpha(1) = 1^*$ and $\alpha \circ \sigma = \sigma^* \circ \alpha$.

X.3 The number of partitions of a finite set

As mentioned in Section 4 of Chapter 1, we derive a recurrence relation for the number $b(n)$ of all possible partitions of a set A_n having n elements. The initial four values of $b(n)$ are

$$b(1) = 1,\ b(2) = 2,\ b(3) = 5,\ b(4) = 15,\ \ldots\ldots, \tag{X.3.1}$$

For example, in the case $A_4 = \{a, b, c, d\}$, there are the following equivalence relations (that is, partitions) of A_4:

$$\{a \sim b \sim c \sim d\}, \{a \sim b \sim c, d\}, \{a \sim b \sim d, c\}, \{a \sim c \sim d, b\}, \{b \sim c \sim d, a\},$$

$$\{a \sim b, c, d\}, \{a \sim c, b, d\}, \{a \sim d, b, c\}, \{b \sim c, a, d\}, \{b \sim d, a, c\}, \{c \sim d, a, b\},$$

$$\{a \sim b, c \sim d\}, \{a \sim c, b \sim d\}, \{a \sim d, b \sim c\}, \{a, b, c, d\}.$$

The sequence (X.3.1) appears in many other connections and is known as the **Bell sequence** referring to **Eric Temple Bell (1883 - 1960).**

We now prove the following statement:

If we define the empty set $A_0 = \emptyset$ to have one (trivial) partition, that is, $b(0) = 1$, then

$$b(n+1) = \sum_{t=0}^{n} \binom{n}{t} b(t) \text{ for all } n \in \mathbb{N}_0, \tag{X.3.2}$$

where $\binom{n}{t} = \dfrac{n!}{t!(n-t)!}$ for $t = 0, 1, \ldots, n$ are the binomial coefficients.

To prove the formula (X.3.2) we proceed as follows : Let $A_{n+1} = \{a_0, a_1, \ldots, a_n\}$. We denote by $\mathcal{R}(A_{n+1})$ the collection of all partitions of the set A_{n+1}.

Now, for every $1 \leq t \leq n+1$, we denote by $\mathcal{R}_t(A_{n+1})$ the set of **all partitions** of the set A_{n+1} such that in each of these partitions, the element a_0 belongs to a subset that has exactly t elements. Thus

$$\{\mathcal{R}_t(A_{n+1}) \mid t = 1, 2, \ldots, n+1\} \text{ is a partition of the set } \mathcal{R}(A_{n+1}).$$

Consequently, $\mathcal{R}_1(A_{n+1})$ are all partitions of the set A_{n+1} that contain the one-element subset $\{a_0\}$, and their number is exactly $b(n) = \binom{n}{n} b(n)$. Similarly, $\mathcal{R}_2(A_{n+1})$ are all partitions of the set A_{n+1} that contain the subsets $\{a_0, a_{k_1}\}, 1 \leq k_1 \leq n$, and their number is obviously $nb(n-1) = \binom{n}{1} b(n-1) = \binom{n}{n-1} b(n-1)$. In general, $\mathcal{R}_{t+1}(A_{n+1})$ are all partitions of the set A_{n+1} that contain the subsets $\{a_0, a_{k_1}, a_{k_2}, \ldots, a_{k_t}\}$ of $t+1$ elements, and their number is clearly $\binom{n}{t} b(n-t) = \binom{n}{n-t} b(n-t)$. In particular, the number of elements in $\mathcal{R}_n(A_{n+1})$ is equal to $\binom{n}{1} b(1)$ and, evidently, there is only one partition that contains the entire set A_{n+1} and that is counted, as we agreed, by $\binom{n}{0} b(0)$.

Using the formula (X.3.2), we readily determine that

$$b(5) = \binom{4}{0} \times 1 + \binom{4}{1} \times 1 + \binom{4}{2} \times 2 + \binom{4}{3} \times 5 + \binom{4}{4} \times 15 = 1 + 4 + 12 + 20 + 15 = 52$$

and similarly

$$b(6) = \mathbf{1} \times 1 + \mathbf{5} \times 1 + \mathbf{10} \times 2 + \mathbf{10} \times 5 + \mathbf{5} \times 15 + \mathbf{1} \times 52 = 1 + 5 + 20 + 50 + 75 + 52 = 203.$$

The coefficients of these sums form **the rows of the Yang Hui - Pascal triangle** (see Definition IV.3.1). Another proof of formula (X.3.2) is given in Example II.2.5; see equation (II.2.15) and related exercises.

X.4 Lattices of equivalences, all subsets of a universe

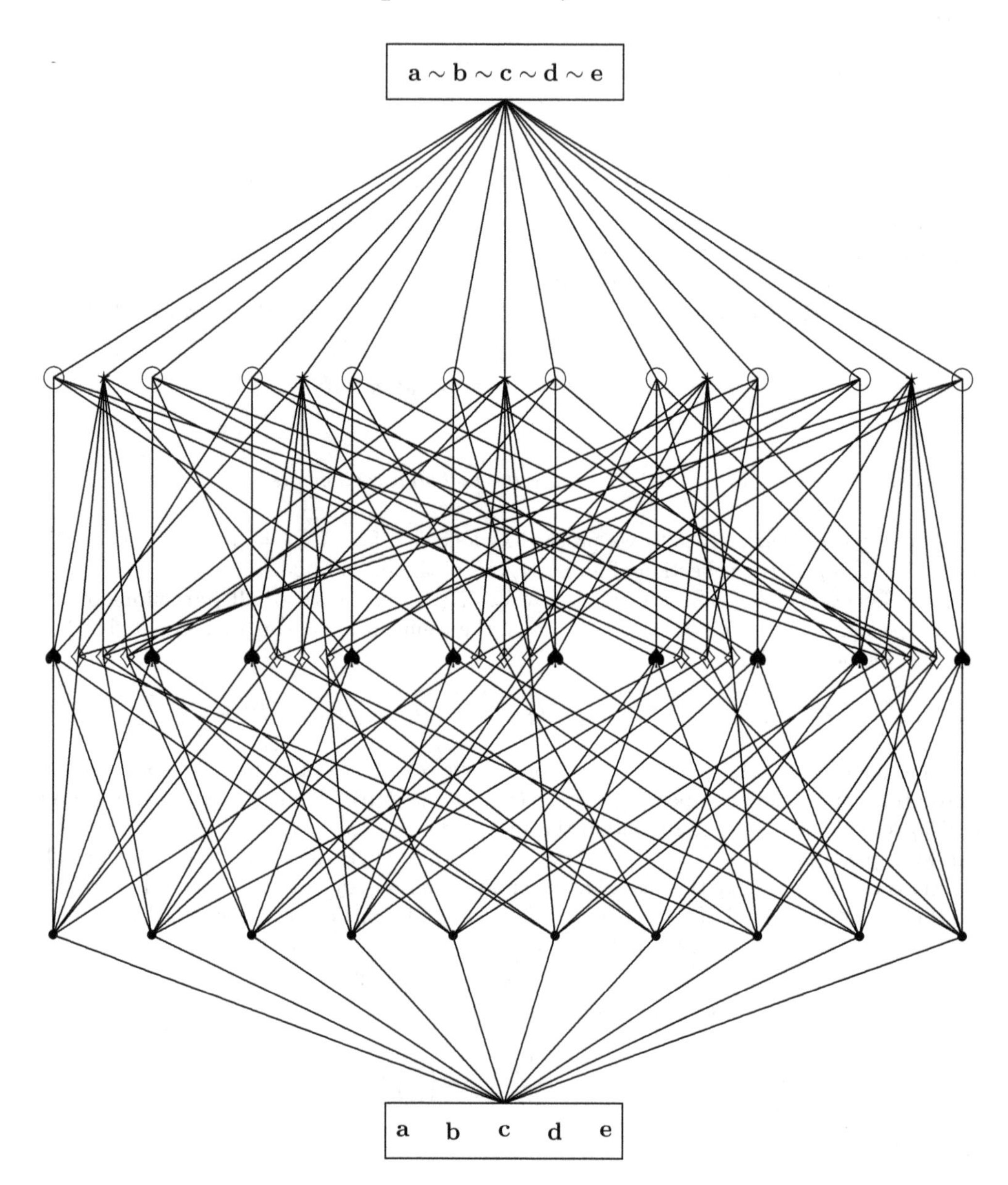

Figure X.4.1 Lattice of equivalences on a set of five elements

In Figure X.4.1 the symbol ◯ denotes equivalences of the form $\{\mathbf{x} \sim \mathbf{y} \sim \mathbf{u},\ \mathbf{v} \sim \mathbf{w}\}$, the symbol $\star$ equivalences of the form $\{\mathbf{x} \sim \mathbf{y} \sim \mathbf{u} \sim \mathbf{v},\ \mathbf{w}\}$, the symbol ♠ equivalences of the form $\{\mathbf{x} \sim \mathbf{y} \sim \mathbf{u},\ \mathbf{v},\ \mathbf{w}\}$, the symbol $\diamondsuit$ equivalences of the form $\{\mathbf{x} \sim \mathbf{y},\ \mathbf{u} \sim \mathbf{v},\ \mathbf{w}\}$, and the symbol $\bullet$ equivalences of the form $\{\mathbf{x} \sim \mathbf{y},\ \mathbf{u},\ \mathbf{v},\ \mathbf{w}\}$.

A very important partially ordered set which forms a lattice, is the collection $\mathcal{P}(U)$ of all subsets of a universe U as introduced in Section 1 of Chapter 1. Here, for A and B of $\mathcal{P}(U)$,

$$A \vee B = A \cup B \text{ and } A \wedge B = A \cap B.$$

In this lattice, for every element A, there exists its (unique) **complement** $A^* = U \backslash A$ defined by the properties

$$A \vee A^* = U \text{ and } A \wedge A^* = \emptyset.$$

Moreover, this lattice is **distributive**, that is, for arbitrary A, B, C of $\mathcal{P}(U)$,

$$A \cap (B \cup C) = (A \cap B) \cup (A \cap C) \text{ and } A \cup (B \cap C) = (A \cup B) \cap (A \cup C).$$

We remark that the mapping $D : \mathcal{P}(U) \to \mathcal{P}(U)$ defined by $D(A) = A^*$ reflects the **duality** of $\mathcal{P}(U)$ expressed earlier in the form of **de Morgan rules**:

$$D(A \cap B) = D(A) \cup D(B) \text{ and } D(A \cup B) = D(A) \cap D(B).$$

The inclusion $A \subseteq B$ implies the inclusion $D(A) \supseteq D(B)$ and also $D^2 = D \circ D$ is the identity mapping of the set $\mathcal{P}(U)$.

Remark X.4.1. We show that the (finite) lattice $\mathcal{R}(A)$ of all partitions of a finite set A has the property that, for every partition $\mathcal{P}$ there exists a partition $\mathcal{Q}$ such that $\mathcal{P} \vee \mathcal{Q}$ is the one-element partition $\mathbf{1} = \{A\}$ and $\mathcal{P} \wedge \mathcal{Q}$ is the "finest" partition $\mathbf{0} = \{\ \{a\}\ \mid a \in A\}$. Such a "pseudocomplement" $\mathcal{Q}$ is not unique.

Given $\mathcal{P} = \{A_1, A_2, \dots, A_d\}$, we assume that the numbers n_r of elements of the subsets A_r satisfy $n_1 \le n_2 \le \cdots \le n_d$. We now create a $d \times n_d$ scheme (matrix)

$$\begin{array}{ccccccccccc}
a_{11} & a_{12} & \dots & a_{1n_1} & \star & \star & \star & \star & \star & \star & \star \\
a_{21} & a_{22} & \dots & a_{2n_1} & \dots & a_{2n_2} & \star & \star & \star & \star & \star \\
a_{31} & a_{32} & \dots & a_{3n_1} & \dots & a_{3n_2} & \dots & a_{3n_3} & \star & \star & \star \\
\dots & \dots & \dots & \dots & \dots & \dots & \dots & \dots & \dots & \star & \star \\
\dots & \dots & \dots & \dots & \dots & \dots & \dots & \dots & \dots & \dots & \star \\
a_{d1} & a_{d2} & \dots & a_{dn_1} & \dots & a_{dn_2} & \dots & a_{dn_3} & \dots & \dots & a_{dn_d}
\end{array}$$

in which the rows list the elements of the components $A_r = \{a_{ri} \mid 1 \le i \le n_r\}$, $1 \le r \le d$, of the partition $\mathcal{P}$. Now, define $\mathcal{Q}$ to consist of n_d subsets of A whose elements are listed as columns of the above scheme: $\mathcal{Q} = \{B_1, B_2, \dots, B_{n_d}\}$, where B_s is the set of all a_{js}, $1 \le s \le n_d$.

It can be shown that the partitions $\mathcal{P}$ and $\mathcal{Q}$ of Remark X.4.1 satisfy

$$\mathcal{P} \vee \mathcal{Q} = \mathbf{1} \text{ and } \mathcal{P} \wedge \mathcal{Q} = \mathbf{0},$$

and we encourage the reader to prove this.

Further, it can be shown for a set A with more than 3 elements, there are partitions $\mathcal{P}_1, \mathcal{P}_2, \mathcal{P}_3$ in the lattice $\mathcal{R}(A)$ such that

$$\mathcal{P}_1 \wedge (\mathcal{P}_2 \vee \mathcal{P}_3) \ne (\mathcal{P}_1 \wedge \mathcal{P}_2) \vee (\mathcal{P}_1 \wedge \mathcal{P}_3),$$

that is, the lattice $\mathcal{R}(A)$ is not distributive.

X.5 Finite automata

The concept of a monoid permits the presentation of the algebraic theory of **finite automata** (finite state machines). We sketch some of the basic facts. Given a monoid $M = (M, \circ)$ with the identity element 1, we need also to consider its **submonoids** $N = (N, \circ)$, that is, subsets of M that contain $x \circ y$ with elements $x, y \in N$, as well as the identity element 1.

Given a finite set, an **alphabet** $Z = \{z_1, z_2, \dots, z_n\}$ of **letters (symbols)**, we denote by $F(Z)$ the monoid of all **words** (strings of letters):

$$F(Z) = \{x = z_{j_1} z_{j_2} \dots z_{j_r} \mid z_{j_t} \in Z, 1 \le t \le r\}$$

with the operation of **word concatenation** (juxtaposition):

For $x = z_{j_1} z_{j_2} \dots z_{j_r}$ and $y = z_{k_1} z_{k_2} \dots z_{k_s}$, $x \circ y = z_{j_1} z_{j_2} \dots z_{j_r} z_{k_1} z_{k_2} \dots z_{k_s}$.

The identity element is the **"empty word"** $\omega : \omega \circ x = x \circ \omega = x$ for all $x \in F(Z)$. The monoid $F(Z)$ is the **free monoid** over the basis Z.

Now let $(M, \circ)$ be an arbitrary monoid and $B \subseteq M$ its **generating system.** Thus, every element $a \in M$ can be expressed in the form

$$a = b_1 \circ b_2 \circ \dots \circ b_d, \quad \text{where } b_t \in B \text{ for } 1 \le t \le d.$$

This expression is, in general, not unique. As we have done previously, we denote by $F(B)$ the free monoid over the basis B and call it a **free cover** of the monoid M. For each element $x = b_{j_1} b_{j_2} \dots b_{j_d}$, where $b_{j_t} \in B, 1 \le t \le d$, we define the image

$$\varphi(x) = b_{j_1} \circ b_{j_2} \circ \dots \circ b_{j_d}$$

in the monoid M. The mapping φ is called the **canonical homomorphism** $\varphi : F(B) \to M$ and its **kernel** $\mathrm{Ker}\varphi = \{x \in F(B) \mid \varphi(x) = \omega\}$ is a submonoid of M.

For a given set S, we denote by $\mathrm{Hom}(S, S)$ the monoid of all mappings of S into itself with the operation of composition of mappings. The **image** of the element $s \in S$ in the mapping $\alpha : S \to S$ will be denoted by $\alpha(s)$. Thus, the mapping $\alpha \circ \beta : S \to S$ for $\alpha : S \to S$ and $\beta : S \to S$ is defined, in agreement with the usual notation for finite automata, by $(\alpha \circ \beta)(s) = \beta(\alpha(s))$. This explains why some authors use the notation $(s)\alpha$ instead of $\alpha(s)$. The identity mapping is the identity element of this monoid $1 : S \to S$, that is, $1(s) = s$ for all $s \in S$.

The monoid $\mathrm{Hom}(S, S)$ contains a submonoid $G(S)$ of all bijective mappings of the set S onto S. Hence, for every element $\alpha \in G(S)$, there is the **inverse element** $\beta \in G(S)$, usually denoted by α^{-1}, and satisfying $\alpha \circ \beta = \beta \circ \alpha = 1$. The submonoid $G(S)$ is thus a group, the so-called **symmetric (permutation) group** of the set S. In the important case, when the set is finite and has n elements, then $\mathrm{Hom}(S, S)$ is finite and has n^n elements, while the group $G(S)$ has $n!$ elements.

Finally, we define a finite automaton.

Definition X.5.1. *Let S be a finite set, $s_0 \in S$ and $K \subseteq S$. Let $B = \{\alpha_1, \alpha_2, \dots, \alpha_n\} \subseteq \mathrm{Hom}(S, S)$ and A_B the submonoid of the monoid $\mathrm{Hom}(S, S)$ generated by the subset B. Then the triple (A_B, s_0, K) is said to be a finite automaton,*

$$\mathbf{Tr}(A_B, s_0, K) = \{x \in A_B \mid x(s_0) \in K\}$$

the **trace** *of the automaton* (A_B, s_0, K) *and*

$$\mathbf{Lang}(A_B, s_0, K) = \{y \in F(B) \mid y(s_0) \in K\}$$

the **language** *of the automaton* (A_B, s_0, K).

Here $F(B)$ is the free cover of the monoid A_B. With reference to Section 5 of Chapter VIII, we can describe the (regular) language of a finite automaton in terms of the homomorphism $\varphi : F(B) \to A_B$, namely

$$\mathbf{Lang}(A_B, s_0, K) = \varphi^{-1}(\mathbf{Tr}(A_B, s_0, K)).$$

We note that in the literature, a finite automaton is usually presented as a quintuple (S, Z, s_0, K, π), where S is a finite non-empty **set of states**, Z an **alphabet**, $s_0 \in S$ the **initial state**, $K \subseteq S$ the set of the **final states** and $\pi : S \times Z \to S$ the **state-transition function**. In our definition, $Z = B$ and $\pi(s, v) = \varphi(v)(s)$. In the case that the set S has only a few elements, it is convenient to represent a finite automaton as a graph. The mapping $\pi(s, z)$ is described by the vertices representing the states s and $s' = \pi(s, z)$ together with the arrow $s \longrightarrow s'$ "decorated" by the symbol $z \in Z$.

We illustrate this situation with two examples.

Example X.5.1. Let $S = \{s_0, s_1, s_2, s_3\}, K = \{s_0\}$ and $B = \{\alpha_1, \alpha_2\}$, where

$$\alpha_1 : (s_0, s_1, s_2, s_3) \mapsto (s_2, s_3, s_0, s_1) \text{ and } \alpha_2 : (s_0, s_1, s_2, s_3) \mapsto (s_1, s_0, s_3, s_2).$$

Here, $\alpha : (s_0, s_1, s_2, s_3) \mapsto (\alpha(s_0), \alpha(s_1), \alpha(s_2), \alpha(s_3))$ and thus

$$A_B = \{1, \alpha_1, \alpha_2, \alpha_3 = \alpha_1 \circ \alpha_2 = \alpha_2 \circ \alpha_1 : (s_0, s_1, s_2, s_3) \mapsto (s_3, s_2, s_1, s_0)\}$$

is a commutative group of order 4 with the multiplication table

$\circ$	1	α_1	α_2	α_3
1	1	α_1	α_2	α_3
α_1	α_1	1	α_3	α_2
α_2	α_2	α_3	1	α_1
α_3	α_3	α_2	α_1	1

that is, the Klein four-group V_4. Moreover, we see that $\mathbf{Tr}(A_B, s_0, K) = \{\mathbf{1}\}$. This implies immediately that $\mathbf{Lang}(A_B, s_0, K) = \text{Ker } \varphi$, and thus

$$\mathbf{Lang}(A_B, s_0, K) = \{x \in F(B) \mid \text{the numbers of symbols } \alpha_1 \text{ and } \alpha_2 \text{ in the string } x \text{ are even}\}.$$

A graphical representation of this automaton is as follows:

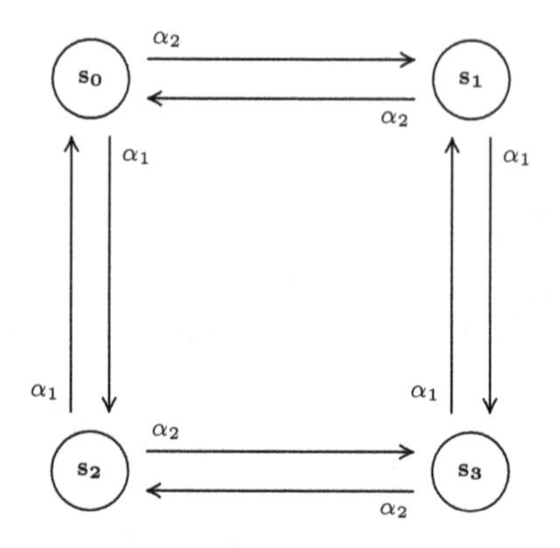

Example X.5.2. Let $S = \{s_0, s_1, s_2\}$, $K = \{s_2\}$ and $B = \{a = \alpha_0, b = \alpha_1\}$, where

$$a : (s_0, s_1, s_2) \mapsto (s_1, s_1, s_1) \text{ and } b : (s_0, s_1, s_2) \mapsto (s_2, s_2, s_0).$$

Then

$$A_B = \{1, a, b, c = a \circ b, d = b \circ b, e = a \circ b \circ b\},$$

where $c\colon (s_0, s_1, s_2) \mapsto (s_2, s_2, s_2)$, $d\colon (s_0, s_1, s_2) \mapsto (s_0, s_0, s_2)$ and $e\colon (s_0, s_1, s_2) \mapsto (s_0, s_0, s_0)$, is a semigroup with the multiplication table

$\circ$	1	a	b	c	d	e
1	1	a	b	c	d	e
a	a	a	c	c	e	e
b	b	a	d	c	b	e
c	c	a	e	c	c	e
d	d	a	b	c	d	e
e	e	a	c	c	c	e

Thus, $\mathbf{Tr}(A_B, s_0, K) = \{b, a \circ b\}$ and $\mathbf{Lang}(A_B, s_0, K) = \varphi^{-1}(b, a \circ b)$. Furthermore, we have $\varphi^{-1}(b) = \{b^{2k+1} \mid k \geq 0\}$. The inverse image $\varphi^{-1}(a \circ b)$ can be described recurrently: $x \in \varphi^{-1}(a \circ b)$ if and only if $ax \in \varphi^{-1}(a \circ b)$ and as long as $x \neq b^{2k+1}$, then also $bx \in \varphi^{-1}(a \circ b)$ if and only if $x \in \varphi^{-1}(a \circ b)$. Hence

$$\mathbf{Lang}(A_B, s_0, K) = \{b, ab, aab, bab, bbb, aaab, abab, abbb, baab, bbab, aaaab,$$
$$aabab, aabbb, abaab, abbab, baaab, babab, babbb, bbaab, bbbab, bbbbb, \dots\},$$

that is, the language consists of the words x that are the concatenations of finite numbers of sequences of r letters "a" and $2s+1$ letters "b", $r \geq 0, s \geq 0$:

$$x = (a^{r_1}b^{2s_1+1})(a^{r_2}b^{2s_2+1})\cdots(a^{r_k}b^{2s_k+1}),\ r_i \geq 0,\ s_j \geq 0,\ 1 \leq i,j \leq k.$$

A graphical representation of this automaton is as follows.

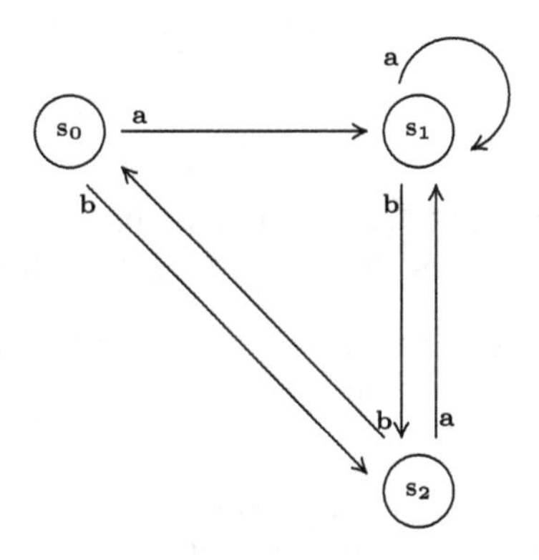

X.6 Categories

The quotation: *Mathematics is the art of giving the same name to different things* - as opposed to the quotation: *Poetry is the art of giving different names to the same thing* - is attributed to **Henry Poincaré**. *Mathematicians do not study objects, but relations between objects.* Thus, mathematicians are free to replace some objects by others so long as the relations between them remain unchanged. In contemporary mathematics they employ the language of **category theory** that emphasizes this point of view.

The language of **categories**, introduced in the middle of the last century by **Saunders MacLane (1909 - 2005)** and **Samuel Eilenberg (1913 - 1998)** became indispensable for present-day mathematics. It has penetrated scientific literature, monographs, and, little by little, mathematical textbooks.

Definition X.6.1. *A* **category** $\mathcal{C}$ *consists of a class of* **objects** $\mathrm{Ob}\mathcal{C}$ *and, for each pair of objects* $X, Y \in \mathrm{Ob}\mathcal{C}$*, of a set* $\mathrm{Mor}(X,Y)$ *of* **morphisms** $f : X \to Y$ *from the object* X *to the object* Y *subject to the following conditions:*

(1) *for every triplet of objects* $X, Y, Z \in \mathrm{Ob}\mathcal{C}$ *and every pair of morphisms* $f : X \to Y$ *and* $g : Y \to Z$*, there is a uniquely defined morphism* $g \circ f : X \to Z$ *which is called the* **composition** *or* **product** *of the morphisms* f *and* g*;*

(2) *composition of morphisms is associative, that is, for every triplet of morphisms*

$$X \xrightarrow{f} Y \xrightarrow{g} Z \xrightarrow{h} W\ ,\ h \circ (g \circ f) = (h \circ g) \circ f.$$

(3) *for every object* $X \in \mathrm{Ob}\mathcal{C}$*, there exists a morphism* $1_X \in \mathrm{Mor}(X,X)$ *such that* $f \circ 1_X = f$ *and* $1_X \circ g = g$ *for all morphisms* $f : X \to Y$ *and* $g : Z \to X$*.*

It is easy to see that a morphism 1_X with the above properties is unique. It is called the **identity morphism** of the object X.

Example X.6.1. (Examples of categories)

(1) The category $\mathcal{P}(U)$ of all subsets of a given universe U. The objects of this category are the subsets $X, Y, \ldots$ of the universe U, the morphisms of $\mathrm{Mor}(X, Y)$ are all mappings from X to Y, that is, $\mathrm{Mor}(X, Y) = \mathrm{Hom}(X, Y)$. A natural extension of this category is the category $\mathcal{S}$ of all sets. Here however, we face questions concerning the foundations of the theory of sets: the objects of this category do not form a set, but a class of all sets.

(2) The category $\mathrm{Vec}_{\mathbf{F}}$: The objects are all finite-dimensional vector spaces over a given field F, the morphisms are the linear transformations between the vector spaces.

(3) Every monoid $(M, \circ)$ can be considered as a category. This category has a single object a and $M = \mathrm{Mor}(a, a)$. The composition of morphisms naturally coincides with the semigroup operation $\circ$.

(4) Every partially ordered set $(S, \leq)$ can be viewed as a category. Here, the objects are the elements of the set S and, for every pair $x, y \in S$, $\mathrm{Mor}(x, y)$ contains a single element $f_{x,y}$ if $x \leq y$ and $\mathrm{Mor}(x, y) = \emptyset$ otherwise. The composition of the morphisms is defined in a natural manner.

(5) The category Mat_F: The objects are all natural numbers and $\mathrm{Mor}(m, n)$ consists of all $n \times m$ matrices with entries from a field F. The composition of morphisms is the usual product of matrices. A verification of all conditions is, as in all preceding examples, straightforward.

(6) The path category $\mathcal{C}_\Gamma$ of an oriented graph $\Gamma = (\mathbf{V}, \mathbf{H})$: The objects are the vertices of the graph, that is, $\mathrm{Ob}\mathcal{C}_\Gamma = \mathbf{V}$ and, for every pair $i, j \in \mathbf{V}$, $\mathrm{Mor}(i, j)$ is the set of all paths from i to j, that is, $\mathrm{Mor}(i, j) = \{\omega \in \mathcal{S}(\Gamma) \mid t(\omega) = i, h(\omega) = j\}$. Of course, $1_i = \epsilon_i$ for every $i \in \mathbf{V}$.

In Chapters VIII and IX we considered the category of groups, the category of commutative (abelian) groups, the category of rings, etc.

In a similar way as morphisms serve to compare and describe relations between objects of a given category, covariant and contravariant functors serve to describe relations between categories.

Definition X.6.2. *Let $\mathcal{C}_1$ and $\mathcal{C}_2$ be categories. A* **covariant functor** *F from $\mathcal{C}_1$ to $\mathcal{C}_2$ is a mapping $F : \mathcal{C}_1 \to \mathcal{C}_2$ assigning to every object $X \in \mathrm{Ob}\mathcal{C}_1$ an object $F(X) \in \mathrm{Ob}\mathcal{C}_2$ and to every morphism $f : X \to Y$ in $\mathcal{C}_1$ a morphism $F(f) : F(X) \to F(Y)$ in $\mathcal{C}_2$ in such a way that*

(1) *$F(1_X) = 1_{F(X)}$ for every $X \in \mathrm{Ob}\mathcal{C}_1$ and*

(2) *$F(g \circ f) = F(g) \circ F(f)$ for every $f : X \to Y$ and $g : Y \to Z$ in $\mathcal{C}_1$.*

A **contravariant functor** *G from $\mathcal{C}_1$ to $\mathcal{C}_2$ is a mapping $G : \mathcal{C}_1 \to \mathcal{C}_2$ assigning to every object $X \in \text{Ob}\mathcal{C}_1$ an object $G(X) \in \text{Ob}\mathcal{C}_2$ and to every morphism $f : X \to Y$ in $\mathcal{C}_1$ a morphism $G(f) : G(Y) \to G(X)$ in $\mathcal{C}_2$ in such a way that*

(1) *$G(1_X) = 1_{G(X)}$ for every $X \in \text{Ob}\mathcal{C}_1$ and*

(2) *$G(g \circ f) = G(f) \circ G(g)$ for every $f : X \to Y$ and $g : Y \to Z$ in $\mathcal{C}_1$.*

Example X.6.2. (1) The **identity functor** $F : \mathcal{C} \to \mathcal{C}$ defined by $F(X) = X$ for all objects of $\mathcal{C}$ and $F(f) = f$ for all morphisms of $\mathcal{C}$.

(2) The **forgetful functor** from the category $\text{Vec}_{\mathbf{F}}$ of the finite-dimensional vector spaces over a given field F to the category $\mathcal{S}$ of sets assigning to a vector space X_F the set X (without any structure) and to a linear transformation $T : X_F \to Y_F$ the mapping $T : X \to Y$.

(3) The **Hom functor** $H_A : \mathcal{C} \to \mathcal{S}$ from a category $\mathcal{C}$ to the category of sets $\mathcal{S}$. Here A is a fixed (chosen) object of the category $\mathcal{C}$. For $X \in \mathcal{C}$, define $H_A(X) = \text{Mor}(A, X)$ and for $f : X \to Y$, $H_A(f) : \text{Mor}(A, X) \to \text{Mor}(A, Y)$ assigning to every $g : A \to X$ the morphism $f \circ g : A \to Y$.

Using the language of functors, we can express an **equivalence of two categories.** This topic is however well beyond the scope of this book. Nevertheless, here we present a simple example illustrating such an equivalence.

Example X.6.3. Let $\text{Vec}_{\mathbf{F}}$ be the category mentioned in Example X.6.1. Thus the objects of $\text{Vec}_{\mathbf{F}}$ are additive (abelian) groups X subject to action by elements of the field $\mathbf{F}$ as displayed at the beginning of VI.3. The morphisms are linear transformations $f : X \to Y$, that is, mappings satisfying

$$f(x_1 + x_2) = f(x_1) + f(x_2) \quad \text{and} \quad f(xa) = f(x)a.$$

Objects of the category $\text{Vec}_{\mathbf{F}}$ are the **representations** of the field F. There is a **unique** (indecomposable) object $F \in \text{Vec}_{\mathbf{F}}$ such that every object of $\text{Vec}_{\mathbf{F}}$ is a (finite direct) sum of copies of F. This is an expression of the fact that every finite-dimensional vector space over a field has a basis (as we know from linear algebra).

Let $\mathcal{K}_{Mat_{n\times n}(\mathbf{F})}$ be the category of all $Mat_{n\times n}(\mathbf{F})$-modules X, that is, of all additive (abelian) groups subject to action by elements of the ring $Mat_{n\times n}(\mathbf{F})$ of all $n\times n$ matrices over the field $\mathbf{F}$, together with mappings $f : X \to Y$ satisfying all of the above listed properties, where elements $a, b \in F$ are replaced by matrices $A, B \in Mat_{n\times n}(\mathbf{F})$. The elements of $\mathcal{K}_{Mat_{n\times n}(\mathbf{F})}$ are representations of $Mat_{n\times n}(\mathbf{F})$.

The categories $\text{Vec}_{\mathbf{F}}$ and $\mathcal{K}_{Mat_{n\times n}(\mathbf{F})}$ are equivalent. This, in particular, means that there is a unique (indecomposable) object $\mathbf{F}_n$ in $\mathcal{K}_{Mat_{n\times n}(\mathbf{F})}$ such that every object of this category is a (direct) sum of several copies of $\mathbf{F}_n$, in the same way as every vector space over $\mathbf{F}$ is a (direct) sum of copies of the (one-dimensional) vector space $\mathbf{F}$.

X.7 Discrete Fourier transform

Inspired by Example VII.4.2, we describe a general situation of (homogeneous) commutative semisimple algebras. Such algebras appear in parts (v), (vi), (vii) and (viii) of this example. We begin with a formal definition.

Definition X.7.1. *An n-dimensional homogeneous commutative F-algebra A is a vector space of all n-tuples of elements from a field F with multiplication defined by*

$$\mathbf{a}\cdot\mathbf{b} \;=\; (a_1b_1, a_2b_2, \ldots, a_nb_n) \text{ for all } \mathbf{a} = (a_1, a_2, \ldots, a_n),\ \mathbf{b} = (b_1, b_2, \ldots, b_n).$$

Thus $A = (A; +, \cdot)$ is a ring with the zero element $\mathbf{0} = (0, 0, \ldots, 0)$, the identity element $\mathbf{1} = (1, 1, \ldots, 1)$ and a canonical copy of the field F : The subalgebra $\{(a, a, \ldots, a) \mid a \in F\}$ is identified with the field F. We emphasize the fact that both addition and multiplication is done coordinatewise. This will be recorded by saying that the F-algebra A is a **direct sum** of n copies of the field F and by writing

$$A \;=\; F_1 \oplus F_2 \oplus \cdots \oplus F_n \;=\; \bigoplus_{t=1}^{n} F_t \text{ with all } F_t \simeq F. \tag{X.7.1}$$

Note that A may also be written as a cartesian product $A = F_1 \times F_2 \times \cdots \times F_n$; we have used that notation in the above mentioned examples.

We leave it to the reader to describe all (multiplicative) invertible elements and all nontrivial divisors of zero in the algebra A defined in (X.7.1), as well as the lattice of all subalgebras of A.

Let $f \in F[x]$ be a monomial polynomial over F that factors into linear factors : $f(x) = (x - a_1)(x - a_2)\ldots(x - a_n)$ with **distinct** $a_t, 1 \leq t \leq n$ from F. Consider the quotient ring

$$A = F[x]/\langle f(x)\rangle.$$

Theorem X.7.1. *The ring A is an n-dimensional homogeneous commutative semisimple F-algebra of the form* (X.7.1), *where the identity element 1_t of the subalgebra F_t is the class $\overline{q}_t$ defined by the Lagrange basic polynomial*

$$q_t(x) \;=\; \prod_{\substack{0 \leq k \leq n \\ k \neq t}} \frac{(x - a_k)}{(c_t - a_k)}\,.$$

Thus

$$\overline{q}_t^{\,2} = \overline{q}_t,\ \overline{q}_r \cdot \overline{q}_s = \overline{f} = 0 \text{ for all } t \text{ and } r \neq s.$$

Moreover, for every $\overline{g} \in A$,

$$\overline{g} = \sum_{t=1}^{n} g(a_t) \cdot 1_t \;\Big(= \sum_{t=1}^{n} \overline{g(a_t)\, q_t(x)}\;\Big)\,.$$

Proof. All of this is a straightforward application of Theorem VII.2.4 (with the index set $\{1, 2, \ldots, n\}$). One should consider all polynomials as polynomial functions of $P(F)$ (see the text following Exercise VII.1.3). □

Example X.7.1 (Discrete Fourier transform). An important application is the choice of the polynomial $f(x) = x^n - 1 \in \mathbb{C}[x]$, where $n \in \mathbb{N}$. Here,

$$x^n - 1 = \prod_{i=0}^{n-1}(x - \omega^i), \quad \text{where} \quad \omega = e^{\frac{2\pi i}{n}}.$$

Defining the map $\Phi : \mathbb{C}[x] \to \mathbb{C}^n$ by

$$\Phi(g) = \big(g(1),\, g(\omega),\, g(\omega^2),\, \ldots,\, g(\omega^{n-1})\big)$$

we obtain a homomorphism from $\mathbb{C}[x]$ to $\underbrace{\mathbb{C} \times \mathbb{C} \times \cdots \times \mathbb{C}}_{n-\text{times}}$. The kernel of this homomorphism is the ideal generated by the polynomial $f(x) = x^n - 1$. Thus Φ induces an isomorphism $\bar{\Phi}$ between the $\mathbb{C}$-algebras $\mathbb{C}[x]/\langle(x^n - 1)\rangle$ and $\mathbb{C}^n$:

$$\bar{\Phi}(\overline{g}) = \big(g(1),\, g(\omega),\, g(\omega^2),\, \ldots,\, g(\omega^{n-1})\big)\ .$$

This isomorphism is called the discrete Fourier transform after **Joseph Fourier (1768 - 1830)**. In this case, we can also very simply express the inverse $\bar{\Phi}^{-1}$ of the isomorphism $\bar{\Phi}$, the **Fourier inversion**.

Theorem X.7.2. *The Fourier inverse is given by*

$$\bar{\Phi}^{-1}[(c_0, c_1, c_2, \ldots, c_{n-1})] = \overline{a_0 + a_1x + a_2x^2 + \cdots + a_{n-1}x^{n-1}},$$

where $n \in \mathbb{N}$ and

$$a_r = \frac{1}{n}\sum_{t=0}^{n-1} c_t\omega^{-rt}. \tag{X.7.2}$$

Proof. Our proof consists of an application of elementary linear algebra. We start with an element $\overline{g} \in \mathbb{C}[x]/\langle(x^n - 1)\rangle$ that is represented by the (unique) polynomial of degree smaller than n, say by $g(x) = a_0 + a_1x + a_2x^2 + \cdots + a_{n-1}x^{n-1}$. Thus

$$\bar{\Phi}(\overline{g}) = (c_0 = \sum_{t=0}^{n-1} a_t,\ c_1 = \sum_{t=0}^{n-1} a_t\omega^t,\ c_2 = \sum_{t=0}^{n-1} a_t\omega^{2t},\ \cdots,\ c_{n-1} = \sum_{t=0}^{n-1} a_t\omega^{(n-1)t}).$$

Since $\bar{\Phi}$ is an isomorphism, to determine the inverse $\bar{\Phi}^{-1}$ we just need to solve the system of linear equations

$$\begin{pmatrix} 1 & 1 & 1 & 1 & \ldots & 1 \\ 1 & \omega & \omega^2 & \omega^3 & \ldots & \omega^{n-1} \\ 1 & \omega^2 & \omega^4 & \omega^6 & \ldots & \omega^{(n-1)2} \\ \ldots & \ldots & \ldots & \ldots & \ldots & \ldots \\ 1 & \omega^r & \omega^{2r} & \omega^{3r} & \ldots & \omega^{(n-1)r} \\ \ldots & \ldots & \ldots & \ldots & \ldots & \ldots \\ 1 & \omega^{n-1} & \omega^{2(n-1)} & \omega^{3(n-1)} & \ldots & \omega^{(n-1)^2} \end{pmatrix} \cdot \begin{pmatrix} a_0 \\ a_1 \\ a_2 \\ a_3 \\ \ldots \\ \ldots \\ a_{n-1} \end{pmatrix} = \begin{pmatrix} c_0 \\ c_1 \\ c_2 \\ c_3 \\ \ldots \\ \ldots \\ c_{n-1} \end{pmatrix}, \tag{X.7.3}$$

which has a unique solution. To verify that the solution is given by (X.7.2), we calculate the r-th row:

$$\sum_{t=0}^{n-1} \omega^{rt} a_t = \sum_{t=0}^{n-1} \omega^{rt} \sum_{k=0}^{n-1} \frac{1}{n} c_k \omega^{-kt} = \frac{1}{n} \sum_{k=0}^{n-1} c_k \sum_{t=0}^{n-1} \omega^{(r-k)t} = c_r .$$

Since $\omega^n = 1$, this is a consequence of the simple fact concerning geometric progressions, namely that

$$\sum_{t=0}^{n-1} \omega^{ts} = \frac{\omega^{ns}-1}{\omega^s - 1} = 0 \text{ for } s \neq tn,\ t \in \mathbb{Z}.$$

Of course, in the case that $s = tn$, the sum equals n.

□

Remark X.7.1. We observe that the formula (X.7.2) is just expressing the fact that the matrix

$$\begin{pmatrix} 1 & 1 & 1 & 1 & \dots & 1 \\ 1 & \omega^{-1} & \omega^{-2} & \omega^{-3} & \dots & \omega^{-(n-1)} \\ 1 & \omega^{-2} & \omega^{-4} & \omega^{-6} & \dots & \omega^{-(n-1)2} \\ \dots & \dots & \dots & \dots & \dots & \dots \\ 1 & \omega^{-r} & \omega^{-2r} & \omega^{-3r} & \dots & \omega^{-(n-1)r} \\ \dots & \dots & \dots & \dots & \dots & \dots \\ 1 & \omega^{-(n-1)} & \omega^{-2(n-1)} & \omega^{-3(n-1)} & \dots & \omega^{-(n-1)^2} \end{pmatrix}$$

is an n-multiple of the inverse of the matrix of the system (X.7.3).

Remark X.7.2. The discrete Fourier transform has a number of important applications. Since it is possible to compute both the transform and its inverse rather quickly, and since the operations in $\mathbb{C}^n$ are immediate, one of the application is to compute products of very large numbers or polynomials. Example X.7.2 indicates how such a process works.

Example X.7.2. Take $n = 5$ and $g_1(x) = x^3 - x,\ g_2(x) = x^2 + x + 1$. Thus

$$\begin{aligned} \bar{\Phi}(\overline{g_1}) &= (0,\ -\omega+\omega^3,\ \omega-\omega^2,\ -\omega^3+\omega^4,\ \omega^2-\omega^4), \\ \bar{\Phi}(\overline{g_2}) &= (3,\ 1+\omega+\omega^2,\ 1+\omega^2+\omega^4,\ 1+\omega+\omega^3,\ 1+\omega^3+\omega^4) \end{aligned}$$

and hence

$$\bar{\Phi}(\overline{g_1 g_2}) = (0,\ 1-\omega-\omega^2+\omega^4,\ 1-\omega^2+\omega^3-\omega^4,\ 1-\omega+\omega^2-\omega^3,\ 1+\omega-\omega^3-\omega^4).$$

From here, $a_0 = \frac{1}{5}(4-\omega-\omega^2-\omega^3-\omega^4) = \frac{1}{5}(5-1-\omega-\omega^2-\omega^3-\omega^4) = 1$. Furthermore, $a_1 = \frac{1}{5\omega^4}(1+\omega+\omega^2+\omega^3+\omega^4-5\omega^4) = -1$. Similarly, $a_2 = -1$, $a_3 = 0$ and $a_4 = 1$. Thus

$$\begin{aligned} g_1(x)\, g_2(x) &= \bar{\Phi}^{-1}(0,\ 1-\omega-\omega^2+\omega^4,\ 1-\omega^2+\omega^3-\omega^4,\ 1-\omega+\omega^2-\omega^3,\ 1+\omega-\omega^3-\omega^4) \\ &= 1 - x - x^2 + x^4. \end{aligned}$$

In order to achieve a greater understanding of $\bar{\Phi}$ and its inverse $\bar{\Phi}^{-1}$, we conclude this book by inviting the reader in the last exercise to determine simple explicit formulas for $\bar{\Phi}$ and $\bar{\Phi}^{-1}$ for some small values of n.

Exercise X.7.1. Give simple explicit formulas for $\bar{\Phi}$ and $\bar{\Phi}^{-1}$ when $n = 3$, 4 and 6.

ANSWERS TO SELECTED EXERCISES

I.7.6	$\sum_{d^2\mid n}\mu(d)g(n/d^2) = \sum_{d^2\mid n}\mu(d)\sum_{e^2\mid n/d^2} f((n/d^2)/e^2)$ $= \sum_{k^2\mid n}\sum_{de=k}\mu(d)f(n/d^2e^2) = \sum_{k^2\mid n} f(n/k^2)\sum_{d\mid k}\mu(d)$ $= \sum_{k^2\mid n, k=1} f(n/k^2) = f(n)$
II.3.4	$a_{1t} = \frac{1}{6}\left(-14t^3 + 111t^2 - 271t + 204\right)$
II.4.1(i)	$\phi(n) = \frac{2n-1}{n+1}\,,\ \psi(n) = 3\,\frac{n-2}{n+1}$
II.5.2(i)	$X_n = (1+\sqrt{10})^n + (1-\sqrt{10})^n$
II.5.2(ii)	$X_n = \frac{A^n}{2^{n-1}}$
II.5.2(iii)	$X_n = A^{n-1}$
II.7.3	$1000 - 790 - 300 - x + 110 + 35 + 15 - 10 = 0$, and thus $x = 60$
II.8.1	$n\,2^{n-1}$
II.8.2	$\frac{n}{3n+1},\ \frac{n+2}{2(n+1)}$
II.8.3	$\frac{3}{4} - \frac{2n+3}{2(n+1)(n+2)}$
II.8.13	$2\binom{n-1}{0} + 9\binom{n-1}{1} + 15\binom{n-1}{2} + 11\binom{n-1}{3} + 3\binom{n-1}{4}$
II.8.15	$1\binom{n-1}{0} + 6\binom{n-1}{1} + 12\binom{n-1}{2} + 10\binom{n-1}{3} + 3\binom{n-1}{4}$
II.8.17(ii)	$\mu(n) = \frac{1}{24}\left(n^4 - 6n^3 + 23n^2 - 18n + 24\right)$
III.5.3	$x = 42 + 71t,\ y = -163 - 277t$. For $t = -1,\ x + y = 85$
III.7.1	$(3 - 33t, 1 + 8t, -1 + 6t)$
III.7.2	$(2q, 45 + 5t - q, 90 + 13t, 120 + 17t)$
III.7.3	$x = 99 + 210t,\ x_1 = 10 + 21t,\ x_2 = 33 + 70t,\ x_3 = 14 + 30t$
III.9.10(i)	$x = 2 + 15t$
III.9.10(ii)	$x = 13 + 85t$
III.9.12	$x = 122 + 13\cdot 14\cdot 107\cdot t$

IV.6.1 The least solution is $x = 649,\ y = 180$

IV.8.7 Expressions are successively $..010101|_2, ..101010|_2, ..101011|_2, ..010110|_2$

IV.8.9 There are two values of $\sqrt{2}:\ \ldots 245450454|_7$ and $\ldots 421216213|_7$

IV.8.10(ii) There are two solutions: $\ldots 302002110|_5$ and $\ldots 142442340|_5$

IV.8.11 Again, there are two values of $\sqrt{-1}:\ \ldots 12013233|_5$ and $\ldots 32431212|_5$

V.2.6 $z = 2 - 2b + bi$

V.3.6(i) $1 - 2i,\ 3 + i$

V.3.6(ii) $1 - 2i,\ -2 + 3i$

V.3.6(iii) $\pm\frac{1}{2}\,(\sqrt{3} + i),\ \pm\frac{1}{2}\,(\sqrt{3} - i)$

V.3.6(iv) $\pm\frac{1}{2}\,(\sqrt{3} + i),\ \pm\frac{1}{2}\,(1 + \sqrt{3}\,i),\ \pm\frac{1}{2}\,(\sqrt{3} - i),\ \pm\frac{1}{2}\,(1 - \sqrt{3}\,i)$

V.3.6(v) $1,\ -1 \pm 2i$

VI.2.10 $D_2 \simeq \mathbb{Z}_2[x]/\langle x^2 + 1\rangle;\ [2 + i]\,[3 + i] = [0]$ in D_5

VII.1.5 $c = [5],\ d = [2]$

VII.1.6 $a^3 - ab + c = c^2 - 4a^2 d = 0$

VII.1.7 $y = x^2 + 2x + 1,\ a = 2,\ b = 9,\ c = 4;$ the roots are $-1 \pm 2\,i$ and $-1 \pm \frac{\sqrt{2}}{2}\,i$

VII.2.3(i) $(x + [2])\,(x^3 + [3]x^2 + [2]x + [2]) = x^4 + [3]x^2 + x + [4]$

VII.2.3(ii) $x\,(x + [1])\,(ax^2 + ([2]a + [3])x + ([2]a + [1])),\ a \in \mathbb{Z}_5$

VII.2.5(i) $n = 13k + 3$ or $n = 13k + 9$

VII.2.5(ii) $[2]x^3 + [2]x^2,\ [2]x^3 + [2]x,\ [2]x^2 + [2]x,\ 0$

VII.2.6 $a_n = \sum_{k=0}^{t} (-1)^k \binom{t}{k} d_{t-k}$

VII.3.8(i) $35\,790\,267,\ 150$

VII.3.8(ii) $x,\ x + [1];\ x^2 + x + [1];\ x^3 + x^2 + [1];\ x^3 + x + [1]$

VII.3.8(iii) $x^2 + 2$

VII.3.8(iv) $x^2 + 2 = 4\,f(x) + 2\,(2x + 1)\,g(x)$

VII.3.12 Over $\mathbb{C}$,

$$r(x) = \frac{1}{8}\left[\frac{1}{x+1} + \frac{1}{x-1} - \frac{1}{x+i} - \frac{1}{x-i} + \frac{i}{x+\frac{\sqrt{2}}{2}+\frac{\sqrt{2}}{2\,i}}\right]$$

$$-\frac{1}{8}\left[\frac{i}{x+\frac{\sqrt{2}}{2}-\frac{\sqrt{2}}{2\,i}} + \frac{i}{x-\frac{\sqrt{2}}{2}+\frac{\sqrt{2}}{2\,i}} - \frac{i}{x-\frac{\sqrt{2}}{2}-\frac{\sqrt{2}}{2\,i}}\right];$$

over $\mathbb{R}$,

$$r(x) = \frac{1}{8}\left[\frac{\sqrt{2}}{x^2+\sqrt{2}x+1} - \frac{\sqrt{2}}{x^2-\sqrt{2}x+1} - \frac{2x}{x^2+1} + \frac{1}{x+1} + \frac{1}{x-1}\right];$$

over $\mathbb{Q}$,

$$r(x) = \frac{1}{8}\left[\frac{-4x}{x^4+1} - \frac{2x}{x^2+1} + \frac{1}{x+1} + \frac{1}{x-1}\right]$$

VII.6.3(i) $(a)\ \dfrac{E_2}{E_3}$, $(b)\ \dfrac{E_2^2 - 2E_1E_3}{E_3^2}$, $(c)\ E_1^3 - 3E_1E_2 + 3E_3$,
$(d)\ E_1^4 - 4E_1^2E_2 + 4E_1E_3 + 2E_2^2$

VII.6.3(ii) $5pq$

VII.6.3(iii) $x^3 + 16x^2 + 61x - 22$; in general,
$x^3 + 2ax^2 + (a^2 + b)x + (ab - c)$

VII.6.4 $a = 1,\ b = -2,\ c = 0$

VII.7.1 $\frac{2n+5}{5}\binom{n+4}{4}$

VII.7.4 If for example $f(x) = a_0 + a_1x + a_2x^2 + a_3x^3$, then
$g(x) = b_0 + b_1x + b_2x^2 + b_3x^3 + b_4x^4$ with $b_0 = 0$,
$b_1 = a_0 + \dfrac{a_1}{2} + \dfrac{a_2}{6},\ b_2 = \dfrac{a_1}{2} + \dfrac{a_2}{2} + \dfrac{a_3}{4}, b_3 = \dfrac{a_2}{3} + \dfrac{a_3}{2},\ b_4 = \dfrac{a_3}{4}$

VIII.3.5 S_3 is generated by $a = \begin{pmatrix} 1 & 2 & 3 \\ 2 & 3 & 1 \end{pmatrix}$ and $b = \begin{pmatrix} 1 & 2 & 3 \\ 2 & 1 & 3 \end{pmatrix}$, $GL_2(GF_2)$ is generated by $A = \begin{pmatrix} 1 & 1 \\ 1 & 0 \end{pmatrix}$ and $B = \begin{pmatrix} 1 & 1 \\ 1 & 0 \end{pmatrix}$, and D_6 is generated by the triangle symmetries $\alpha : ABC \to BCA$ and $\beta : ABC \to BAC$.
The maps $a \to A, b \to B$ and $A \to \alpha, B \to \beta$ define isomorphisms.

VIII.3.9 The automorphism group $\mathbf{Aut}(C(n))$ is isomorphic to the multiplicative group $\mathbf{U}(\mathbb{Z}/n\mathbb{Z})$ of invertible elements of $\mathbb{Z}/n\mathbb{Z}$.
It is an abelian group of order $\varphi(n)$, where φ is Euler's function

VIII.3.27 $\mathcal{D}_{2n}$ is indecomposable if and only if $n \neq 2k$ with k odd.
Compare Exercise VIII.7.18

IX.2.5 Consider the map
$x \to -10x + 5 :\ x \oplus y = x + y - 5,\ x \otimes y = \frac{1}{10}[-xy + 5(x + y) + 25]$

IX.3.6 $d(-8 + 5i, 7 + 9i) = 1; 1 = (-8 + 5i)(-2 + 5i) + (7 + 9i)(4 + 2i)$

IX.4.8 $(\alpha^4 + 1)^{-1} = \alpha^3 + \alpha^2 + \alpha + 1$ in $\mathbb{Q}(\alpha)$

IX.4.15 $x^4 + 4$ splits completely in $\mathbb{Q}(i)$ as
$x^4 + 4 = (x + (1 + i))(x - (1 + i))(x + (1 - i))(x - (1 - i))$.
There is no field F strictly between $\mathbb{Q}$ and $\mathbb{Q}(i)$ as the degree of $\mathbb{Q}(i)$ is a prime, namely 2.

X.7.1 For $n = 3$:
$\bar{\Phi}(\overline{a_0 + a_1x + a_2x^2}) = (a_0 + a_1 + a_2,\ a_0 + a_1\,\omega + a_2\,\omega^2,\ a_0 + a_1\,\omega^2 + a_2\,\omega)$;
$\bar{\Phi}^{-1}[(c_0, c_1, c_2)] = \frac{1}{3}\ \left[(c_0 + c_1 + c_2) + (c_0 + c_1\,\omega^2 + c_2\,\omega)x + (c_0 + c_1\,\omega + c_2\,\omega^2)x^2\right]$

Index

CPSIA information can be obtained
at www.ICGtesting.com
Printed in the USA
BVHW091316270620
582061BV00010B/36